D1271204

RADIATION SHIELDING AND DOSIMETRY

RADIATION SHIELDING AND DOSIMETRY

A. EDWARD PROFIO
Professor of Nuclear Engineering
University of California, Santa Barbara

A WILEY-INTERSCIENCE PUBLICATION
JOHN WILEY & SONS, New York • Chichester • Brisbane • Toronto

Library of Congress Cataloging in Publication Data

Profio, A. Edward, 1931-
Radiation shielding and dosimetry.

"A Wiley-Interscience publication."
Includes index.
1. Shielding (Radiation). 2. Radiation—Dosage.
3. Radiation dosimetry. 4. Radiology, Medical—
Safety measures. I. Title.
TK9152.P74 621.48'32 78-15649
ISBN 0-471-04329-X

Printed in the United States of America

10 9 8 7 6 5 4 3 2 1

To
THEOS J. THOMPSON
Man and Mentor

PREFACE

This book is intended as a text and reference for nuclear engineers, radiation protection technologists, and medical physicists. It treats the design of shields for radioactive sources, X-ray machines, low energy accelerators, and nuclear reactors. It is also concerned with the prediction and measurement of dose in the body from external and internal sources, and the biological effects of ionizing radiation.

The book is written at the graduate and upper-division undergraduate university level. It assumes familiarity with the concepts of modern physics and mathematics through the calculus and differential equations. However, emphasis is placed on physical understanding and applications rather than theory and mathematical development. A previous course in nuclear measurements and nuclear physics would be helpful. Many engineers do not have this background, however, and pertinent information on interactions of ionizing radiation with matter and radiation detection is included. For greater detail, the reader may consult the author's previous book, *Experimental Reactor Physics*.

Chapter 1 introduces the subject of ionizing radiation and attenuation in matter. Sources of ions, electrons, photons, and neutrons are discussed in Chapter 2, and interactions in Chapter 3. Methods for computing the transport of radiation are treated in Chapter 4, with emphasis on Monte Carlo, discrete ordinates, and kernel techniques. Definitions of "dose" and methods of measuring or calculating fluence and dose are covered in Chapter 5, while biological effects of ionizing radiation and radiation protection standards are the subject of Chapter 6. Shield engineering and materials are discussed in Chapter 7, which also includes examples of shield designs. Chapter 8 treats dose in the body from external photon and neutron beams, as well as dose from inhaled, ingested, or injected radionuclides.

I wish to thank Darcy Radcliffe for an outstanding job in typing and illustrating the lecture notes from which this book was derived, Tod Still for preparing the camera-ready copy for publication, editor Beatrice Shube for her able assistance and unfailing encouragement throughout the writing of two books, and my family, Janet, Chris, Claudia, and Suzy, for their love.

Santa Barbara, California
December 1978

A. Edward Profio

CONTENTS

RADIATION SHIELDING AND DOSIMETRY

1

INTRODUCTION

Radiation shielding is the discipline concerned with the transport and interaction of ionizing radiation in matter. The term "shielding" implies the deliberate introduction of material between the radiation source and an object to reduce the radiation intensity and hence damage to the object. However, attenuation and absorption of radiation are also important in radiotherapy and radiation processing, where the aim is to deliver a precise amount or "dose" of radiation to a tumor or sample. In this book we consider the penetration and effects of neutrons, photons, and charged particles in various materials, including the human body.

"Dosimetry" is a conventional term for characterization of the damaging potential of ionizing radiation. Biological damage is related to the energy deposited by ionization in unit mass of tissue (the absorbed dose), modified by factors such as the microscopic spatial distribution of the ionizations, the concentration of oxygen, and the rate of energy deposition. Neutron damage in metals is related to displacements of atoms and transmutations by nuclear reactions. Embrittlement of steel is often correlated with the fluence of neutrons with energies above 1 MeV. In calculations and some experiments the energy spectrum of the flux density is obtained and then convoluted with an energy-dependent "damage function" to get a measure of the damage potential. We discuss methods of calculating and measuring the spectrum and "dose."

Figure 1.1 illustrates a simplified reactor shielding problem involving determination of the thicknesses X and Y required to reduce the neutron and γ-ray doses at D to less than regulated limits. The source region (reactor core) is indicated by S. Region 1 is iron, and region 2 is concrete.

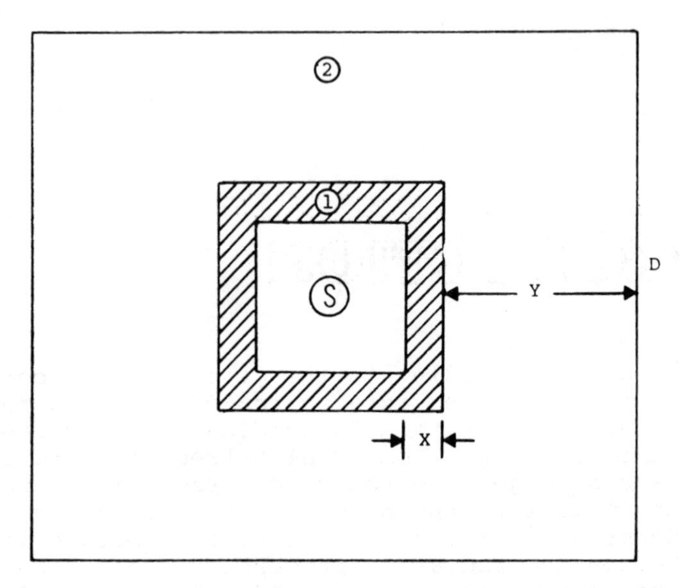

Fig. 1.1. Simplified reactor shield problem. Region S contains radiation source, region 1 contains iron, and region 2 contains concrete. Determine thicknesses X and Y for specified dose at D.

A radiation shielding and dosimetry problem may be divided into eight steps:

1. Specify the source, including strength (particles emitted per second), energy spectrum, angular distribution, spatial distribution, and time dependence unless a constant source is assumed.

2. Specify the source-shield-detector geometry, material compositions, and densities for all materials. In composite or heterogeneous shields or in the body, a somewhat simplified geometrical configuration is often necessary and adequate.

3. Obtain the interaction cross-sections or attenuation coefficients, secondary-radiation production cross-sections if pertinent, and other radiological data for the materials involved, and reduce the data to a form suitable for the calculation of radiation transport.

4. Specify the location, energy, and angular response of one or more detectors, according to the purpose of the calculation. Specify a response

function or damage function for calculation of "dose" at pertinent locations such as accessible regions immediately outside a shield. Decide whether point, surface, or volume detectors are appropriate for averaging in the problem.

5. Decide on the computer program or other calculational method, based on the complexity of the problem and the accuracy required in the results. Starting from the source, calculate the transport of radiation through the shield and any intervening space, and calculate the response (e.g., absorbed dose) in each detector to the required precision.

6. Compare the calculated response to limiting criteria, for example, the maximum dose permitted by regulations. If unacceptable, modify the design (e.g., increase thickness of a shield layer or perhaps substitute another material) and iterate until the dose or other response is within limits. Response can be scaled linearly with source strength as long as the material properties are unchanged.

7. Other constraints such as cost should be considered in the choice of materials and their fabrication into the shield. Shields may be optimized on cost, thickness, or mass. Overdesigned shields (giving lower dose than necessary) may be "safer," but there is a penalty. A thinner shield may be desired to reduce the size and cost of surrounding structures or the length of piping and so forth within the shield. Less massive shields are desired for transportation of radioactive materials and for mobile sources. In critical applications such as space vehicles, small mass overrides all other criteria.

8. Verify the performance of the shield or the dose in the body or sample by actual measurement at startup or first use. If this is not feasible, make the measurements in a shield mockup or a phantom representing the body or sample.

The source is generally given, although part of the problem in radiotherapy and radiation processing is to select the type and energy of radiation required to deliver a certain dose over a given volume with reasonable uniformity without exceeding another specified dose in surrounding tissue or components. The types of sources we are concerned with are radioactive materials emitting α and β particles, γ rays, and neutrons; X-ray machines and medium- or low-voltage accelerators generating X-rays, electrons, and hydrogen or helium ions with energies below 20 to 30 MeV; nuclear-fission reactors as sources of neutrons and γ rays; and nuclear-fusion devices as sources of neutrons. Particles from high-energy accelerators, space radiation, and direct radiation from nuclear weapons lie outside our scope. Radionuclides that are mixed with air or water or are deposited on the ground are considered. Internal sources from radionuclides inhaled, ingested, or injected (as in nuclear medicine) are also treated. Table 1.1 summarizes important properties of ionizing particles and electromagnetic radiation (X-ray and γ-ray photons).

In nuclear medicine and radiology, and in evaluating the dose at depth for radiation protection purposes, it is necessary to investigate the attenuation and absorption of radiation in body tissues and organs. Because of differences in size and development, it is usual to specify a standard or reference model (sometimes for different ages or sex). The most elaborate models include the size and approximate shape of many organs. Bone and soft tissue are always differentiated, and soft tissue may be specified in terms of density and content of fat, muscle, or other watery tissue. Figure 1.2 shows a model used in calculations of organ dose from internally deposited radionuclides. In some calculations the body is approximated by ellipsoidal or other simple shapes and by a homogenized composition, or even by just water or plastic. Anthropomorphic phantoms are available commercially with a human skeleton, plastic approximating soft-tissue composition and density, and low-density spaces for the lungs.

An accuracy of 30% in dose calculations may be considered good except in radiation therapy, where an accuracy of better than 10% may be needed to achieve the desired therapeutic response (destruction of all tumor cells) while not damaging surrounding normal tissues excessively.

Specification of the shield geometry, composition, and density is the essence of shield design. Experience is needed to design efficient shields. Shielding of fission fragments, ions, and electrons is easy because of their small range in solid or liquid matter (micrometers to millimeters), although generation of penetrating X-rays (bremsstrahlung) by high-energy electrons must be considered. X-Rays and γ rays are attenuated by dense and preferably high-atomic-number materials, such as lead or iron, or by concrete.

TABLE 1.1. Properties of Ionizing Radiations

Type	Particle	Symbol	Charge[a]	Rest Mass (u)[b]	Half-life[c]
Ions	Proton	p	+e	1.007	stable
	Deuteron	d	+e	2.014	stable
	Triton	t	+e	3.015	12.3 a
	Alpha	α	+2e	4.002	stable
	Fission fragment	f	~+20e	~ 100	varies
Electrons	Negatron	β^-, e^-	-e	4.586×10^{-4}	stable
	Positron	β^+, e^+	+e	same	stable
Neutral	Photon	γ	0	0	stable
	Neutron	n	0	1.009	11.7 min

[a] One electronic charge, $e = 1.60210 \times 10^{-19}$ coulomb (C).

[b] One unified mass unit, $u = 1.6604 \times 10^{-24}$ gram (g)
$= 9.31478 \times 10^{8}$ electron volt (eV).

[c] Abbreviations for units of time used in tables in the present work are as follows: s = second, min = minute, h = hour, d = day, a = year.

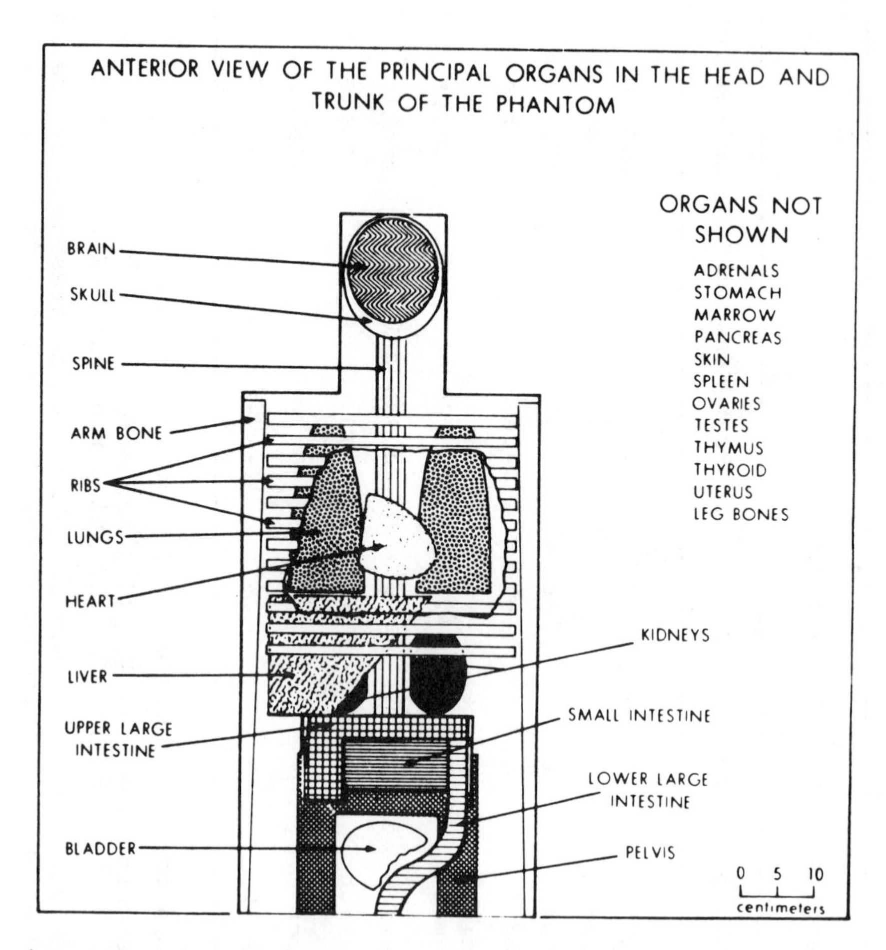

Fig. 1.2. Anatomical model for calculation of dose to an organ from γ-ray-emitting radionuclide deposited in another organ (Society of Nuclear Medicine MIRD Pamphlet 5, 1969).

Thermal neutrons are easily absorbed, especially by certain elements such as boron and cadmium. In materials such as water, plastics, or concrete, fast neutrons are slowed down by collisions with hydrogen. A problem with neutron shields is production of secondary γ rays by neutron capture or inelastic scattering. Additional material has to be included to attenuate these γ rays.

Practical shields usually contain access holes, penetrations for piping and electrical cables, and the like. Manufacturing tolerances and allowances for expansion may leave gaps that permit streaming of γ rays and neutrons, shortcircuiting part of the shield. Shield analysis proceeds by the steps outlined. Shield synthesis involves engineering and economic sense, reiterated calculations, and dose limits or other design criteria. The iteration process is aided if the reduction in dose for unit increase in layer thickness is known.

The interactions of radiation with matter are described by cross-sections. The cross section gives the probability per target atom and unit flux density of incident radiation that a given interaction will occur. Cross sections depend on the type of incident particle and its energy. Differential scattering cross sections also involve the angle of scattering. Cross sections are also defined for energy transfer and production of secondary radiation, such as neutron capture or inelastic-scattering γ rays. The cross section is a function of the energy of the neutron, the energy of the γ ray, and the angular distribution of the γ rays unless isotropic emission is assumed. Obtaining the cross sections in sufficient detail is only part of the job. It is also necessary to reduce them to a form acceptable to the radiation transport computer program or other calculational procedure. This often involves obtaining "group" cross sections, averaged over a certain energy interval, plus group-to-group transfer cross sections and some representation of angular distributions (for example, in Legendre polynomial expansions as discussed later). In the kernel method, linear attenuation coefficients and buildup factors are used. The linear attenuation coefficient is the sum of the products of the atom densities and corresponding total cross section for each element. The buildup factor corrects dose to include scattered radiation and is a function of depth of penetration in the medium as well as source energy and angular distribution. Other material properties are used in approximate methods for neutron calculations. Transmission of charged particles may be handled simply by range and stopping power (dE/dx) concepts, or by more elaborate methods that give emergent flux densities and spectra.

The type of detector used in most calculations gives the "dose" or other response in terms of

$$D(\underline{r}) = \int h(E)\, \phi(\underline{r},E)\, dE$$

where h(E) is the energy-dependent response function or damage function and $\phi(\underline{r},E)$ is the flux density spectrum at position $\underline{r}$. Another detector may give the flux density spectrum, the current density spectrum, or even the angular and energy distribution of the flux density. The energy-integrated flux density (or "flux") is easily obtained with h(E) = 1. Sometimes the uncollided flux density (or flux density spectrum, or dose) and the scattered flux density (or spectrum, or dose) are calculated separately. Any of these quantities may be calculated at a given point or averaged over a given surface or volume.

With the availability of large, fast, digital computers it has become possible to calculate radiation transport in three-dimensional source-shield-detector configurations, including essentially all materials and interactions exactly, using the Monte Carlo or stochastic procedure. However, if good statistical precision is to be achieved, the calculations can become quite lengthy and expensive, especially for thick or complex shields. The difficulty is compounded if differential information is needed, such as energy or angular distributions. Also, the proper application of Monte Carlo methods to deep penetrations requires some expertise in selection of variance-reduction parameters at this stage of development. Discrete-ordinates transport programs allow a nearly exact calculation of differential distributions and doses in one-dimensional geometries (slabs, spheres, or infinite cylinders) at reasonable cost. Two-dimensional discrete-ordinates programs are available but require very large computer memories and long running times in deep penetration shielding problems. Thus approximate but faster and less costly procedures are often used where they can obtain results of adequate accuracy for the purpose. These methods include the kernel-buildup factor method for γ rays and the removal-diffusion method for fast neutrons. Diffusion theory alone, which is often quite successful for thermal reactor criticality calculations, is not suitable for calculations of the anisotropic flux density of fast neutrons in a shield.

Dose to occupational workers and the public is limited by federal and state regulations, based on biological effects of radiation and recommendations of expert groups such as the National Council on Radiation Protection and Measurements (NCRP) and the International Commission on Radiological Protection (ICRP). Dose or response criteria for other applications are not as well formalized, but radiation effects and standards are discussed later.

Verification of shield design and dose to tissues or specimens requires measurements of dose or flux density and spectra. Measurements of cross sections and other data such as reflection or albedo factors also require detectors, spectrometers, or dosimeters of various kinds. These devices and associated electronic instrumentation are discussed later.

2

SOURCES

2.1 RADIONUCLIDES

The decay of radioactive nuclei may be accompanied by emission of α particles, positive or negative electrons, γ or X-rays, or fission fragments and neutrons.

Radioactive Decay Laws

The probability of decay, that is, the probability of spontaneous transformation of one nuclide into another, is independent of age and environment. The mean activity or decay rate

$$A = \lambda N \quad \text{transformations-s}^{-1} \tag{2.1}$$

where N is the number of nuclei at a given time and the decay constant $\lambda(s^{-1})$ is the probability that one nucleus will decay in unit time. The SI unit of activity is simply transformations per second (s^{-1}). Other units are the becquerel (Bq) and curie (Ci):

$$1 \text{ Bq} = 1.0 \quad \text{transformations-s}^{-1} \tag{2.2a}$$

$$1 \text{ Ci} = 3.7 \times 10^{10} \text{ transformations-s}^{-1} \tag{2.2b}$$

exactly. The number of particles emitted per second, the source strength, is equal to the decays per second times n, the number of particles of that type emitted per decay, on the average; hence

$$S_i = n_i A \tag{2.3}$$

where the subscript i refers to particles of a given type, such as electrons or γ-rays.

The activity is equal to the rate of depletion of the original number of radioactive nuclei; hence

$$\frac{-dN}{dt} = \lambda N \tag{2.4}$$

Integrating with an initial number N_0 nuclei,

$$A(t) = \lambda N(t) = \lambda N_0 e^{-\lambda t} = A_0 e^{-\lambda t} \tag{2.5}$$

The reciprocal of the decay constant is the mean life, τ, the time for the activity to decay to $(1/e) = 0.368$ of the initial value. A more common measure is the half-life, the time for the activity to decay to one-half of the initial value,

$$T_{1/2} = \frac{\ln 2}{\lambda} = \frac{0.693}{\lambda} \tag{2.6}$$

The product nucleus may also be radioactive. If the decay series is

$$N_a \rightarrow N_b \rightarrow N_c \quad \text{(stable)}$$

the differential equation for the number of daughter or b nuclei is

$$\frac{dN_b}{dt} = \lambda_a N_a - \lambda_b N_b \tag{2.7}$$

where the first term on the right-hand side is the rate of production from the transformation of parent nuclei and the second term is the rate of depletion by radioactive decay of the daughter nuclei. With the initial conditions $A_a(0) = A_{a0}$ and $A_b(0) = A_{b0}$ the activity of the daughter is

$$A_b(t) = A_{b0} \exp -\lambda_b t$$
$$+ A_a(t) \frac{\lambda b}{\lambda_b - \lambda_a}\left(1 - \exp -(\lambda_b - \lambda_a)t\right) \tag{2.8}$$

while $A_a(t)$ is given by Eq. (2.5) using A_{a0} and λ_a.

If the daughter has a longer half-life than the parent (or $\lambda_b < \lambda_a$), the second term in Eq. (2.8) indicates a buildup followed by a decay that approaches the decay of type b nuclei alone. If the daughter has a shorter half-life than the parent (or $\lambda_b > \lambda_a$), the daughter activity eventually follows the rate of decay of the parent, a condition termed "transient equilibrium." Then the ratio of daughter:parent activity is $\lambda_b/(\lambda_b - \lambda_a)$. Details of series decay may be found in Evans [1].

Modes of Decay and Spectra

The important modes of radioactive decay for radiation shielding and dosimetry are summarized in Table 2.1. The decay scheme, half-life, and other useful information may be found in the *Table of Isotopes* [2]. Properties of certain nuclides are given in Table 2.2.

α Decay is characterized by emission of one or more α particles ($^4He^{++}$ ions). If the daughter nucleus is left in an excited state, deexcitation and return to the ground state (or lower excited states) is accompanied by emission of γ-rays. Figure 2.1 is a decay diagram for the α decay of ^{220}Rn. Decay to the 0.550-MeV excited state of ^{216}Po occurs in 0.07% of the transformations, whereas decay to the ground state occurs in 99.93% of the transformations. The transition energy $E_t = Q_\alpha - E_s$, where E_s is the energy above the ground state. The transition energy is shared by the α particle and the recoil nucleus (^{216}Po),

$$E_\alpha + E_R = E_t \tag{2.9}$$

Conservation of momentum requires

$$E_\alpha = E_t \frac{M_0}{M_\alpha} \tag{2.10}$$

where the reduced mass

$$M_0 = \frac{M_\alpha M_R}{M_\alpha + M_R} \tag{2.11}$$

The energies of the α particles are then $E_{\alpha 1} = 5.7470$ MeV, with a recoil ^{216}Po ion energy of 0.1064 MeV and

TABLE 2.1. RADIOACTIVE DECAY MODES

Name (Symbol)	Product Nucleus	Principal Radiation	Other Radiation
Alpha (α)	$Z - 2$ $A - 4$	α line	γ line (often)
Negatron beta (β^-)	$Z + 1$ A	β^- continuous	γ line (often)
Positron beta (β^+)	$Z - 1$ A	β^+ continuous	γ line (often) 0.511-MeV annihilation
Electron capture (EC)	$Z - 1$ A	neutrino (not observed)	X or Auger e^- line (daughter)
Internal conversion (IC)	Z A	e^- line	X or Auger e^- line (parent)
Isomeric transition (IT)	Z A	γ line	
Spontaneous fission (SF)	~1/2 Z ~1/2 A	2 fission fragments	n continuous γ continuous

TABLE 2.2. RADIONUCLIDE SOURCES

Nuclide (Decay)	$T_{1/2}$	Particles			Photons	
		n/dis	E_m (MeV)	$\bar{E}$ (MeV)	n/dis	E (MeV)
^{3}H (β^-)	12.3 a	1.00	0.018	0.0057	0.00	
^{14}C (β^-)	5730.0 a	1.00	0.156	0.0493	0.00	
^{16}N (β^-)	7.2 s	0.26	10.4		0.05	7.11
		0.68	4.27		0.69	6.13
		0.05	3.31		0.01	2.75
^{22}Na (EC, β^+)	2.60 a	0.906	0.546	0.216	1.00	1.275
		0.094		EC	1.81	0.511
^{24}Na (β^-)	15.0 h	1.00	1.392	0.555	1.00	2.754
					1.00	1.368
^{32}P (β^-)	14.3 d	1.00	1.710	0.695	0.00	
^{35}S (β^-)	87.0 d	1.00	0.167	0.049	0.00	
^{55}Fe (EC)	2.70 a	Auger 5 to 6 keV			0.26	0.006
^{57}Co (EC)	270.0 d	0.01	0.129	IC	0.10	0.136
		0.02	0.115	IC	0.86	0.122
		0.68	0.007	IC	0.10	0.014
^{60}Co (β^-)	5.26 a	1.00	0.313	0.094	1.00	1.332
					1.00	1.173
^{85}Kr (β^-)	10.7 a	1.00	0.672	0.246	0.004	0.514

^{90}Sr (β^-)	28.1 a	1.00	0.546	0.196	0.00	
^{90}Y (β^-)	64.0 h	1.00	2.27		0.00	
^{99m}Tc (IT)	6.0 h	0.09 0.01	0.119 0.138	IC IC	0.88	0.140
^{131}I (β^-)	8.06 d	0.90	0.606	0.192	0.07 0.82	0.637 0.364
^{137}Cs (β^-)	30.0 a	0.054 0.946	1.176 0.514	0.427 0.175	see ^{137m}Ba	
^{137m}Ba (IT)	2.55 m	0.08	0.624	IC	0.90	0.662
^{147}Pm (β^-)	2.62 a	1.00	0.224	0.070	0.00	
^{210}Po (α)	138.0 d	1.00	5.305			
^{220}Rn (α)	55.0 s	0.0007 0.9983	5.747 6.287	α_1 α_2	0.0007	0.550
^{241}Am (α)	458.0 a	0.85 0.13 0.32	5.484 5.442 0.039	α α IC	0.38	0.06

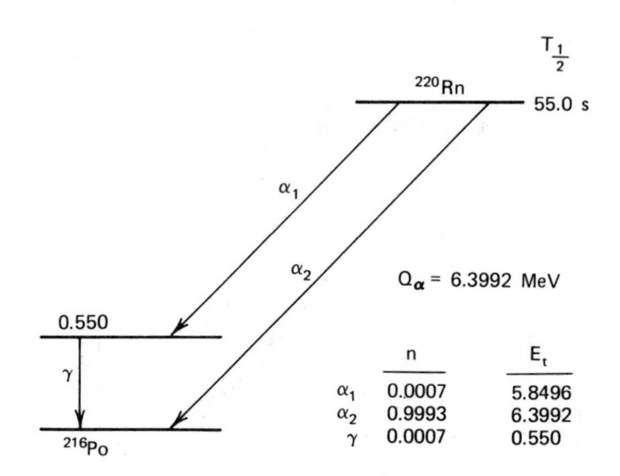

Fig. 2.1. Decay scheme of radon-220

$E_{\alpha 2}$ = 6.2870 MeV with E_{R2} = 0.1164 MeV. The energy of the γ ray is 0.550 MeV, with recoil negligible even for light nuclei (the spectra are shown in Figure 2.2). Emission is isotropic.

In β^- decay a neutron in the nucleus transforms into a proton, negative electron, and a neutrino (actually an antineutrino), $n \rightarrow p + e^- + \bar{\nu}$. The proton stays in the nucleus hence the atomic number of the daughter nucleus is one greater than the atomic number of the parent. The neutrino and electron escape. The neutrino interacts hardly at all with matter and is not significant in damage or heating. However, its influence is felt because the available transition energy is shared between the electron and the neutrino (the share of the residual nucleus is negligible because the other particles are so much lighter). The shape of the electron (β particle) spectrum depends on the transition energy and hence maximum energy available to the electron, E_{max} (when the neutrino carries off no energy), the atomic number Z, and the change in nuclear spin and parity [3]. A transition between states where the spin changes by one unit (or none, except 0 → 0) and there is no change in parity is called "allowed." Many β decays are of this type. Some examples of allowed spectra for the decays of the pure β emitters ^3H, ^{14}C, and ^{32}P are shown in Figure 2.3. In radiation shielding and dosimetry it is seldom necessary to know the exact

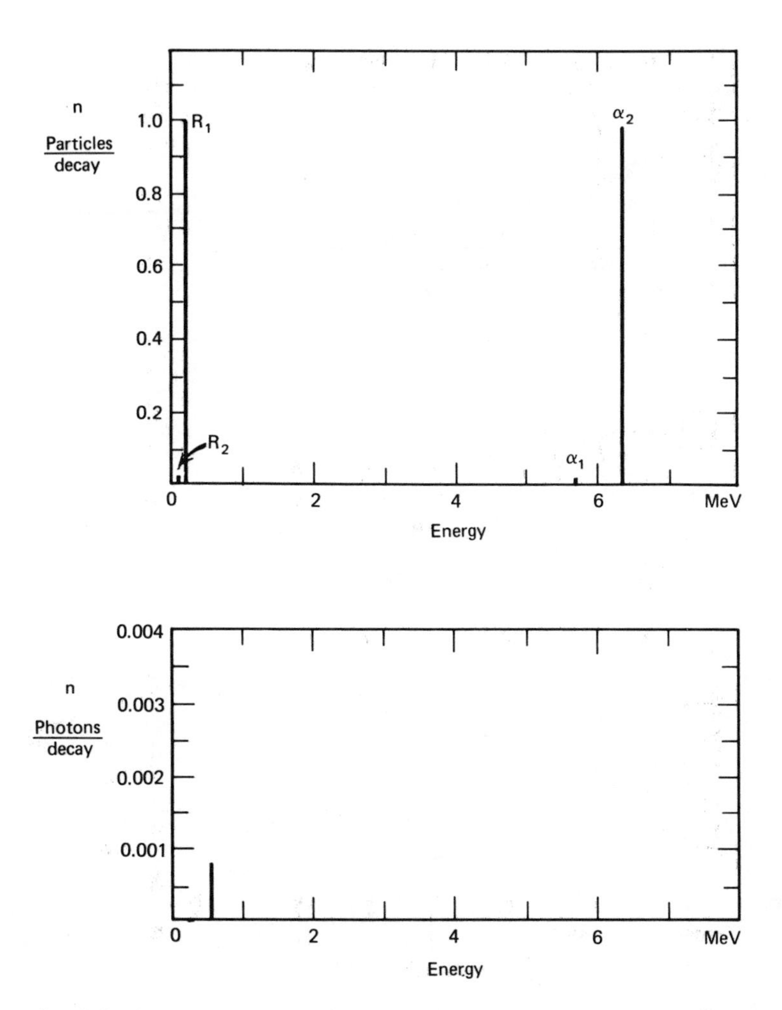

Fig. 2.2. Ion spectrum (α particles and recoil nuclei) and γ-ray spectrum in decay of radon-220.

shape of the spectrum, and most problems can be solved knowing only the maximum energy and the average energy of the β particles. The maximum energy $E_{max} = Q_{\beta^-} - E_s$, where E_s is the energy of any excited state; the maximum penetration or range is a function of E_{max}. Since the range of β particles is small in solids or liquids, it is usually a good approximation to assume the energy is deposited at the point of emission. Then the average

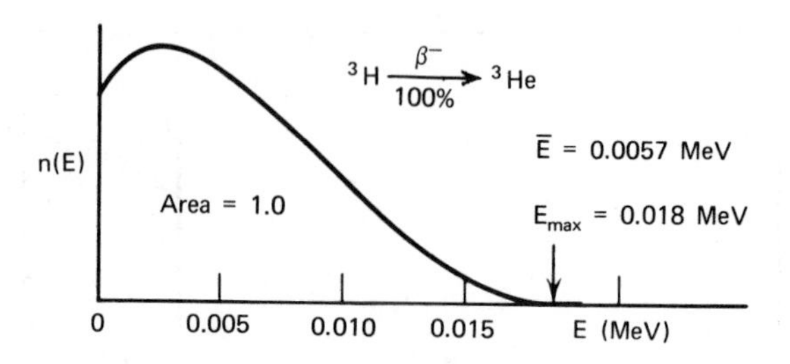

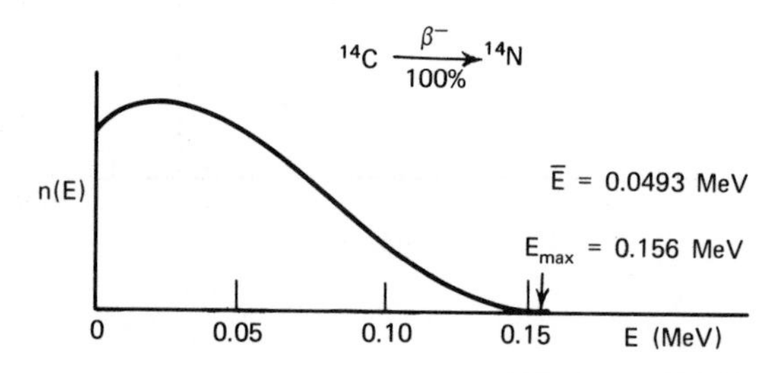

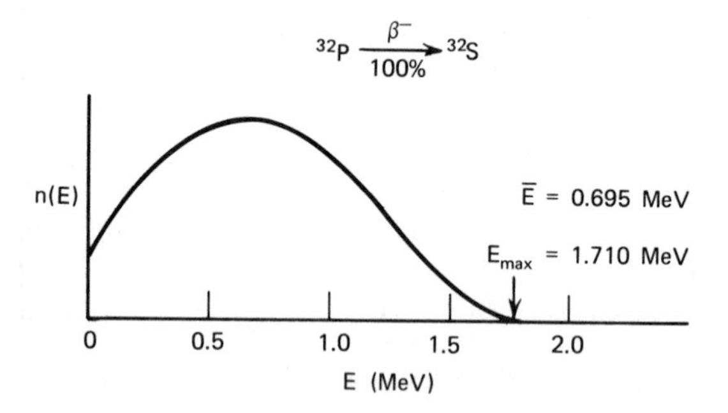

Fig. 2.3. Spectrum of β particles in negatron β decay n(E) electron/MeV per decay, as a function of energy, for ^{3}H, ^{14}C, and ^{32}P.

energy absorbed is equal to the average energy of the β-particle spectrum, $\bar{E}$. Graphs of $\bar{E}/E_{max}$ as a function of E_{max} and the Z of the daughter nuclide, for allowed and forbidden transitions, have been published [4].

Many beta decays are followed by deexcitation of an excited state with emission of γ rays. Figure 2.4 shows the decay scheme of cobalt-60 and the spectrum

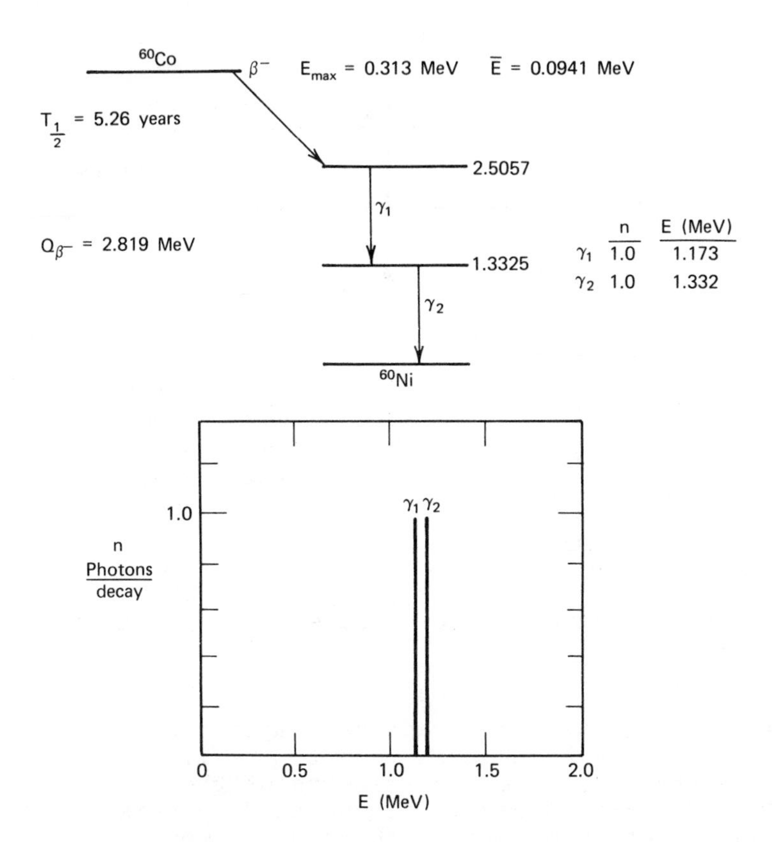

Fig. 2.4. Decay scheme and γ-ray spectrum for negatron decay of cobalt-60.

of γ rays from the decay. Except for nuclei aligned in a magnetic field, β-particle and γ-ray emission is isotropic.

In β^+ decay, a proton is transformed into a neutron, positive electron, and neutrino, $p \rightarrow n + e^+ + \nu$. The atomic number of the daughter is one less than the atomic number of the parent. When a balance is made on the rest masses of the parent and daughter atoms (including the orbital electrons for neutrality), it is found that rest energy equivalent to two electron masses, or 1.022 MeV total, must come out of the transition energy. Hence positron decay is energetically impossible unless the transition energy is at least 1.022 MeV. Figure 2.5 illustrates the position decay

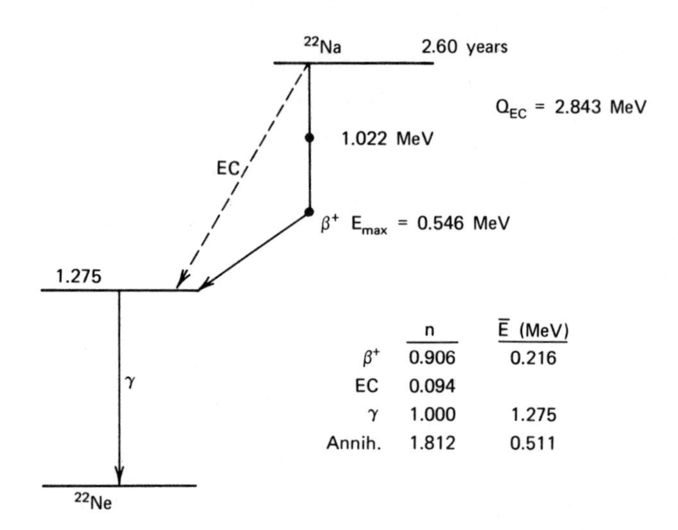

Fig. 2.5. Positron β decay and electron capture decay of ^{22}Na.

of sodium-22. Electron capture competes with positron decay. In ^{22}Na both modes lead to the 1.275-MeV excited state in ^{22}Ne and emission of the 1.275-MeV γ ray. Positron decay is accompanied by emission of two annihilation quanta, each of 0.511 MeV, when the positron slows down and combines with a free electron in matter. The spectrum of positrons is somewhat different from the spectrum of negatrons because of the influence of the nuclear Coulomb field but otherwise follows the same theory. The maximum and average energies are often sufficient.

Quantum mechanics tells us that there is a finite probability that an atomic orbital electron will find itself in the nucleus. A K-shell electron, or sometimes an L-shell electron, may be captured by the nucleus. In electron capture decay a proton is transformed into a neutron and a neutrino is emitted. The neutrino is practically undetectable but X-rays or Auger (ohzhay) electrons are emitted as the vacancy in the K or L shell is filled. The X-ray energies are characteristic of transitions in the daughter nucleus (Z - 1). In Auger electron emission, the atomic transition energy that would otherwise be imparted to an

X-ray is instead transferred to an outer-shell electron, which in turn is ejected. The ratio of X-ray photons emitted per inner shell vacancy is called the "fluorescence yield." X-ray emission is more probable in heavy elements, while Auger electron emission is favored in light elements. In the electron capture decay of ^{22}Na, Auger electrons are emitted with an average energy of 0.0008 MeV. Electron capture competes with β^+ decay and is more probable for large Z and small transition energy.

Internal conversion is a competing process to radiation of a γ ray and is highly probable at large Z and small transition energy. The energy that would otherwise go to a γ ray is imparted to an atomic electron (K, L, or perhaps M shell). The energy of the electron is

$$E_{e^-} = E_t - E_B \tag{2.12}$$

where E_B is the binding energy or K-, L-, or M-shell absorption-edge energy in the parent atom. The ratio of internal conversion electrons to γ-ray photons is the internal conversion coefficient

$$\alpha = \alpha_K + \alpha_L + \ldots + = \frac{N_e}{N_\gamma} \tag{2.13}$$

Sometimes the percentage of photons or electrons is given instead of the ratio. Note that internal conversion electrons are monoenergetic, unlike the continuous-spectrum β particles. After ejection of the inner-shell electron, the binding energy E_B is rapidly released as X-rays (characteristic of transitions between atomic levels in the parent atom) Auger electrons, or both. Internal conversion is negligible in ^{60}Co but is significant in heavier elements.

An example of isomeric transition is seen in Figure 2.6. The isomers are ^{137m}Ba and ^{137}Ba. The transition occurs from a relatively long-lived excited state, long enough in fact that ^{137m}Ba can be chemically separated from the ^{137}Cs and treated as a distinct radioactive substance. Barium-137m decays with emission of a 0.6616-MeV γ ray. Because of internal conversion, internal conversion electrons, X-rays, and Auger electrons are also emitted as shown in Figure 2.6. If chemical separation is not carried out, it is

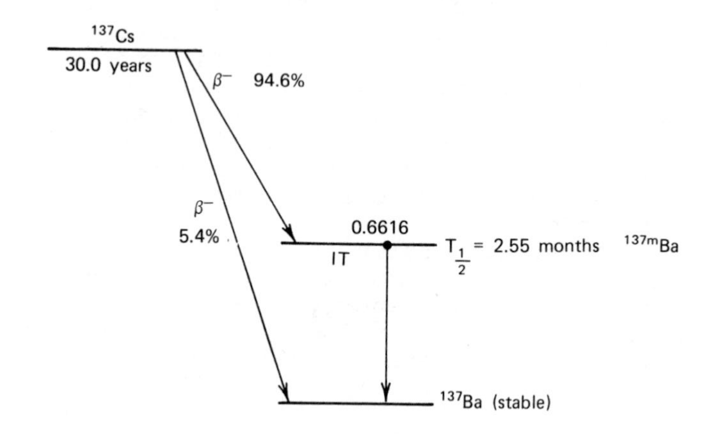

Fig. 2.6. Negatron decay of ^{137}Cs and isomeric transition decay of ^{137m}Ba. Decay of ^{137m}Ba (isomeric transition):

Radiation	n (number/dis)	E(Mev)
γ	0.8981	0.6616
K IC e^-	0.0820	0.6241
L IC e^-	0.0147	0.6560
M IC e^-	0.0049	0.6605
K X-rays	0.0728	0.0348 average
L X-rays	0.0133	0.0044
Auger	0.0133	0.001-0.0263

conventional to refer to the 0.6616-MeV γ ray from ^{137}Cs, as in other β decays followed by γ-ray emission.

Spontaneous fission is a mode of radioactive decay in which a heavy nucleus splits roughly in half (actually with a wide range of light and heavy fission fragments) and emits a continuous spectrum of neutrons and γ rays. Spontaneous fission usually competes with α decay, and the rate of depletion is given by the combined decay constant $\lambda = (\lambda_\alpha + \lambda_{SF})$. Neutrons from spontaneous fission of ^{238}U, ^{240}Pu, or ^{242}Pu are sometimes significant in shutdown reactors or fuel elements, but the source strength is small compared to prompt fission neutrons in the operating reactor.

Neutron Sources

Californium-252 is a spontaneous fission nuclide used as a laboratory source of neutrons in encapsulated sources up to several milligrams. Properties of ^{252}Cf are summarized in Table 2.3. The neutron spectrum is plotted in Figure 2.7 and the spectra of neutrons, prompt fission γ rays, and equilibrium fission-product γ rays are given in Table 2.4 [5]. In addition, there are 2.78×10^9 photons/second per gram (s-g) of ^{252}Cf of 0.043 MeV energy, and 2×10^9 photons/s-g of 0.1-MeV energy emitted in α decay, but most of these low energy γ rays will be absorbed in the source capsule.

Neutrons are also produced by bombardment of beryllium by α particles from radionuclides, or by bombardment of beryllium or heavy water (deuterium oxide, D_2O) by energetic γ rays. The reactions are

$$^{9}Be + \alpha \rightarrow n + {}^{12}C^* + 5.704 \text{ MeV}$$

$$^{9}Be + \gamma \rightarrow n + {}^{8}Be - 1.666 \text{ MeV}$$

$$^{2}H + \gamma \rightarrow n + {}^{1}H - 2.225 \text{ MeV}$$

The various (α,n) sources differ according to the α-particle emitter chosen. The most common ones are ^{241}Am, ^{239}Pu, and ^{210}Po. Characteristics of these sources are summarized in Table 2.5. The Am-Be and Po-Be sources are made by mixing fine powders of beryllium metal with polonium metal or americium oxide. The Pu-Be source is an intermetallic compound, $PuBe_{13}$. The 4.43 MeV γ ray comes from deexcitation of $^{12}C^*$. Theoretically, the neutron spectrum could be calculated from the α-particle energy, the Q value, and properties of beryllium. The spectrum is not monoenergetic because of loss of α-particle energy by ionization and integration over all angles of incidence. Because of uncertainties in particle size and mixing, it is usual to use measured spectra [6,7]. Neutron spectra are plotted in Figure 2.8.

The γ-ray energy in a photoneutron source must be higher than the Q value. The most commonly used radionuclide is ^{124}Sb, which has a half-life of 60.4 d and emits a 1.692-MeV γ ray in 48% of the decays, as well

TABLE 2.3. PROPERTIES OF CALIFORNIUM-252

Neutron emission rate	2.34×10^{12} n/s-g
	4.4×10^{9} n/s-Ci
Neutrons per fission	3.76
Average n energy	2.35 MeV
Effective half-life	2.65 a
α Half-life	2.73 a
SF Half-life	85.5 a
Average α energy	6.12 MeV
Total heating	38.5 W/g
n Dose rate at 1 m	2.2×10^{3} rem/h-g
γ Dose rate at 1 m	1.6×10^{2} rad/h-g

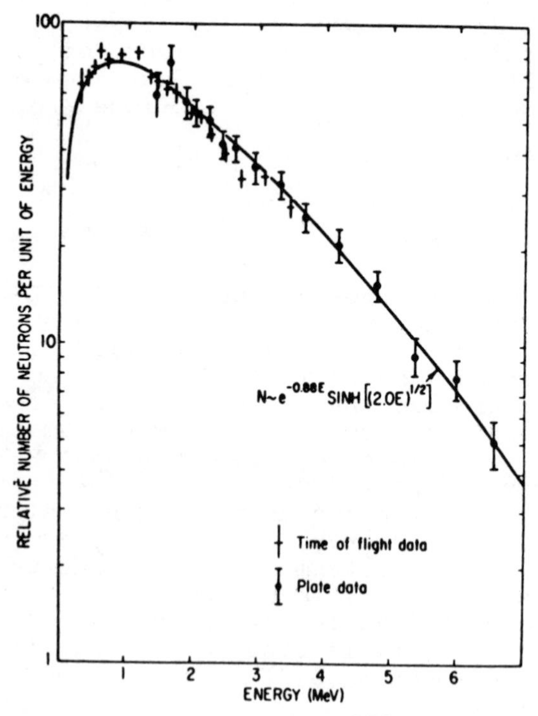

Fig. 2.7. Neutron spectrum of ^{252}Cf spontaneous fission (after Californium-252 Symposium, CONF-681032; from A. Edward Profio, *Experimental Reactor Physics*, Wiley, New York, 1976, p. 125).

TABLE 2.4. NEUTRON AND γ-RAY SPECTRUM OF ^{252}Cf

Neutrons

Energy (MeV)	Neutrons/s-g
0.0 to 0.5	2.8×10^{11}
0.5 to 1.0	3.7×10^{11}
1.0 to 2.0	7.6×10^{11}
2.0 to 3.0	4.6×10^{11}
3.0 to 4.0	2.8×10^{11}
4.0 to 5.0	1.6×10^{11}
5.0 to 6.0	5.6×10^{10}
6.0 to 7.0	4.0×10^{10}
7.0 to 8.0	1.3×10^{10}
8.0 to 10.0	9.9×10^{9}
10.0 to 13.0	2.2×10^{9}

γ Rays (photons/s-g)

Energy (MeV)	Prompt Fission	Fission Product	Total
0.0 to 0.5	3.3×10^{12}	1.3×10^{12}	4.6×10^{12}
0.5 to 1.0	1.7×10^{12}	4.0×10^{12}	5.7×10^{12}
1.0 to 1.5	7.7×10^{11}	9.1×10^{11}	1.7×10^{11}
1.5 to 2.0	4.2×10^{11}	3.5×10^{11}	7.7×10^{11}
2.0 to 2.5	2.2×10^{11}		
2.5 to 3.0	1.1×10^{11}		
3.0 to 3.5	5.6×10^{10}		
3.5 to 4.0	3.0×10^{10}		
4.0 to 4.5	1.7×10^{10}		
4.5 to 5.0	8.2×10^{9}		
5.0 to 5.5	4.9×10^{9}		
5.5 to 6.0	1.8×10^{9}		
6.0 to 6.5	1.0×10^{9}		

TABLE 2.5. PROPERTIES OF α/NEUTRON SOURCES

	^{241}Am-Be	^{239}Pu-Be	^{210}Po-Be
Yield (n/s-Ci)	2.2×10^6	1.7×10^6	2.5×10^6
Grams emitter/Ci	0.3	16	2.2×10^{-4}
Half-life	458 a	24360 a	138 d
Approximate mℓ/Ci	3	12	0.1
Average n energy (MeV)	4.5	3.2	4.2
n Dose rate (mrem/h-Ci at 1 m)	1.7	1.3	1.9
γ Energies (MeV)	0.06, 4.43	4.43	0.8, 4.43
γ Dose rate (mR/h-Ci at 1 m)	10, 1[a]	0.08	0.006, 0.11
Heating (W/Ci, no fission)	0.033	0.031	0.032

[a]Approximately 0.03 with capsule attenuation.

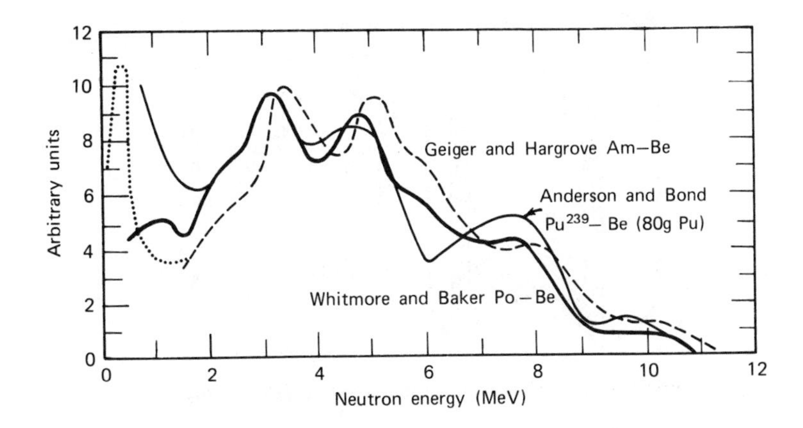

Fig. 2.8. Neutron spectra of α/n sources (from DePangher and Tochilin [7]).

as several other γ rays. The energy of the neutron emitted by an ^{124}Sb-Be source is 26 keV, with some spread because of scattering. The yield depends on the thickness of the beryllium shell surrounding the γ-ray source. For a thickness of 2 cm, practical yield is about 10^7 n/s per curie. However, the neutrons are accompanied by a very intense γ-ray field. The ^{124}Sb γ-ray energy is too low for the ^{2}H(γ,n) reaction. High-energy γ or X-rays are found in reactors and high-energy electron accelerators, as is discussed later.

Encapsulation

The α-particle emitters and other emitters with activities higher than a few microcuries are normally encapsulated to prevent spread of radioactive contamination. Sources such as ^{252}Cf, the (α,n) sources, and other hazardous sources are doubly encapsulated in refractory metals to protect against corrosion, fire, or leakage. Although radioactive emission is isotropic, the capsule and self-absorption may introduce anisotropy. Some capsule designs are shown in Figure 2.9.

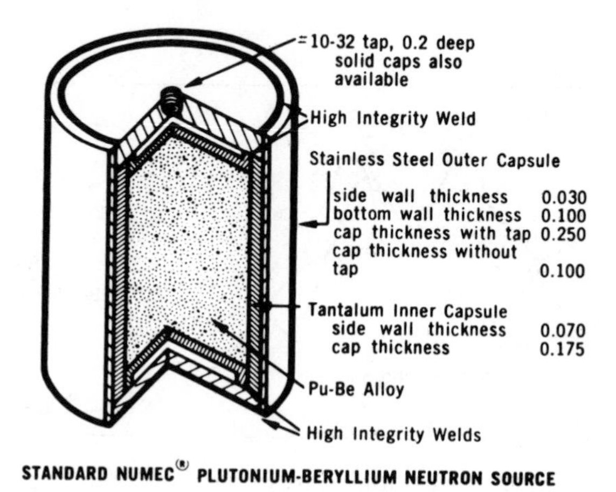

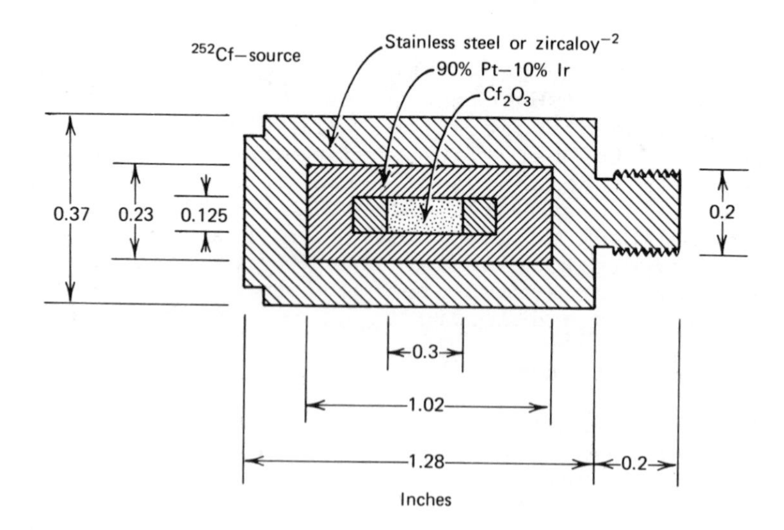

Fig. 2.9. Designs of radioactive source capsules, (from A. Edward Profio, *Experimental Reactor Physics*, Wiley, New York, 1976).

2.2 X-RAY MACHINES AND ACCELERATORS

In X-ray machines and accelerators, charged particles are accelerated to high energies by electromagnetic fields. The charged particle (electron or ion) may be extracted or may strike a target, generating secondary radiations. Interactions of charged particles with matter are discussed in Chapter 3. Here we are concerned with the method of acceleration and characteristics of the radiations emitted from practical targets.

X-Ray Machines

X-Rays are generated by acceleration of electrons and impingement on a heavy metal target, usually tungsten, in a vacuum tube such as shown in Figure 2.10. The electrons are produced by thermionic emission from a hot filament and are then focused and accelerated by a potential applied between the cathode and anode. The anode consists of a tungsten plate embedded in a cooled copper backing. At higher electron beam power the anode may be rotated to help dissipate the heat. The X-rays are emitted in all directions, but only those along a certain direction are used, exiting through a window. The projected area of the X-ray beam at the anode is small, on the order of mm^2, but the beam then diverges with a cone half-angle on the order of 10 degrees. The tube is inserted in a housing filled with

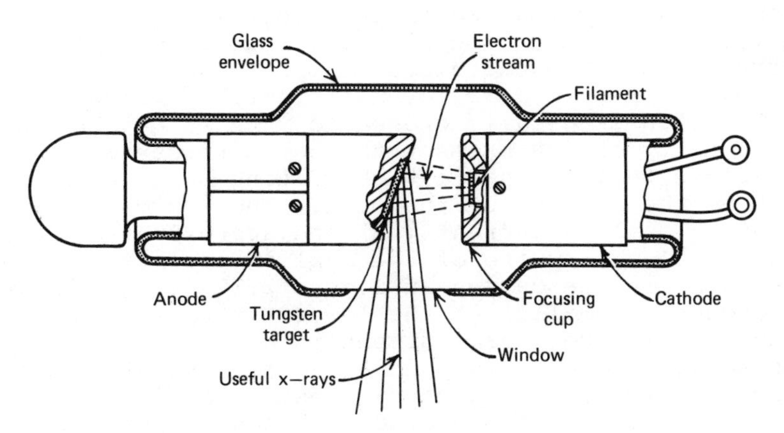

Fig. 2.10. Cross section of X-ray tube.

oil for electrical insulation and heat transfer. The housing reduces the radiation outside the beam to a low, but not negligible, dose rate.

The accelerating potential is provided by a transformer-rectifier type of power supply. Full-wave rectification is usually provided but not much filtering, so the voltage, electron-beam current and X-ray intensity vary, as shown in Figure 2.11, for single-phase alternating current (ac) power. With three-phase power and additional rectifiers, the voltage and current ripple may be reduced to a few percent. Both the X-ray spectrum and the intensity are functions of the voltage.

The X-rays are generated by two processes: bremsstrahlung on deceleration of the electrons and "fluorescence" on filling of vacancies in atomic shells produced by electron impact. Bremsstrahlung production is discussed in Chapter 3. The spectrum is continuous from zero to the maximum energy of the electrons, which is equal to the electron charge times the maximum voltage applied between anode and cathode. Generally we refer to the bremsstrahlung spectrum by the maximum or peak voltage in kilovolts peak, kVp. Superimposed on the continuous spectrum are lines corresponding to the characteristic X-rays of the target, when the incident electron energy exceeds the binding energy of the atomic electron in the K, L, or M shell. The spectrum is modified by attenuation in the window and any additional material between the anode and object. Metal absorbers or filters are often added to attenuate very low-energy X-rays that contribute to skin dose, for example, but not to a useful X-ray image. Thus the spectrum is hard to calculate. Approximate spectra are shown in Figure 2.12 for different kVp values.

Tungsten is usually chosen for the target because the intensity of the bremsstrahlung increases with Z and because tungsten has a high melting point and can withstand the heating by the beam. The source strength is proportional to current (mA) and approximately proportional to the square of the applied potential. Thus the area under the curves in Figure 2.12 (total number of photons) is smaller at the lower kVp values, even though the electron beam current is constant. The fluorescent X-rays corresponding to filling of vacancies in the K shell do not appear at less than 70 kVp in tungsten. The energies of the K_α lines (transitions between L shell and K shell) occur at 57 keV, and the

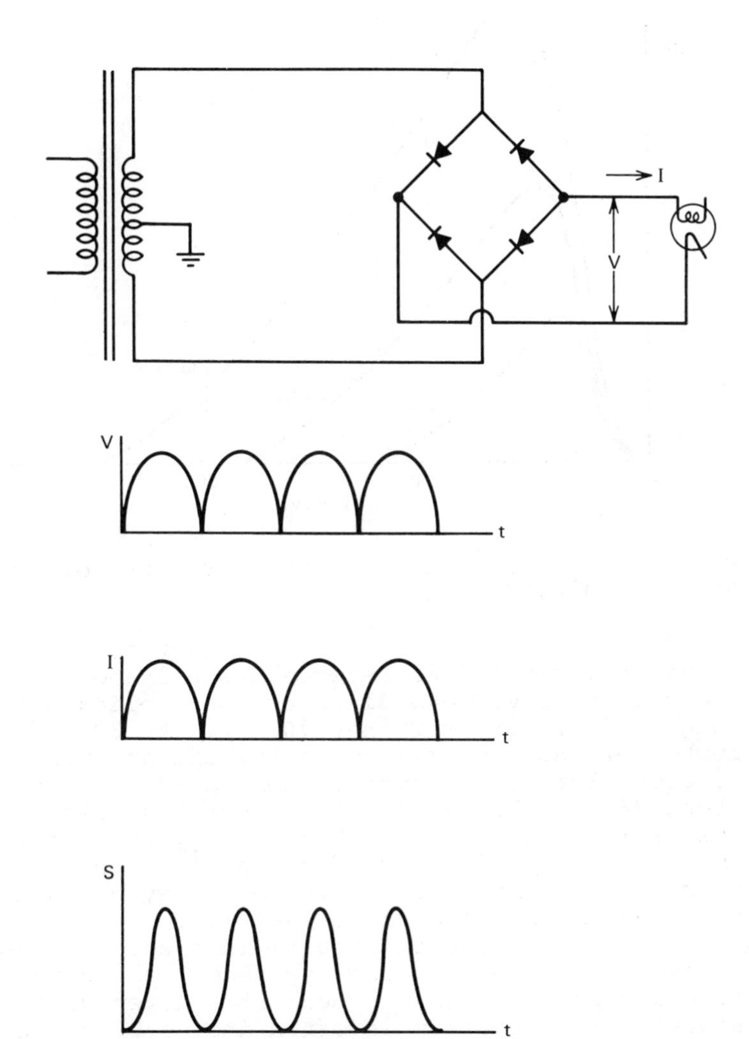

Fig. 2.11. Full wave rectifier circuit for X-ray tube and waveforms of voltage, current, and X-ray source strength.

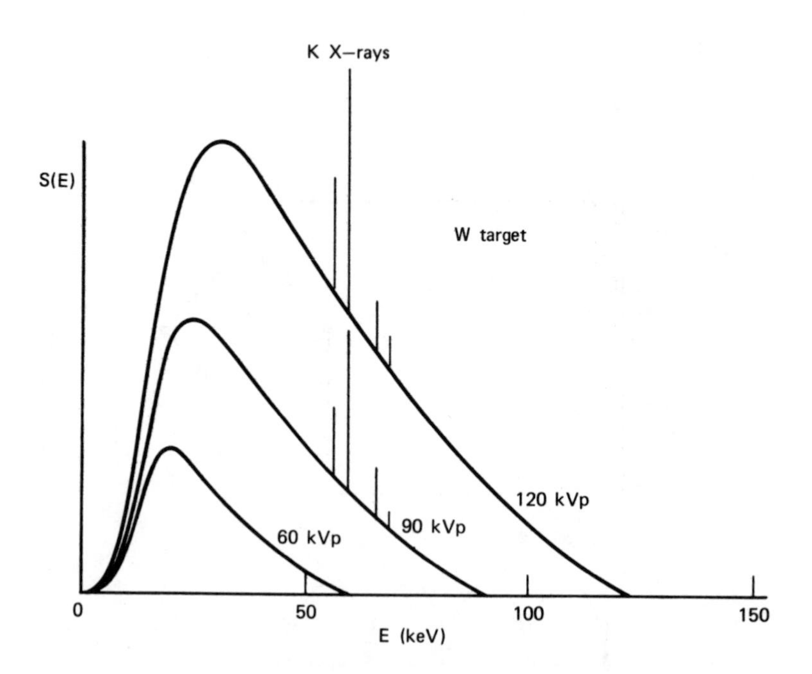

Fig. 2.12. Spectrum of X-rays for peak voltage of 60, 90, or 120 kilovolts.

K_β lines (transitions between M shell and K shell) occur at 69 keV, regardless of the applied potential as long as it is greater than 70 kVp. At 80 kVp the number of photons emitted in the characteristic lines is about 10% of the total, whereas at 150 kVp it is about 28% of the total. At potentials below 200 kVp, over 99% of the electron-beam energy appears as heat (mostly by way of ionization) and less than 1% as X-rays. Additional information on X-ray production may be found in the literature [8-9-10].

Because of the difficulty in predicting spectra and source strengths, one may need to measure the spectrum and strength in the beam at a given distance. However, the spectrum is often indicated by the half-value layer (HVL) or thickness of some absorber (e.g., Al or Cu) required to reduce the dose rate by half (see Section 3.4). The strength is usually given in terms of the dose rate at a fixed distance for a given kVp and mA beam current. Relative exposures or doses are referred to kVp, filtration, and milliamperes times seconds of beam-on time (mAs).

Cockcroft-Walton Accelerator

The Cockcroft-Walton accelerator consists of a source of ions or electrons, accelerating and focusing electrodes in a vacuum system, voltage-multiplying direct-current (dc) high-voltage supply, and target such as shown schematically in Figure 2.13. The high-voltage supply is basically a transformer-rectifier circuit with n rectifiers and n capacitors that are charged to peak potential V and then discharged in series to provide peak potential nV at the terminal when unloaded. A typical value for nV is 200 to 300 kV. The high-voltage circuit and accelerator terminal and column may be insulated by air or transformer oil. The electron source is basically a heated filament as in the X-ray tube. Ion sources use radiofrequency fields or

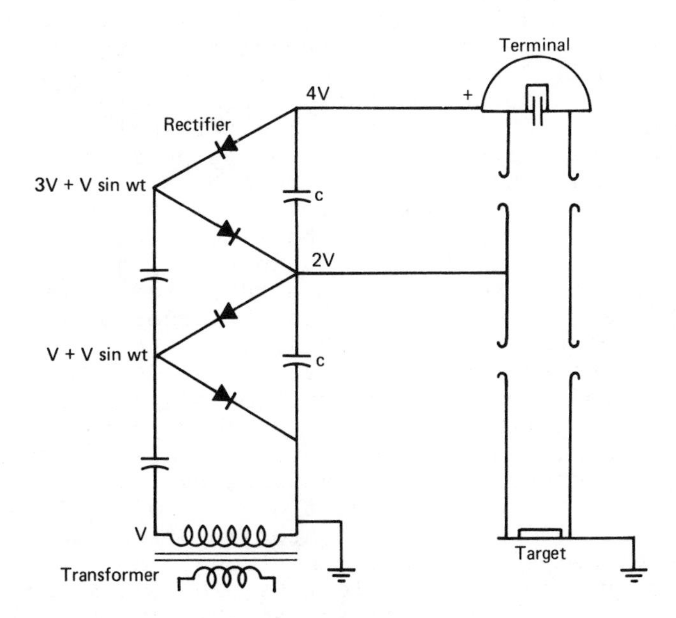

Fig. 2.13. Cockcroft-Walton voltage multiplier circuit and accelerating column.

cold-cathode discharges to generate a low density, low-energy plasma in hydrogen, deuterium, or helium gas. The corresponding protons, deuterons, or helium ions are extracted by an electric field from a small-diameter hole or canal. The extracted beam is accelerated through the potential nV and focused on the target. Typical current at the target is 1 to 2 mA, and the focal spot diameter may be 1 to 2 cm. The Cockcroft-Walton accelerator is often used to generate neutrons by either the $^{3}H(d,n)^{4}He$, or $^{2}H(d,n)^{3}He$ reaction. These and other neutron-producing reactions are discussed later. In addition to neutrons, there may be X-rays (bremsstrahlung) generated by secondary electrons ejected from the target or any parts intercepting some beam and accelerated back to the ion source in the terminal. Accelerated electrons, of course, produce bremsstrahlung and characteristic X-rays in the target itself. The beam energy is usually controlled to 1%.

Van de Graaff Accelerator

The main features of a Van de Graaff accelerator are shown in Figure 2.14. The HV potential at the terminal is generated by mechanical transfer of charge on an insulating belt driven by a motor. The charge is sprayed on by a corona discharge from needle points and removed by collector points in the terminal. The charge then distributes itself on the outside of a polished metal sphere or hemisphere housing the electron or ion source. Potentials of 3 MV are typical, but single-stage accelerators have been built with a potential of 10 MV. Insulation is by SF_6 gas in a pressure vessel. Beam energy is regulated to about 0.1%, and target currents are on the order of 0.1 mA. Higher ion energies can be achieved in the tandem Van de Graaff. Negative ions are accelerated to a positively charged terminal. A fraction of these are then converted to positive ions by passage through a thin foil. The positive ions are then accelerated from terminal to ground, doubling the energy. However, the beam current attainable is on the order of 0.01 mA.

The Van de Graaff accelerator is often used to generate almost-monoenergetic neutrons. Some useful reactions are listed in Table 2.6. The cross sections for these reactions are plotted against incident ion energy in Figure 2.15. The maximum in the T(d,n) cross section occurs at 105 keV. The deuteron reactions are

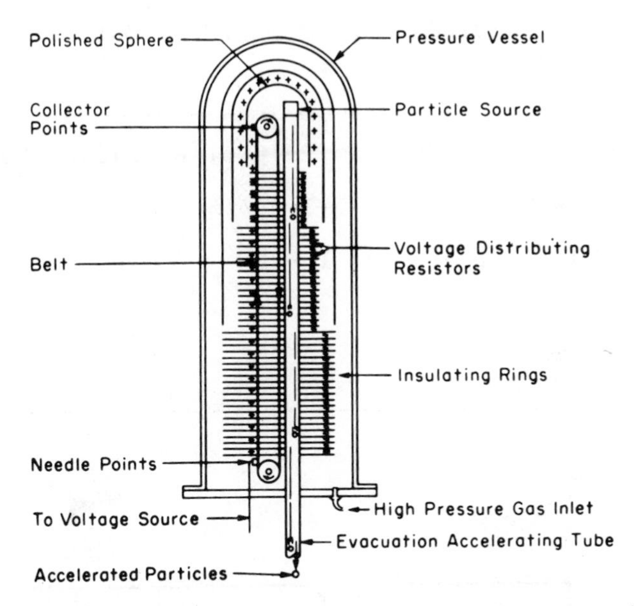

Fig. 2.14. Diagram of Van de Graaff accelerator (from A. Edward Profio, *Experimental Reactor Physics*, Wiley, New York, 1976, p. 182).

exothermic and occur even at very low energies. The proton reactions are endothermic, and the cross section exhibits a threshold.

The neutron energy depends on the Q value, incident ion energy E_1 and mass M_1, target nucleus mass M_2 (assumed at rest), residual nucleus mass M_4, and angle of emission θ_L relative to the incident ion direction. If the mass of the neutron is M_3, kinematics gives, for the neutron energy E_3,

$$\sqrt{E_3} = u \pm \sqrt{u^2 + w} \tag{2.14}$$

where

$$u = \frac{\sqrt{M_1 M_3 E_1}}{M_3 + M_4} \cos \theta_L \tag{2.15}$$

TABLE 2.6. NEUTRON-PRODUCING REACTIONS FOR ION ACCELERATORS

Reaction	$T(p,n)^3He$	$^7Li(p,n)^7Be$	$D(d,n)^3He$	$T(d,n)^4He$
Q Value (MeV)	-0.764	-1.646	3.266	17.586
Threshold (MeV)	1.019	1.882		
Minimum E_n (MeV, 0°)	0.0639	0.0294	2.448	14.05

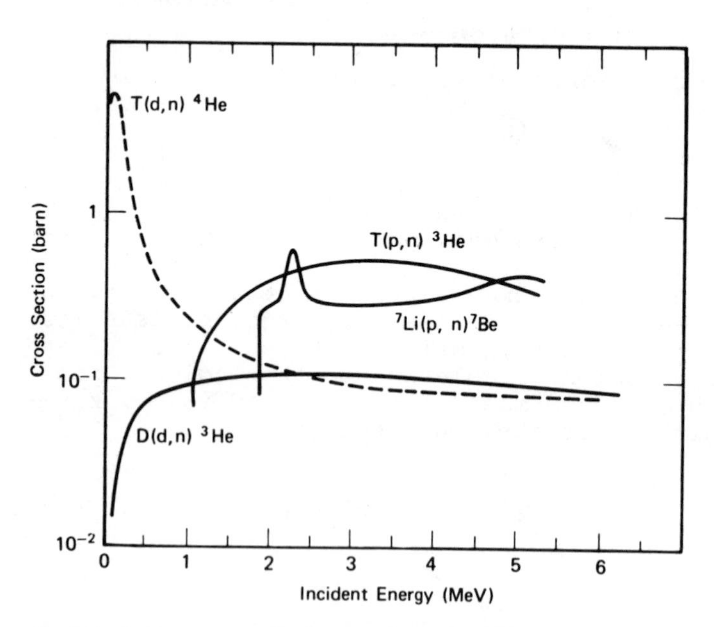

Fig. 2.15. Cross sections for neutron-producing reactions (from A. Edward Profio, *Experimental Reactor Physics*, Wiley, New York, 1976, p. 197).

and

$$w = \frac{M_4 Q + E_1(M_4 - M_1)}{M_3 + M_4} \tag{2.16}$$

The reaction is possible when $\sqrt{E_3}$ is real, positive. The neutron energy at 0° and 150° is plotted as a function of ion energy in Figure 2.16. The emission rate also varies with angle. Details are given in Marion and Fowler [11]. The T(d,n) reaction is nearly isotropic below 200 keV and not very anisotropic at higher energies. The D(d,n) neutrons are forward

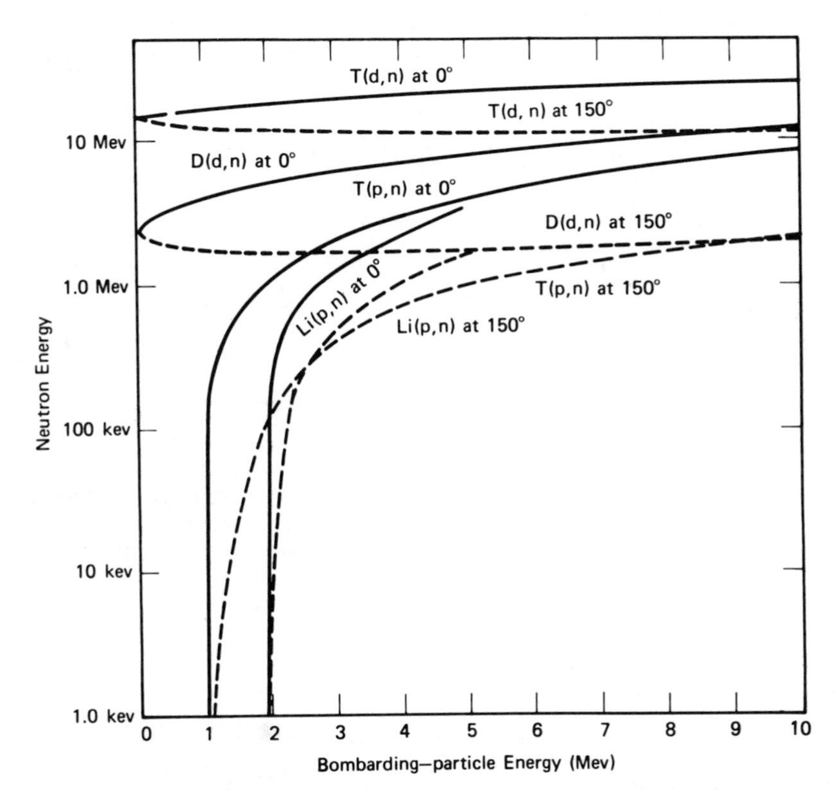

Fig. 2.16. Neutron energies as function of incident ion energy and neutron-emission angle (from A. Edward Profio, *Experimental Reactor Physics*, Wiley, New York, 1976, p. 196).

peaked (0°), as are the T(p,n) and ^{7}Li(p,n) neutrons, and emission rates at 0° may be used for shield design (see Tables 2.7 and 2.12).

TABLE 2.7. PRACTICAL YIELDS OF TARGETS

Reaction (Target)	Yield (n/s-sr at 0°) (Energy) per μA
T(p,n) (2.5 mg/cm^2 TiT_x)	6×10^6 (1.5 MeV)
	1×10^7 (2.5 MeV)
^{7}Li(p,n) (Thick)	9×10^6 (2.0 MeV)
	~ 10^8 (2.5 MeV)
D(d,n) (Thick TiD)	5.5×10^4 (150 keV)
(D_2 Gas 1.9 cm NTP)	2×10^7 (2.0 MeV)
T(d,n) (Thick TiT_x)	1×10^7 (150 keV)
	2×10^7 (≥ 500 keV)

Ions lose energy by ionization in the target material; hence the energy and angular distributions as well as the reaction cross section should be averaged over ion energy. The yield (neutrons per incident ion) is

$$Y = N \int_{E_1}^{E_2} \frac{\sigma(E)}{dE/dx} \, dE \approx N \frac{\sigma(E_1)\Delta E}{(dE/dx)_{E_1}} \qquad (2.17)$$

where N is the atomic density, E_1 and E_2 the ion energies entering and leaving the target or active region, σ the cross section for the nuclear reaction, and dE/dx the rate of energy loss by ionization in the target material, at energy E. The approximation in Eq. (2.4) applies when the target is "thin," that is, where energy loss ΔE is small compared to E_1. Targets for production of monoenergetic neutrons have to be thin, whereas thick targets are used when maximum neutron production is desired. Tritium or deuterium targets for the Van de Graaff accelerator may be gaseous, contained in a tube perhaps 5 mm in diameter and 5 cm in length, with a thin window at one end to admit the beam. Lithium targets are usually vacuum evaporated, and the ΔE may be 50 keV or so. Solid targets for the Cockcroft-Walton and Van de Graaff accelerators are made of titanium tritide or deuteride on a copper backing. Diameter is usually 2.5 cm and thickness of the active layer some 0.25 to 2.5 mg/cm^2. The composition of titanium tritide or deuteride is variable, with a maximum corresponding to $TiH_{1.5}$. The thickness may be specified in mg/cm^2 of the hydride, or in terms of cm^3 of gas at normal temperature and pressure (NTP), absorbed per unit area of target surface. The maximum loading available corresponds to 1 Ci/cm^2 of tritium. The rate of energy loss of deuterons in $TiT_{1.5}$ is plotted in Figure 2.17 [12,13].

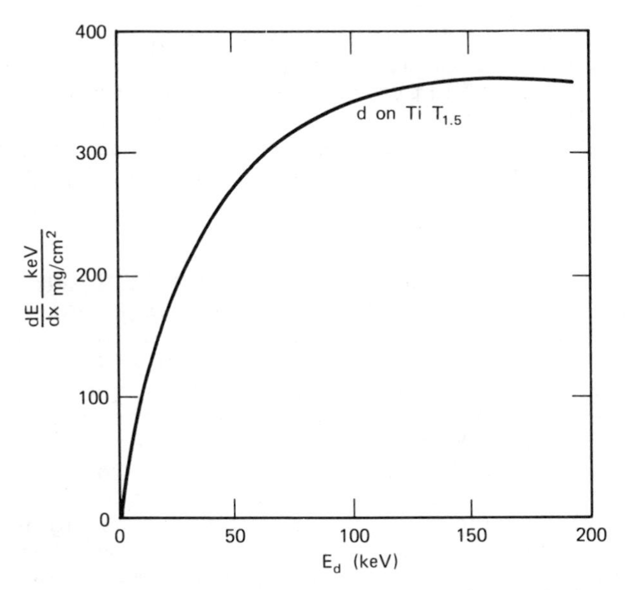

Fig. 2.17. Rate of energy loss dE/dx of deuterons in a titanium tritide target as a function of deuteron energy.

Often practical, measured yields are given in terms of n/s [or n/s-steradian (sr) at a specified angle], per microampere (μA) of deuteron or proton current, as a function of ion energy [14]. Some typical values are listed in Table 2.7 for fresh targets. The yield of TiT_x decreases with irradiation, perhaps because of sputtering. Half-life is about 2.7 mAh/cm^2.

Cyclotron

The features of a cyclotron may be seen in Figure 2.18. The ions are generated in a source at the center of a vacuum chamber installed between the poles of a large electromagnet and are accelerated as they cross the gap between two hollow, D-shaped electrodes. The magnetic field constrains the ions to move in a circular orbit between accelerations and also provides a focusing action. The D-to-D potential is on the order of 100 kV, but the ions can be accelerated to 20 MeV or more on repeated traversals of the gap. The frequency of the RF field applied to the Dees is adjusted to $eB/2\pi m$, where e is the charge and m the mass of the ion, and B is the magnetic field strength. The beam current impinging on a target installed within the magnetic field may be about 1 mA. If the ions are extracted to an external target, the beam current may be only 10 to 100 μA. The most prolific neutron producing reaction is $^{9}Be(d,n)^{10}B$. The yield is plotted as a function of energy in Figure 2.19. The neutron spectrum is shown in Figure 2.20 for three deuteron energies [7].

Electron Linear Accelerator

Principal features of the electron LINAC (linear accelerator) are shown in Figure 2.21. A large pulsed current of electrons is generated in an oxide-coated hot cathode, and accelerated to some 200 keV in a Cockcroft-Walton accelerator. A radio frequency (rf) buncher may be employed to increase current and improve energy homogeneity. The electrons enter the LINAC proper at near the speed of light. The accelerating section consists of a waveguide loaded with iris diaphragms and fed by a klystron tube operating usually at either 1300 MHz or 2856 MHz. A traveling electromagnetic wave is made to propagate down the waveguide, with the electric field along the axis and a phase velocity matching the velocity of the electrons. When electrons are injected in proper phase they are

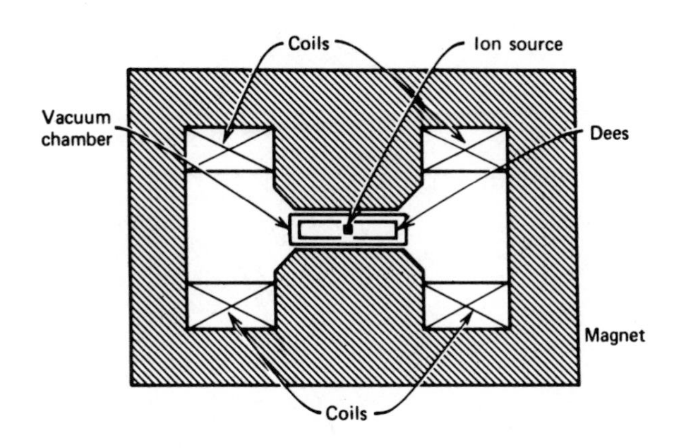

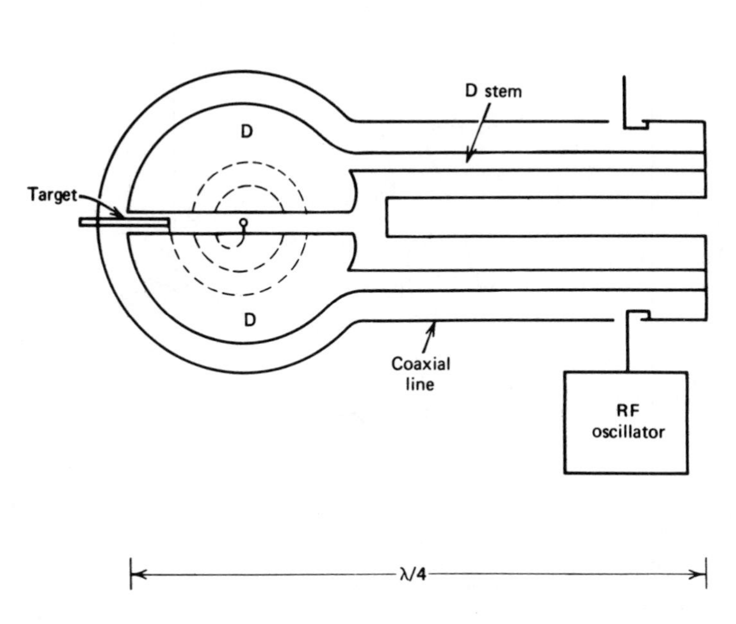

Fig. 2.18. Schematic diagram of cyclotron (from A. Edward Profio, *Experimental Reactor Physics*, Wiley, New York, 1976, p. 184).

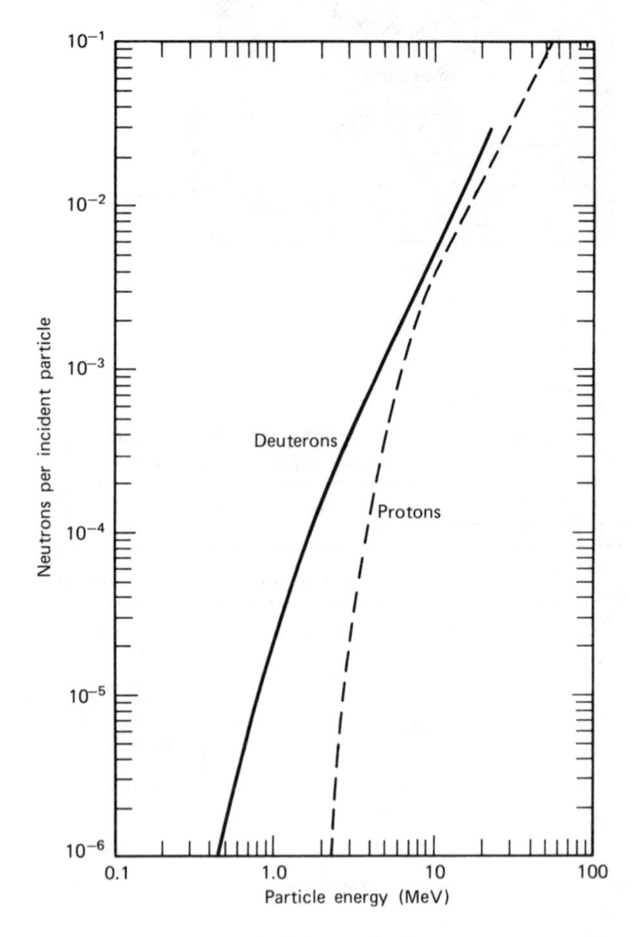

Fig. 2.19. Neutron yields of deuterons and protons on a thick beryllium target (from DePangher and Tochilin [7]).

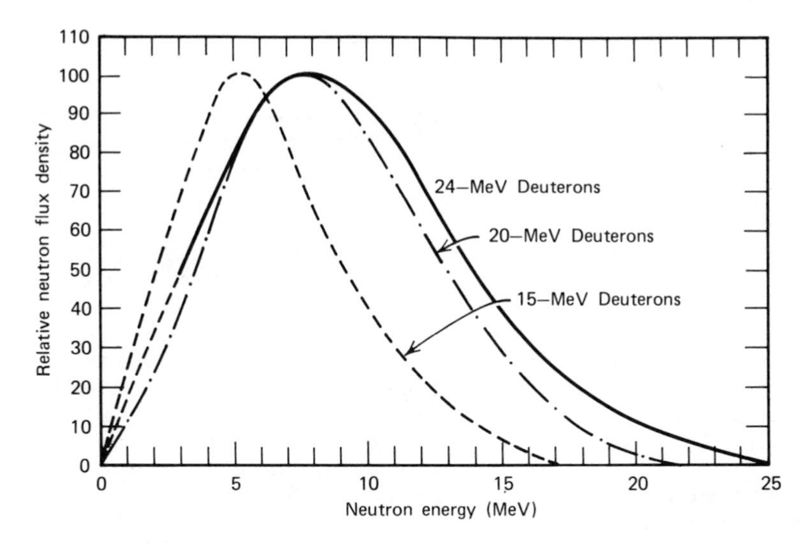

Fig. 2.20. Neutron spectra for deuterons on beryllium (from DePangher and Tochilin [7]).

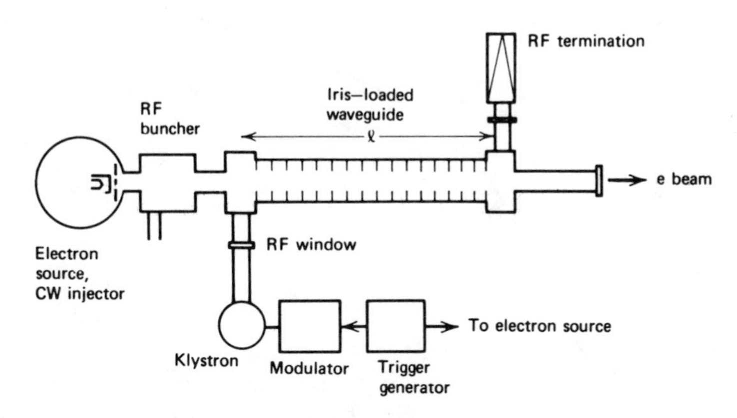

Fig. 2.21. Schematic diagram of electron linear accelerator (from A. Edward Profio, *Experimental Reactor Physics*, Wiley, New York, 1976, p. 187).

accelerated, generally to between 4 MeV and 10 MeV in a single-stage machine. To reach higher energies (e.g., 30 to 100 MeV), waveguide and klystron stages are installed in tandem, with the electrons eventually emerging in a well-collimated beam at the far end. Energy homogeneity is about 3%. The accelerator, including klystrons, is pulsed repetitively from a few pulses per second to 360 pulses/s. Pulse widths may be a few nanoseconds to a few microseconds. Peak beam current may be several amperes, but the average current is on the order of 1 mA.

Bremsstrahlung is generated when the electrons strike a heavy metal target such as tungsten or depleted uranium. Photoneutrons and photofission neutrons may also be generated. Figure 2.22 plots the total neutron yield from a thick target of depleted uranium, as a function of electron energy (the uranium acts as its own electron-bremsstrahlung converter). The yield is given in terms of neutrons/second per megawatt of beam power, that is, the beam energy (MeV) times the beam current (amperes). The spectrum is broad; Figure 2.23 shows the measured spectrum from a 7.63-cm-diameter sphere of depleted uranium (designed with a reentrant hole and diameter for nearly isotropic emission), cooled by a thin stream of water both

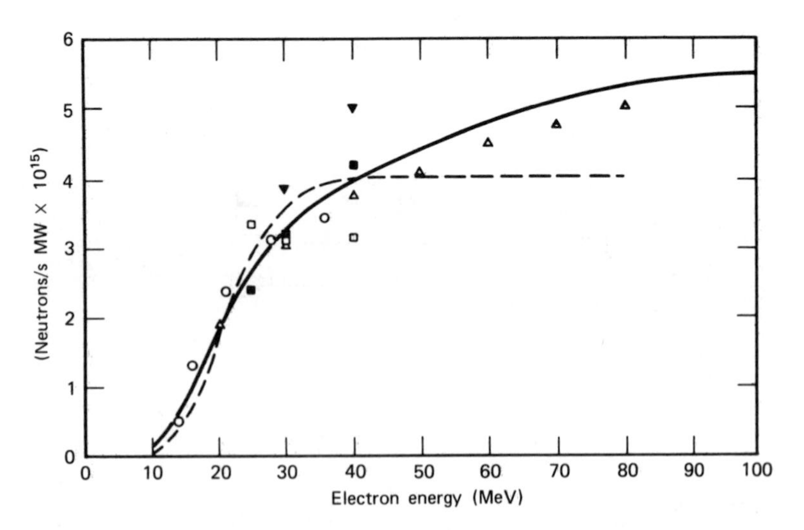

Fig. 2.22. Neutron yield from thick target of depleted uranium (after D. E. Groce, GA-8087).

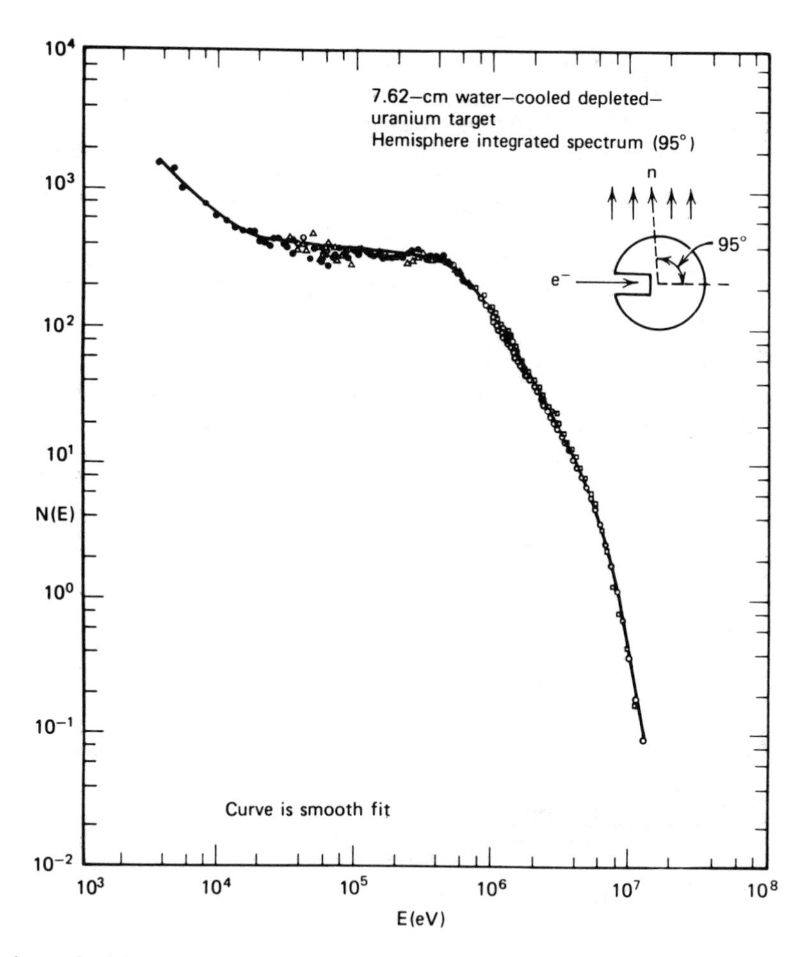

Fig. 2.23. Neutron spectrum of water-cooled, depleted uranium target irradiated with 28-MeV electrons (after A. Edward Profio, *Experimental Reactor Physics*, Wiley, New York, 1976, p. 200).

outside and inside the hole and irradiated with a beam of 28-MeV electrons. The rise below 1 MeV is from scattering in the water.

Sealed-tube Neutron Generator

A sealed-tube accelerator, suitable for generating 14-MeV neutrons by the D-T reaction, is shown in Figure 2.24. There is no vacuum pump; hence the Phillips ion gage type ion source and the accelerating gap are at the same pressure (0.01 to 0.02 torr). The gap must be small (1 to 2 cm). Pressure is controlled by releasing deuterium or a deuterium-tritium mixture from the replenisher or by absorbing gas in the getter. The target contains tritium or a deuterium-tritium mixture. By applying a voltage pulse of -120 kV, a beam current of several hundred milliamperes may be obtained, resulting in a neutron generation rate of 10^{13} n s^{-1} during the pulse. The pulse width may be 2.5 μs and the repetition rate 10 pulses per second; hence the average rate is only about 10^{8} n s^{-1}. However, similar devices with cooled targets are being designed for a continuous output of 5×10^{12} n s^{-1}.

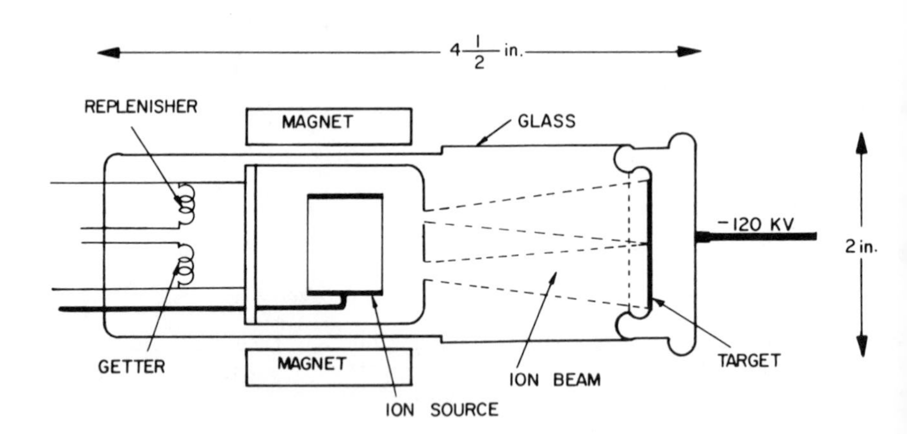

Fig. 2.24. Sealed-tube neutron generator.

2.3 FUSION DEVICES

Thermonuclear fusion reactors do not yet exist, but many experimental devices are in operation or planned. It is expected that the first fusion reactors will use the $T(d,n)^4He$ reaction because the ignition temperature is lowest. The reaction occurs in a plasma with an ion temperature on the order of 10 to 100 keV in different designs. The products are 14-MeV neutrons, which easily escape from the plasma (ion density on the order of 10^{14} cm^{-3}), and 3.6 MeV α particles that may be trapped in the plasma by a magnetic field, for a time. In addition, there will be a flux of low energy ions and X-rays incident on the "first wall" or inner boundary of the plasma container. The problems associated with the heating and damage to the first wall lie outside our scope. However, the penetration of the neutrons and interactions in the "blanket" (lithium-containing layer designed for recovery of the neutron kinetic energy as heat and for breeding of tritium by neutron reactions in lithium), and in the shield for the magnet and personnel, are of interest. Laser-pellet fusion devices do not require a magnet, but a blanket and shielding are still necessary.

2.4 NUCLEAR REACTORS

Nuclear-fission reactors are sources of prompt fission neutrons, delayed neutrons, prompt-fission γ rays, fission fragments, and β particles and γ rays from decay of the radioactive fission products and neutron-activation products. In operation, the most important radiations from the reactor core are the prompt-fission neutrons and prompt-fission γ rays. Delayed neutrons do not present a shielding problem because of their small yield (< 1% of the total) and low average energy (< 1 MeV). Fission fragments are stopped within the fuel material or cladding. The intensity of the prompt-fission γ rays is much higher than that of the fission product or activation product γ rays during operation. However, the radioactive products are significant when they are no longer contained within the reactor core shield, and these are considered here.

Prompt-Fission Neutrons

The source strength is given by the fission rate times ν, the prompt neutron multiplicity or average number of neutrons per fission. The value depends somewhat on the energy of the neutron causing fission, E_n,

$$\nu(E_n) = \nu_0 + \left(\frac{d\nu}{dE_n}\right) E_n \qquad (2.18)$$

Table 2.8 lists these parameters for common fissionable nuclides [15]. If the flux density is known, the fission rate may be obtained

$$R(\text{fissions/cm}^3\text{s}) = \phi(E)\ \Sigma_f(E)\ dE \qquad (2.19)$$

and Σ_f is the macroscopic fission cross section.

TABLE 2.8. PROMPT-NEUTRON MULTIPLICITY

Nuclide	E_n (MeV)	ν_0	$d\nu/dE$
^{235}U	0 to 1	2.43	0.066
	> 1	2.35	0.015
^{239}Pu	0 to 1	2.87	0.138
	> 1	2.91	0.133

The fission rate may be normalized to the total power. Although there is some variation according to the contribution of fission-product decay, activation-product decay, and escape of neutrons and γ rays, a value of 3.1×10^{10} fissions/joule is often used, corresponding to 194 MeV/fission.

Measured-fission neutron spectra are shown in Figure 2.25 [16]. The ordinate is $\chi(E)$, the fraction of the total number of fission neutrons between E and (E + dE), with neutron energy in MeV. The solid line is a fit with a "Maxwellian" spectrum of the form

$$N(E) = \frac{2}{\sqrt{\pi}\ T^{3/2}} E^{1/2} \exp \frac{-E}{T} \qquad (2.20)$$

where T is an effective "temperature" in energy units related to the average energy for this spectrum,

$$\overline{E} = \frac{3T}{2} = 1.94 \text{ MeV} \qquad (2.21)$$

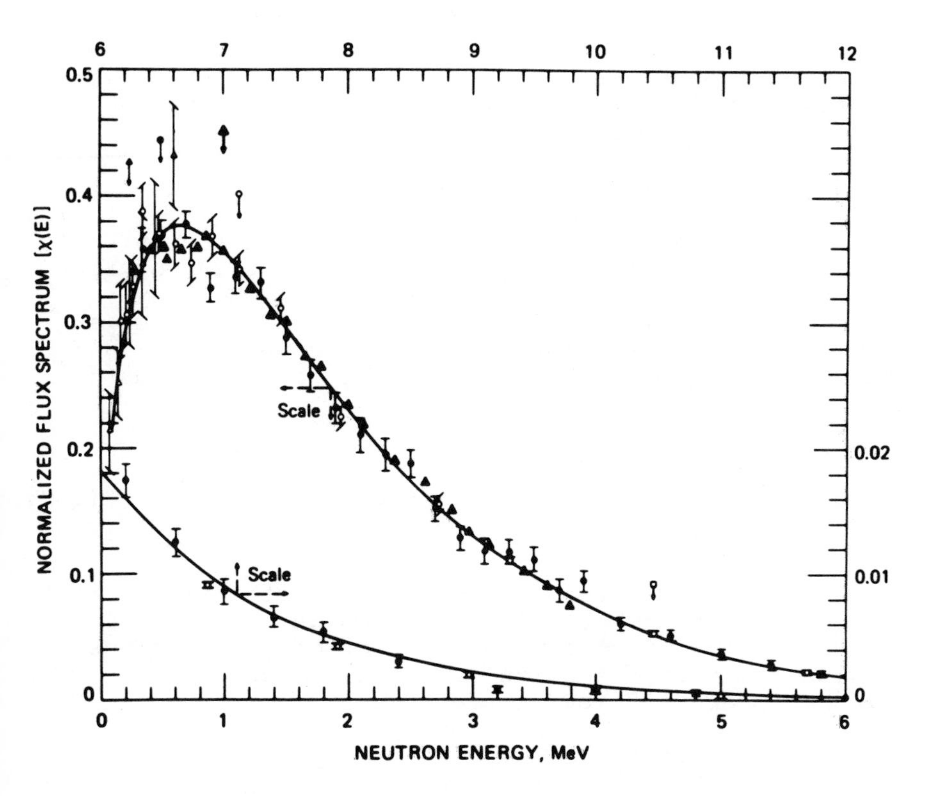

Fig. 2.25. Measured prompt-neutron spectrum from thermal-neutron fission of ^{235}U (after Grundl [16]).

Data for other nuclides and incident neutron energies are sparse. The measured spectrum for thermal neutron fission of ^{239}Pu can be fitted by a Maxwellian with $\overline{E}$ = 2 MeV. A semiempirical relationship between ν and $\overline{E}$ may be used to extrapolate to higher energy,

$$\overline{E} = 0.74 + 0.65\sqrt{\nu + 1}\ \text{MeV} \tag{2.22}$$

Other representations have been used for the spectrum from thermal neutron fission of ^{235}U [17]. The most accurate fit to the earlier measurements is given by Cranberg,

$$\chi(E) = 0.453 \exp \frac{-E}{0.965} \sinh (2.29E)^{1/2} \qquad (2.23)$$

in terms of neutrons/MeV at energy E(MeV), normalized to unity. This relationship is used to obtain the spectrum in Table 2.9.

Prompt-fission Gamma Rays

The prompt-fission γ-ray spectrum, as measured by Peele and Maienschein [18], is shown in Figure 2.26 for thermal neutron fission of ^{235}U. Other fission γ-ray spectra are similar [19]. The two lines are upper and lower error bounds. The Gaussian-like peaks indicate energy resolution. The curve is fit to within 10 to 20% by the empirical formulas

$$\begin{array}{lll} 6.6 \text{ photons/fission-MeV} & 0.1 < E < 0.6 \text{ MeV} & \\ 20.0 \exp(-1.78 E) & 0.6 < E < 1.5 \text{ MeV} & (2.24) \\ 7.2 \exp(-1.09 E) & 1.5 < E < 10.5 \text{ MeV} & \end{array}$$

These expressions are used to calculate the N(E) versus E values given in Table 2.10. Emission is isotropic, and the spatial distribution is that of the fissions, as with the prompt fission neutrons. Total energy release is about 7 MeV/fission.

Fission-product γ Rays

Although the decay γ-ray and β-particle spectra can be calculated for each of the fission-product nuclides taking into account fission yield, buildup, and series decay, this is complicated. For reactor-core shielding calculations, it is more convenient to use a composite γ-ray spectrum corresponding to equilibrium or at least long irradiation of the fuel (β particles are easily absorbed in the fuel and may be neglected except for their heating effect, or when the fission products are released from the fuel). Table 2.11 gives the spectrum averaged in fairly broad energy intervals [20].

Fission Distribution

The spatial distribution of the fission rate in a modern power reactor is complicated by the heterogeneity of the fuel and moderator (in a thermal reactor)

TABLE 2.9. NEUTRON SPECTRUM FROM THERMAL-NEUTRON FISSION OF ^{235}U

E (MeV)	$\chi(E)$ n/MeV	E (MeV)	$\chi(E)$ n/MeV
0.00	0.000	7.0	8.77×10^{-3}
0.25	0.290	7.5	6.01×10^{-3}
0.50	0.347	8.0	4.10×10^{-3}
0.75	0.358	8.5	2.79×10^{-3}
1.00	0.347	9.0	1.88×10^{-3}
1.25	0.325	9.5	1.27×10^{-3}
1.50	0.298	10.0	8.56×10^{-4}
1.75	0.268	10.5	5.74×10^{-4}
2.00	0.239	11.0	3.84×10^{-4}
2.50	0.184	11.5	2.56×10^{-4}
3.00	0.138	12.0	1.70×10^{-4}
3.50	0.102	13.0	7.48×10^{-5}
4.00	0.0738	14.0	3.26×10^{-5}
4.50	0.0528	15.0	1.41×10^{-5}
5.00	0.0375	16.0	6.07×10^{-6}
5.50	0.0263	17.0	2.59×10^{-6}
6.00	0.0184	18.0	1.10×10^{-6}
6.50	0.0127		

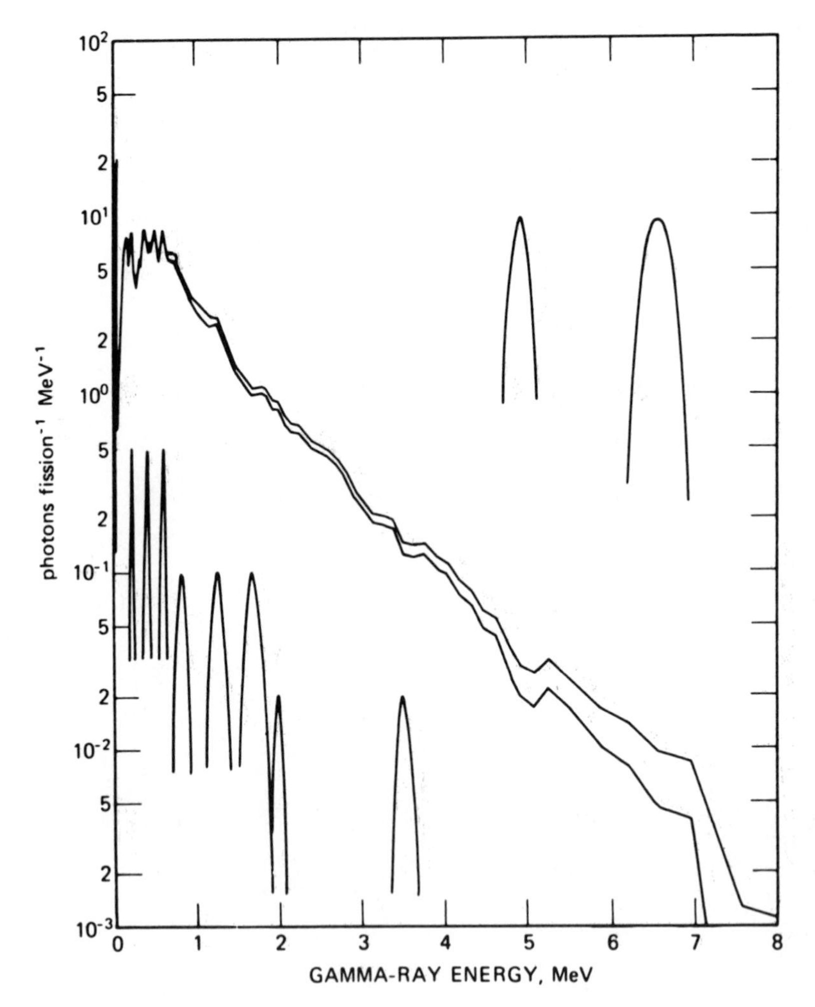

Fig. 2.26. Prompt-fission γ-ray spectrum for thermal-neutron fission of ^{235}U (from A. Edward Profio, *Experimental Reactor Physics*, Wiley, New York, 1976, p. 97).

TABLE 2.10. PROMPT-FISSION γ-RAY SPECTRUM OF $^{235}U(n_{th},F)$

E (MeV)	N(E) $\frac{\text{Photons/MeV}}{\text{fission}}$
0.00	6.60
0.25	6.60
0.50	6.60
0.75	5.32
1.00	3.41
1.25	2.18
1.50	1.40
1.75	1.07
2.00	0.814
2.50	0.472
3.00	0.274
3.50	0.159
4.00	9.20×10^{-2}
4.50	5.33×10^{-2}
5.00	3.09×10^{-2}
5.50	1.79×10^{-2}
6.00	1.04×10^{-2}
6.50	6.03×10^{-3}
7.00	3.50×10^{-3}
7.50	2.03×10^{-3}
8.00	1.18×10^{-3}
8.50	6.82×10^{-4}
9.00	3.95×10^{-4}
9.50	2.29×10^{-4}
10.00	1.33×10^{-4}
10.50	7.71×10^{-5}

TABLE 2.11. EQUILIBRIUM FISSION-PRODUCT γ-RAY SPECTRUM $^{235}U(n_{th},F)$

Energy Interval (MeV)	Effective Energy (MeV)	MeV/ Fission	Photons/ Fission
0.1 to 0.4	0.4	0.645	1.61
0.4 to 0.9	0.8	3.87	4.84
0.9 to 1.35	1.3	0.645	0.496
1.35 to 1.8	1.7	1.06	0.624
1.8 to 2.2	2.18	0.677	0.311
2.2 to 2.6	2.5	0.290	0.116
> 2.6	2.8	0.032	0.011
	Total	7.219	

or fuel and coolant (in a fast reactor) and by the spatial variation of the isotopic composition of the fuel as a result of burnup, buildup of fission-product nuclides (affecting the neutron flux density), arrangements of control rods, and differing enrichment in different sections of the core (e.g., for power flattening). It is the province of the reactor-core analyst to specify the fission rate or power distribution for the shield analyst, but the latter investigator should be aware that the distributions are not as simple as the bare or reflected flux density and power distributions usually given in elementary reactor-theory texts. Figure 2.27 is an example for the radial distribution in a pressurized water reactor (PWR) for two different fuel reloading schemes: OI refers to out/in and LBP to lumped burnable poison. In the OI scheme the fuel in an outer zone of the core is moved closer to the center and replaced by fresh fuel in the outer zone while fuel near the center is discharged. In the LBP scheme fresh fuel replaces spent fuel in a kind of checkerboard pattern throughout the core [21].

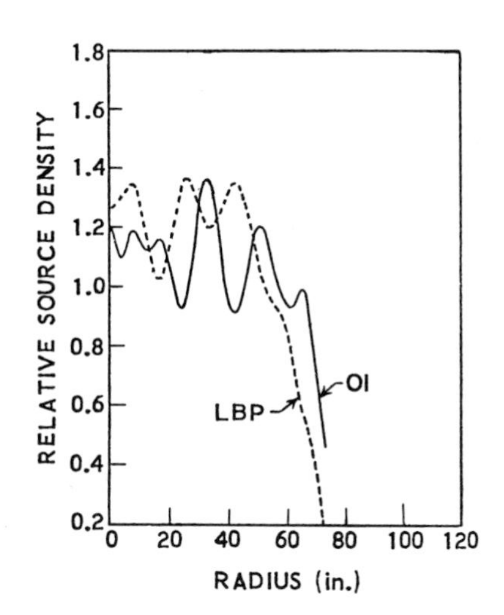

Fig. 2.27. Fission-rate distribution in PWR core for OI and LBP methods of refueling (after Sapyta and Simmons [6], Fig. 3).

Activation Products

Neutron activation products are negligible in shielding the core during operation but are important in shielding pipes and vessels or other equipment containing coolant. In the liquid metal fast breeder reactor the sodium is activated, decaying by β^- with emission of 1.37 and 2.75 MeV γ rays.

$$^{23}\text{Na} + {}^{1}\text{n} \rightarrow {}^{24}\text{Na} \xrightarrow[15\ \text{h}]{\beta^-} {}^{24}\text{Mg (stable)}$$

In water-cooled reactors the fast neutron reaction

$$^{16}\text{O} + {}^{1}\text{n} \rightarrow {}^{1}\text{p} + {}^{16}\text{N} \xrightarrow[7.2\ \text{s}]{\beta^-} {}^{16}\text{O (stable)}$$

is significant, especially in the boiling water reactor because of the short delay between core and steam turbine and other equipment external to the reactor shield. A 6.13-MeV γ ray is emitted in 69% of the decays.

The rate of production of radioactive atoms N_2 is

$$R = N_1 \int \sigma(E)\ \phi(E)\ dE \quad \text{atoms cm}^{-3}\ \text{s}^{-1} \qquad (2.25)$$

where the atomic density of target atoms

$$N_1 = \frac{N_A \rho}{A} \quad \text{atoms cm}^{-3} \qquad (2.26)$$

with $N_A = 6.02 \times 10^{23}$ atoms/gram-mole as the Avogadro number, A the atomic weight, ρ (g/cm³) the density of target atoms, σ the cross section for activation, and ϕ the neutron flux density. If the production rate is constant, the differential equation for the buildup is

$$\frac{dN_2}{dt} = R - \lambda N_2 \qquad (2.27)$$

The solution for the activity, for zero N_2 atoms at $t = 0$, is

$$A = \lambda N_2 = R\left(1 - e^{-\lambda t_r}\right) \text{ dis s}^{-1}\ \text{cm}^{-3} \qquad (2.28)$$

The decay constant is $\lambda = \tau^{-1}$ and t_r is the irradiation time. The buildup curve is plotted in Figure 2.28. After the end of irradiation, the activity decays as

$$A = A_0\ e^{-\lambda t_d} \qquad (2.29)$$

where A_0 is the activity at the end of the irradiation time and t_d is the decay time. The decay curve is plotted in Figure 2.28 for A_0 equal to the equilibrium or saturated activity $A_{sat} = R$, corresponding to large t_r.

Because the coolant circulates, it is necessary to integrate the production rate over the core or region where the neutron flux density is appreciable, and to correct for decay while outside the region. Let us

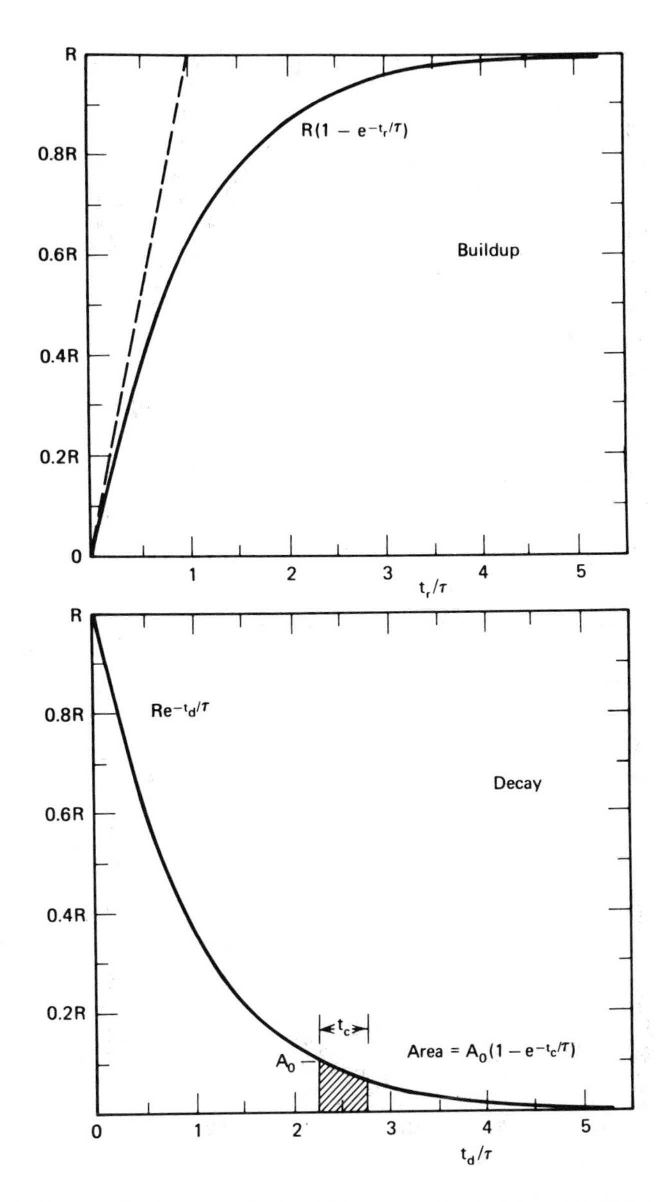

Fig. 2.28. Universal buildup and decay curves for constant production rate R (from A. Edward Profio, *Experimental Reactor Physics*, Wiley, New York, 1976, p. 142).

approximate the situation by assuming slug flow and follow a parcel of coolant (e.g., 1 cm^3) as it is irradiated for a time t_r at an effective flux density ϕ_{eff} with a neutron-spectrum averaged cross section $\bar{\sigma}$. The circulation time, that is, the time for the parcel to return to the entrance of the core, is T. Then at equilibrium, after many cycles, the activity at the exit of the core is

$$A = N\bar{\sigma}\ \phi_{eff} \left(\frac{1 - e^{-\lambda t_r}}{1 - e^{-\lambda T}} \right) \tag{2.30}$$

in disintegrations/second per cm^3. To correct for decay, the activity per cm^3 at other locations in the coolant loop must take into account the transit time in pipes and holdup time in vessels.

2.5 SOURCE DESCRIPTIONS

The source strengths (particles or photons per second) may be a fundtion of the position coordinates, energy, angle, and time,

$$S(\underline{r},E,\underline{\Omega},t) \tag{2.31}$$

where $\underline{r}$ is the position vector or coordinates (x,y,z as in Figure 2.29 for rectangular coordinate geometry, r,ϕ,z in cylindrical geometry, or ρ,θ,ϕ in spherical geometry), E is the energy, $\underline{\Omega}$ is the unit direction vector, and t is the time.

The dependence on position is straightforward for point sources. It is sometimes convenient to represent the source as a line, with strength S_l per unit length (e.g., photons s^{-1} m^{-1}). Other sources may be represented as a plane or other surface with a strength S_a per unit area (e.g., photons s^{-1} m^{-2}). Finally, one may have a volume source S (photons s^{-1} m^{-3}). Source strength may vary with position on the line, surface, or in the volume.

Sources emitting particles or photons at a few discrete energies may be specified by giving the number or particles s^{-1} emitted at each of the energies. A continuous spectrum is given in terms of S(E) particles s^{-1} MeV^{-1}. Thus S(E)dE is the number of particles per second emitted with energies between E and (E + dE).

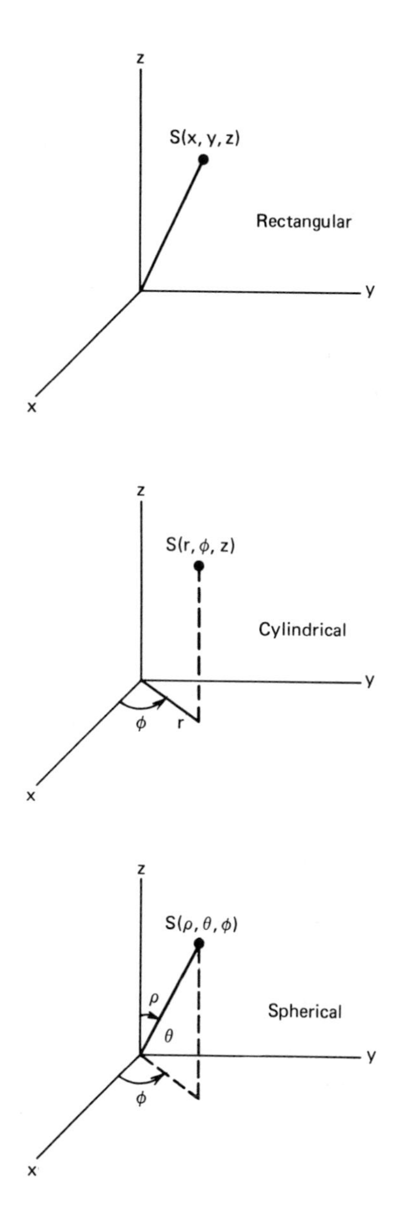

Fig. 2.29. Source-position description in rectangular, cylindrical, and spherical coordinate geometries.

The distribution may be expressed by either a mathematical formula, a table of point values of E and the corresponding S(E) together with an interpolation law (usually linear) to obtain values between the tabulated points, or in the multigroup form. In the multigroup representation the source strength is tabulated as a "group source" for each group number g from 1 to the total number of groups G, usually with group 1 as the highest energy; hence

$$S_g \, \Delta E_g = \int_{E_{g+1}}^{E_g} S(E) \, dE \quad \text{particles s}^{-1} \tag{2.31}$$

The group width $\Delta E_g = E_{g+1}$ need not be constant, but it may be chosen according to the rapidity of variation of of S(E) with E or to correspond to the group structure chosen for the cross sections as discussed in Chapter 3. Representations of source spectra are illustrated in Figure 2.30.

The angular distribution is often isotropic; that is, an equal number of particles per second is emitted in each differential solid angle $d\Omega$(steradian), regardless of direction. The solid angle subtended at a point by an area dA projected on a sphere of radius r around that point is $d\Omega = (dA/r^2)$ steradians. The area of a sphere is $4\pi r^2$; hence there are 4π steradians in a sphere. An isotropic source with strength S particles per second emits $S/4\pi$ particles s^{-1} sr^{-1} in all directions. If the source is anisotropic, it is necessary to define the direction unit vector with reference to certain fixed directions. This is usually done by specifying the polar angle θ and azimuthal angle ϕ as shown in Figure 2.31. The source strength may be given as $S(\underline{\Omega})$ particles s^{-1} sr^{-1} in $d\Omega$ about $\underline{\Omega}$. Let us suppose the differential solid angle is defined by polar angle θ to $(\theta + d\theta)$ and azimuthal angle ϕ to $(\phi + d\phi)$. Then the area projected on a sphere of unit radius is $dA = \sin\theta \, d\theta \, d\phi$, and $d\Omega = dA$ since $r^2 = 1$. For a point source located at the center of the sphere, $S(\underline{\Omega}) \sin\theta \, d\theta \, d\phi$ particles s^{-1} pass through dA, and the number of source particles per second emitted between angle θ_1 and θ_2, and ϕ_1 and ϕ_2 is

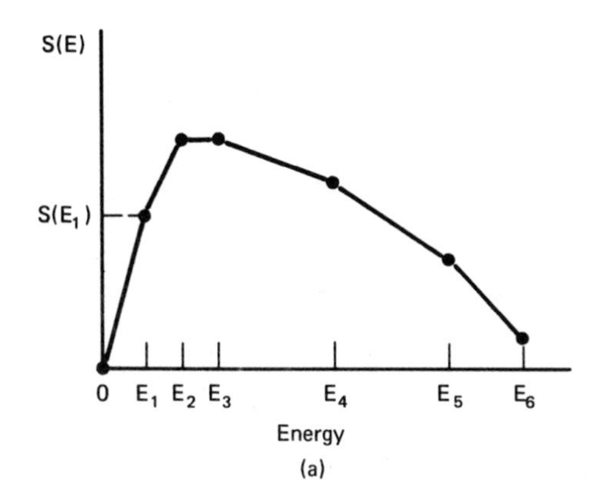

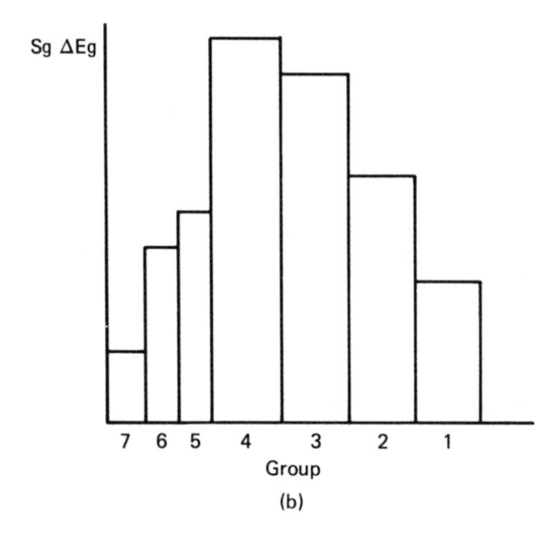

Fig. 2.30. Representation of source spectrum by (a) energy and source strength per unit energy with linear interpolation or (b) group source strength for unequal group widths.

$$N = \int_{\phi_1}^{\phi_2} \int_{\phi_2}^{\phi_2} S(\underline{\Omega}) \sin\theta \, d\theta \, d\phi \qquad (2.32)$$

Very often the angular distribution is symmetric about an axis, such as along the beam of an ion accelerator. It thus makes sense to align the reference direction u_{ref} along the beam, in the direction of travel of the ions, as shown in Figure 2.32. Emission is independent of azimuthal angle (equally probable for all ϕ) but is a function of polar angle. Then if the angular distribution is expressed in terms of the angle, $S(\theta)$, the number of source particles per second emitted between θ_1 and θ_2 is

$$N = \int_{0}^{2\pi} \int_{\theta_1}^{\theta_2} S(\theta) \sin\theta \, d\theta \, d\phi$$

$$= 2\pi \int_{\theta_1}^{\theta_2} S(\theta) \sin\theta \, d\theta \qquad (2.33)$$

Another convenient representation is in terms of the cosine of the polar angle, $\mu = \cos\theta$. Then because $d\mu = \sin\theta \, d\theta$, we have

$$N = 2\pi \int_{\mu_1}^{\mu_2} S(\mu) \, d\mu \qquad (2.34)$$

If the energy and angle are uncorrelated, one could express the source distribution as

$$S(\underline{r},E,\underline{\Omega},t) = S_1(r,E,t)F(\underline{\Omega}) \qquad (2.35)$$

However, there is often a 1:1 relationship between angle and the energy of the particle emitted at that angle. We thus need a table such as Table 2.12, which gives the neutron yield of a thick titanium deuteride

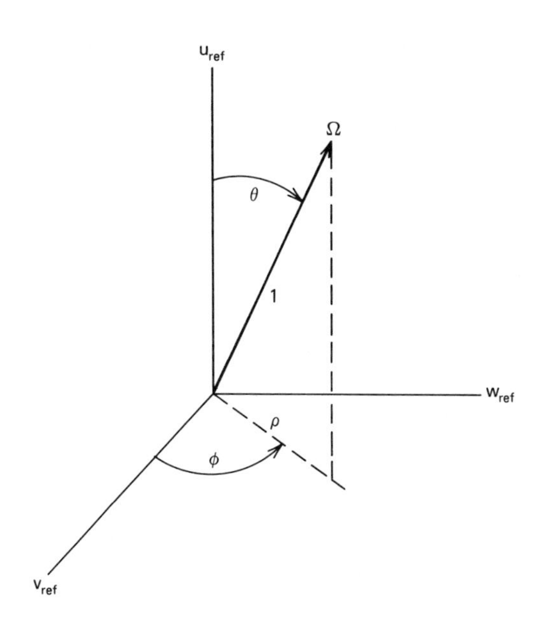

Fig. 2.31. Description of unit direction vector by polar angle θ with reference to fixed direction u_{ref} and azimuthal angle ϕ with reference to reference direction v_{ref}.

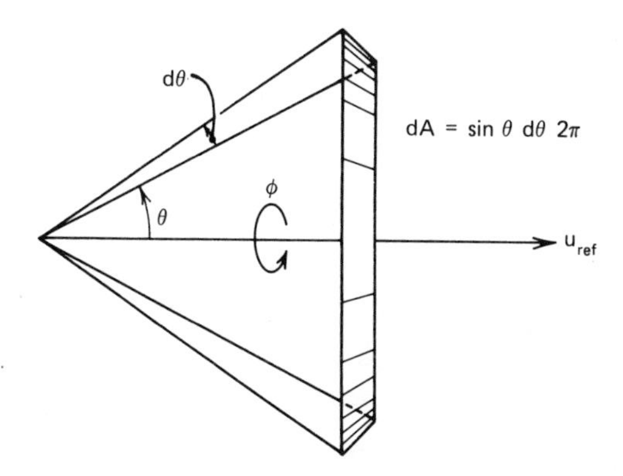

Fig. 2.32. Angular distribution with azimuthal symmetry.

TABLE 2.12. YIELD AS A FUNCTION OF ANGLE AND NEUTRON ENERGY AT THAT ANGLE FOR THICK TITANIUM HYDRIDE TARGETS, E_d = 150 keV

Angle	D(d,n) Reaction		T(d,n) Reaction	
(°)	E_n (MeV)	Y(n s^{-1} sr^{-1} μA^{-1})	E_n (MeV)	Y(n s^{-1} sr^{-1} μA^{-1})
0	2.80	5.5×10^4	14.75	1.0×10^7
20	2.75	5.1×10^4	14.70	9.9×10^6
40	2.70	4.1×10^4	14.60	9.8×10^6
60	2.62	2.9×10^4	14.40	9.7×10^6
80	2.52	2.2×10^4	14.20	9.5×10^6
100	2.41	2.1×10^4	13.95	9.3×10^6
120	2.25	2.2×10^4	13.70	9.1×10^6
140	2.15	2.9×10^4	13.45	9.0×10^6
160	2.05	3.7×10^4	13.42	8.8×10^6
180	2.00	4.1×10^4	13.40	8.7×10^6

target bombarded by 150-keV deuterons, as a function of angle to the deuteron beam, as well as the corresponding neutron energy at that angle [22].

The source strength may be given as a function of time. However, in some applications the quantity of interest is a response integrated over time. Hence the source output

$$Q(\underline{r},E,\underline{\Omega}) = \int_0^T S(\underline{r},E,\underline{\Omega},t)\ dt \qquad (2.36)$$

may be used, with units of particles or photons MeV^{-1} sr^{-1}. Because of the linearity in most shielding configurations, a response such as dose per particle has the same numerical value as a response rate such as dose per second, normalized to a source strength of one particle per second.

If the source is distant or outside the shield but little attenuated, it may be convenient to replace the actual source by an equivalent surface source at the shield. For example, in Figure 2.33 the incident

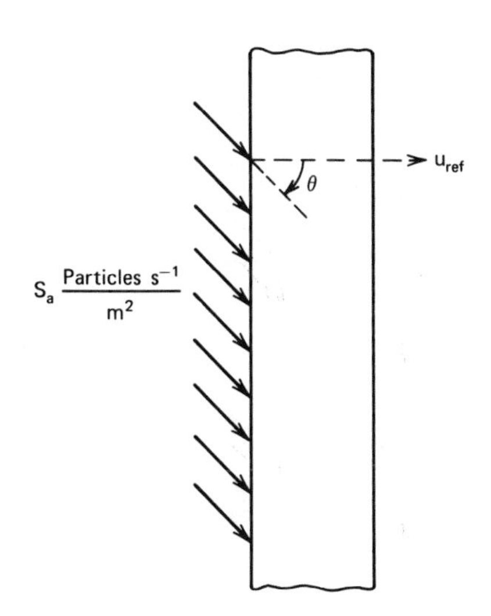

Fig. 2.33. Parallel beam of particles incident at angle θ to a normal of a surface, with S_a particles per second crossing 1 m^2 of surface.

radiation is represented by a parallel beam with strength S_a particles s^{-1} crossing 1 m^2 of surface at an angle θ. Often θ is 0°; that is, the beam is perpendicular to and directed into the surface.

REFERENCES

1. R. D. Evans, *The Atomic Nucleus*, McGraw Hill, New York, 1955.
2. C. M. Lederer, J. M. Hollander, and I. Perlman, *Table of Isotopes*, 6th ed., Wiley, New York, 1967.
3. K. Siegbahn, Ed., *Alpha-, Beta-, and Gamma Ray Spectroscopy*, North-Holland Publishing Company, Amsterdam, 1965.
4. L. T. Dillman and F. C. Von der Lage, "Radionuclide Decay Schemes and Nuclear Parameters for Use in Radiation-Dose Estimation," nm/mird Pamphlet No. 10, Society of Nuclear Medicine, New York, 1975.
5. D. H. Stoddard and H. E. Hootman, "^{252}Cf Shielding Guide," Report DP-1246, Savannah River Lab., NTIS, Springfield, Va., 1971.
6. R. S. Caswell, "Review of Measurements of Absolute Neutron Emission Rates and Spectra from Neutron Sources," in *Neutron Sources and Applications*, CONF-710402, p. I-53 (1971).
7. J. De Pangher and E. Tochilin, F. H. Attix and E. Tochilin, Eds., in *Radiation Dosimetry*, Vol. III, 2nd ed., Academic Press, New York, 1969, pp. 300-302, 309-358.
8. M. M. Ter-Pogossian, *The Physical Aspects of Diagnostic Radiology*, Harper and Row, New York, 1967.
9. W. J. Meredith and J. B. Massey, *Fundamental Physics of Radiology*, Williams and Wilkins, Baltimore, 1972.
10. W. R. Hendee, *Medical Radiation Physics*, Yearbook Medical Publishers, Chicago, 1970.
11. J. B. Marion and J. L. Fowler, Eds., *Fast Neutron Physics*, Part I, Interscience, New York, 1960, pp. 73-176.
12. J. H. Ormrod, *Nucl. Instrum. Methods*, 95, 49-51 (1971).
13. E. M. Gunnersen and G. James, *Nucl. Instrum. Methods*, 8, 173-184 (1960).
14. E. A. Burrill, "Neutron Production and Protection," High Voltage Engineering Company, Burlington, Mass.

15. G. R. Keepin, *Physics of Nuclear Kinetics*, Addison-Wesley, Reading, Mass., 1965, p. 54.
16. J. A. Grundl, "Fission-Neutron Spectra," in *Neutron Standards and Flux Normalization*, Proc. Symp. Argonne Natl. Lab., 1970, CONF-701002.
17. E. D. Arnold, "Prompt Fission Neutrons," in R. G. Jaeger, Ed., *Engineering Compendium on Radiation Shielding*, Vol. I, Springer-Verlag, New York, 1968, pp. 68-72.
18. R. W. Peele and F. C. Maienschein, "Prompt Gamma Rays from Thermal Neutron Fission of ^{235}U," *Nucl. Sci. Eng.*, 30, 485 (1970); also report ORNL-4457.
19. V. V. Verbinski et al., "Measurements of Prompt Gamma Rays from Thermal Neutron Fission of ^{235}U and ^{239}Pu, and Spontaneous Fission of ^{252}Cf," DASA Report 2234 (GA-9148), 1969.
20. *Reactor Physics Constants*, 2nd ed., ANL-5800 (1963), p. 630.
21. J. J. Sapyta and G. L. Simmons, "Shielding Design and Analysis Methods for Pressurized Water Reactors," *Nucl. Technol.*, 26, 508-515 (1975).
22. J. D. Seagrave, "D(D,n)$\overline{He}^3$ and T(d,n)He^4 Neutron Source Handbook," Los Alamos Report LAMS-2162 (1958).

PROBLEMS

1. What evidence is there that the decay constant λ is independent of age, temperature, pressure, and other parameters?

2. Derive Eq. (2.8). Plot on semilog paper the parent and daughter activities for 1-Ci parent and 0-Ci daughter at t = 0.

$$^{99}Mo \xrightarrow{\beta^-} {}^{99m}Tc \xrightarrow{\beta^-} {}^{99}Tc$$

$$T_{1/2} \qquad 67\ h \qquad 6\ h$$

3. Plot on semilog paper the parent and daughter activities

$$^{131}\text{Te} \xrightarrow[30\ \text{h}]{\beta^-} {}^{131}\text{I} \xrightarrow[8.05\ \text{d}]{\beta^-} {}^{131}\text{Xe}$$

and the atoms of ^{131}Xe as a function of time for 1 Ci of ^{131}Te and 0.1 Ci ^{131}I at t = 0.

4. Given that 99.6% of the decays of ^{22}Na proceed by positron emission and the remainder by electron capture, what is the partial half-life for positron decay, if the total half-life is 2.60 years? What is $Q_{\beta+}$ for ^{22}Na. Sketch the electron and photon spectra for ^{22}Na. How many X-rays are emitted per decay, assuming K-shell electron capture? Are their energies characteristic of ^{22}Na or ^{22}Ne?

5. Calculate the heating rate per gram of ^{60}Co, assuming that all of the β particle energy and 5% of the γ-ray energy are absorbed.

6. Explain the difference between isomeric transition and internal conversion. Is it possible for both processes to occur in the same nuclide?

7. Draw a diagram of a Cockcroft-Walton accelerator for acceleration of positive ions and explain how the voltage-multiplying circuit operates. Discuss the advantages and disadvantages of the Van de Graaff accelerator for neutron production. Look up how the cyclotron works, including the principle of magnetic resonance, and explain why the cyclotron is not suitable for electrons.

8. Calculate the energy of a neutron emitted at 90° in the D(d,n)^{3}He reaction for E_d = 400 keV, for the T(d,n)^{4}He reaction with E_d = 150 keV, and for the ^{7}Li(p,n)^{7}Be reaction with E_p = 2.0 MeV.

9. What is the fraction of neutrons with energies greater than 5 MeV for thermal neutron fission of ^{235}U? What is the fraction above 10 MeV?

10. A 10-g sample of cobalt is irradiated for 30 days in a thermal-neutron flux density of 10^{12}n/cm^{2}s,

and then allowed to decay for 1 day. What is the activity of ^{60}Co? The cross section for the $^{59}Co(n,\gamma)^{60}Co$ reaction is 37.2 b (barns).

11. A radionuclide is produced at a constant rate of 10^3 nuclei per second, for 3 half-lives. It then decays for 2 half-lives. How many nuclei decay in the next half-life?

3 INTERACTIONS

3.1 FLUX DENSITY AND CROSS SECTION

A particle field may be specified by the differential flux-density $\psi(\underline{r},E,\underline{\Omega},t)$ particles cm^{-2} s^{-1} MeV^{-1} sr^{-1}. The concept is illustrated in Figure 3.1. The number of particles at $\underline{r}$ and at time t, with energies between E and E+dE, moving in differential solid angle $d\Omega$ about direction $\underline{\Omega}$, that cross differential area $dA_{\perp}$ (normal to $\underline{\Omega}$) in unit time is $\psi(\underline{r},E,\underline{\Omega},t)\ dA\ dE\ d\Omega$. Note that the reference area is always oriented perpendicular to the direction of motion.

The integral of the differential flux density over angle is the scalar flux density,

$$\phi(\underline{r},E,t) = \int \psi(\underline{r},E,\underline{\Omega},t)\ d\Omega \tag{3.1}$$

Because the reference area rotates to be normal to $\underline{\Omega}$, the scalar flux density may be identified with the number of particles entering a differential sphere, per unit time, divided by the cross-sectional area of the sphere, as shown in Figure 3.2. The sphere presents the same cross-sectional area regardless of the direction of incidence. It follows that the scalar flux density (often called simply the "flux density") can be measured by an isotropic detector, that is, one that responds equally regardless of the direction of incidence. The detector does not have to be spherical as long as it is "thin," that is, absorbing or scattering only a small fraction of the incident particles. In most situations one can assume atoms and nuclei

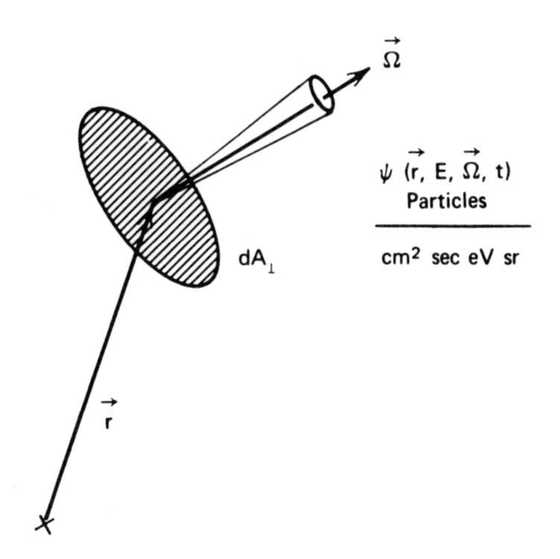

Fig. 3.1. Definition of differential flux density (from A. Edward Profio, *Experimental Reactor Physics*, Wiley, New York, 1976, p. 39).

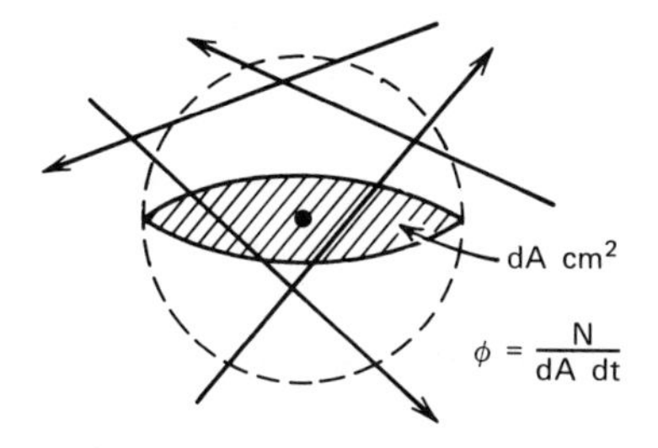

Fig. 3.2. Scalar flux density is equal to N particles in dt second entering differential sphere, divided by cross-sectional area dA.

interact equally to all directions of incidence. The relationship between flux density and number density (particles cm^{-3}) can be seen in Figure 3.3, where we consider all particles to be moving in the same direction, with speed v cm s^{-1}. Consider a cylindrical volume with base area 1 cm^2 and length vt cm. Then in t seconds the number of particles crossing the 1 cm^2 (normal to the direction of the particles) is the volume of this cylinder, vt cm^3 times the number density, n particles cm^{-3}; hence in 1 s we have

$$\phi = nv \text{ particles cm}^{-2}\text{ s}^{-1} \tag{3.2}$$

The scalar flux density has also been identified as the total track length or distance traveled in 1 s, by all the particles in 1 cm^3. There are n particles in 1 cm^3 and each travels v cm in 1 s, so the total track length per cm^3 is nv. Although derived for a monodirectional, single-speed beam, the results apply to particles with any velocity.

Sometimes the energy flux density (intensity) is used instead of the particle flux density,

$$I(\underline{r},E,t) = E\phi(\underline{r},E,t) \tag{3.3}$$

The power density

$$I(\underline{r},t) = \int E\phi(\underline{r},E,t)\, dE \text{ MeV cm}^{-2}\text{ s}^{-1} \tag{3.4}$$

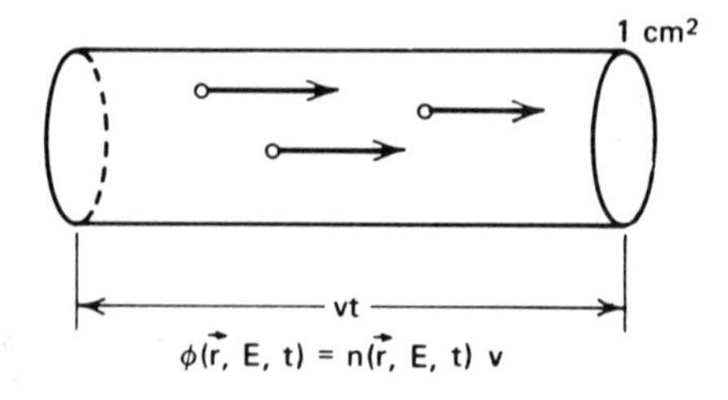

Fig. 3.3. Relationship between number density and flux density (from A. Edward Profio, *Experimental Reactor Physics*, Wiley, New York, 1976, p. 40).

where

$$1 \text{ MeV s}^{-1} = 1.6012 \times 10^{-13} \text{ W}$$

The time-integrated particle flux density is called the "fluence,"

$$\Phi(\underline{r},E) = \int \phi(\underline{r},E,t)\, dt \quad \text{particles cm}^{-2} \qquad (3.5)$$

The energy fluence

$$F(\underline{r}) = \int I(\underline{r},t)\, dt \quad \text{MeV cm}^{-2} \qquad (3.6)$$

The rate of nuclear or atomic reactions or interactions can be expressed by the yield, Y, or by the cross section, σ. Consider a beam of particles incident on a thin foil as seen in Figure 3.4. The number of particles incident per second is ϕA. Almost always we find the reaction rate R is proportional to the rate of incidence, as would be expected if the foil does not change under irradiation, and collisions of incident particles with each other are negligible. Then the yield of the reaction is

$$Y = \frac{R}{\phi A} \; \frac{\text{reactions/s}}{\text{particles/s}} \qquad (3.7)$$

In general, the yield depends on the type and energy of the incident particle, the composition and thickness δx of the foil, and sometimes other variables such as the temperature (for slow-neutron reactions).

It is desirable to eliminate the dependence on the thickness and area and refer reaction rates per cm^3 to a property of the target atom or nucleus, and incident particle. Experimentally, we find this simplification is possible except for coherent scattering of X-rays or slow neutrons from crystals, where the interaction is best described in terms of a plane wave of radiation incident on the whole crystal.

Assuming the foil is thin so all nuclei or atoms react independently and are exposed to the same flux density (hence negligible shadowing of one atom by another and negligible attenuation of the beam), we can

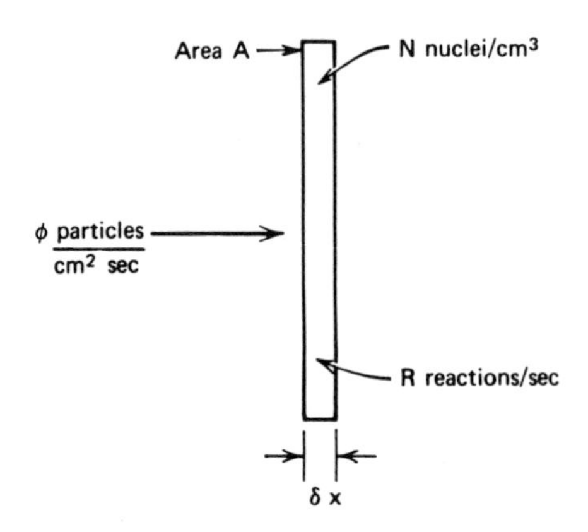

Fig. 3.4. Reaction rate in a thin foil (from A. Edward Profio, *Experimental Reactor Physics*, Wiley, New York, 1976, p. 92).

define the microscopic cross section for incident particle of energy E,

$$\sigma(E) = \frac{R(E)}{\phi A\ \delta x\ N}\ \text{cm}^2 \text{ per atom} \tag{3.8}$$

where R(E) reactions/s is the reaction rate for particles of energy E incident. The atom density

$$N = \frac{N_A \rho}{W}\ \text{atoms cm}^{-3} \tag{3.9}$$

where Avogadro's number $N_A = 6.023 \times 10^{23}$ atoms/mole, ρ is the density (g/cm^3), and W is the atomic weight. The dimensions of σ are cm^2; hence the term "cross section," but the common unit is

$$1 \text{ barn} = 10^{-24} \text{ cm}^2$$

If we have a compound or mixture, one can use the macroscopic cross section

$$\Sigma(E) = \underset{i}{S} N_i \sigma_i(E) \text{ cm}^{-1} \quad (3.10)$$

where $\underset{i}{S}$ indicates the sum over the atoms or nuclei of type i. In γ-ray interactions, Σ is called the "linear attenuation coefficient," $\mu(\text{cm}^{-1})$.

Although the total cross section (or collision cross section or interaction cross section) is a property of the incident particle as well as the nucleus or atom and is a function of energy, one can visualize it as an effective projected area to point particles, as shown in Figure 3.5. The number of nuclei is NAx; hence the projected area of the target nuclei is NAxσ. Thus the fraction of area obscured by nuclei is NAxσ/A. With ϕA particles/second incident, the number of collisions per second is ϕA(NAσx/A) = Nσ ϕ(Ax) = Nσ ϕV, where V is the volume.

The total or collision cross section gives the rate of interaction by all processes, including absorption and scattering; hence for incident particles of energy E_0 we have

$$\sigma(E_0) = \sigma_a(E_0) + \sigma_s(E_0) \quad (3.11)$$

The absorption cross section is sufficient to describe the rate of disappearance of incident particles by capture or other absorption processes, but a differential scattering cross section is needed to define the energy and angular distribution of the particles after scattering. Let us assume an incident particle of energy E_0 and direction $\underline{\Omega}_0$ as shown in Figure 3.6. After scattering the energy is E and the direction is $\underline{\Omega}$. Then

$$\frac{d\sigma_s(E_0 \rightarrow E, \underline{\Omega}_0 \rightarrow \underline{\Omega})}{dE\, d\Omega} \text{ cm}^2 \text{ MeV}^{-1} \text{ sr}^{-1}$$

is the cross section for scattering into solid angle dΩ about direction $\underline{\Omega}$, and into energy interval dE about energy E. For unpolarized radiation and isotropic medium the cross section is independent of any fixed direction in space and is azimuthally symmetric, depending only on the angle θ between the scattered and incident directions, where $\cos\theta = (\underline{\Omega}_0 \cdot \underline{\Omega})$. The

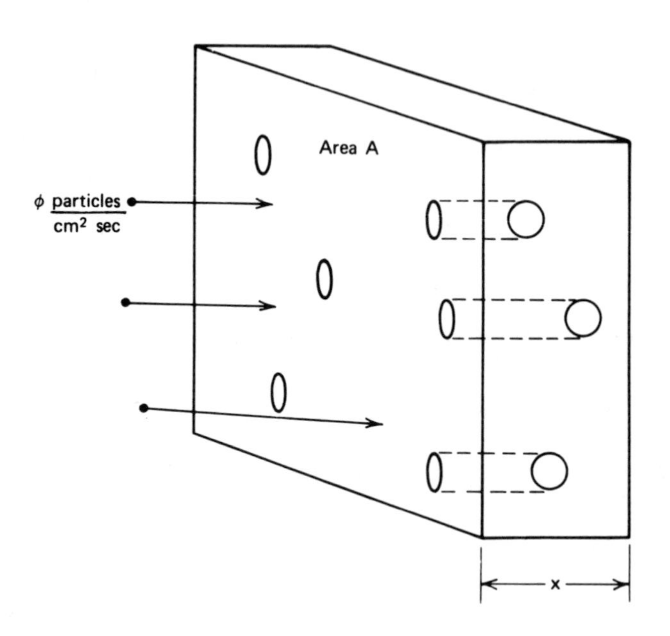

Fig. 3.5. Collision cross section as effective projected area of nucleus for point particles incident. Number of nuclei, NAx; projected area per nucleus, σ; projected area, NAxσ; collisions per second, (ϕA) $(NAx\sigma/A) = N\sigma\ \phi\ (Ax)$ (from A. Edward Profio, *Experimental Reactor Physics*, Wiley, New York, 1976, p. 43).

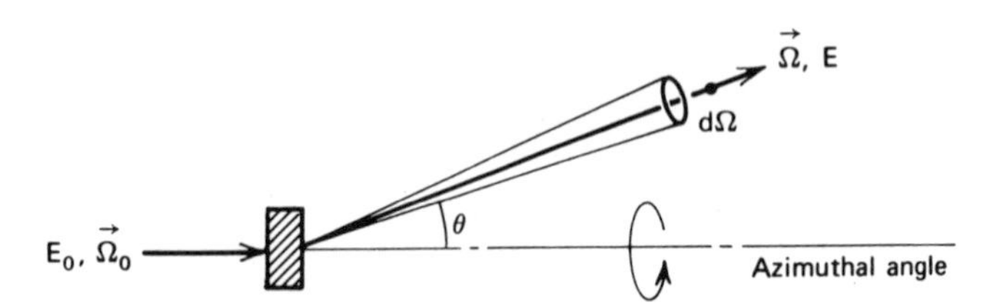

Fig. 3.6. Geometry for definition of the differential scattering cross section (from A. Edward Profio, *Experimental Reactor Physics*, Wiley, New York, 1976, p. 96).

integral over scattered angle and energy is the scattering cross section,

$$\sigma_s(E_0) = \iint \frac{d\sigma_s(E_0 \rightarrow E, \underline{\Omega}_0 \rightarrow \underline{\Omega})}{dE\, d\Omega}\, dE\, d\Omega \qquad (3.12)$$

Similar cross sections may be defined for the energy and angular distribution of secondary particles, such as the production of γ rays by interactions of neutrons,

$$\frac{d\sigma(E_n \rightarrow E_\gamma, \underline{\Omega}_n \rightarrow \underline{\Omega}_\gamma)}{dE_\gamma\, d\Omega_\gamma}$$

Other "cross sections" are defined for energy transfer, as is discussed later.

3.2 INTERACTIONS OF IONS

The principal interactions of ions (mass $M \gg$ electron mass m_0) in the energy range of interest (roughly 0.1 MeV to 20 MeV) are: (1) inelastic Coulomb scattering from atomic electrons, resulting in ionization and excitation of the atom but little change in direction of the ion, (2) charge exchange, that is, capture and loss of electrons, (3) elastic Coulomb scattering on the nucleus (Rutherford scattering), and (4) nuclear reactions.

The most important mechanism for energy loss is ionization and excitation. As indicated, the heavy ion is little deflected by the interaction with electrons, and the track or path of the ion is usually straight. Rutherford scattering is infrequent for the hydrogen and helium ions but can be significant for fission fragments and other highly charged ions. Rutherford scattering can produce a large deflection and energy loss. Capture and loss of electrons, reducing the effective charge of the ion, is significant for the slowly moving fission fragments and for the hydrogen and helium ions below roughly 1 MeV. Nuclear reactions are very infrequent at energies below 20 MeV and can be neglected in considering energy loss or energy deposition. However, they are significant in accelerators in producing neutrons and other secondary particles, as discussed in Chapter 2.

Ionization and Excitation

The theory of energy loss by ionization and excitation is developed by Bichsel [1] and by Evans [2]. Essentially, one considers the Coulomb force between the ion and an atomic electron as the ion travels by, and calculates the momentum and energy transferred to the electron as a function of the impact parameter b (closest distance of approach in absence of an interaction). The energy-transfer cross section for losing energy E to (E + dE) is $d\sigma = (2\pi\ b\ db)$ in terms of the corresponding impact parameter. The smallest value of b is limited by the uncertainty principle, and the largest value of b is limited by an energy transfer too small to excite an electron to a higher energy state. The binding of the electrons is represented by a mean ionization and excitation potential I (eV), a function of atomic number Z of the element that has to be determined experimentally. Other corrections involve relativistic effects at higher energies and corrections for binding in different shells. The effect of NZ electrons/cm^3 averaged over all impact parameters can then be expressed in terms of the stopping power or linear rate of energy loss,

$$\left(\frac{dE}{dx}\right)_{ion} = \text{const}\ \frac{z^2 NZ}{v^2}\left(\ln\frac{2m_0 v^2}{I}\right)\ \text{MeV/cm} \qquad (3.13)$$

where z is the charge of the ion (in units of the electron charge), v is the speed of the ion, and m_0 is the mass of the electron. The stopping power does not depend on the mass M of the ion, but v^2 is proportional to E/M.

For the same material, the stopping power of deuterons, tritons, ^{3}He ions and α particles can be obtained from the stopping power for protons (evaluated at E/A, where A is the ratio of ion mass to proton mass) by multiplying by z^2:

$$\left(\frac{dE}{dx}\right)_d = \left(\frac{dE}{dx}\right)_p \quad \text{at}\ \frac{E_d}{2.00} \qquad (3.14a)$$

$$\left(\frac{dE}{dx}\right)_t = \left(\frac{dE}{dx}\right)_p \quad \text{at}\ \frac{E_t}{2.99} \qquad (3.14b)$$

$$\left(\frac{dE}{dx}\right)_{^3He} = 4\left(\frac{dE}{dx}\right)_p \quad \text{at } \frac{E_{^3He}}{2.99} \tag{3.14c}$$

$$\left(\frac{dE}{dx}\right)_{\alpha} = 4\left(\frac{dE}{dx}\right)_p \quad \text{at } \frac{E_{\alpha}}{3.97} \tag{3.14d}$$

These relationships do not apply to helium ions below about 2 MeV because of charge exchange.

Instead of the linear stopping power, it is often convenient to tabulate or plot the mass stopping power

$$\frac{S}{\rho} = \frac{1}{\rho}\left(\frac{dE}{dx}\right) = z^2 \frac{Z}{W} f(V^2, I) \quad \frac{\text{MeV}}{\text{g/cm}^2} \tag{3.15}$$

Because Z/W is approximately constant (except for hydrogen) and the functional dependence on I is logarithmic, the mass stopping power varies only slowly with Z. Values of I and ℓn I are listed in Table 3.1 for several elements [1]. Chemical binding and density have little effect on stopping power at the energies of interest, and the stopping power of a compound or mixture may be found by summing the stopping powers of the constituent elements at the proper atom densities. Mass stopping powers of water and lead for protons are listed in Table 3.2 and plotted in Figure 3.7 [1].

It should be noted that ionization and excitation of atoms is only the first step in producing what may be the ultimate effect of significance, such as biological damage. Except for some chemical-bond energies, most of the ionization and excitation energy eventually is degraded to heat. The effects of ionizing radiation on matter are discussed in Chapter 6.

Charge Exchange

An ion passing through matter captures and loses electrons in thousands of exchanges with the atoms. The capture and loss cross sections are strong functions of the speed, with capture predominating at small speed v. Thus the mean charge and the rate of energy loss decrease and the stopping power deviates from the $1/v^2$ law at low energies. Because the charge exchange is different for helium and hydrogen ions, the ratio of stopping power for say α particles is less than 4 at α particle energies below 2 MeV, as shown in Figure 3.8.

TABLE 3.1. MEAN IONIZATION AND EXCITATION POTENTIAL

Element	Z	Z/W	I (eV)	ℓn I
H	1	0.992	18	2.890
He	2	0.500	42	3.738
Be	4	0.444	64	4.159
C	6	0.500	78	4.357
N	7	0.500	78	4.357
O	8	0.500	100	4.605
Al	13	0.482	164	5.100
Si	14	0.498	170	5.136
Ar	18	0.451	184	5.215
Cu	29	0.456	322	5.775
Ge	32	0.441	350	5.858
Pb	82	0.396	820	6.709
U	92	0.386	900	6.802

Rutherford Scattering

The differential cross section for elastic nuclear Coulomb scattering (Rutherford scattering) is given by

$$\frac{d\sigma}{d\Omega} = \frac{b^2}{16} \frac{1}{\sin^4(\theta/2)} \tag{3.16}$$

where $b = 2Zze^2/M_0v^2$, with e as the electron charge and θ the angle of scattering in the center-of-mass system (CMS). The reduced mass

TABLE 3.2. MASS STOPPING POWER (MeV per g/cm^2) FOR PROTONS

E (MeV)	Water	Lead	E (MeV)	Water	Lead
1.0	271.4	62.87	5.0	81.3	26.88
1.1	253.4	60.15	5.5	75.4	25.42
1.2	238.0	57.71	6.0	70.4	24.12
1.3	224.5	55.52	6.5	66.1	22.98
1.4	212.7	53.52	7.0	62.3	21.95
1.5	202.2	51.75	7.5	58.9	21.03
1.6	192.9	50.20	8.0	56.0	20.20
1.7	184.5	48.73	8.5	53.3	19.45
1.8	176.9	47.33	9.0	50.9	18.75
1.9	170.0	46.01	9.5	48.7	18.11
2.0	163.6	44.76	10.0	46.8	17.53
2.2	152.4	42.45	11.0	43.3	16.48
2.4	142.8	40.57	12.0	40.4	15.56
2.6	134.4	38.87	13.0	37.8	14.76
2.8	127.1	37.31	14.0	35.6	14.05
3.0	120.6	35.91	15.0	33.7	13.41
3.2	114.8	34.67	16.0	31.9	12.84
3.4	109.6	33.52	17.0	30.4	12.32
3.6	104.9	32.45	18.0	29.0	11.85
3.8	100.6	31.48	19.0	27.8	11.42
4.0	96.7	30.57	20.0	26.6	11.02
4.5	88.3	28.57			

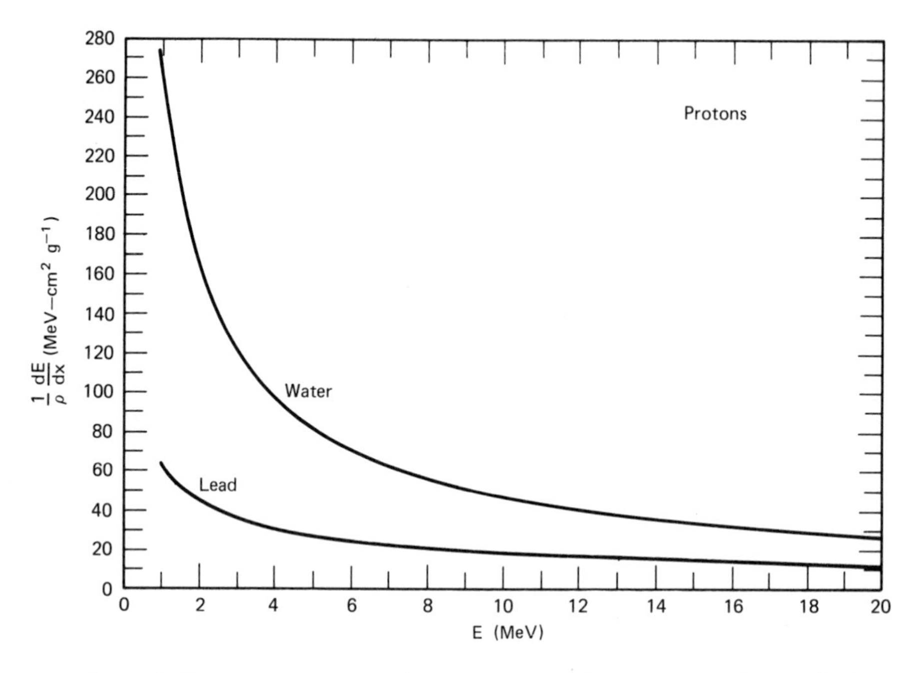

Fig. 3.7. Mass stopping power of water and lead for protons.

$$M_0 = \frac{M_1 M_2}{M_1 + M_2} \qquad (3.17)$$

with M_1 as the mass of the incident ion and M_2 the mass of the target nucleus. Rutherford scattering is predominantly forward but can give a large deflection. The maximum energy loss occurs for 180° scattering. Kinematics of elastic scattering gives for the scattered ion energy

$$E_2 = \left(\frac{r - 1}{r + 1}\right)^2 E_1 \qquad (3.18)$$

where E_1 is the incident energy and R is the ratio of the mass of the target nucleus to the mass of the ion.

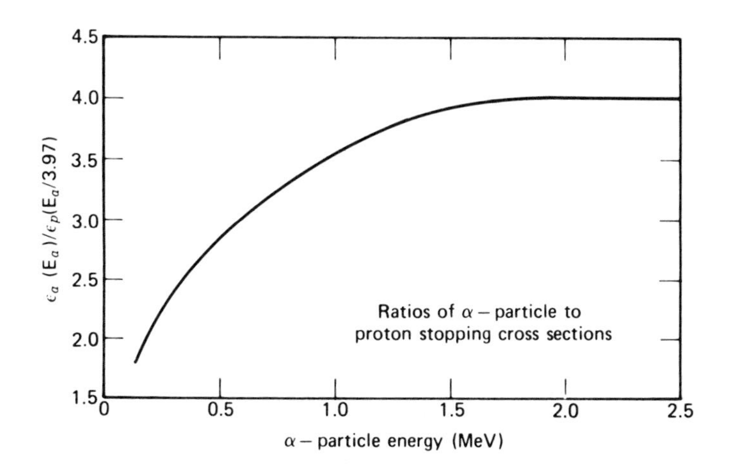

Fig. 3.8. Effect of charge exchange on ratio of stopping power for α particles to stopping power for protons at same speed (from A. Edward Profio, *Experimental Reactor Physics*, Wiley, New York, 1976, p. 50).

For α particles on lead, $E_2 = 0.93\ E_1$ for 180° scattering, whereas for α particles on aluminum, $E_2 = 0.55\ E_1$.

Range

The discussion of stopping power has already assumed that the thousands of individual collisions, mostly with small energy loss and deflection, can be modeled by a continuous slowing down approximation (CSDA). If one calculates the distance the ion travels from source energy E_0 to zero energy by the CSDA model, the CSDA range

$$R(E_0) = \int_{E_0}^{0} \frac{dE}{(dE/dx)} = \int_{E_0}^{0} \text{const}\ \frac{M}{z^2}\ \frac{v^3\ dv}{NZ\ \ell n\left(2m_0 v^2/I\right)} \qquad (3.19)$$

where $E = (Mv^2/2)$. The effect of charge exchange can be taken into account by using an effective z, or if z is allowed to stand for the initial charge, one can add a correction term. Then the CSDA range can be expressed as

$$R(E_0) = \frac{M}{z^2} g(v_0, Z, I) + C \qquad (3.20)$$

The range for protons in various materials is plotted in Figures 3.9 and 3.10. The range is listed in Table 3.3 for water and lead as a function of proton kinetic energy T or E. The range for other hydrogen ions may be found by multiplying M/z^2 by the proton range evaluated at the same initial speed v_0. The difference in range of helium ions compared to protons, because of the difference in charge exchange, is about 0.25 mg/cm^2 or 0.2 cm of air. Hence we have for the same material

$$R_d(E_0) = 2R_p \quad \text{at } \frac{E_0}{2} \qquad (3.21a)$$

$$R_t(E_0) = 2R_p \quad \text{at } \frac{E_0}{3} \qquad (3.21b)$$

$$R_{^3He}(E_0) = \frac{3}{4} R_\alpha \quad \text{at } \frac{4}{3} E_0 \qquad (3.21c)$$

$$R_\alpha(E_0) \simeq R_p \frac{E_0}{4} + 0.25 \text{ mg/cm}^2 \qquad (3.21d)$$

In the absence of data on a certain material, the approximate Bragg-Kleeman rule is useful,

$$\frac{R_1}{R_0} \simeq \frac{\rho_0}{\rho_1} \frac{\sqrt{W_1}}{\sqrt{W_0}} \qquad (3.22)$$

where R_1 is the range in the material, R_0 is a known range in a standard material, ρ is the density, and W is the atomic weight.

Straggling

The crow-flight distance an ion travels in matter (distance from source to stopping point) is slightly different from the CSDA range because of multiple slight deflections or Rutherford scattering. The flux density/thickness curve, shown in Figure 3.11, does not drop to zero abruptly because of these fluctuations in the paths

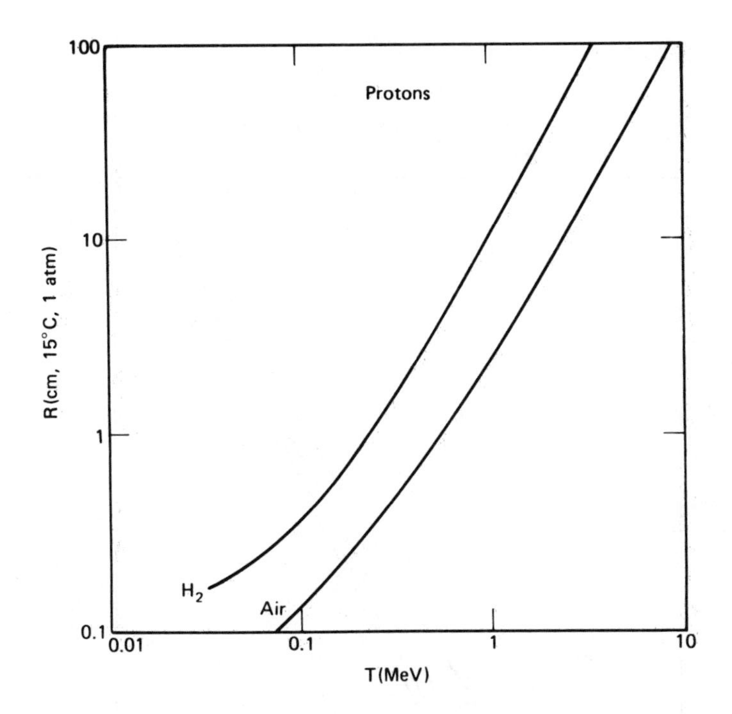

Fig. 3.9. Range of protons in air and hydrogen gas as function of kinetic energy (from A. Edward Profio, *Experimental Reactor Physics*, Wiley, New York, 1976, p. 52).

of individual particles. Thus we define a mean range R at the 50% point on the flux density vs distance curve, or integral number curve (number of ions that travel beyond distance x). The derivative with respect to x is the differential number curve shown in Figure 3.11, giving the number of ions that stop between x and (x + dx). It is approximately Gaussian with a standard deviation σ (not to be confused with a cross section). For historical reasons the width is expressed by the range-straggling parameter $\alpha = (2\sigma)^{1/2}$. The range straggling is exaggerated in Figure 3.11; it is about 1.5% of the mean range for α particles and 3% of the mean range for protons. Other definitions of range may be found, especially

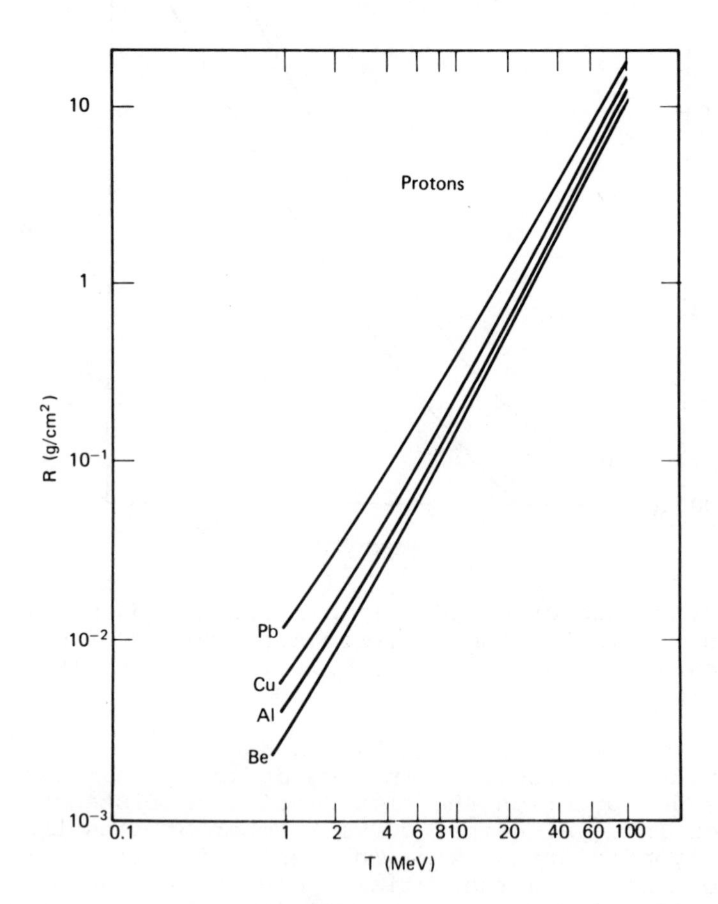

Fig. 3.10. Range of protons in beryllium, aluminum, copper, and lead in terms of mass thickness ρx (g/cm^2) as function of kinetic energy (from A. Edward Profio, *Experimental Reactor Physics*, Wiley, New York, 1976, p. 53).

TABLE 3.3. PROTON CSDA RANGE (g/cm^2) IN WATER AND LEAD

E (MeV)	Water	Lead	E (MeV)	Water	Lead
1.0	0.0039	0.0116	4.6	0.0319	0.1068
1.2	0.0046	0.0149	4.8	0.0342	0.1139
1.4	0.0055	0.0185	5.0	0.0367	0.1213
1.6	0.0065	0.0224	6.0	0.0499	0.1606
1.8	0.0076	0.0265	7.0	0.0650	0.2042
2.0	0.0088	0.0308	8.0	0.0820	0.2517
2.2	0.0101	0.0354	9.0	0.1008	0.3031
2.4	0.0114	0.0402	10.0	0.1213	0.3583
2.6	0.0129	0.0453	11.0	0.1435	0.4172
2.8	0.0144	0.0505	12.0	0.1675	0.4797
3.0	0.0160	0.0560	13.0	0.1931	0.5457
3.2	0.0177	0.0617	14.0	0.2203	0.6151
3.4	0.0195	0.0675	15.0	0.2492	0.6880
3.6	0.0213	0.0736	16.0	0.2798	0.7642
3.8	0.0233	0.0799	17.0	0.3119	0.8437
4.0	0.0253	0.0863	18.0	0.3445	0.9265
4.2	0.0274	0.0929	19.0	0.3808	1.0125
4.4	0.0296	0.0998	20.0	0.4176	1.1016

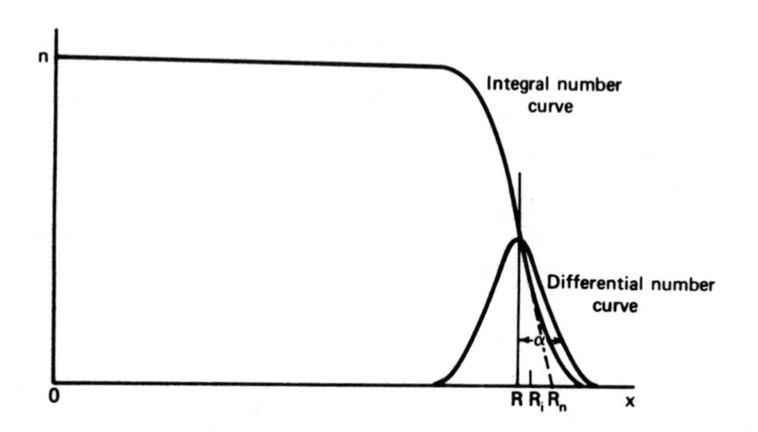

Figure 3.11. Integral and differential number versus distance curves for ions, with mean range R, extrapolated number range R_n, extrapolated ionization range R_i, and range straggling parameter a. The straggling is exaggerated (from A. Edward Profio, *Experimental Reactor Physics*, Wiley, New York, 1976, p. 54).

in experimental papers. If the integral number is extrapolated as shown, one obtains the extrapolated number range R_0. If ionization density is measured instead, one obtains a slightly different measure, the extrapolated ionization range R_i. All are very close, and it is clear that ions are stopped by a thickness of material only a few percent larger than the mean range. For a thickness of material less than the mean range, nearly all the ions penetrate, but with less than the source energy and an energy straggling of about 3%.

Energy Distributions

An ion of a certain energy E will travel a mean distance R(E) before being stopped, regardless of its initial energy. Thus range-energy tables can be used to construct the energy/"residual range" and energy/distance curve such as shown in Figure 3.12.

One could also plot the rate of energy loss or stopping power dE/dx against residual range or x, instead of energy because energy and distance are related. Figure 3.13 shows a measured Bragg ionization curve for ^{210}Po α particles (initial energy 5.25 MeV) in air. The ordinate is the relative specific ionization with a peak of 6600 ion-electron pairs generated

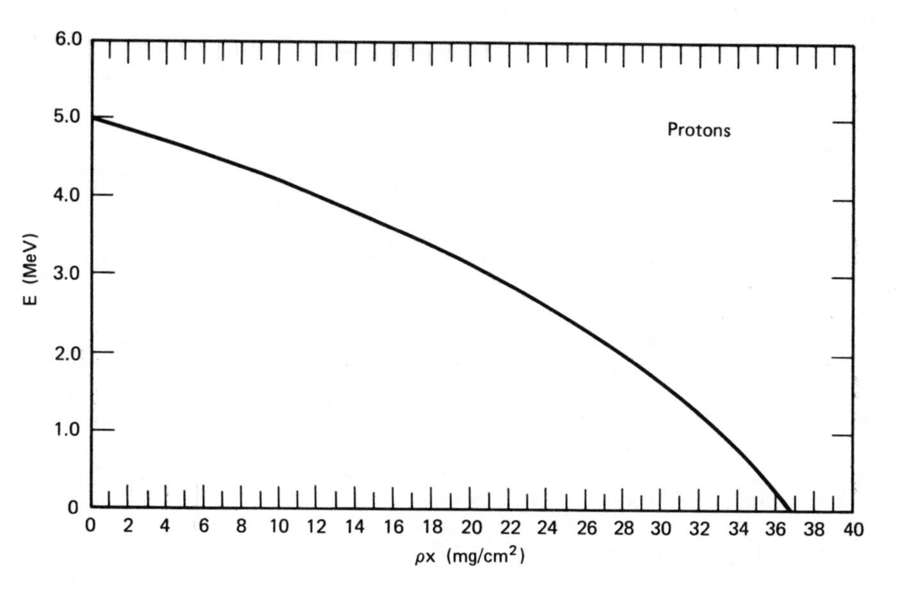

Fig. 3.12. Energy versus distance for 5-MeV protons in water.

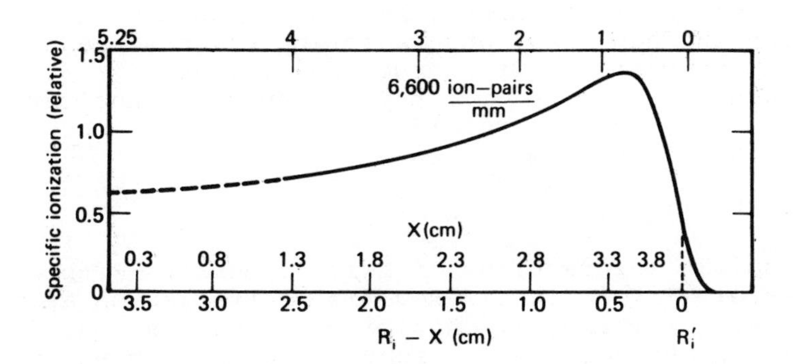

Fig. 3.13. Bragg ionization curve for ^{210}Po α particles in air, 15°C, 1 atm (from A. Edward Profio, *Experimental Reactor Physics*, Wiley, New York, 1976, p. 55).

per mm of distance. The specific ionization is proportional to dE/dx because the energy required to produce one ionized atom is almost constant ($\sim$ 34 eV in air).

Fission Fragments

Fission fragments experience more frequent Rutherford scatterings than hydrogen and helium ions because of the large charge (about 20 initially), but the path is still fairly straight. The rate of energy loss and specific ionization are greatest at the beginning of the track, decreasing with distance because of electron pickup (charge exchange). Ranges of the average light and heavy fission fragments are listed in Table 3.4 for several materials. The initial energy of the average light fragment is about 100 MeV, and the initial energy of the average heavy fragment is about 70 MeV.

Additional data on interactions of ions with matter may be found in Marion and Young [3].

Shielding of Ions

Because of the small range of ions in matter, shielding is simple. α Particles from radioactive nuclides do not penetrate even the dead layer of the skin. However, the energy deposited by ions is very important when they are generated within tissue, for example α particles inhaled or ingested in the body, or recoil protons from elastic scattering of fast neutrons. Radiation damage to metals is caused by displacement of atoms induced by neutron scattering, and by hydrogen or helium produced in (n,p) or (n,α) reactions. Energy deposition is discussed in Chapter 5, and radiation effects are discussed in Chapter 6. Of course, any secondary radiations such as neutrons or γ rays produced by nuclear reactions will require shielding.

3.3 INTERACTIONS OF ELECTRONS

The principal interactions of swift electrons with matter are: (1) inelastic Coulomb scattering on atomic electrons, resulting in ionization and excitation, (2) elastic nuclear Coulomb scattering, (3) emission of electromagnetic radiation (bremsstrahlung) on deceleration or acceleration, mostly in the nuclear Coulomb field, and (4) annihilation of a positron or negatron.

TABLE 3.4. MEAN RANGE OF FISSION FRAGMENTS

Material[a]	Light	Heavy
Air (cm)	2.54	1.95
H_2 (cm)	2.11	1.77
Ar (cm)	2.39	1.94
Al (mg/cm^2)	4	3
U (mg/cm^2)	11	9

[a]Gases at 15°C, 1 atm.

Ionization and excitation are important throughout the Z range and E range of interest. Elastic nuclear Coulomb scattering is much more important for electrons than ions, resulting in a very crooked path with large straggling in range and energy and a high probability of being scattered through greater than 90° (backscattering). Radiation of X-rays (bremsstrahlung) is most significant at large Z and E. Annihilation of a positron usually occurs after it has been slowed to thermal energies and is characterized by emission of two 0.511-MeV γ rays (annihilation quanta) at 180° to each other.

Ionization and Excitation

Ionization and excitation proceeds by the same Coulomb interaction as for ions [1]. However, the electron is relativistic, approaching the speed of light c, even at keV energies. Also, as the incident and ionization electrons cannot be differentiated, the faster electron emerging from the interaction is arbitrarily taken to be the incident one; hence the maximum energy transfer is E/2. The stopping power or rate of energy loss depends on the density of the medium because of polarization by the moving electron. With these modifications, the stopping power [2]

$$\left(\frac{dE}{ds}\right)_{ion} = 0.511 \times 10^{-24} \frac{NZ}{\beta^2} \left[\ln \beta\left(\frac{E + 0.511}{I}\right)\left(\frac{E}{0.511}\right)^{1/2} - \frac{1}{2}\beta^2\right] - \delta \frac{\text{MeV}}{\text{cm}} \qquad (3.23)$$

where s refers to distance along the path, NZ is the electron density (electrons/cm^3), $\beta = (V/c)$, where V is the speed of the incident electron, E is the kinetic energy, I is the mean ionization and excitation potential as given in Table 3.1, and δ corrects for the density of the medium (zero for gases). The stopping power first decreases with increasing energy, reflecting the $1/\beta^2$ or $1/V^2$ dependence, then increases at higher energies because of the relativistic effect. In a solid or liquid the rise is not as rapid as in gases because of the polarization effect. Because of the crooked path, the rate of energy loss along the path is not directly comparable to the loss of energy in foils. The collision or ionization stopping power of water is plotted in Figure 3.14 for electrons and protons [4]. The rate of energy loss by ionizing collisions and by radiation is plotted for electrons in lead in Figure 3.15 and listed in Table 3.5 [5]. The radiation yield is the fraction of the electron's initial kinetic energy that is radiated as bremsstrahlung ("braking" or deceleration radiation) in a thick target.

Scattering

Elastic nuclear Coulomb or Rutherford scattering, modified by relativistic and quantum-mechanical effects, is discussed by Evans [6]. The effect is to make the path of an electron quite crooked, so that an increment of path length cannot be equated with an increment on in range dx or thickness of absorber as with ions. Thus most results on the number and energy spectrum of

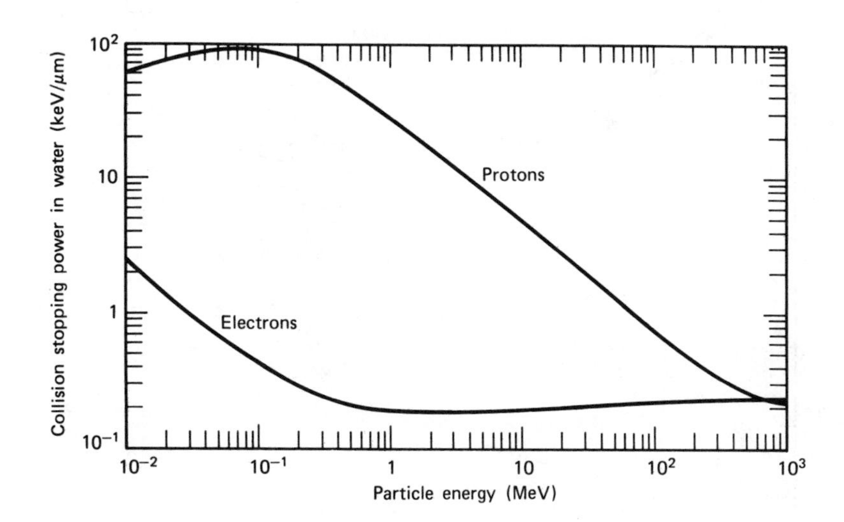

Fig. 3.14. Ionization-collision stopping power for electrons and protons in water (after Ref. 4).

electrons as a function of thickness are obtained experimentally, although calculations have been performed [5].

Backscattering, or reemergence from the surface of incidence, is significant in some counting experiments. The fraction of incident electrons that are backscattered increases with the thickness of material up to a thickness equal to about 0.2 of the range. The fraction backscattered also increases with the atomic number Z. For a thick, high-Z material such as lead, the number backscattered into a detector on the source side of the slab may equal the number reaching the detector directly from the source. Figure 3.16 shows the dependence of the thick-slab backscattering factor

$$f_b = \frac{\text{counts with slab}}{\text{counts without slab}} \tag{3.24}$$

as a function of Z. Backscattering is almost independent of the energy of the electrons.

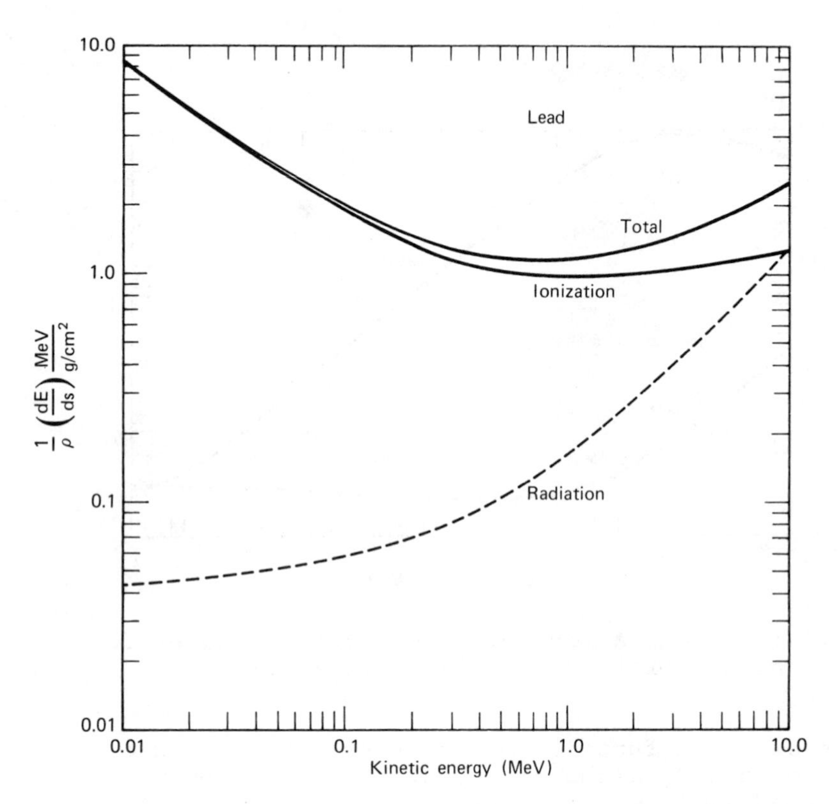

Fig. 3.15. Mass stopping power of electrons in lead.

Bremsstrahlung

According to classical electromagnetic theory, a charge should radiate when it is decelerated or accelerated. The acceleration produced by a nucleus of charge Ze on a particle of charge ze and mass M is proportional to Zze^2/M. The amplitude of the electromagnetic wave is proportional to the acceleration. The intensity is proportional to the square of the (amplitude, times ze) hence proportional to $Z^2z^4e^6/M^2$. The proportionality holds also in quantum theory. Radiative losses are small for heavy particles (ions) and in low-Z materials. Actually, because of quantum effects the electron does

TABLE 3.5. MASS STOPPING POWER OF ELECTRONS IN LEAD

Kinetic Energy (MeV)	Ionization ($MeV-cm^2/g$)	Radiation ($MeV-cm^2/g$)	Total ($MeV-cm^2/g$)	Radiation Yield
0.01	8.42	4.51(-2)[a]	8.46	3.4(-3)
0.02	5.45	4.62(-2)	5.50	5.2(-3)
0.05	3.00	5.26(-2)	3.05	9.8(-3)
0.10	1.96	5.94(-2)	2.02	1.7(-2)
0.20	1.39	7.25(-2)	1.46	2.8(-2)
0.40	1.11	9.62(-2)	1.20	4.7(-2)
0.60	1.03	1.19(-1)	1.15	6.2(-2)
0.80	1.01	1.42(-1)	1.15	7.5(-2)
1.00	1.00	1.66(-1)	1.17	8.7(-2)
2.00	1.04	2.80(-1)	1.32	1.3(-1)
4.00	1.11	5.21(-1)	1.63	2.0(-1)
6.00	1.16	7.68(-1)	1.92	2.5(-1)
8.00	1.19	1.02	2.21	3.0(-1)
10.00	1.22	1.28	2.49	3.4(-1)
20.00	1.29	2.61	3.91	4.7(-1)

[a] Read 4.51×10^{-2}.

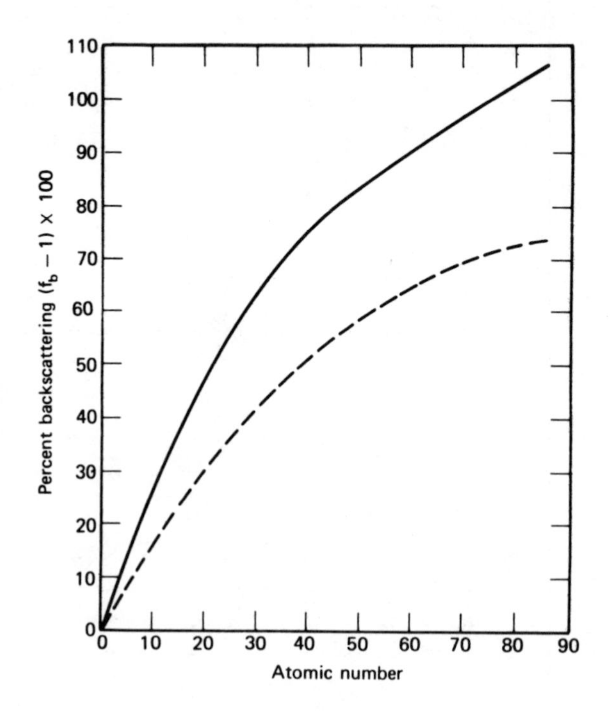

Fig. 3.16. Saturation backscattering of β particles plotted as function of atomic number of backscattering material. Solid line represents data for source directly on scatterer and broken line, source on a thin film over scatterer (after E. P. Steinberg, ANL-5622; from A. Edward Profio, *Experimental Reactor Physics*, Wiley, New York, 1976, p. 60).

not radiate in every nuclear encounter. When it does, the photon may carry off any energy between 0 and

$$(h\nu)_{max} = E_0 \tag{3.25}$$

where E_0 is the initial kinetic energy of the electron. The electron may lose energy between E_0 and 0, the nucleus receiving negligible energy.

In a thin target, that is, one where the electron loses very little energy on the average, the bremsstrahlung intensity spectrum is approximately uniform between 0 and E_0 [7,8]; hence the number of photons per unit energy is approximately proportional to $(h\nu)^{-1}$ between 0 and E_0. In a thick target, the electron loses energy by ionization as well as radiation, and the number versus photon energy spectrum is enhanced at low energies. Figure 3.17 illustrates measured and calculated bremsstrahlung spectra for collimated electron beams of 5.3 MeV and 10.0 MeV incident on a gold-tungsten target 0.2 radiation lengths thick [9,10]. The radiation length is the thickness required to reduce the electron's kinetic energy by a factor of $(1/e) = 0.368$ by radiation. At low photon energies the bremsstrahlung spectrum is modified by attenuation in the target material and housing (see Figure 3.18).

The angular distribution of bremsstrahlung is forward-peaked. Figure 3.18 plots the measured "exposure" rate in roentgens/min at 1 m (approximately proportional to intensity), per milliampere of current in a collimated electron beam incident on a thick tungsten target (thickness slightly greater than the range), for a few electron energies in the megaelectron volt region [11]. The distribution is more isotropic at low energies.

The rate of loss of electron kinetic energy by radiation is given by [12]

$$\left(\frac{dE}{ds}\right)_{rad} = N(E + m_0c^2)\ \sigma_0\ Z^2\ \overline{B}\ \ \text{MeV/cm} \qquad (3.26)$$

where N is the density of nuclei per cm^3, $m_0c^2 = 0.511$ MeV, and E is in MeV, $\sigma_0 = 0.580 \times 10^{-27}$ cm^2/nucleus, and $\overline{B}$ is a photon-energy averaged factor from the theory, plotted in Figure 3.19 as a function of electron kinetic energy for H_2O, Cu, and Pb. At low energies, $E << m_0c^2$, $\overline{B}$ is 5.33 for all materials. The rate of energy loss per unit mass thickness ρx (g/cm^2) in lead, where $\rho = 11.35$ g/cm^3 and $Z = 82$, is given in Table 3.5 and plotted in Figure 3.15 [5].

Even pure β emitters are sources of electromagnetic radiation because of the bremsstrahlung in the surrounding material. The fraction of the initial kinetic energy that is converted into bremsstrahlung in a thick target is approximately [7]

$$\frac{I}{E} \simeq 0.0007ZE \qquad (3.27)$$

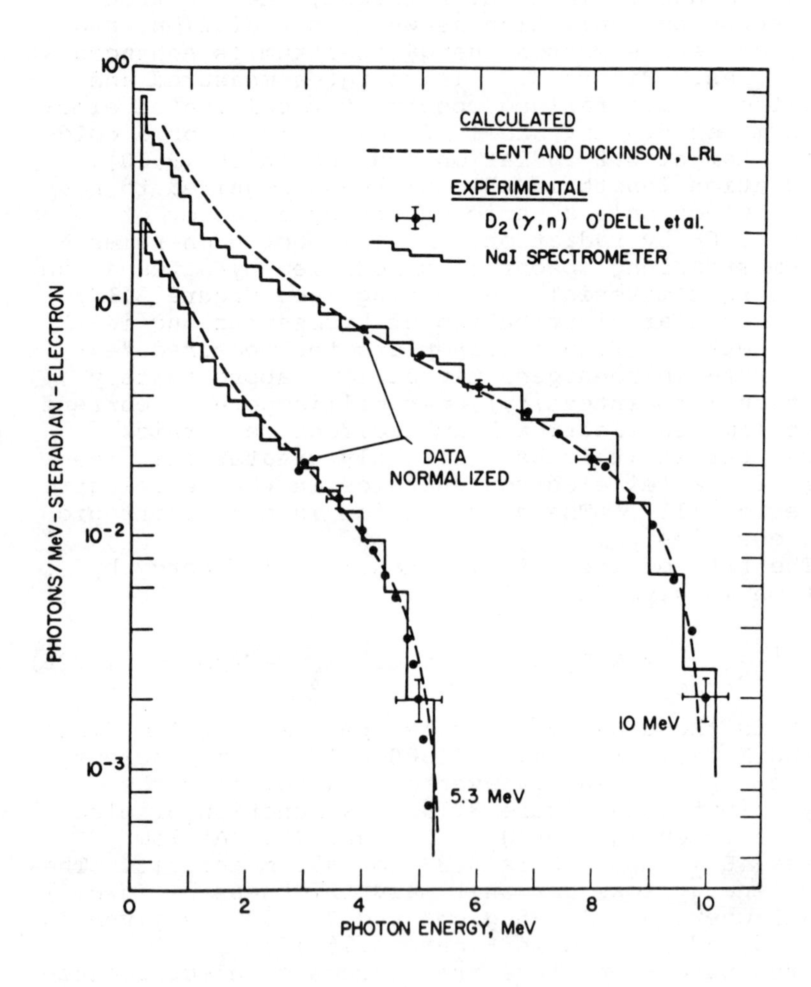

Fig. 3.17. Comparison of measured and calculated thick-target bremsstrahlung spectra shapes produced by 5.3- and 10-MeV electrons. The NaI spectrometer measurements and calculations of Lent and Dickinson have been normalized to D_2 (γ,n) data at points indicated (after Sandifer and Taherzadeh [9], from A. Edward Profio, *Experimental Reactor Physics*, Wiley, New York, 1976, p. 63).

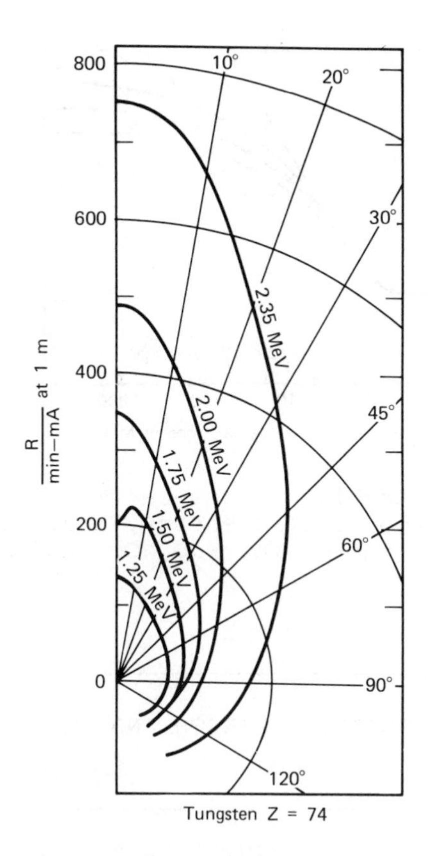

Fig. 3.18. Measured exposure rate as function of angle, for various initial energy electron beams on a tungsten target slightly thicker than the range.

where E is in MeV. To reduce the bremsstrahlung from β sources, it is advantageous to shield them with a low-Z material such as plastic, unless radiation damage or thermal stability dictates a metal such as aluminum. Table 3.5 gives the radiation yield in lead. At 1 MeV, for example, 8.7% or 87 keV will be radiated as bremsstrahlung in a thick target. Table 3.6 lists the radiation length and the critical energy, or energy where the rates of energy loss by radiation and by ionization are equal, for several materials.

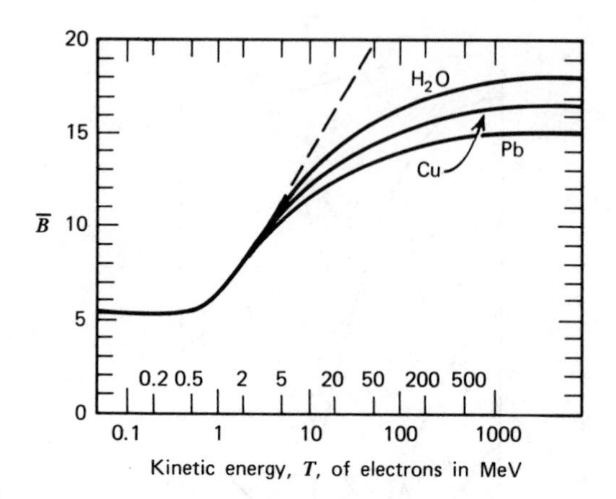

Fig. 3.19. Bremsstrahlung cross-section factor $\overline{B}$ as function of electron kinetic energy, including screening effect at high energy for H_2O, Cu, and Pb (after Evans [2]).

TABLE 3.6. RADIATION LENGTH AND CRITICAL ENERGY

Material	Z	Radiation Length (g/cm^2)	Critical Energy (MeV)
Hydrogen	1	58.0	340.0
Carbon	6	43.0	103.0
Aluminum	13	24.0	87.0
Iron	26	14.0	47.0
Lead	82	5.8	6.9

Range and Straggling

The mean energy of electrons passing through a slab of material is decreased, but there is large straggling in energy as well as in range. Figure 3.20 plots the number of primary electrons transmitted by a slab of material, per electron perpendicularly incident on the slab, for monoenergetic electrons and for β particles whose maximum energy is equal to the energy of the monoenergetic electrons. Empirical expressions have been obtained for monoenergetic electrons [12]. The shape depends on the details of the detector and geometry, but the latter half of the transmission curve for monoenergetic electrons tends to be linear, extrapolating to a practical range R. The transmission curve for β particles is approximately exponential but has a finite range also nearly equal to R. Measurements of range R in aluminum are plotted in Figure 3.21. Within 0.5 to 3 MeV the data are represented to within $\pm 5\%$ by

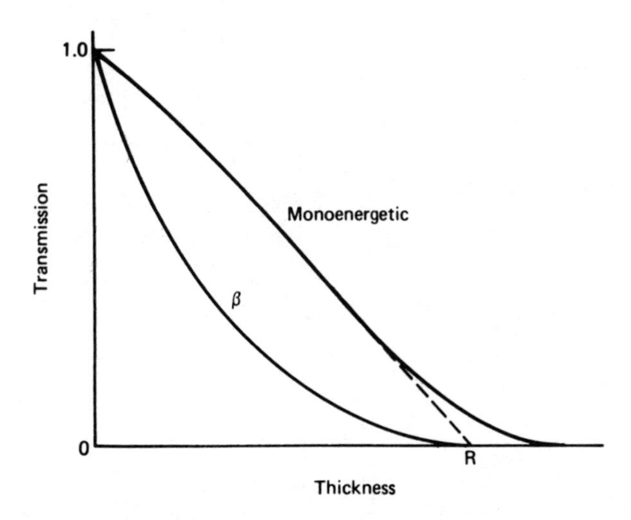

Fig. 3.20. Transmission of monoenergetic electrons and continuous-spectrum β rays through matter; R is extrapolated or practical range for monoenergetic electrons or maximum range for β rays (from A. Edward Profio, *Experimental Reactor Physics*, Wiley, New York, 1976, p. 64).

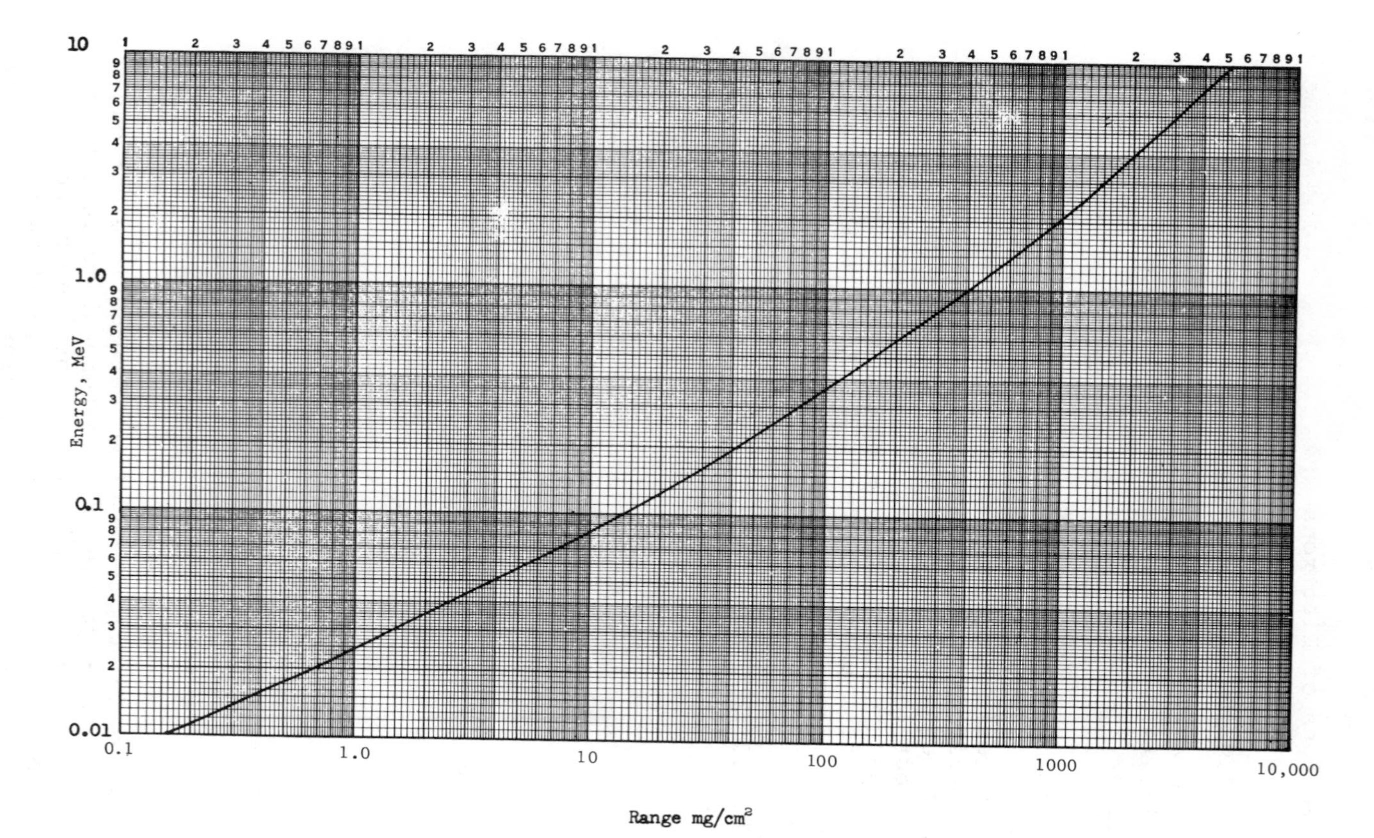

Fig. 3.21. Range-energy relationship for β particles (data from *Radiological Health Handbook*).

$$R(g/cm^2) = 0.52E(MeV) - 0.09 \quad (3.28)$$

The range expressed in g/cm^2 has been found to be nearly independent of Z; hence the data may be used for other materials. Because of straggling, the thickness of the material should be about 25% greater than R to be certain of stopping all electrons.

Shielding of Electrons

The range of medium or low-energy electrons in solids is only a few millimeters, and shielding does not present a serious problem as long as low-Z materials are used to minimize production of the more penetrating bremsstrahlung. However, the dose to the skin can be significant with external β-particle emitters, and dose to organs containing β emitters must always be considered. Energy deposition by electrons is considered in Chapter 5. The electrons may be generated by γ-ray interactions as well, as discussed in Section 3.4.

3.4 INTERACTIONS OF PHOTONS

The most important interactions of γ-ray or X-ray photons with matter are summarized in Table 3.7. The photoelectric effect, Compton effect, and pair production are most significant in attenuation of γ rays, heating, and radiation damage. Rayleigh scattering is often neglected because the photon is only deflected through a small angle and dose not lose energy. However, it does remove photons from a narrow collimated beam and may be included in the total interaction cross section. The nuclear photoeffect is small but is a source of neutrons, as discussed in Chapter 2.

The photoelectric effect predominates at low energies and high Z values. Pair production predominates at high energies and high Z values. The Compton effect is the main interaction at moderate energies ($\sim$1 MeV) for all Z and is significant even at low energies for small Z.

Photoelectric Effect

In the photoelectric effect the γ ray imparts all its energy to an electron bound in the atom. The electron uses part of its energy to overcome the binding energy E_B, with the remainder appearing as kinetic energy of

TABLE 3.7. PHOTON INTERACTIONS

Process	Interacts with	Products	Cross-section Dependence	Remarks
Photoelectric effect	Atom	e^- X, Auger	$\sim Z^4 (h\nu)^{-3}$	$E = h\nu - E_B$
Pair production	Nucleus	e^-, e^+ annih.	$\sim Z^2 (h\nu)^n$	$E_+ + E_- =$ $h\nu - 1.02$ MeV
Compton effect	Free electron	e^-, γ	$\sim Z (h\nu)^{-m}$	$E = 0$ to $h\nu/(1 + (1/2\alpha))$
Rayleigh scattering	Bound electron	γ	$Z^2 (h\nu)^{-k}$	Coherent, small angle
Nuclear photoeffect	Nucleus	p, n, F, etc.	Threshold ~ 2 to 20 MeV, peak 10 to 30 MeV	< 10% of electronic cross sections

the electron,

$$T = h\nu - E_B \tag{3.29}$$

Momentum is conserved by recoil of the atom, which is essential to the process. The energy of the photoelectron is independent of angle. At low energies the electron tends to be ejected at 90° and at high energies the angle approaches 0°, but the angle is seldom of concern [13,14].

The greatest transfer of energy occurs with the most tightly bound electrons, and a K electron is usually ejected if this is energetically possible. The vacancy is rapidly filled by transitions from the next shell, accompanied by emission of either a characteristic X-ray or an Auger electron (as in electron capture). The ratio of X-ray photons to vacancies is called the "fluorescence yield." Fluorescence yield is nearly unity for high-Z elements and nearly zero for low-Z elements. The binding energy is in the keV range for high-Z elements (e.g., 88 keV for the K-shell in lead) and in the eV range for low-Z elements. Although it is possible for X-rays emitted near the surface to escape, it is usually a good approximation in shielding and dosimetry to assume all of the photon energy, $h\nu$ is absorbed at the point of the photoelectric interaction.

The photoelectric cross section for lead is plotted in Figure 3.22. The cross section for the photoelectric effect goes approximately as

$$\sigma_{pe} = \text{const}\ \frac{Z^4}{(h\nu)^3}\ \text{cm}^2/\text{atom} \tag{3.30}$$

But the cross section drops abruptly at the K edge because below this there is insufficient energy to overcome the binding and the K-shell electrons no longer participate. The cross section drops again as the energy goes below the binding energies of the L-shell subshells.

Pair Production

In some interactions of a photon with the nuclear Coulomb field, a negatron-positron pair is formed and the photon disappears. Momentum is conserved by the nuclear recoil. The photon energy appears as the

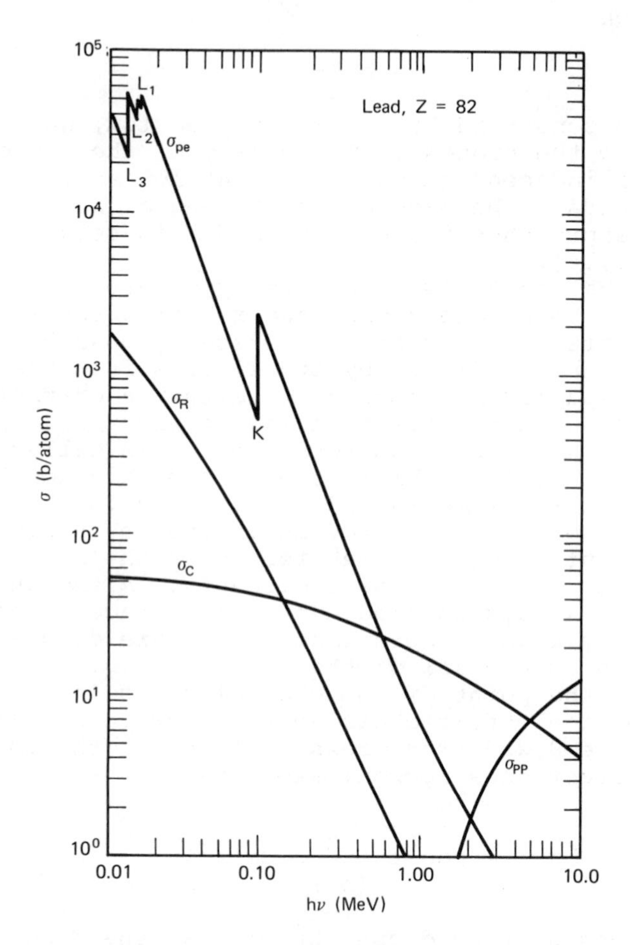

Fig. 3.22. Photon-interaction cross sections in lead.

total energy (kinetic plus rest energy) of the pair,

$$h\nu = (E_- + m_0c^2) + (E_+ + m_0c^2) \tag{3.31}$$

The photon must have an energy of at least $2m_0c^2$ to create the negatron and positron, hence the threshold for pair production is 1.022 MeV. The kinetic energy $h\nu - 2m_0c^2$ is distributed between the negatron and positron [13]. The energy spectra are essentially identical, and the electron spectrum seldom matters as the interactions of negatrons and positrons are almost the same and their ranges are small. The principal difference between negatrons and positrons is that the positron, on slowing down and combining with a negatron, releases two 0.511-MeV annihilation photons, isotropically and back-to-back. Thus all of the incident photon's energy is released rapidly, although it is possible for the 0.511-MeV annihilation photons to travel some distance, and they should be followed in the overall γ-ray transport problem.

The cross section for pair production increases with energy above the threshold, roughly as $(h\nu)^n$, where n varies from 2 to 1. The cross section is proportional to Z^2 except above 5 MeV, where the cross section increases less rapidly because the pair is created at some distance from the nucleus, and screening by the inner atomic electrons slightly reduces the effective Z. The pair production cross section for lead is plotted in Figure 3.22.

Compton Effect

The Compton effect involves the interactions of a photon with a free electron at rest, that is, where the incident photon energy is large compared with the binding energy. The interaction may be treated as an elastic collision of the photon, with energy $h\nu$ and momentum $h\nu/c$, with the electron. As shown in Figure 3.23, the photon is deflected through an angle θ and emerges with a lower energy $h\nu'$ and momentum $h\nu'/c$. The electron emerges at angle ψ with kinetic energy E and momentum p [14,15]. Using conservation of momentum and energy, with the relativistic relationship $pc = [E(E + 2m_0c^2)]^{1/2}$, we obtain

$$\frac{h\nu'}{h\nu} = \frac{1}{1 + \alpha(1 - \cos\theta)} \qquad (3.32)$$

where $\alpha \equiv h\nu/m_0c^2$, the incident photon energy in units of the rest energy of the electron, 0.511 MeV. This relationship is plotted in Figure 3.24. It is interesting to note that for $\alpha >> 1$, the energy after deflection through 180° approaches 0.25 MeV, and the energy after scattering through 90° approaches 0.5 MeV.

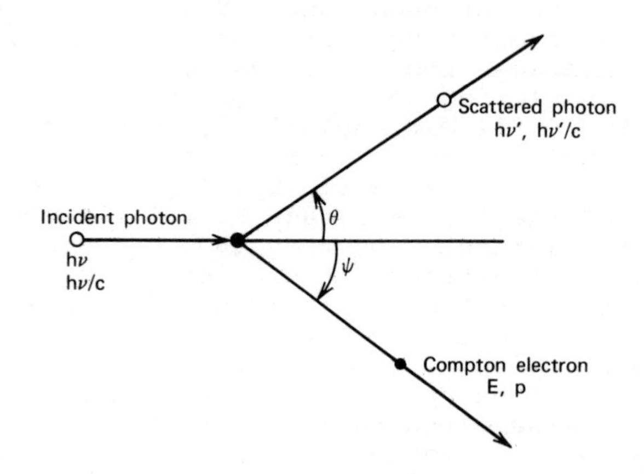

Fig. 3.23. Diagram of Compton interaction.

The kinetic energy imparted to the electron is

$$E = h\nu - h\nu' = h\nu \frac{\alpha(1 - \cos\theta)}{1 + \alpha(1 - \cos\theta)} \tag{3.33}$$

The maximum value of E corresponds to the minimum value of $h\nu'$; hence 180° scattering of the photon, and

$$E_{max} = \frac{h\nu}{1 + (1/2\alpha)} \tag{3.34}$$

is called the "Compton edge."

The differential cross section per electron, for the number of photons scattered into solid angle $d\Omega = 2\pi \sin\theta \, d\theta$, as derived theoretically by Klein and Nishina [16], is

$$\frac{d\sigma_e}{d\Omega} = r_0^2 \left[\frac{1}{1 + \alpha(1 - \cos\theta)}\right]^2 \left(\frac{1 + \cos^2\theta}{2}\right) \left\{1 + \frac{\alpha^2(1 - \cos\theta)^2}{(1 + \cos^2\theta)[1 + \alpha(1 - \cos\theta)]}\right\} \tag{3.35}$$

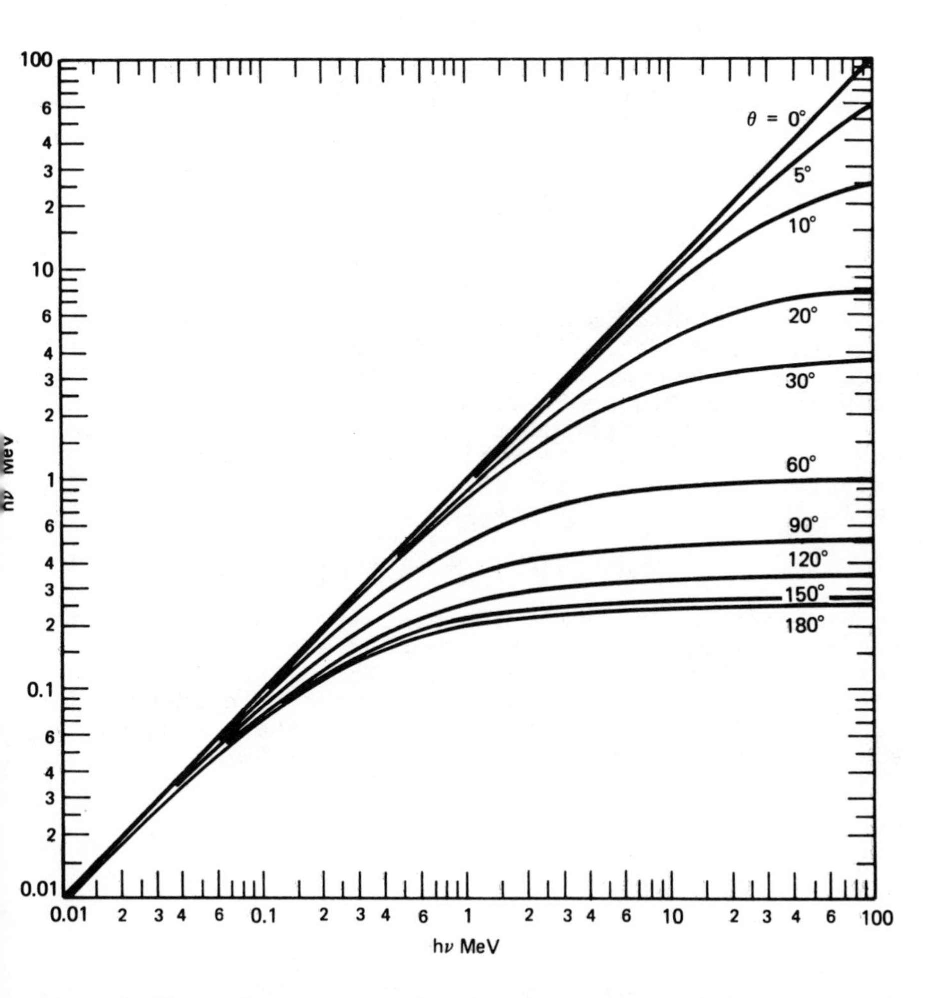

Fig. 3.24. Photon-scattered energy hν' as function of incident energy hν and angle of scattering.

where the classical radius of the electron $r_0 = 2.818 \times 10^{-13}$ cm. The differential cross section is plotted in Figure 3.25 for several values of hν. The Compton interaction cross section per electron corresponds to Eq. (3.35) integrated over θ. It is

$$\sigma_e = 2\pi r_0^2 \left\{ \frac{1+\alpha}{\alpha^2} \left[\frac{2(1+\alpha)}{1+2\alpha} - \frac{1}{\alpha} \ln(1+2\alpha) \right] + \frac{1}{2\alpha} \ln(1+2\alpha) - \frac{1+3\alpha}{(1+2\alpha)^2} \right\} \tag{3.36}$$

cm^2/electron, and is listed in Table 3.8. The cross section for the atom is Z times the cross section per electron. The Compton cross section for lead (Z = 82) is plotted in Figure 3.22.

The angular distribution for the Compton electron can be derived from the Klein-Nishina formula and the relationship between the photon angle and electron angle,

$$\cot \psi = (1+\alpha) \tan \frac{\theta}{2} \tag{3.37}$$

As θ goes from 0° to 180°, ψ goes from 90° to 0°. The angular distribution of the Compton electron is even more forward-peaked than the angular distribution of the photons but is seldom of concern. More interesting is the differential cross section for the number of electrons with kinetic energy E to (E + dE),

$$\frac{d\sigma_e}{dE} = \frac{\pi r_0^2}{\alpha^2 m_0 c^2} \left\{ 2 + \left(\frac{E}{h\nu - E} \right)^2 \left[\frac{1}{\alpha^2} + \frac{h\nu - E}{h\nu} - \frac{2}{\alpha} \frac{h\nu - E}{E} \right] \right\} \tag{3.38}$$

The energy spectrum is plotted in Figure 3.26 for a few initial photon energies. The average energy imparted to the electron is given in Table 3.8 along with the fraction of the incident photon carried off by the

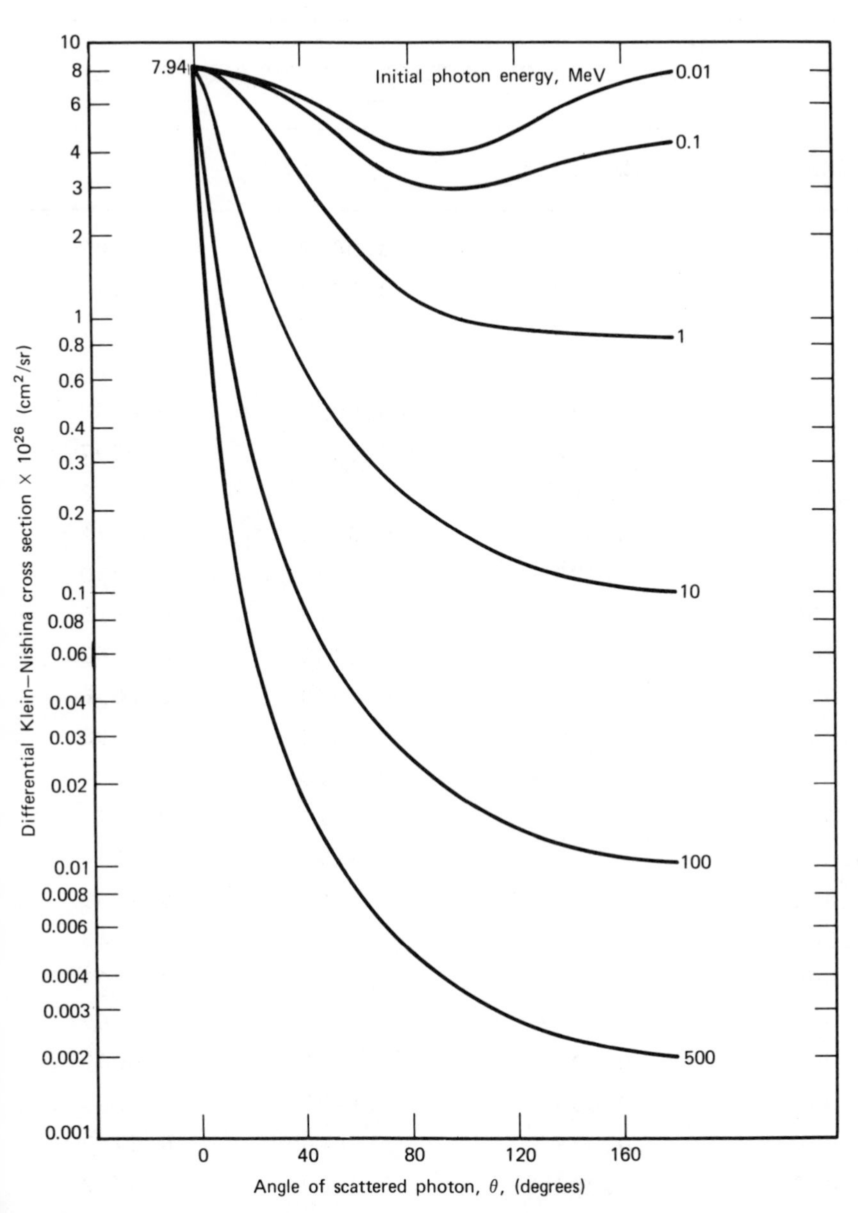

Fig. 3.25. Differential cross section for number of photons Compton scattered into unit solid angle (after Nelms [16]).

TABLE 3.8. COMPTON CROSS SECTION AND AVERAGE ENERGY

$h\nu$ (MeV)	σ_e (b/electron)	E_{av} (MeV)	$f_a = E_{av}/h\nu$
0.010	0.640	0.0002	0.0187
0.015	0.629	0.0004	0.0277
0.020	0.618	0.0007	0.0361
0.030	0.598	0.0016	0.0520
0.040	0.579	0.0027	0.0667
0.050	0.562	0.0040	0.0807
0.060	0.546	0.0056	0.0938
0.080	0.517	0.0094	0.1171
0.100	0.493	0.0138	0.1380
0.150	0.444	0.0272	0.1815
0.200	0.406	0.0432	0.2162
0.300	0.354	0.0809	0.2696
0.400	0.317	0.1240	0.3098
0.500	0.290	0.1710	0.3424
0.600	0.268	0.2210	0.3675
0.800	0.235	0.3270	0.4089
1.000	0.211	0.4400	0.4399
1.500	0.172	0.7420	0.4948
2.000	0.146	1.0610	0.5307
3.000	0.115	1.7310	0.5769
4.000	0.096	2.4280	0.6069
5.000	0.083	3.1400	0.6280
6.000	0.073	3.8640	0.6440
8.000	0.060	5.3380	0.6672
10.000	0.051	6.8350	0.6835

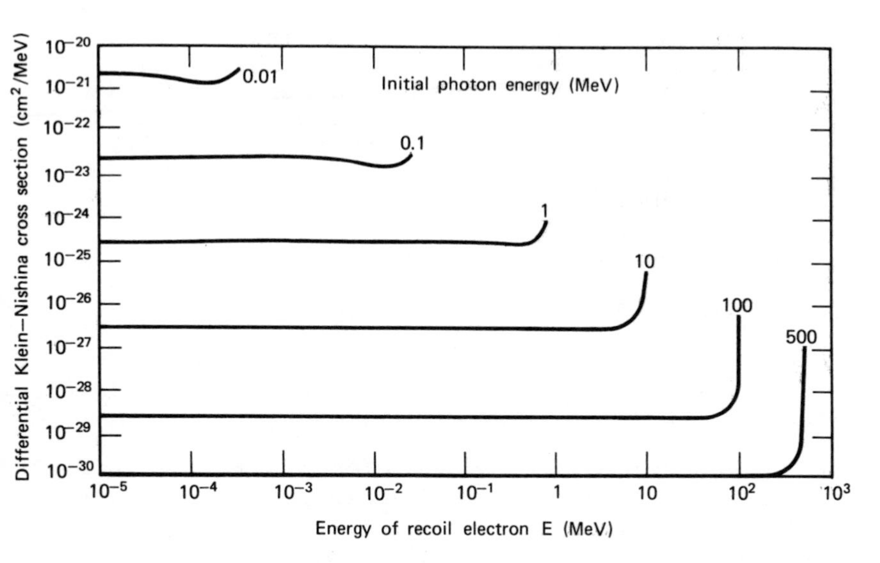

Fig. 3.26. Differential cross section for Compton electron receiving energy E to (E + dE) (after Nelms [16]).

electron, on the average. This fraction, f_a times the interaction cross section, is sometimes called the "Compton absorption cross section." Useful graphs of the energy-angle relationships and cross sections are given in Nelms [17].

At very low energies and in high-Z elements, where atomic binding cannot be neglected, the cross section for incoherent scattering with energy loss deviates from the Klein-Nishina formula. Corrections are expressed as "form factors." However, in this region the scattering is usually overwhelmed by the photoelectric absorption and corrections are often ignored.

Rayleigh Scattering

Rayleigh or coherent scattering is scattering from bound electrons, where the atom as a whole recoils. The energy is essentially unchanged. The radiation from all bound electrons in an atom interferes coherently, with the result that Rayleigh scattering is sharply at 0°, especially at low Z and high energy. Rayleigh scattering deflects photons from a very well-collimated beam but has little effect in the broad-beam situations usually encountered in shielding. Rayleigh scattering cross section varies as Z^2 and decreases with increasing energy approximately as $(h\nu)^{-k}$, where k is about 3, as seen in Figure 3.22 for lead. Cross sections for photon interactions are tabulated by Hubbell [18].

Attenuation of Photons

The total linear attenuation coefficient

$$\mu = \sum_i N_i(\sigma_{pe} + \sigma_{pp} + \sigma_C + \sigma_R)_i \ \text{cm}^{-1} \tag{3.39}$$

where the summation is over the elements i in a compound or mixture. In terms of the coefficients for different interactions

$$\mu = \mu_{pe} + \mu_{pp} + \mu_C + \mu_R \tag{3.40}$$

For broad-beam attenuation, coherent scattering σ_R or μ_R may be omitted. Linear-attenuation coefficients for water (1 g cm^{-3}) are plotted in Figure 3.27 (without coherent scattering). The mass-attenuation coefficient is μ/ρ. Total mass-attenuation coefficients for some common materials are given in Table 3.9 [18].

Let us now consider the attenuation of a neutral particle beam (photons or neutrons) in a slab, such as shown in Figure 3.28. For neutrons, the attenuation coefficient is the total macroscopic cross section, Σ. The total interaction rate is also the rate of removal of particles from the beam, hence the change in flux density

$$d\phi(x,E_0) = -\mu\phi(x,E_0)\,dx \tag{3.41}$$

where the zero subscript reminds us that we are considering only the particles that have the initial energy, E_0, and direction. The attenuation coefficient is evaluated at E_0. Integration gives

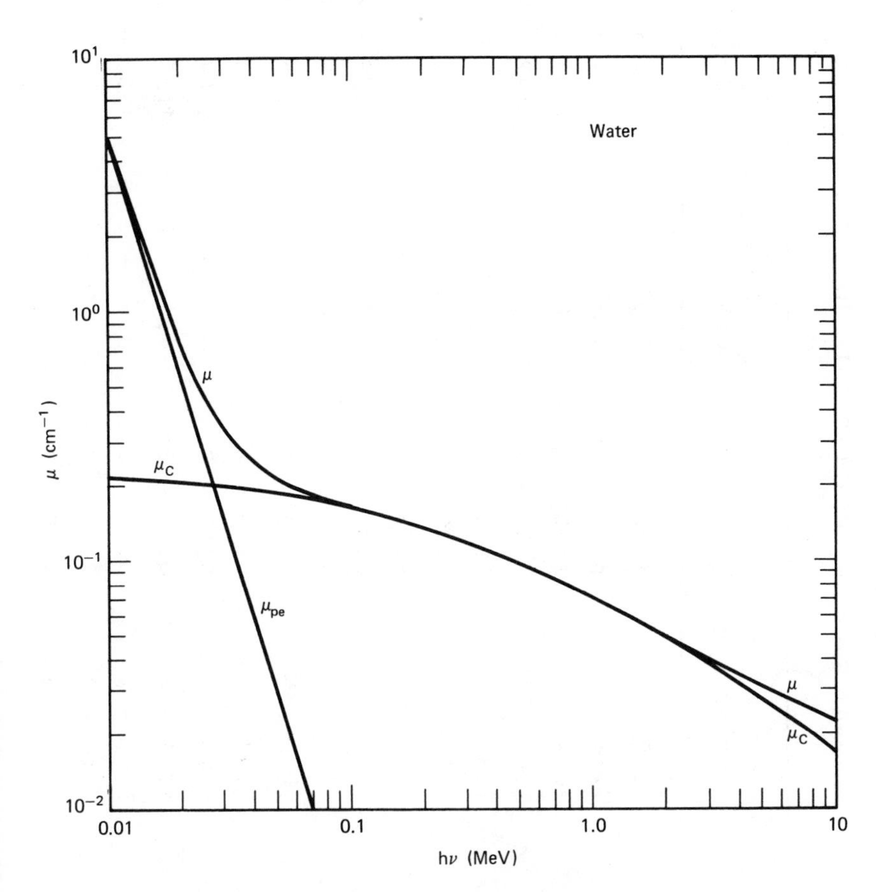

Fig. 3.27. Attenuation coefficients for water (1 g cm^{-3}), without coherent scattering.

TABLE 3.9. MASS-ATTENUATION COEFFICIENTS (WITHOUT COHERENT)

$h\nu$ (MeV)	μ/ρ (cm^2/g) Water	Air	Concrete	Iron	Lead
0.010	4.99 + 0[a]	4.82 + 0	2.65 + 1	1.72 + 2	1.28 + 2
0.020	7.11 - 1	6.91 - 1	3.45 + 1	2.51 + 1	8.34 + 1
0.050	2.14 - 1	1.96 - 1	3.61 - 1	1.84 + 0	7.22 + 0
0.100	1.68 - 1	1.51 - 1	1.70 - 1	3.42 - 1	5.23 + 0
0.200	1.36 - 1	1.23 - 1	1.25 - 1	1.39 - 1	9.45 - 1
0.300	1.18 - 1	1.06 - 1	1.07 - 1	1.07 - 1	3.83 - 1
0.400	1.06 - 1	9.53 - 2	9.58 - 2	9.21 - 2	2.20 - 1
0.500	9.67 - 2	8.70 - 2	8.73 - 2	8.29 - 2	1.54 - 1
0.600	8.95 - 2	8.05 - 2	8.07 - 2	7.62 - 2	1.20 - 1
0.800	7.86 - 2	7.07 - 2	7.09 - 2	6.65 - 2	8.56 - 2
1.000	7.07 - 2	6.36 - 2	6.37 - 2	5.96 - 2	6.90 - 2
1.500	5.75 - 2	5.18 - 2	5.19 - 2	4.87 - 2	5.10 - 2
2.000	4.94 - 2	4.45 - 2	4.48 - 2	4.25 - 2	4.50 - 2
3.000	3.97 - 2	3.58 - 2	3.65 - 2	3.62 - 2	4.16 - 2
4.000	3.40 - 2	3.08 - 2	3.19 - 2	3.31 - 2	4.14 - 2
5.000	3.03 - 2	2.75 - 2	2.90 - 2	3.14 - 2	4.24 - 2

6.000	2.77 - 2	2.52 - 2	2.70 - 2	3.05 - 2	4.34 - 2
8.000	2.43 - 2	2.23 - 2	2.45 - 2	2.98 - 2	4.59 - 2
10.000	2.22 - 2	2.04 - 2	2.31 - 2	2.98 - 2	4.84 - 2

[a] Read 4.99×10^0

[b] Weight fractions: H(0.0056), O(0.8791), Na(0.0171), Mg(0.0024), Al(0.0456), Si(0.3158), S(0.0012), K(0.0192), Ca(0.0826), Fe(0.0122). Density 2.2 to 2.4 g/cm^3.

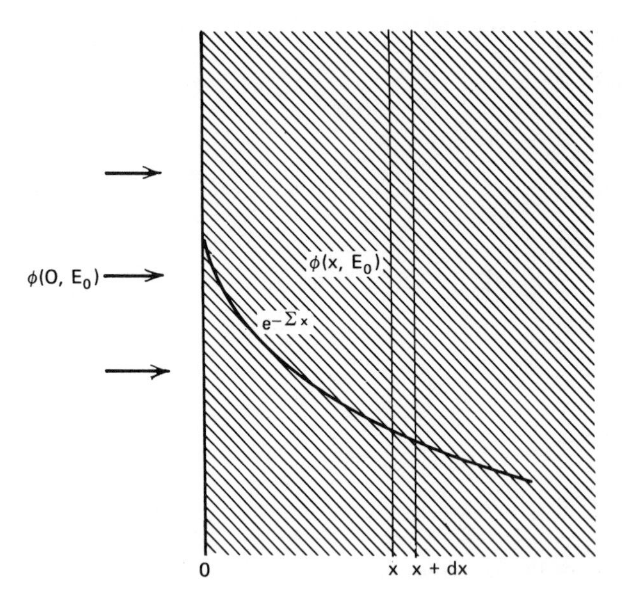

Fig. 3.28. Attenuation of a neutral-particle beam (from A. Edward Profio, *Experimental Reactor Physics*, Wiley, New York, 1976, p. 49).

$$\phi(x,E_0) = \phi(O,E_0)\ e^{-\mu x} \tag{3.42}$$

Note that $e^{-\mu x}$ is the probability that a particle travels a distance x without interaction, and μdx is the probability that a particle then interacts in dx. Thus the mean distance to interaction, or mean free path (distance) between interactions, is

$$\lambda = \frac{\int_0^\infty x\ e^{-\mu x}\ \mu dx}{\int_0^\infty e^{-\mu x}\ \mu dx} = \mu^{-1}\ \text{cm} \tag{3.43}$$

The half-value layer (HVL) is the thickness required to attenuate the particle flux density to half the incident flux density, $e^{-\mu x} = 0.5$; hence

$$\text{HVL} = -\ell n \frac{0.5}{\mu} = \frac{0.693}{\mu} \tag{3.44}$$

The HVL as a function of photon energy is plotted in Figure 3.29 for aluminum (Z = 13), copper (Z = 29), and lead (Z = 82).

In practice, the attenuation of γ rays (or neutrons) is not calculated so simply. The scattered particles also contribute to the flux density and dose. Transport of particles is discussed in Chapter 4. Energy absorption and dose are discussed in Chapter 5.

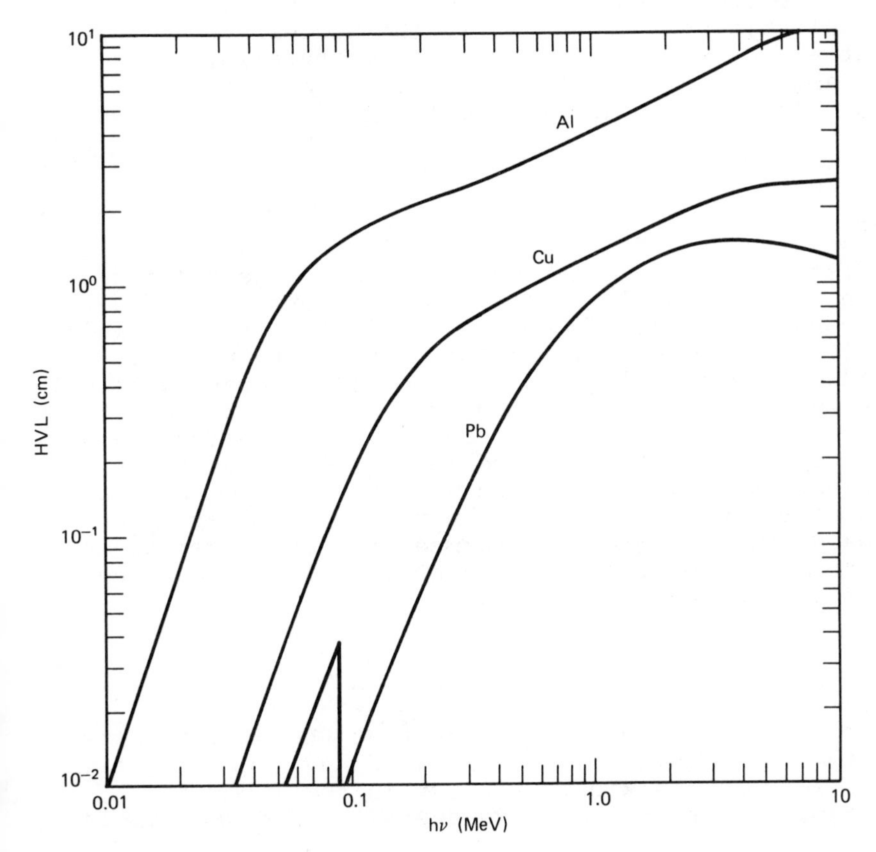

Fig. 3.29. Half-value layer as function of photon energy.

3.5 INTERACTIONS OF NEUTRONS

The interactions of charged particles and γ rays with matter proceed mainly by way of the electromagnetic force, and the cross sections vary relatively smoothly with atomic number. In contrast, the interactions of neutrons with matter occur mainly through the strong nuclear force, with each nuclide different, but general trends follow mass number because of the charge independence of the nuclear force. There are also regularities at proton and neutron numbers corresponding to filled nuclear shells, the so-called magic nuclides. The principal interactions of the neutron with matter in the energy range of interest (<20 MeV) are: (1) elastic scattering (n,n), (2) nuclear inelastic scattering (n,n'), (3) radiative capture (n,γ), (4) absorption with charged particle emission (e.g., n,p), (5) fission (n,F), and (6) (n,2n) and similar reactions.

Elastic Scattering

Elastic scattering predominates in the lighter nuclei and at low energy. It occurs by both direct interaction with the nuclear potential and formation of a compound nucleus. Potential scattering is essentially constant with energy, and the cross section is equal to a few barns for slow and intermediate energy neutrons. The compound nucleus formed from nuclide ^{A}Z and neutron is excited with an energy equal to the binding energy of the final neutron in nuclide ^{A+1}Z, plus the kinetic energy available in the CMS, as shown in Figure 3.30 for ^{12}C. The binding energy is the energy released when a nucleus is formed from its constituent protons and neutrons and may be calculated from the mass difference,

$$B_n = c^2(M_A + m_n - M_{A+1}) \tag{3.45}$$

From measured masses one finds that the binding energy of the last neutron in ^{13}C is 4.95 MeV. The relationship between the kinetic energy in the laboratory system, E_L, and the kinetic energy in the CMS, E_C is

$$E_L = \frac{M_A + m_n}{M_A} E_C \tag{3.46}$$

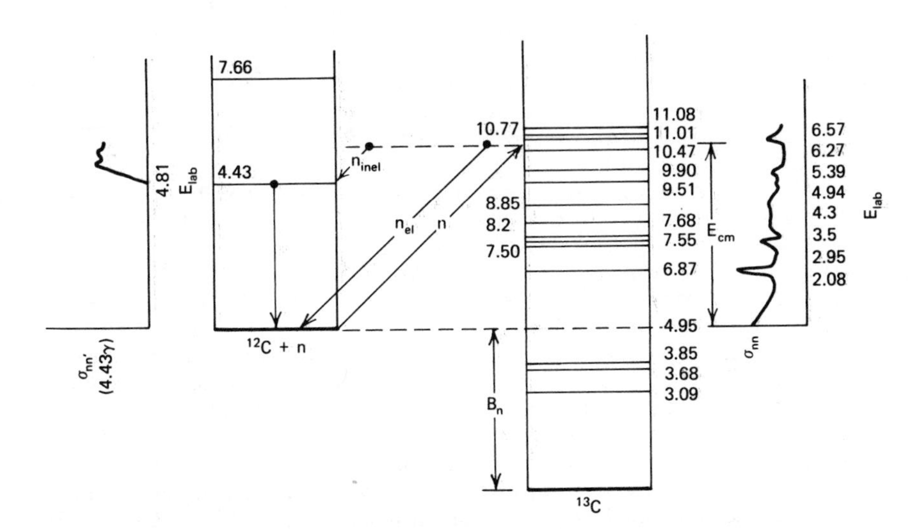

Fig. 3.30. Energy-level diagram for elastic and inelastic scattering in ^{12}C (data from T. Lauritsen and F. Ajzenberg-Selove, "Energy Levels of Light Nuclei," May 1962, NAS-NRS Nuclear Data Sheets). All energies are in megaelectron volts.

The excitation energy is then

$$E_x = B_n + E_C \tag{3.47}$$

When the excitation energy is equal to the energy of an excited level in the compound nucleus, the cross section exhibits a maximum, or resonance, as illustrated for $^{12}C + n \rightarrow {}^{13}C^* \rightarrow {}^{12}C + n_{e\ell}$ in Figure 3.30. The next level above 4.95 MeV is at 6.87 MeV. Thus a resonance in the compound elastic scattering cross section occurs at

$$E_L = \frac{12 + 1.009}{12} (6.87 - 4.95) = 2.08 \text{ MeV} \tag{3.48}$$

In ^{13}C and other light nuclei the levels are widely spaced (average level spacing $\overline{D}$ = 0.1 to 1 MeV), and compound elastic scattering occurs at MeV energies with widely spaced, and relatively broad resonances. The average level spacing decreases with A and E_x.

In the heavier nuclei the spacing becomes so small that the resonances are unresolvable and tend to overlap because of Doppler broadening; hence the cross section appears to vary smoothly with energy in the kilo- and megaelectron volt regions.

In elastic scattering there is no loss of energy in the CMS, but the neutron does lose energy in the laboratory system because of the recoil of the nucleus. The kinematics of the two-body collision (not necessarily elastic) are discussed with reference to the diagrams in Figure 3.31. In the laboratory system a particle of mass M_1, speed v_1, and kinetic energy E_1 is incident on a nucleus of mass M_2, which we assume is free and initially at rest. The particles may combine briefly to form a compound nucleus or may exist in some other intermediate state, but the details do not matter unless energy is gained or lost while in that state. After the collision, in general, a lighter particle of mass M_3, speed v_3, and kinetic energy E_3 emerges at laboratory-system angle θ_L to the original direction. A residual nucleus or heavy particle of mass M_4, speed v_4, and kinetic energy E_4 emerges at angle ψ_L. The difference between the combined kinetic energies of the products and reactants is, by definition, the nuclear disintegration energy

$$Q = E_3 + E_4 - E_1 \tag{3.49}$$

since we assumed $E_2 = 0$. The value Q is positive for exoergic reactions such as some charged-particle reactions and fission, zero for elastic scattering, and negative for endoergic reactions, including inelastic scattering.

Applying conservation of momentum parallel and perpendicular to the initial direction and solving for Q, we find

$$Q = E_3\left(1 + \frac{M_3}{M_4}\right) - E_1\left(1 - \frac{M_1}{M_4}\right) - \frac{2\sqrt{M_1E_1M_3T_3}}{M_4} \cos \theta_L \tag{3.50}$$

Usually we observe the lighter particle M_3 (e.g., the scattered neutron) at a given angle θ_L, and inquire about the energy E_3, given the incident laboratory-system energy, E_1. Considering Eq. (3.50) as a quadratic in E_3, we find

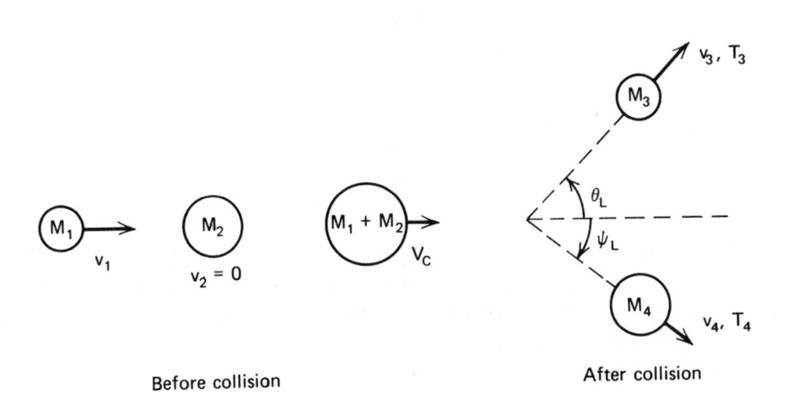

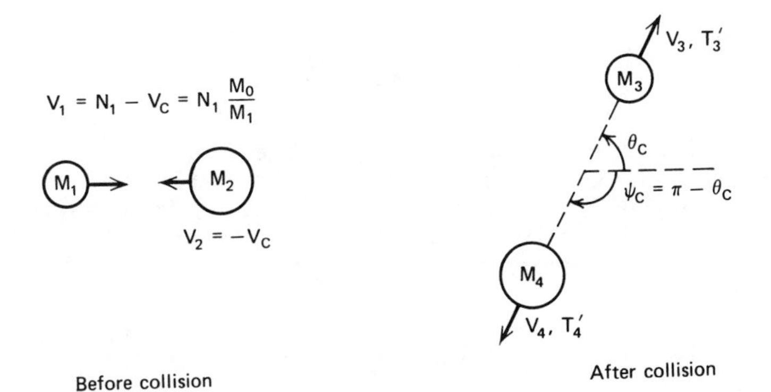

Fig. 3.31. Nuclear collisions in (a) laboratory system and (b) CMS frames; T is kinetic energy, E (from A. Edward Profio, *Experimental Reactor Physics*, Wiley, New York, 1976, p. 100).

$$\sqrt{E_3} = u \pm \sqrt{u^2 + w} \tag{3.51}$$

where

$$u = \frac{\sqrt{M_1 M_3 E_1}}{M_3 + M_4} \cos \theta_L \tag{3.52}$$

and

$$w = \frac{M_4 Q + E_1 (M_4 - M_1)}{M_3 + M_4} \tag{3.53}$$

The reaction is possible when $\sqrt{E_3}$ is real, positive.

The collision may also be examined in the CMS, where the total momentum, by definition, is zero. The velocity of the CMS is

$$V_C = v_1 \frac{M_0}{M_2} \tag{3.54}$$

where the reduced mass

$$M_0 = \frac{M_1 M_2}{M_1 + M_2} \tag{3.55}$$

is 0.5 to 1.0 times the mass of the lighter particle. After collision, the particle M_3 is deflected through the CMS angle θ_C, and M_3 and M_4 recede from each other in opposite directions as shown in Figure 3.31. The total kinetic energy of the reactants,

$$(E_0)_{in} = \frac{1}{2} M_1 V_1^2 + \frac{1}{2} M_2 V_2^2 = \frac{1}{2} M_0 v_1^2 = E_1 \frac{M_1}{M_1 + M_2} \tag{3.56}$$

The total kinetic energy of the products follows from Eq. (3.49).

The relationship between the laboratory-system and CMS angles may be seen from the velocity diagram (Figure 3.32). By the law of sines,

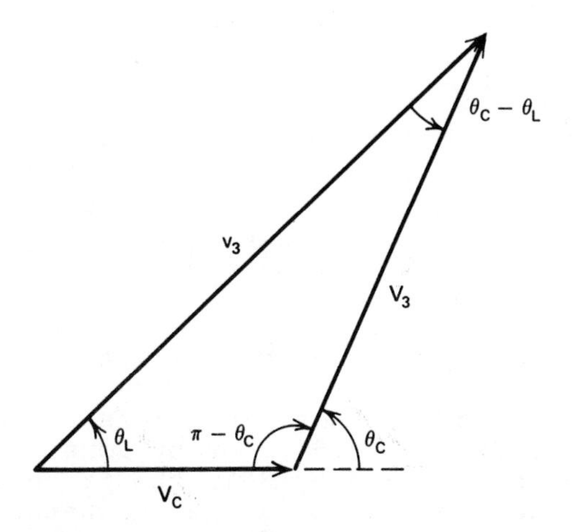

Fig. 3.32. Velocity diagram for relating laboratory systems and CMS (from A. Edward Profio, *Experimental Reactor Physics*, Wiley, New York, 1976, p. 113).

$$\frac{\sin(\theta_C - \theta_L)}{\sin \theta_L} = \frac{V_C}{V_3} = x \tag{3.57}$$

where

$$x^2 = \frac{M_1 M_3}{M_4(M_3 + M_4 - M_1)} \left(1 + \frac{M_3 + M_4}{M_3 + M_4 - M_1} \frac{Q}{E_1}\right)^{-1} \tag{3.58}$$

Cross sections are nearly always tabulated or plotted against the incident laboratory energy, but the angular distribution may be given as a function of either θ_C or θ_L. The number of particles scattered into a given solid angle must be the same in both systems,

$$\left(\frac{d\sigma}{d\Omega}\right)_C d\Omega_C = \left(\frac{d\sigma}{d\Omega}\right)_L d\Omega_L \tag{3.59}$$

The ratio of solid angles is

$$G = \frac{\sin \theta_L \, d\theta_L}{\sin \theta_C \, d\theta_C} = \frac{\sin^2 \theta_L}{\sin^2 \theta_C} \cos(\theta_C - \theta_L)$$

$$= \left[1 - x^2 \sin^2 \theta_L\right]^{1/2}$$

$$\left[x \cos \theta_L + \left(1 - x^2 \sin^2 \theta_L\right)^{1/2}\right]^{-2} \tag{3.60}$$

Tables have been prepared to facilitate the transformation [19].

The energy of the heavy particle E_4, may be found from Eqs. (3.49) and (3.51), or by considering the particle to be the same as 3 in the preceding equations because it does not matter actually whether the heavy or light particle is labeled 3 or 4, as long as we are consistent.

The equations are simplified when $M_3 = M_1$, as in neutron scattering. For elastic scattering (and to a very good approximation for inelastic scattering), $M_4 = M_2$. Let us define $A = (M_2/M_1) \simeq (M_4/M_1)$ and note that for neutron scattering, $A \geq 1$. Then only one real, positive root is obtained for E_3. Thus

$$x^{-1} = \frac{V_3}{V_C} = A \sqrt{1 + \frac{A + 1}{A} \frac{Q}{E_1}} = AB \tag{3.61}$$

which defines B. Also,

$$V_C = \frac{V_1}{A + 1} \tag{3.62}$$

Applying the law of cosines to the velocity diagram in Figure 3.30, and noting $\cos(\pi - \theta_C) = -\cos \theta_C$, we obtain

$$E_3 = \frac{E_1}{(A + 1)^2} \left(1 + A^2B^2 + 2AB \cos \theta_C\right) \tag{3.63}$$

and

$$\cos \theta_L = \frac{1 + AB \cos \theta_C}{\left(1 + A^2B^2 + 2AB \cos \theta_C\right)^{1/2}} \tag{3.64}$$

For elastic scattering of neutrons, $Q = 0$; hence $B = 1$. From Eq. (3.63) the smallest scattered energy occurs when $\cos \theta_C = -1$, and

$$(E_3)_{min} = rE_1 \tag{3.65}$$

where

$$r = \left(\frac{A - 1}{A + 1}\right)^2 \tag{3.66}$$

The maximum fractional loss of energy is

$$\frac{E_1 - (E_3)_{min}}{E_1} = 1 - r \tag{3.67}$$

For hydrogen, $A = 1$ and all of the neutron's energy can be lost in a single collision. For ^{12}C, on the other hand, $r = 0.716$ and the maximum fractional energy loss is 0.284. Hydrogen has a relatively large cross section for neutrons and, because of the large energy loss per collision, is favored as a major element in neutron-shielding materials, such as water, polyethylene (CH_2) and other organics, and as a constituent of concrete.

The distribution of scattered neutron energies between E_1 and rE_1 depends on the angular distribution of scattering. In hydrogen, scattering is isotropic in the CMS up to about 10 MeV, and most other elements scatter isotropically up to energies on the order of 0.1 MeV. Hence in isotropic scattering, all solid angles are equally probable, and

$$\left(\frac{d\sigma}{d\Omega}\right)_C = \frac{\sigma_s}{4\pi} \text{ cm}^2 \text{ sr}^{-1} \tag{3.68}$$

where σ_s is the elastic scattering cross section. The angular distribution in the laboratory system follows from Eqs. (3.59) and (3.60) with $A = 1$ for hydrogen and

B = 1 for elastic scattering in Eqs. (3.61) and (3.64). We obtain, with $\mu_L = \cos \theta_L$,

$$\left(\frac{d\sigma}{d\Omega}\right)_L = \frac{\sigma_s}{4\pi} 4\mu_L \quad cm^2 \ sr^{-1} \qquad 0 \le \mu_L \le 1 \tag{3.69}$$

$$= 0 \qquad -1 < \mu_L < 0$$

The differential cross section for scattering from incident energy E_1 to energy E_3 may be obtained from Eq. (3.63) and

$$\frac{d\sigma(E_1 \rightarrow E_3)}{dE_3} = \frac{d\sigma(E_1,\theta_C)}{d\Omega_C} \frac{d\Omega_C}{dE_3} \tag{3.70}$$

with $d\theta_C = 2\pi \sin \theta_C \, d\theta_C = 2\pi \, d(\cos \theta_C)$,

$$\frac{d\sigma(E_1 \rightarrow E_3)}{dE_3} = \frac{\sigma_s(E_1)}{E_1} \frac{(A + 1)^2}{4A} = \frac{\sigma_s(E_1)}{(1 - r)E_1} \tag{3.71}$$

from E_1 to rE_1 and zero elsewhere. The recoil nucleus has energy $E_4 = (E_1 - E_3)$. The maximum energy is

$$(E_4)_{max} = (1 - r)E_1 \tag{3.72}$$

and the average energy is

$$\bar{E}_4 = \frac{2A}{(A + 1)^2} E_1 \tag{3.73}$$

All these relationship apply to isotropic elastic scattering in the CMS.

The angular distribution of scattering may be anisotropic, especially for higher energies and heavier elements. The angular distributions for ^{12}C at several energies are plotted in Figure 3.33 as a function of $\cos \theta_C$.

Cross sections for elastic scattering, $\sigma_s = \sigma_{nn}$ are plotted in Figure 3.34 for ^{12}C, along with other

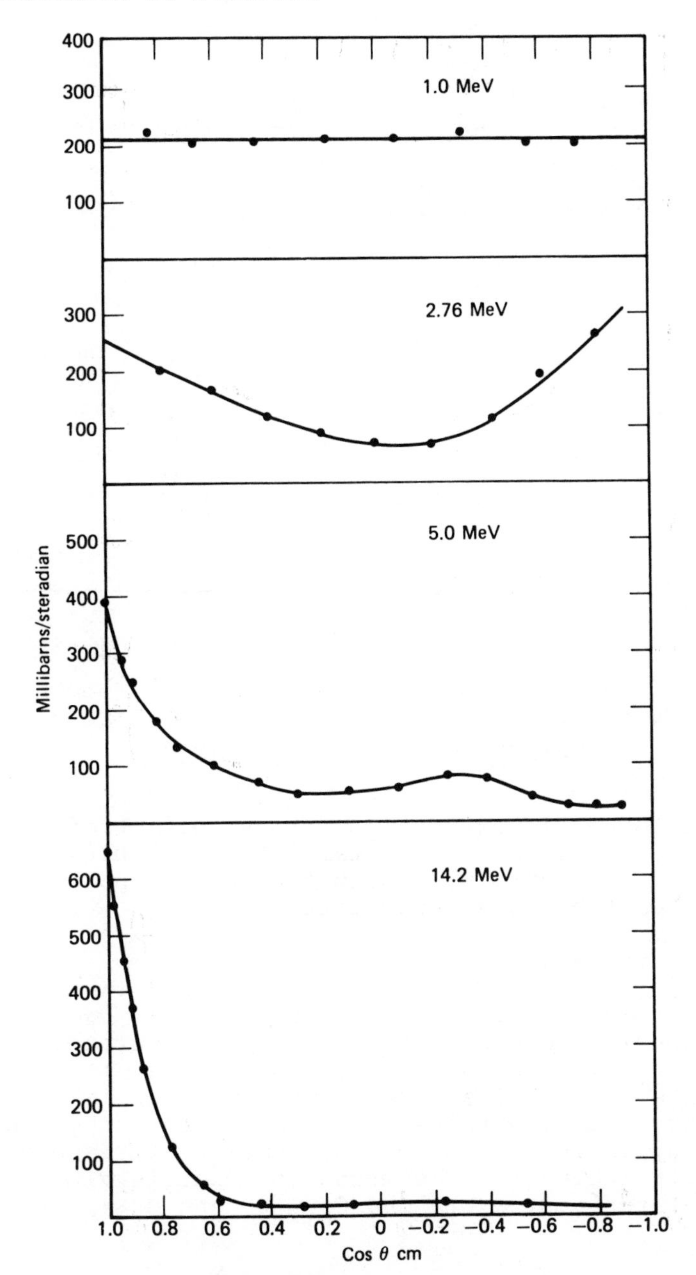

Fig. 3.33. Angular distributions of fast-neutron scattering in ^{12}C (after BNL-400, 2nd Ed., Vol. I, Oct. 1962) (from A. Edward Profio, *Experimental Reactor Physics*, Wiley, New York, 1976, p. 87).

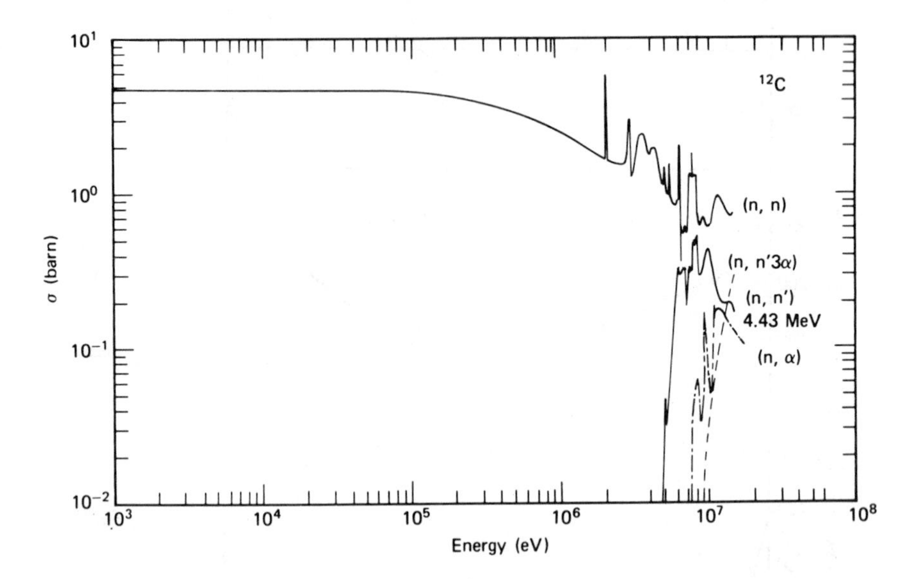

Fig. 3.34. Neutron cross sections for ^{12}C (from A. Edward Profio, *Experimental Reactor Physics*, Wiley, New York, 1976, p. 76).

reactions to be discussed. The cross section of hydrogen is very important and is tabulated in Table 3.10 and plotted in Figure 3.35 with the charged-particle-emitting reactions discussed later.

Inelastic Scattering

Nuclear inelastic scattering is energetically possible when the nucleus can be left in an excited state, as shown for n_{inel} in Figure 3.30. The excited state decays with emission of a γ ray. The emission is usually prompt, but isomeric transitions can occur. The smallest energy in the laboratory system, the threshold energy, for inelastic scattering to occur is given by Eq. (3.35) for $u^2 + w = 0$ and $\theta_L = 0$, and $M_3 \simeq M_4$; hence

$$(E_1)_{threshold} = -Q \frac{M_3 + M_4}{M_4} \tag{3.74}$$

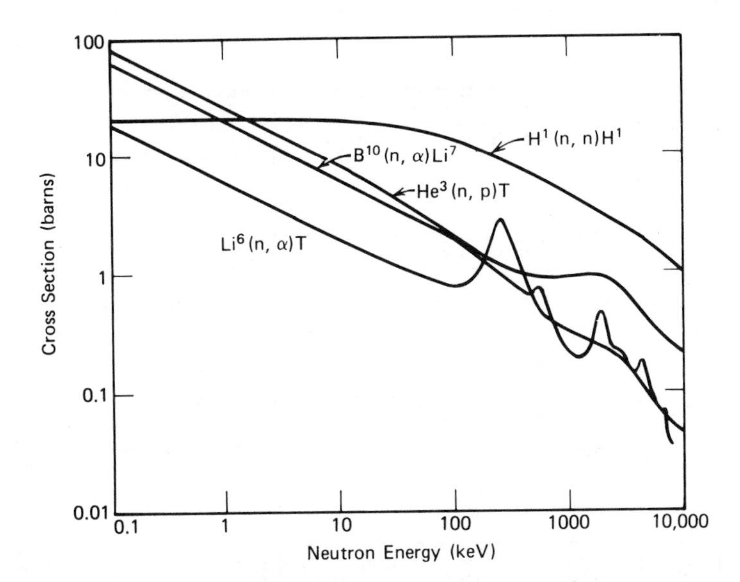

Fig. 3.35. Cross sections for neutron reactions yielding charged particles suitable for neutron detection (after J. B. Marion and F. C. Young, *Nuclear Reaction Analysis Graphs and Tables*, North-Holland, Amsterdam, and Wiley, New York, 1968)(from A. Edward Profio, *Experimental Reactor Physics*, Wiley, New York, 1976, p. 91).

In ^{12}C the first excited state is at 4.43 MeV. The neutron may be inelastically scattered from this level when the CMS energy exceeds $-Q$ = 4.43 MeV or when the laboratory energy exceeds

$$(E_1)_{threshold} = 4.43 \frac{1.009 + 12.0}{12.00} = 4.81 \text{ MeV} \qquad (3.75)$$

As the energy increases, inelastic scattering becomes possible with the nucleus left in a higher excited state, for example, 7.66 MeV in ^{12}C. Cross sections may be defined for inelastic scattering from each of the levels and for the combined effect from all levels. When the levels can be resolved, the scattered neutron

TABLE 3.10. CROSS SECTION FOR ELASTIC SCATTERING IN HYDROGEN

E(eV)	σ(b)	E(eV)	σ(b)	E(MeV)	σ(b)	E(MeV)	σ(b)
0.0010	20.449	1.50 + 5	10.965	1.00	4.261	6.50	1.3290
0.0253	20.449	1.60 + 5	10.673	1.10	4.051	6.80	1.2990
100.0000	20.449	1.70 + 5	10.398	1.20	3.868	7.00	1.2690
1.00 + 3[a]	20.329	1.80 + 5	10.140	1.30	3.706	7.50	1.2010
2.00 + 3	20.198	1.90 + 5	9.898	1.40	3.561	8.00	1.1390
3.00 + 3	20.068	2.00 + 5	9.671	1.50	3.429	8.50	1.0830
4.00 + 3	19.941	2.20 + 5	9.258	1.60	3.309	9.00	1.0320
5.00 + 3	19.815	2.40 + 5	8.892	1.70	3.198	9.50	0.9859
6.00 + 3	19.691	2.60 + 5	8.562	1.80	3.097	10.00	0.9432
8.00 + 3	19.448	2.80 + 5	8.262	1.90	3.003	10.50	0.9035
1.00 + 4	19.213	3.00 + 5	7.987	2.00	2.915	11.00	0.8665
1.50 + 4	18.651	3.20 + 5	7.734	2.20	2.759	11.50	0.8323
2.00 + 4	18.126	3.40 + 5	7.501	2.40	2.622	12.00	0.8005
2.50 + 4	17.634	3.60 + 5	7.284	2.60	2.501	12.50	0.7710
3.00 + 4	17.172	3.80 + 5	7.083	2.80	2.392	13.00	0.7433
3.50 + 4	16.737	4.00 + 5	6.897	3.00	2.293	13.50	0.7173
4.00 + 4	16.327	4.20 + 5	6.725	3.20	2.203	14.00	0.6929
4.50 + 4	15.941	4.40 + 5	6.565	3.40	2.120	14.50	0.6698
5.00 + 4	15.575	4.60 + 5	6.415	3.60	2.043	15.00	0.6480
5.50 + 4	15.228	4.80 + 5	6.275	3.80	1.973		
6.00 + 4	14.900	5.00 + 5	6.143	4.00	1.907		
6.50 + 4	14.587	5.50 + 5	5.845	4.20	1.845		
7.00 + 4	14.291	6.00 + 5	5.584	4.40	1.788		
7.50 + 4	14.008	6.50 + 5	5.354	4.60	1.734		
8.00 + 4	13.738	7.00 + 5	5.148	4.80	1.683		

8.50 + 4	13.481	7.50 + 5	4.964	5.00	1.635
9.00 + 4	13.235	8.00 + 5	4.797	5.20	1.589
9.50 + 4	12.999	8.50 + 5	4.645	5.40	1.547
1.00 + 5	12.774	9.00 + 5	4.506	5.60	1.506
1.10 + 5	12.351	9.50 + 5	4.378	5.80	1.467
1.20 + 5	11.964			6.00	1.430
1.30 + 5	11.607			6.20	1.395
1.40 + 5	11.275			6.40	1.362

[a] 1.00 + 3 reads 1.00×10^3.

energy can be calculated from Eq. (3.35) and the angular distribution. Inelastic scattering can be isotropic or anisotropic. At higher energies, levels may not be resolved and the neutron energy-angle distribution is described by a formula or table, usually specified in the laboratory frame. Inelastic scattering is especially important in heavy elements and at high energies and can give a significant loss in energy as well as generating γ rays in shields containing iron and other heavy materials. γ-Ray production data for iron and lead are given in Table 3.11.

Radiative Capture

Radiative capture is possible in all nuclides but predominates at low energies and in heavy nuclides. The cross section generally exhibits reasonances, and usually goes as

$$\sigma(n,\gamma) = \frac{\text{const}}{v} \tag{3.76}$$

below the lowest energy resonance. The energy available is the binding energy of the last neutron in the product (compound) nucleus, plus the kinetic energy in the CMS. However, instead of emitting a neutron as in scattering, one or more γ rays are emitted. Figure 3.36 illustrates the energy-level diagram for radiative capture of thermal neutrons in ^{12}C. Deexcitation proceeds by transition to the ground state in 69% of the captures and by transition to the 3.680 MeV state and then to the ground state in 31% of the captures. Hence γ rays of 4.947 MeV ($n = 0.69$), 1.267 MeV ($n = 0.31$), and 3.680 MeV ($n = 0.31$) are emitted following thermal neutron capture in ^{12}C. The cross section for radiative capture in ^{12}C is very small even at thermal energies and is not plotted in Figure 3.34. It happens that the product, ^{13}C, is stable, but the product in many radiative capture reactions is radioactive because of the excess neutron, decaying usually by negatron emission with release of delayed γ rays in many instances.

The thermal neutron-capture cross section for hydrogen is $1/v$ and equal to 0.332 b at 2200 m/s. It emits a 2.23-MeV γ ray in 100% of the captures. The thermal-capture cross section for iron is 2.55 b, and for pure lead it is 0.17 b. The complex γ-ray spectrum for iron is given in Table 3.12. Pure lead emits a 7.37-MeV γ ray (95%) and a 6.74-MeV γ ray (5%).

TABLE 3.11. INELASTIC GAMMA RAY PRODUCTION IN IRON AND LEAD

Nuclide (Abundance)	E_γ (MeV)	σ(b) at E_n 2 MeV	4 MeV	14 MeV
^{54}Fe	1.13		0.3	
(5.8%)	1.41	0.9	0.9	0.4
^{56}Fe	0.845	1.0	1.1	0.9
(91.2%)	1.24	0.01	0.02	0.03
	1.81		0.01	0.01
	2.12		0.01	0.01
^{206}Pb	0.54		1.5	
(25.1%)	0.80		2.3	
^{207}Pb	0.57		1.9	
(21.7%)	0.89		0.8	
	1.08		0.3	
	1.77		0.6	
	2.12		0.3	
	2.64		0.3	
^{208}Pb	0.58		0.6	
(52.3%)	0.86		0.2	
	2.61		1.1	

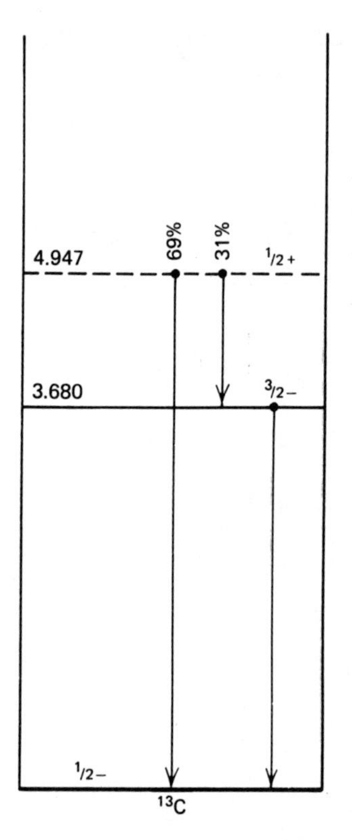

Fig. 3.36. Energy-level diagram for thermal-neutron radiative capture in ^{12}C (data from Bartholomew [26]). γ Rays of total energy $(E_B + E_n) \simeq E_B$ are emitted, in this case a 4.95-MeV γ ray in 69% of the captures, and a 1.27-MeV plus a 3.68-MeV γ ray in 31% of the captures (from A. Edward Profio, *Experimental Reactor Physics*, Wiley, New York, 1976, p. 89).

Capture γ rays are often the major contributor to the dose in thick shields. Some neutron shields contain boron to reduce the generation of penetrating γ rays from hydrogen and other constituents.

Charged-particle Emission

Absorption with emission of charged particles, such as the (n,p) and (n,α) reactions, is possible when the

TABLE 3.12. THERMAL-CAPTURE γ-RAY SPECTRA FOR IRON

E_γ (MeV)	Number / 100 Captures	E_γ (MeV)	Number / 100 Captures
0 to 0.25	1.7	5.50	0.8
0.50[a]	16.2	5.75	0.5
0.75	6.3	6.00	9.2
1.00	3.7	6.25	8.8
1.25	3.6	6.50	0.8
1.50	3.7	6.75	0.3
1.75	15.2	7.00	0.2
2.00	2.4	7.25	0.3
2.25	2.8	7.50	5.0
2.50	1.5	7.75	53.5
2.75	3.2	8.00	0.0
3.00	2.2	8.25	0.1
3.25	2.4	8.50	0.1
3.50	5.8	8.75	0.0
3.75	1.0	9.00	0.7
4.00	2.1	9.25	0.0
4.25	5.1	9.50	4.2
4.50	2.8	9.75	0.0
4.75	0.6	10.00	0.0
5.00	2.7	10.25	0.1
5.25	0.5	10.50	0.0

[a] Upper energy, interval 0.25 to 0.50 MeV.

excitation energy is sufficient to give a significant probability of penetrating the Coulomb barrier. The cross section rises rapidly from the threshold energy, and tends to level off as the barrier penetrability approaches unity but may decrease again because of competing reactions or because the cross section for formation of the compound nucleus decreases. In a few light nuclei the (n,p) or (n,α) reactions are exoergic and thus may occur even with thermal neutrons. These reactions are summarized in Table 3.13. As indicated in the $^{10}B(n,\alpha)^7Li$ reaction, γ rays may be emitted as well as charged particles, when the final nucleus is left in an excited state. The cross sections and branching probabilities apply to thermal neutron absorption. The energies are Q values; energy is shared between the reaction products.

TABLE 3.13. CHARGED-PARTICLE REACTIONS WITH THERMAL NEUTRONS

Reaction	σ_a (barns)
$^3He + n \rightarrow p + {}^3H + 0.764$ MeV	5327 ± 10
$^6Li + n \rightarrow \alpha + {}^3H + 4.786$ MeV	940 ± 4
$^{10}B + n \rightarrow \alpha + {}^7Li + 2.79$ MeV (6.1%)	3837 ± 9
$\alpha + {}^7Li^* + 2.31$ MeV (93.9%)	
$^7Li^* \rightarrow {}^7Li + 0.478$ MeV γ	
$^{14}N + n \rightarrow p + {}^{14}C + 0.626$ MeV	1.81 ± 0.05

In other nuclides the (n,p) and (n,α) reactions are endoergic, occur only above a threshold, and are characteristic of fast-neutron reactions along with inelastic and elastic scattering, and other reactions such as (n,2n). The $^{12}C(n,n'3\alpha)$ reaction is a special case. It could be considered as inelastic scattering because a neutron of lower energy in the CMS is emitted, but inelastic scattering is usually reserved as a term for the $(n,n'\gamma)$ reaction.

Cross sections for some common reactions with charged particle emission, used in neutron detectors, are plotted in Figure 3.35 together with the elastic scattering cross section for hydrogen. There are no inelastic scattering or charged-particle-emitting absorption reactions in hydrogen.

Fission and Other Reactions

Fission has been discussed already as a source in Chapter 2. Shields do not generally contain fissionable material.

At high energies, the excited nucleus remaining after emission of one neutron may emit additional neutrons instead of charged particles or γ rays; thus (n,2n) and (n,3n) reactions may occur. As these reactions are always endoergic, they are characterized by a threshold and release a broad spectrum of low energy neutrons. The actual spectrum may be known or may be approximated by an "evaporation" model spectrum of the form

$$n(E) = \text{const}\ \frac{E}{T^2} \exp \frac{-E}{T} \qquad (3.77)$$

where T is a nuclear "temperature" expressed in energy units. It is usually assumed that emission is isotropic in the laboratory system.

Sources of Nuclear Data

Nuclear level diagrams are published in Landölt-Bornstein [20] and the *Table of Isotopes* [21].

Plots and references for measured neutron cross sections are compiled in the "barn book," Brookhaven National Laboratory Report BNL-325, which is periodically updated [22]. Measured angular distributions are given [23] in BNL-400. However, a computer-oriented system, SCISRS [24], has been implemented to store and retrieve experimental data, on request to the National Neutron Cross Section Center at Brookhaven. A bibliography of cross-section measurements [25] is published periodically by the European Nuclear Energy Agency. Consistent Q values and binding energies have been published [26], and data on radiative capture γ-ray spectra may be found in the journal *Nuclear*

Data [27] and in a report by Orphan, Rasmussen, et al. [28].

Measured nuclear data are still incomplete, imprecise, or discordant for many nuclides and energies. The data must be evaluated to select the best fit, decide between conflicting measurements, and fill in gaps. The evaluation is often based on fits and interpolations with nuclear-model calculations, renormalizations of experimental data, and comparisons of calculated spectra and reaction rates with results of integral or bulk experiments. The Evaluated Nuclear Data File/B (ENDF/B) is a computer-based system at the National Neutron Cross Section Center, Brookhaven National Laboratory, and contains evaluated data for most nuclides [29], retrievable on request. One of the purposes of ENDF/B is to serve as a reference-data set for intercomparisons of calculations, but it is updated from time to time (Versions) as better evaluations become available.

Attenuation of Neutrons

The total linear attenuation coefficient for neutrons is the macroscopic total cross section

$$\Sigma_T = \sum_i N_i \sigma_{Ti} \quad \text{cm}^{-1} \tag{3.78}$$

where the summation is over all the nuclides i in the compound or mixture, and σ_T is the sum of the partial cross sections,

$$\sigma_T = \sigma_{nn} + \sigma_{nn'} + \sigma_{n\gamma} + \sigma_{np} + \cdots + \tag{3.79}$$

and so forth.

The flux density of uncollided neutrons follows the same exponential law as uncollided photons [Eq. (3.42)], with $\mu = \Sigma_T$. The mean free path for neutrons is $\mu = \Sigma_T^{-1}$. The mean free path for slow (∿eV) neutrons is small, but intermediate (∿keV) and fast (∿MeV) neutrons may have mean free paths of several centimeters. Scattered neutrons also contribute to the flux density and dose. Transport of neutrons is discussed in Chapter 4. Radiative capture and inelastic scattering generate secondary γ rays, which also have to be attenuated. The neutrons and

γ rays deposit energy in the shield and body, with some damaging effect. Energy absorption is discussed in Chapter 5.

3.6 CROSS-SECTION DESCRIPTIONS

Total cross sections and other cross sections that depend only on the incident energy may be specified by a table of point values of energy and the corresponding cross section, such as Table 3.10. It is assumed that the cross section variès linearly with energy unless a different interpolation law is specified. The behavior at the K and L edges of the photoelectric cross section, and in the total interaction cross section for photons, may be specified by double values of the energy or by energies that lie on either side of the jump and are very close together.

Resonances

The few, broad resonances in the neutron cross sections of light elements can be specified by a table of values, more closely spaced where the cross section varies rapidly. However, in general the shape of a resonance is specified by the single-level Breit-Wigner formula [30]

$$\sigma_{ni}(E) = \pi \lambda\!\!\!^{-2} g_J \frac{\Gamma_n \Gamma_i}{(E - E_0)^2 + (\Gamma/2)^2} \qquad (3.80)$$

where

$\lambda\!\!\!^{-} = \hbar(2M_0E)^{1/2}$ cm, the wavelength in the CMS;
$g_J = (2J + 1)/2(2I + 1)$, a statistical factor for the spins where J is the spin quantum number of the compound nucleus and I the spin of the target nucleus;
Γ_n = neutron width (eV);
Γ_i = width for reactions i (eV);
Γ = total width = $\Gamma_n + \Gamma_i + \cdots +$
E_0 = resonance energy (eV); and
E = neutron energy in the CMS

The total width is the natural line width and is related to the lifetime τ of the compound nucleus by the uncertainty principle, $\tau\Gamma = h$. The Γ is also the full width at half-maximum of the resonant part of the total cross section (any smoothly varying part is added to the resonant part). The ratio Γ_i/Γ expresses the probability for process i (and for J, implicitly). Tabulations of neutron cross-section data give the resonance E_0, Γ_n, Γ_γ (for radiative capture), g (or I or J), and the reduced neutron width

$$\Gamma_n^0 = \frac{\Gamma_n}{\sqrt{E_0}} \tag{3.81}$$

In the heavier nuclei, the radiative capture width Γ_γ is on the order of tens of millielectron volts (meV) and is found to be essentially constant for levels of the same angular momentum or spin, in a given nuclide. The neutron width is on the order of MeV. In light nuclei, Γ_γ is small, Γ_n is on the order of keV, and the resonances are mostly in the scattering cross section.

The resonance formula for radiative capture can be written more compactly by defining the peak cross section,

$$\sigma_0 = 4\pi\lambda_0^2 \, g_J \, \frac{\Gamma_n(E_0)}{\Gamma(E_0)} \tag{3.82a}$$

with

$$4\pi\lambda_0^2 = 2.608 \times 10^6 \, \frac{(A + 1)^2}{A^2 E(\text{eV})} \tag{3.82b}$$

Let

$$y \equiv \frac{2(E - E_0)}{\Gamma(E_0)} \tag{3.82c}$$

and experimentally for s-wave ($\ell = 0$) neutrons,

$$\Gamma_n = \Gamma_n(E_0) \sqrt{\frac{E}{E_0}} \tag{3.82d}$$

Then

$$\sigma_{n\gamma}(E) = \sigma_0 \sqrt{\frac{E_0}{E}} \, \frac{\Gamma_\gamma}{\Gamma} \, \frac{1}{1 + y^2} \tag{3.83}$$

The shape of a radiative capture resonance is shown in Figure 3.37. For $E << E_0$,

$$\sigma_{n\gamma} = \frac{\text{const}}{\sqrt{E}} \tag{3.84}$$

and the radiative capture cross section varies as 1/v below the lowest energy resonance. At thermal energies, only s-wave capture is possible. However, deviations from the 1/v law may occur because of a negative-energy resonance (i.e., from a level below the binding energy of the neutron) or multichannel processes not described by the Breit-Wigner theory.

For elastic scattering, $\Gamma_i = \Gamma_n$. However, one must also consider the potential scattering and the interference between potential and resonance scattering. To account for the interference, the scattered wave amplitudes are added and the sum squared to obtain the intensity. For s-wave scattering and low energies,

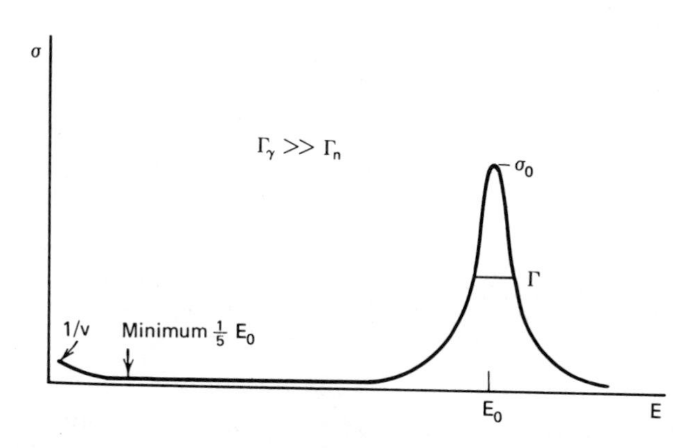

Fig. 3.37. Radiative capture resonance shape (from A. Edward Profio, *Experimental Reactor Physics*, Wiley, New York, 1976, p. 111).

$$\sigma_{nn}(E) = \sigma_0 \sqrt{\frac{E_0}{E}} \frac{\Gamma_n}{\Gamma} \frac{1}{1 + y^2} + \sigma_p$$

(resonance) (potential)

$$+ \left(\sigma_0 \sigma_p g_J \frac{\Gamma_n}{\Gamma} \sqrt{\frac{E_0}{E}}\right)^{1/2} \frac{2y}{1 + y^2} \qquad (3.85)$$

(interference)

The potential scattering σ_p is often identified with σ_{fa}, the free-atom scattering cross section. The shape of an elastic scattering resonance, for $\ell = 0$, is shown [31] in Figure 3.38. For $\ell > 0$, the peak is not as high and the valley is not as deep. The small cross section at the valley, or minimum, can be significant in shielding because neutrons near the energy can penetrate large distances before being scattered.

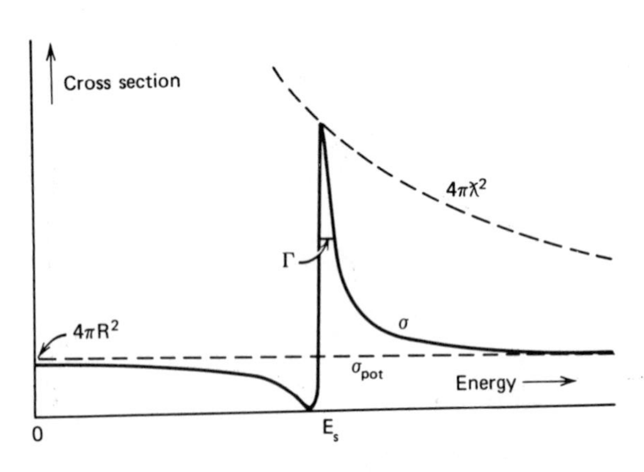

Fig. 3.38. Elastic scattering resonance shape for s-waves, with $\sigma_p = 4\pi R^2$, where R is the nuclear radius (after Blatt and Weisskopf [32]; from A. Edward Profio, *Experimental Reactor Physics*, Wiley, New York, 1976, p. 112).

The resonance formulas apply to the kinetic energy of the neutron-nucleus system. If the nucleus is not at rest but is moving with velocity $\underline{V}$,

$$E = \frac{1}{2} M_0 (\underline{v} - \underline{V})^2 = \frac{M_0}{M}\left(\frac{1}{2} MV^2\right) + \frac{M_0}{M}\left(\frac{1}{2} mv^2\right) - M_0 \underline{v} \cdot \underline{V} \qquad (3.86)$$

The first term is the kinetic energy of the nucleus. For nuclei in thermal agitation at room temperature or so, it is on the order of hundredths of an electron volt and may be neglected compared to resonance energies of several eV or greater. The second term involves the laboratory system to CMS conversion and is handled by redefining E_0 in the laboratory. The final term is the important one and expresses the Doppler effect resulting from the relative motion of neutron and nucleus. As in optics or sound, the effective motion is the "radial" component V_r, the component of $\underline{V}$ parallel to the neutron velocity $\underline{v}$. Thus the Doppler term may be written as

$$-M_0 \sqrt{\frac{2E}{m}}\; V_r$$

where E is the kinetic energy of the neutron measured in the laboratory system and m is the mass of the neutron. For most resonant absorbers or scatterers, the distribution of V_r is given with sufficient accuracy by the one-dimensional Maxwellian distribution,

$$p(V_r) = \sqrt{\frac{M}{2\pi kT^*}}\; \exp - \frac{MV_r^2}{2kT^*} \qquad (3.87)$$

where the effective temperature is T*. For gases, T* = T. For liquids and most solids at room temperature and above, T* is greater than, but approximately equal to the thermodynamic temperature T.

We now wish to define an average or effective cross section $\langle\sigma\rangle$ such that the reaction rate $vN\langle\sigma\rangle$

is equal to the average reaction rate for the actual distribution of relative velocities of neutron and nucleus. The concept of average or effective cross section is frequently used and is good as long as the weighting function is clearly defined. For the resonance cross section,

$$\langle\sigma\rangle = \frac{\int \sigma(V_r)\ p(V_r)\ dV_r}{\int p(V_r)\ dV_r} \tag{3.88}$$

The convolution of the Maxwellian distribution with the resonance cross section has been performed [33], and

$$\langle\sigma_{n\gamma}\rangle = \sigma_0 \sqrt{\frac{E_0}{E}}\ \frac{\Gamma_\gamma}{\Gamma}\ \psi(\theta, y) \tag{3.89}$$

$$\langle\sigma_{nn}\rangle = \sigma_0 \sqrt{\frac{E_0}{E}}\ \frac{\Gamma_n}{\Gamma}\ \psi(\theta, y) + \left(\sigma_0\ \sigma_p\ g_J\ \frac{\Gamma_n}{\Gamma} \sqrt{\frac{E_0}{E}}\right)^{1/2} \chi(\theta, y)$$

$$+\ \sigma_p \tag{3.90}$$

where the symmetric and asymmetric shape functions ψ, χ are functions of the ratio of the natural width Γ to the Doppler width Δ,

$$\theta \equiv \frac{\Gamma}{\Delta} \tag{3.91}$$

and the Doppler width

$$\Delta \equiv \sqrt{\frac{4mkTE_0}{M}} = \sqrt{\frac{4E_0kT}{A}} \tag{3.92}$$

The ψ function is plotted in Figure 3.39 for $\theta = \infty$, 1.58, and 1.12. As the temperature is increased the value in the "wings," say, $y \geq 5$, remains about the same, but the peak is lowered and broadened. It turns out that the area in unchanged. Doppler broadening does not change the total absorption rate if all nuclei

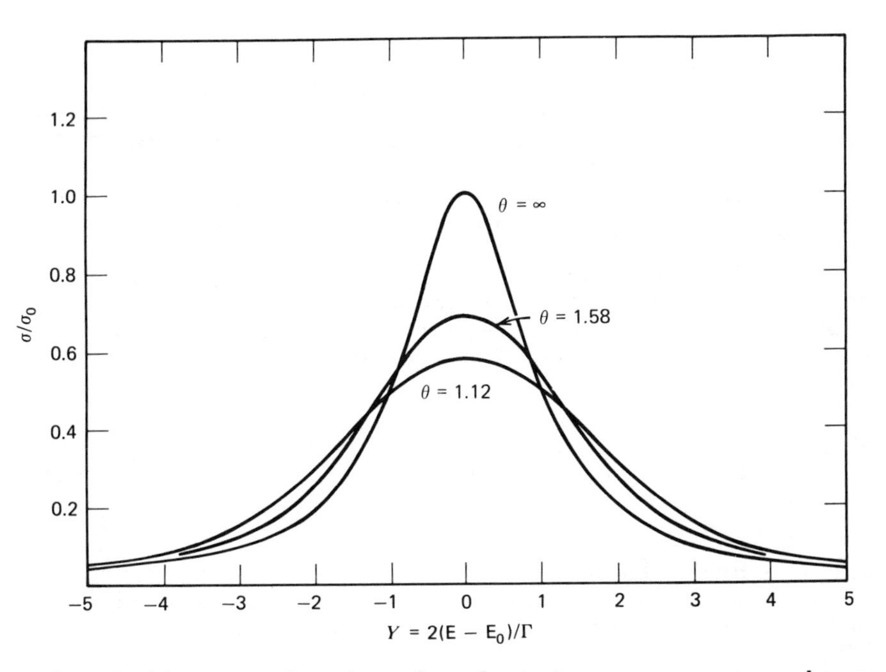

Fig. 3.39. Doppler broadened capture resonance shapes (from A. Edward Profio, *Experimental Reactor Physics*, Wiley, New York, 1976, p. 114).

are exposed to the same flux density, but in heterogeneous or concentrated mixtures of absorber and moderator or scatterer, flux density is decreased by absorption. Doppler broadening and self-shielding are discussed later under multigroup cross sections.

In the unresolved resonance region, generally at the higher energies, parameters for individual resonances can no longer be measured. Scattering and absorption are calculated from the statistical or average properties of many resonances. The radiative capture width Γ_γ is assumed to be the same for unresolved and resolved resonances of the same J. The resonance energies E_0 are replaced by the average level spacing $\overline{D}_\ell$ in a given energy range including many resonances. The strength function is defined by

$$S_\ell = \frac{\overline{\Gamma^0_{n\ell}}}{\bar{D}_\ell} \qquad (3.93)$$

where the numerator is the averaged reduced neutron width for angular momentum ℓ and J. The strength functions for s- and p-wave resonances, S_0 and S_1 are on the order of 10^{-4} for all nuclei. Calculations of Doppler broadening are done as before, except that interference is often neglected. More accurate calculations include the fluctuations in the level spacings and reduced widths about the average. Theoretical expressions are available for these: the Wigner distribution for spacings and the Porter-Thomas (same as χ^2) distribution of widths. The average cross sections are then of the form

$$\langle\sigma_{ni}\rangle = 2\pi^2 \lambda^2 \sum_{\ell,J} g_J \, S_\ell \, \frac{\langle\Gamma_i\rangle}{\langle\Gamma\rangle} \sqrt{E} \qquad (3.94)$$

where the brackets indicate averages. Because definitions differ, the expression for the cross section should be given along with the resonance parameters.

Thermal Cross Section

A neutron with energy less than 1 eV gains as well as loses energy in collisions with nuclei, and the chemical binding of the scatterer can have a significant effect on the thermal neutron spectrum in reactors. However, for shielding purposes one is mainly interested in cross sections for radiative capture or charged-particle reactions, usually 1/v or close to it. The reaction rate for 1/v reactions is proportional to the neutron density, R = const (nv/v) = const n, and does not depend on the spectrum of the neutron flux density. Thermal neutron cross sections are often specified at a reference speed v_0 of 2200 m/s, corresponding to 0.0253 eV, the most probable energy for a Maxwell-Boltzmann spectrum at a temperature T = 293 K. Thermal neutron spectra are discussed in Duderstadt and Hamilton [33]. For small absorption the spectrum is close to the Maxwell-Boltzmann distribution of velocities

$$n(v)dv = \frac{4}{\sqrt{\pi}} \left(\frac{v}{v_0}\right)^2 \exp -\left(\frac{v}{v_0}\right)^2 d\left(\frac{v}{v_0}\right) \qquad (3.95)$$

where the most probable velocity

$$v_0 = \sqrt{\frac{2kT}{m}} \qquad (3.96)$$

The distribution of energies is

$$n(E)dE = \frac{2}{\sqrt{\pi}} \sqrt{\frac{E}{kT}} \exp \frac{-E}{kT} d\left(\frac{E}{kT}\right) \qquad (3.97)$$

where

$$kT(eV) = 8.617 \times 10^{-5} \text{ T kelvin} \qquad (3.98)$$

These distributions are plotted in Figure 3.40. The average energy

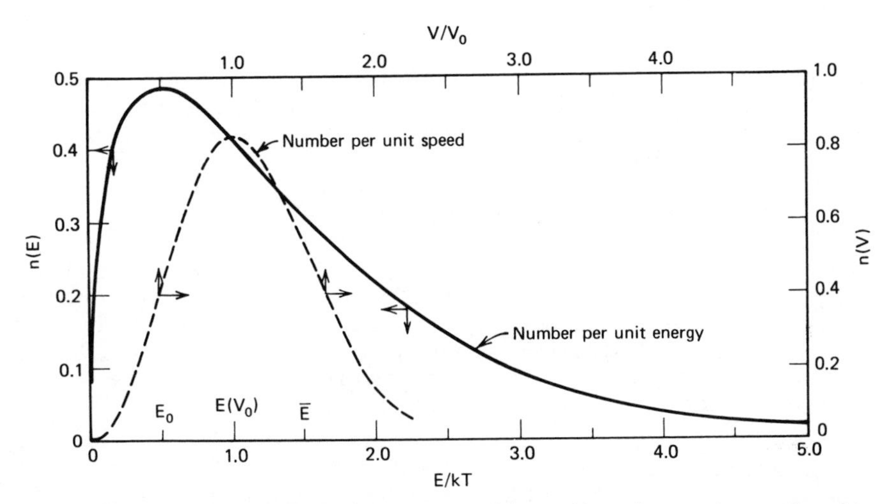

Fig. 3.40. Maxwell-Boltzmann distributions in velocity and energy (from A. Edward Profio, *Experimental Reactor Physics*, Wiley, New York, 1976, p. 115).

$$\bar{E} = \frac{3}{2} kT \tag{3.99}$$

If the cross section is 1/v and is specified at 2200 m/s, the reaction rate is given properly if the flux density is evaluated for a speed of 2200 m/s,

$$R = \phi(2200)\ N\sigma(2200) \tag{3.100}$$

independent of temperature. However, one may have instead ϕ_T, the flux density averaged over the Maxwell-Boltzmann distribution for some temperature T. Then since

$$\phi_T = \frac{2}{\sqrt{\pi}} \sqrt{\frac{T}{293}}\ \phi(2200) \tag{3.101}$$

we have

$$R = \phi_T\ N\ \langle\sigma\rangle_{th} \tag{3.102}$$

where

$$\langle\sigma\rangle_{th} = \frac{\sqrt{\pi}}{2}\ g(T) \sqrt{\frac{293}{T}}\ \sigma(2200\ m/s) \tag{3.103}$$

and the g(T) factor is a correction for non-1/v behavior (g is unity for 1/v absorbers). Thermal scattering cross sections are often considered to be constant in shielding calculations.

Differential Cross Sections

The angular distribution of scattering may be specified by a matrix of $d\sigma_s(E \rightarrow E',\mu)/d\Omega$ against incident energy E, scattered energy E', and the cosine of the scattering angle in the laboratory system, $\mu = \cos\theta_L = \underline{\Omega} \cdot \underline{\Omega}'$. However, this is unwieldy and a common alternative is to expand in Legendre polynomials,

$$\frac{d\sigma_s(E \rightarrow E',\mu)}{d\Omega} = \sum_{\ell=0}^{L} \frac{2\ell + 1}{4\pi}\ \sigma_s(E)\ f_\ell(E \rightarrow E')\ P_\ell(\mu) \tag{3.104}$$

where the orthogonal expansion coefficients are

$$f_\ell(E \to E') = \frac{2\pi}{\sigma_s(E)} \int_{-1}^{1} P_\ell(\mu) \frac{d\sigma_s(E \to E',\mu)}{d\Omega} d\mu \qquad (3.105)$$

and the P_ℓ are the Legendre polynomials as given in Table 3.14.

TABLE 3.14. LEGENDRE POLYNOMIALS

$$P_0(\mu) = 1$$

$$P_1(\mu) = \mu$$

$$P_2(\mu) = \frac{1}{2}(3\mu^2 - 1)$$

$$P_3(\mu) = \frac{1}{2}(5\mu^3 - 3\mu)$$

$$P_4(\mu) = \frac{1}{8}(35\mu^4 - 30\mu^2 + 3)$$

$$P_5(\mu) = \frac{1}{8}(63\mu^5 - 70\mu^3 + 15\mu)$$

$$P_n(\mu) = \frac{1}{2^n n!} \frac{d^n}{d\mu^n} (\mu^2 - 1)^n$$

The expansion is exact if $L = \infty$, but in practice the series is truncated at some low order of L, often 3 or 5. The error introduced by truncation in the differential cross section for elastic scattering in hydrogen, may be seen in Figure 3.41. The $L = 0$ approximation, corresponding to the average cross section and isotropic scattering in the laboratory system, is poor. For $L = 1$ (P_0 and P_1 terms), the approximation is somewhat better. At $L = 2$ the approximation is fairly good and is not improved very much on going to $L = 5$. Neutron transport calculations in a variety of shielding materials such as H_2O, CH_2, LiH, and concrete suggest that a P_3 expansion is sufficient in many shielding problems. Many γ-ray shielding calculations have also been made with $L = 3$. If in doubt, the order of

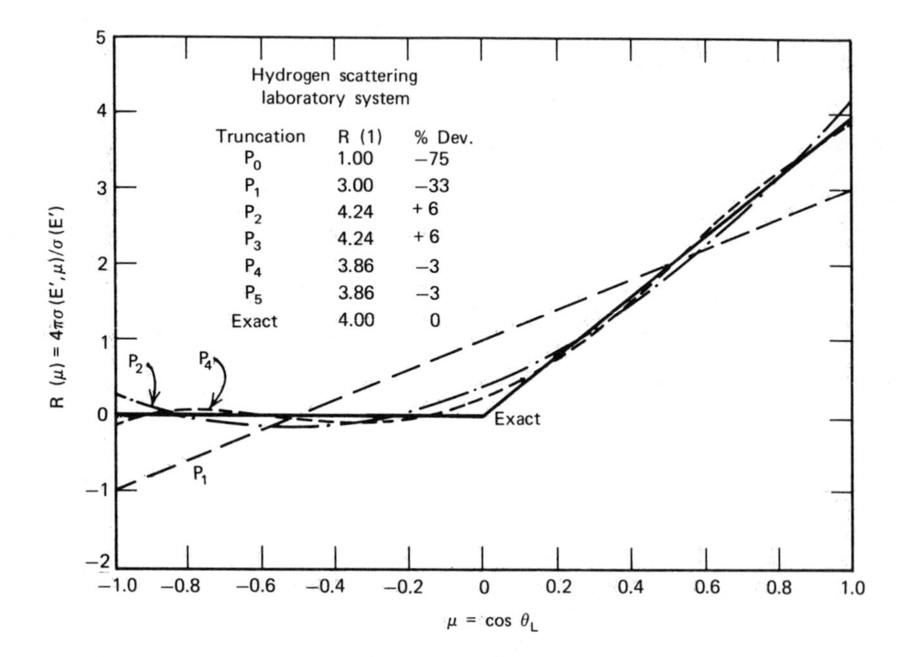

Fig. 3.41. Differential scattering cross section for hydrogen and truncated Legendre polynomial expansions.

expansion can be increased until no significant difference is noted in the results of a shielding calculation similar to the intended application.

Multigroup Cross Sections

Many radiation transport computer codes use the multigroup approximation, in which the continuous variation of the cross section with energy is replaced by a weighted average over the energy interval E_1 to E_2,

$$\sigma_g(E_1,E_2) = \frac{\int_{E_1}^{E_2} \sigma(E)\, W(E)\, dE}{\int_{E_1}^{E_2} W(E)\, dE} \tag{3.106}$$

The weighting function W(E) is usually chosen to approximate the spectrum expected in the application, since then the reaction rate

$$R(E_1,E_2) = N\, \sigma_g(E_1,E_2)\, \overline{\phi_g}\, \Delta E = \int_{E_1}^{E_2} N\, \sigma(E)\, \phi(E)\, dE \tag{3.107}$$

where $\overline{\phi_g}$ is the average flux density in the interval. The averaging may be carried out in two steps, once to generate perhaps 100 to 1000 fine-group averaged cross sections, and again to reduce to perhaps 10 to 100 broad-group cross sections. The weighting spectrum for the fine-group set is supposedly representative of the spectrum in a broad class of transport problems. For example, for shielding fission neutron sources, a fission spectrum joined to a 1/E spectrum at about 100 keV may be appropriate. The more groups and smaller ΔE, the less sensitive is the average or group cross section to the spectrum chosen. The number of groups may be reduced somewhat if the group boundaries (E_1, E_2, etc.) are selected to coincide with thresholds, resonances, or other energies where the cross section varies rapidly. However, as all elements use the same group structure in any given calculation, such selection may not be practical. A common structure for 99 fast-group and one thermal-neutron group cross sections uses equal lethargy intervals (lethargy $u \simeq \ln (E_0/E)$, where E_0 is usually 10 MeV) of 0.1 from $u = -0.4$ (14.92 MeV) to $u = 4.5$ (111.1 keV), and $\Delta u = 0.25$ from $u = 4.5$ to 17.0 (0.414 eV). This is termed the "GAM-II" structure after the cross section preparation code that first used it [35]. For calculations with fusion (14-MeV) sources it would be preferable to include more groups in the MeV region. Because too much computer storage and time are required to perform transport calculations with even a hundred groups, it is desirable to reduce the set to fewer "broad-group" cross sections. The fine groups are collapsed to broad groups using a calculated weighting spectrum typical of the source, composition, and geometry of the intended specific problem. Often an infinite-medium approximation is made, but in fact the spectrum changes with distance from the source and may be modified by leakage. It is usually uneconomical to use spatially dependent cross sections. The weighting spectrum is averaged over space and relatively narrow energy intervals are used where feasible.

The GAM-II structure is quite adequate for a smoothly varying cross section such as the total cross section for hydrogen. The fractional error in a group, comparing the true cross section and the average, is never more than $\pm 5\%$. The uncollided flux-density spectrum in water at 150 cm from a fission-spectrum source is calculated to within a few percent by the group-averaged cross sections, considering only the hydrogen component.

The major difficulty comes in averaging over resonances. Figure 3.42 plots the total cross section for ^{16}O near the 2.37-MeV interference minimum. The solid line shows the variation with energy as given by the ENDF/B library. The horizontal bar illustrates the group cross section in GAM-II fine group number 19, obtained by averaging over the group interval 2.23 to 2.47 MeV, with $W(E) = E^{-1}$ as GAM-II assumes. The average cross section is 13 times the cross section at the minimum. The effect of the approximation on the

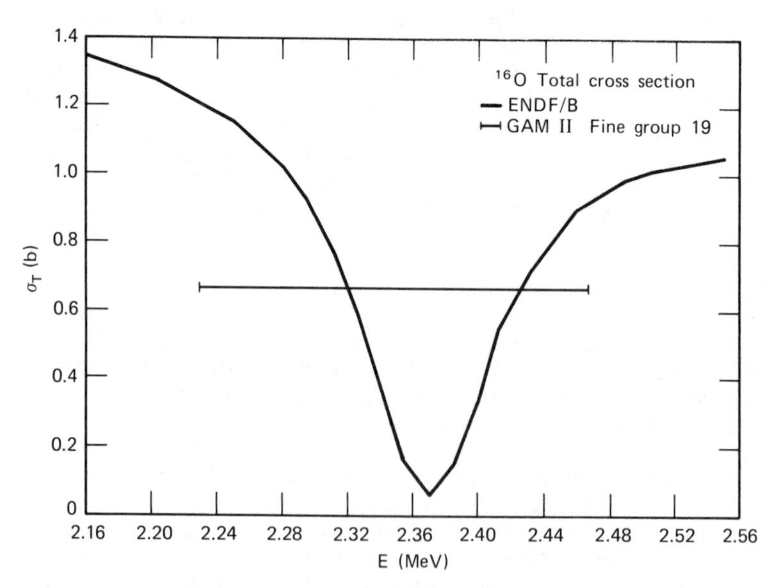

Fig. 3.42. Total cross section of ^{16}O near 2.37-MeV resonance minimum, and group cross section with 1/E weighting.

transmission of uncollided fission-spectrum neutrons for Nd = 2.01 (corresponding to d = 60 cm of oxygen at the density N = 0.0334 atoms/barn-cm in water), is shown in Figure 3.43. The average cross section gives a transmission that is 20% lower than the average of the transmission calculated from the point-by-point data. The problem is that the weighting spectrum goes as exp $[-N\sigma(E)x]$ instead of E^{-1}. One cannot expect the group cross section to give the transmitted spectrum shape within the group, but it should allow one to predict the correct transmission integrated over the group energy interval. This is difficult to achieve even with very fine-group cross sections, and one might better calculate "streaming" through cross-section minima by an analytical or numerical point-by-point method. Fortunately, the influence of the minima is mitigated by mixing with hydrogen or other nuclides.

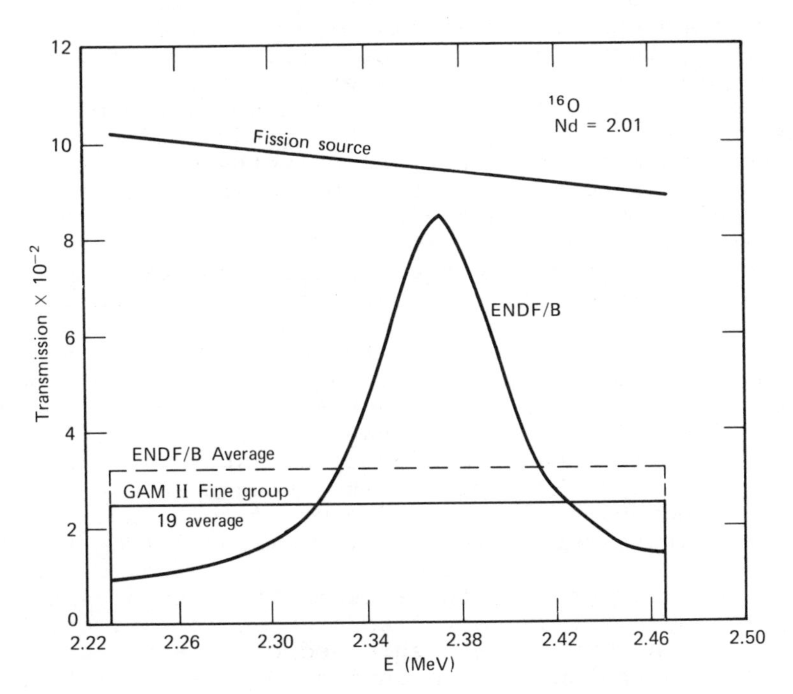

Fig. 3.43. Transmission at the 2.37-MeV resonance minimum in ^{16}O using group average and point-by-point data from ENDF/B for Nd = 2.01.

Radiative capture in resonances is important in the generation of secondary γ rays in substances such as depleted uranium or tungsten, sometimes used as dense γ-ray shielding materials, in combination with hydrogeneous materials for shielding neutrons. The problem is similar to resonance capture in reactor lattices [35] but is complicated when the spatial distribution of captures must be predicted to account for attenuation of the capture γ rays. One may also need the transmitted neutron spectrum for subsequent attenuation in surrounding materials. In a dilute mixture of absorber and Σ_s = const moderator, below the source energy, the spectrum is 1/E, and an "infinitely dilute" cross section can be obtained, evaluated at 0°K but valid at any temperature because Doppler broadening does not affect the total absorption. As the concentration of absorber is increased, the flux density through the resonance is decreased (flux-density depression or energy self-shielding) because of the absorption, but Doppler broadening reduces the self-shielding. If the absorber is physically separated from the moderator, there is spatial self-shielding also. Lahti [36] has analyzed an experiment performed by Antúnez and Neill [37] at General Atomic, where a 3.14-cm slab of depleted uranium, sandwiched between 2.49-cm thick slabs of borated polyethylene, was irradiated with neutrons. The capture in the 3 to 11 eV interval, encompassing the prominent 6.68-eV capture resonance in ^{238}U, was calculated with a code that developed region-averaged cross sections for 10 groups within this interval. Another code made approximate calculations for one-group averages. The 10-group method gave good agreement with an exact transport calculation for the region-averaged capture, but the spatial distribution was somewhat off, leading to underprediction of the capture-γ-ray flux density by 19%, which is not too bad. One-group averages were in error by up to 50% in the capture-rate integrated over the uranium region, and up to a factor of two discordant in the capture-γ-ray flux density emerging from the region. For accurate calculations of spatial distribution of absorptions, there seems to be no alternative to point-by-point or very fine-group calculations.

One can compute self-shielded cross sections, by multiplying the infinite-dilution, 0 K cross section by a self-shielding factor that includes the effects of temperature in Doppler broadening, flux-density depression as a function of scattering and absorption cross sections, and the thickness of a heterogeneous absorber. The Bondarenko approximation [38] is sometimes used in shielding calculations [39]. The homogeneous self-shielding factor

$$f_g \equiv \frac{[\sigma_g]}{\sigma_g} \tag{3.108}$$

where σ_g is the infinite-dilution group cross section and the self-shielded group cross section $[\sigma_g]_j$ is obtained by weighting by a smooth "background" spectrum $\phi_b(E)$ (usually 1/E), a "background" cross section σ_b for the group, and the total cross section of the resonance absorber $\sigma_t(E)$, including Doppler broadening,

$$[\sigma_g]_j = \frac{\int_g \sigma_j(E,T) \frac{\phi_b(E)}{\sigma_t(E,T) + \sigma_b} dE}{\int_g \frac{\phi_b(E)}{\sigma_t(E,T) + \sigma_b} dE} \tag{3.109}$$

where $\sigma_j(E,T)$ is the Doppler broadened cross section for reaction j. The background cross section is taken as the total cross section for all nuclides in the mixture, per atom of absorber,

$$\sigma_b = \frac{\Sigma_g^t}{N_a} \tag{3.110}$$

Thus it is necessary not only to specify the temperature, but also the mixture or at least σ_b for the group. However, the self-shielded cross section and self-shielding factor can be computed with temperature and σ_b as parameters and the proper choice made later in the transport calculation.

Group-to-group-transfer cross sections are derived from the kinematics or expressions for the secondary neutron spectrum by averaging over the scattered energy interval as well as the incident group interval. The angular distribution is handled by a P_ℓ expansion; hence there is a group-to-group transfer matrix or table for each term, or L + 1 altogether. Computer programs such as SUPERTOG [40] prepare the fine-group transfer tables from ENDF/B data. Truncation at a low order in the laboratory system can introduce appreciable errors in the scattered neutron energy distribution. Thus it is better to compute the group-to-group transfer cross sections at high-order P_ℓ expansion in the CMS, transform to the laboratory system, and then truncate at relatively low order (say, P_3) in the laboratory system. Truncation does not affect the values of the lower-order coefficients; hence fine-group cross sections are often prepared to P_8 or P_{12}, whereas the user has the option to simply ignore higher terms when reducing the fine-group set to broad-group cross sections. The averaging spectrum for the broad-group cross sections is often calculated for the same or similar composition and source as the actual shield, but in a simple geometry or infinite-medium approximation. The GAM-II infinite-medium code has been used for shielding work, but to obtain an averaging spectrum, it is probably better to use a discrete-ordinates code such as discussed in Chapter 4.

The energy group structure of the coupled neutron-γ multigroup cross section set DLC-23/CASK (available from the Radiation Shielding Information Center, Oak Ridge) is listed in Table 3.15. This broad-group library has 22 neutron groups from 15 MeV to 0, plus 18 γ-ray groups from 10 MeV to 0. Neutron scattering is handled by P_0 through P_3 cross sections for transfer from one neutron energy group to another, lower energy neutron group. γ-Ray production from neutron inelastic scattering or radiative capture is handled by a cross section for transfer from a neutron group to a γ-ray group. Table 3.16 gives the P_0 through P_3 cross section tables for the first three neutron groups for iron, from the DLC-23 library. The format corresponds to the positions used in the ANISN discrete ordinates program. Position 1 contains the absorption cross section in barns. Position 2 contains $\nu\sigma_f$ (0 since there is no fission in iron). The total cross section appears in position 3. Position 4 contains the within-group scattering cross section, $\sigma_{g\to g}$ where g is the group number. Position 5 contains the cross section for scattering from group (g - 1) into group g. Position 6 contains the cross section for scattering from group (g - 2) into group g, and so forth. Table 3.17 gives the cross sections for Group 24 in the DLC-23 library for iron. Position 1 contains the energy absorption coefficient $\overline{E\sigma_{en}}$. Position 2 is not used. Position 3 contains the total cross section for γ rays. Positions 4 and 5 contain cross sections for

TABLE 3.15. DLC-23/CASK ENERGY GROUP STRUCTURE

Neutron Groups		γ-Ray Groups	
Group	Upper Energy (eV)	Group	Upper Energy (MeV)
1	15.0 + 6[a]	23	10.0
2	12.2 + 6	24	8.0
3	10.0 + 6	25	6.5
4	8.18 + 6	26	5.0
5	6.36 + 6	27	4.0
6	4.96 + 6	28	3.0
7	4.06 + 6	29	2.5
8	3.01 + 6	30	2.0
9	2.46 + 6	31	1.66
10	2.35 + 6	32	1.33
11	1.83 + 6	33	1.0
12	1.11 + 6	34	0.8
13	5.50 + 5	35	0.6
14	1.11 + 5	36	0.4
15	3.35 + 3	37	0.3
16	5.33 + 2	38	0.2
17	1.01 + 2	39	0.1
18	2.90 + 1	40	0.05
19	1.07 + 1		
20	3.06 + 0		
21	1.12 + 0		
22	0.414 + 0		

[a]Read 15.0×10^6.

TABLE 3.16. IRON NEUTRON GROUP CROSS SECTIONS

Position			Group 1		Group 2		Group 3
P_0	1	σ_a	2.022 - 1	σ_a	1.608 - 1	σ_a	1.121 - 1
	2	$\nu\sigma_f$	0.0		0.0		0.0
	3	σ_t	2.620 + 0	σ_t	2.939 + 0	σ_t	3.259 + 0
	4	$\sigma_{1\to1}$	1.267 + 0	$\sigma_{2\to2}$	1.546 + 0	$\sigma_{3\to3}$	1.784 + 0
	5		0.0	$\sigma_{1\to2}$	1.097 - 1	$\sigma_{2\to3}$	1.155 - 1
	6		0.0		0.0	$\sigma_{1\to3}$	2.100 - 2
P_1	1		0.0		0.0		0.0
	2		0.0		0.0		0.0
	3		0.0		0.0		0.0
	4		3.002 + 0		3.881 + 0		4.522 + 0
	5		0.0		1.356 - 2		2.748 - 2
	6		0.0		0.0		0.0
P_2	1		0.0		0.0		0.0
	2		0.0		0.0		0.0
	3		0.0		0.0		0.0
	4		4.250 + 0		5.550 + 0		6.531 + 0
	5		0.0		4.013 + 0		7.055 - 2
	6		0.0		0.0		0.0
P_3	1		0.0		0.0		0.0
	2		0.0		0.0		0.0
	3		0.0		0.0		0.0
	4		4.968 + 0		6.252 + 0		6.858 + 0
	5		0.0		3.535 - 2		3.979 - 2
	6		0.0		0.0		0.0

TABLE 3.17. IRON γ-RAY GROUP CROSS SECTIONS

Position	From	To	Group 24 (8.0 to 6.5 MeV) P_0	P_1	P_2	P_3
1	$\overline{E\sigma_{en}}$		1.465 + 1	0.0	0.0	0.0
2	-		0.0	0.0	0.0	0.0
3	σ_t		2.739 + 0	0.0	0.0	0.0
4	24	24	8.885 - 2	2.653 - 1	4.379 - 1	6.044 - 1
5	23	24	1.258 - 1	3.724 - 1	6.040 - 1	8.116 - 1
6	22	24	1.259 + 0	0.0	0.0	0.0
7	21	24	2.694 - 1	0.0	0.0	0.0
8	20	24	1.635 - 1	0.0	0.0	0.0
9	19	24	9.341 - 2	0.0	0.0	0.0
10	18	24	5.306 - 2	0.0	0.0	0.0
11	17	24	3.046 - 2	0.0	0.0	0.0
12	16	24	1.456 - 2	0.0	0.0	0.0
13	15	24	6.924 - 2	0.0	0.0	0.0
14	14	24	5.642 - 3	0.0	0.0	0.0
15	13	24	1.916 - 3	0.0	0.0	0.0
16	12	24	6.264 - 4	0.0	0.0	0.0
17	11	24	0.0	0.0	0.0	0.0
18	10	24	0.0	0.0	0.0	0.0
19	9	24	0.0	0.0	0.0	0.0
20	8	24	0.0	0.0	0.0	0.0
21	7	24	0.0	0.0	0.0	0.0
22	6	24	0.0	0.0	0.0	0.0
23	5	24	8.919 - 5	0.0	0.0	0.0
24	4	24	2.744 - 2	0.0	0.0	0.0
25	3	24	8.869 - 2	0.0	0.0	0.0
26	2	24	1.104 - 1	0.0	0.0	0.0
27	1	24	1.041 - 1	0.0	0.0	0.0

γ ray scattering. Positions 6 through 27 contain cross sections for production of γ rays by neutrons in groups 22 through 1.

REFERENCES

1. H. Bichsel, "Charged Particle Interactions," in F. H. Attix and W. C. Roesch, Eds., *Radiation Dosimetry*, 2nd ed., Vol. I, Academic Press, New York, 1968, pp. 157-228.
2. R. D. Evans, *The Atomic Nucleus*, McGraw-Hill, New York, 1955, pp. 567-591, 632-671.
3. J. B. Marion and F. C. Young, *Nuclear Reaction Analysis Graphs and Tables*, North-Holland, Amsterdam and Wiley-Interscience, New York, 1968.
4. ICRP Publication 21, "Data for Protection Against Ionizing Radiation from External Sources: Supplement to ICRP Publ. 15," Pergamon Press, Oxford, 1973, p. 42, Figure 1.
5. National Academy of Sciences-National Research Council Publication 1133, "Studies of Penetration of Charged Particles in Matter" (1964). Nuclear Science Series Report No. 39.
6. Evans, *The Atomic Nucleus*, pp. 592-599.
7. Evans, *The Atomic Nucleus*, pp. 600-631.
8. H. W. Koch and J. W. Motz, "Bremsstrahlung Cross Section Formulas and Related Data," *Rev. Mod. Phys.*, 31, 920-955 (1959).
9. C. W. Sandifer and M. Taherzadeh, "Measurement of Linac Thick-Target Bremsstrahlung Spectra Using a Large NaI Scintillation Spectrometer," paper at IEEE Conference on Nuclear and Space Radiation Effects, July 15-18, 1968.
10. A. A. O'Dell, C. W. Sandifer, R. B. Knowlen, and W. D. George, "Measurement of Absolute Thick-Target Bremsstrahlung Spectra," *Nucl, Instrum. Methods*, 61, 340 (1968).
11. W. W. Buechner, R. J. Van de Graaff, E. A. Burrill, and A. Sperduto, *Phys. Rev.*, 74, 1348 (1958).
12. Tatsuo Tabata and Rinsuke Ito, "A Generalized Empirical Equation for the Transmission Coefficient of Electrons," *Nucl. Instrum. Methods*, 127, 429-434 (1975).
13. Evans, *The Atomic Nucleus*, pp. 695-710.

14. R. D. Evans, "X-Ray and γ-Ray Interactions, in Attix and Roesch, *Radiation Dosimetry*, Vol. I, pp. 93-155.
15. Evans, *The Atomic Nucleus*, pp. 672-694.
16. O. Klein and Y. Hishina, *Z. Physik*, 52, 853 (1929).
17. A. T. Nelms, "Graphs of the Compton Energy-Angle Relationship and the Klein-Nishina Formula from 10 keV to 500 MeV," National Bureau of Standards Circular 542 (1953).
18. J. H. Hubbell, "Photon Cross Sections, Attenuation Coefficients, and Energy Absorption Coefficients from 10 keV to 100 GeV," NSRDS-NBS 29 (1969) (available from Superintendent of Documents, U.S. Government Printing Office, Washington, D. C. 20402).
19. J. B. Marion, T. I. Arnette, and H. C. Owens, "Tables for the Transformation Between the Laboratory and Center of Mass Coordinate Systems and for the Calculation of the Energies of Reaction Products," ORNL-2574, Oak Ridge National Laboratories, Tenn., 1959.
20. Landölt-Bornstein, New Series, Vol. 1, *Nuclear Physics*, Part 11, *Energy Levels of Nuclei A = 4 to A = 257*, Springer-Verlag, Berlin, 1961.
21. C. M. Lederer, J. M. Hollander, and I. Perlman, *Table of Isotopes*, 6th ed., Wiley, New York, 1967.
22. S. F. Mughabghab and D. I. Garber, *Neutron Cross Sections*, Vol. 1, *Resonance Parameters*, BNL-325, 3rd ed. (1973). Includes thermal cross sections. Available from National Technical Information Service, Springfield, Va. 22151. D. I. Garber and R. R. Kinsey, *Neutron Cross Sections*, Vol. 2, *Curves*, BNL-325, 3rd ed. (1976). Available from NTIS.
23. D. I. Garber, L. G. Strömberg, M. D. Goldberg, D. E. Cullen, and V. M. May, *Angular Distributions in Neutron-Induced Reactions*," Vols. 1 and 2, BNL-400, 3rd ed. (1970). Available from NTIS.
24. J. M. Friedman and M. Platt, "SCISRS-Sigma Center Information Storage and Retrieval System," BNL-883 (1964).
25. "CINDA (year), An Index to the Literature on Microscopic Nuclear Data," ENEA; U.S. requests may be directed to Division of Technical Information Extension, Oak Ridge, Tenn.
26. J. H. E. Mattauch, W. Thiele, and A. H. Wapstra, *Nucl. Phys.*, 67, 1 (1965).

27. G. A. Bartholonew, *Nucl. Data A*, 3, 367 (1967).
28. V. J. Orphan, N. C. Rasmussen, and T. L. Harper, "Line and Continuum Gamma-Ray Yields from Thermal Neutron Capture in 75 Elements," Report DASA-2570 (GA-10248), 1970.
29. D. Garber, "ENDF/B Summary Documentation," ENDF-201, 2nd ed., National Neutron Cross Section Center, BNL (1975).
30. J. J. Duderstadt and L. J. Hamilton, *Nuclear Reactor Analysis*, Wiley, New York, 1976, pp. 26-32, 48-52.
31. J. M. Blatt and V. F. Weisskopf, *Theoretical Nuclear Physics*, Wiley, New York, 1952, Chapters 13 and 9.
32. T. D. Beynon and I. S. Grant, *Nucl. Sci. Eng.*, 17, 547 (1963); results are quoted in Odette and Doiron, *Nucl. Technology*, 29, 346 (1976).
33. Duderstadt and Hamilton, *Nuclear Reactor Analysis*, Chapter 9.
34. J. Adir and K. D. Lathrop, "Theory of Methods Used in the GGC-4 Multigroup Cross Section Code," Report GA-9021, General Atomic (1968). Includes the updated GAM-II code and a thermal spectrum code.
35. Duderstadt and Hamilton, *Nuclear Reactor Analysis*, Chapter 8 and pp. 427-435.
36. G. P. Lahti, "Resonance Neutron Capture in a Thick Slab of Depleted Uranium," NASA TM X-52602 (1969).
37. H. M. Antúnez and J. M. Neill, *Nucl. Sci. Eng.*, 33, 238 (1968).
38. I. I. Bondarenko, "Group Constants for Nuclear Reactor Calculations," Consultants Bureau Enterprises, New York, 1964.
39. P. D. Soran and D. J. Dudziak, "Application of Bondarenko Formalism to Fusion Reactors," Nuclear Cross Sections and Technology, NBS Special Publication 425, Vol. 2, pp. 722-728 (1975).
40. R. Q. Wright, "SUPERTOG: A Program to Generate Fine Group Constants and P_n Scattering Matrices from ENDF/B," ORNL-TM-2679, Oak Ridge National Laboratories, Tenn. (1969).

PROBLEMS

1. The current density is the number of particles per second crossing unit area (1 cm^2), fixed in space.

Derive the relationship between current density and flux density. What is the current density if the differential flux density is isotropic and the scalar flux density is 10^8 particles cm^{-2} s^{-1}?

2. If the total microscopic cross section for neutrons is 4.2 b for hydrogen and 1.5 b for oxygen, what is the total microscopic cross section for water, density 1 g cm^{-3}?

3. A beam of 10-MeV protons is perpendicularly incident on a slab of lead 0.03 mm thick. What is the energy of the protons transmitted? How thick should the slab be to stop all of the protons?

4. Plot dN/dx, N, E, and dE/dx as a function of x for 10 MeV protons and for 10-MeV α particles in water, where N is the number and E the energy.

5. If it requires 25 eV to produce one ionization in water, how many ionizations are produced per micrometer of path length for 1-MeV electrons? For 10-keV electrons? Suppose a cell is 10 μm in diameter. How does the mean spacing between ionizations compare to the cell dimensions?

6. How thick should a slab of aluminum be to stop β particles of ^{32}P? Would aluminum be a good shield for 1-mCi ^{32}P source? Could you suggest a better shielding material?

7. a. Derive Eq. (3.32) for the energy of the Compton scattered photon.

   b. What is the electron energy and scattered photon energy for 0°, 90°, and 180° Compton scattering of 0.1-MeV and 1.0-MeV photons?

8. Plot the electron and photon spectra for: (a) 2-MeV photons on thin lead slab, (b) 0.662-MeV photons on thin lead slab, and (c) 1-MeV photons on thin water slab.

9. a. Calculate laboratory neutron energy for excitation of the 7.66-MeV level in ^{12}C by inelastic scattering.

b. Plot the spectrum of elastically scattered neutrons in ^{12}C, assuming isotropic scattering in the center of mass.

c. What is the energy of a 6-MeV neutron after inelastically scattering through 90° in ^{12}C?

10. What is the energy of the γ rays from capture of 10-keV neutrons in ^{12}C?

11. What is the flux density of uncollided neutrons transmitted through 50 cm of CH_2 if the incident flux density is 10^8 n/cm^2s and the total mean free path is 5 cm?

4

TRANSPORT

Transport of neutrons and photons is characterized by travel along straight paths (without energy loss) between widely spaced collision points. On collision, the particle may be absorbed or else scattered into a new direction and energy. The Monte Carlo or stochastic method simulates this behavior, and then obtains the flux density by averaging the scores from thousands of individual particle tracks.

A different method is to treat the particle motion as though the neutrons or photons were a gas within the shielding material. The flow of particles into and out of a cell, and the collisions within the cell, are incorporated in the conservation equation known as the Boltzmann transport equation. Various numerical and analytical techniques have been devised to solve the transport equation, such as discrete ordinates, spherical harmonics, and moments. They differ mainly in their handling of the angular flux density. The diffusion theory is an approximation where only the scalar flux density and the current are used. In general, diffusion theory has not been successful in radiation transport problems, where the angular flux density is quite anisotropic, and only the unusual particle penetrates to great depths in a shield.

Another approach to particle transport calculations is the kernel method. The method treats the exponential attenuation of the uncollided neutrons or photons correctly but approximates the behavior of the scattered particles. Other approximate methods include removal-diffusion theory for neutrons and schemes for parameterizing the shape of dose-attenuation curves obtained from experiment or exact calculations.

4.1 MONTE CARLO METHOD

The flow of a Monte Carlo calculation is indicated in Figure 4.1. Usually some 10,000 to 100,000 source particles are generated and their tracks or histories computed. The source routine assigns the initial coordinates of each particle (spatial coordinates x,y,z; direction cosines u,v,w; energy E; time t if time dependence is included). The coordinates are selected from the specified spatial, direction, energy, and time distributions of the source, as discussed later.

The next step is to select the "path length" or distance to the first (or next) collision, in units of mean free paths. The selection is made with the aid of a random number, as explained later.

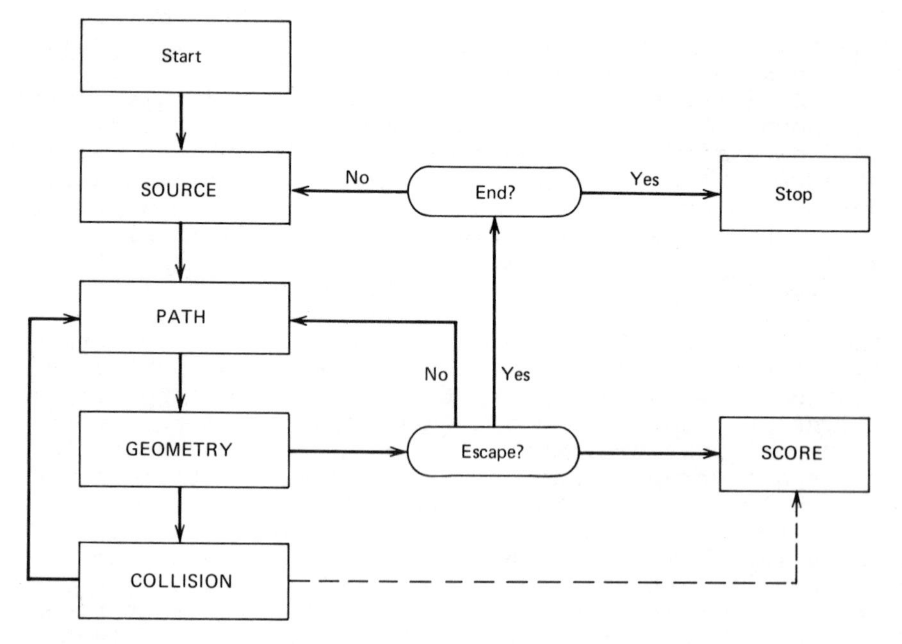

Fig. 4.1. Simplified flow diagram for a Monte Carlo calculation.

The geometry of the source, material zones, and detectors or boundaries must be specified in the input in terms the computer can use. Cross sections must be input for all materials. The geometry-tracking routine computes the spatial coordinates of the collision point, using the mean free path obtained from the cross sections. If a boundary between two materials is crossed, the coordinates are computed at the crossing point. Boundary crossings are usually scored or tallied as part of the output of the computation. If it is an outer boundary, the particle escapes. The program determines if all source particles have been processed; if not, another source particle is initiated.

The particle eventually makes a collision. The collision routine decides whether the particle is scattered or absorbed, and in some programs chooses the nuclide scattering and the type of reaction (e.g., elastic or inelastic). If scattering is selected, the collision routine selects the angle of scattering with the aid of random numbers, computes the scattered energy, and transforms the direction to the coordinate system fixed in the laboratory system. The program then returns to the path-length routine, and the history is continued. In some Monte Carlo programs, "absorption" means termination of the history and a new source particle is started. However, the statistical error or variance in scoring of collisions is reduced by the technique of nonabsorption weighting. A weight, generally 1.0, is assigned to the source particle. Then at each collision the weight W is multiplied by the nonabsorption probability

$$p_{na} = \frac{\Sigma_t - \Sigma_a}{\Sigma_t} \tag{4.1}$$

where Σ_t is the macroscopic total cross section at the incident energy and Σ_a is the macroscopic absorption cross section. The particle is assumed to scatter at every collision but emerges with the weight

$$W' = p_{na}W \tag{4.2}$$

and it is weight, not number of particles, that is scored. The weight is also used in other techniques for reducing variance.

The score in Monte Carlo may be a flux density, collision density, or a response such as dose. The result is obtained by averaging over the scores of many particles, as explained in more detail later. The result cannot be obtained exactly but is subject to a statistical error. In effect, we are performing a computer experiment with a relatively small number of particles. Methods of reducing the statistical error are important.

Random-Number Generator

Selection of the path length, angle of scattering, and many other parameters is made with the aid of a random number. One should be familiar with the way random numbers are generated. One method is to obtain pulses or counts from a truly random source, such as electrical noise or a radioactive substance. However, it is more convenient to calculate them as required. Since the calculation is deterministic, such numbers are often called "pseudorandom". The important point is that a sequence of pseudorandom numbers approximates the properties of true random numbers. Generally we want a floating-point number between 0 and 1, because other distributions can be derived from this one. We want the frequency of values to be equally distributed over the unit interval, and the numbers to be independent of each other (i.e., there should be an equal probability of any number being generated each time the generator is called). Actually, pseudorandom numbers generated on a computer of fixed word length or number of bits cannot be truly independent, but one can avoid patterns such as a short repeating sequence or correlations between terms that lie near each other in the sequence (pairs, triplets, etc.).

Coveyou and MacPherson [1] examined a number of pseudorandom number generators. They recommended the multiplicative congruential generator, where

$$r_i = br_{i-1} \quad \text{(modulo M)} \tag{4.3}$$

where r_i is the i-th random number, r_{i-1} is the previously generated random number (or an initial number), b is the multiplier, and modulo M means the number is the least significant part that fits in the computer of word length $M = 2^k$, where k is the number of bits. The IBM 360 series machines have 32 bits or

31 bits if one is reserved for the sign; hence $M = 2^{31}$. Other computers have $M = 2^{35}$ or higher. It is better to use double precision with the 32-bit machines. The r_i is treated as a binary fraction and is converted to a floating-point fraction for subsequent calculations.

It is important to choose the multiplier b (also termed λ) with great care. The multiplier should not be too small, or have a small number of 1 bits in its binary representation, or be close to a simple rational multiple of M or $\sqrt{M}$. A well-tested and satisfactory multiplier is $b = 5^{15}$. The initial random number should be odd.

Sampling a Distribution

The way in which a random number between 0 and 1 is used to select an event from a discrete probability distribution is explained next. Consider the selection of whether a photon is supposed to interact by the photoelectric effect, Compton effect, or pair production in a particular collision. Let the probabilities be defined by

$$p_1 = \frac{\sigma_{pe}}{\sigma_t} \tag{4.4a}$$

$$p_2 = \frac{\sigma_c}{\sigma_t} \tag{4.4b}$$

$$p_3 = \frac{\sigma_{pp}}{\sigma_t} \tag{4.4c}$$

where $\sigma_t = (\sigma_{pe} + \sigma_c + \sigma_{pp})$.

The discrete probabilities may be represented as shown in Figure 4.2. We also define the cumulative probability distribution

$$P_i = \sum_{i=1}^{j} p_i \tag{4.5}$$

which gives the probability that i is less than j. In the example, $(p_1 + p_2 + p_3) = 1.0$ or in general

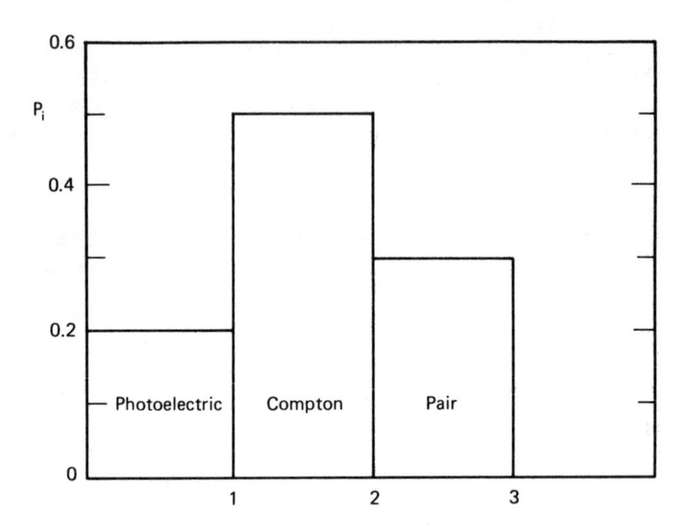

Fig. 4.2. Discrete-probability density function.

$$\sum_{i=1}^{n} p_i = 1.0 \tag{4.6}$$

The cumulative probability distribution may be diagrammed as in Figure 4.3. Now is we generate a random number r ($0 \leq r < 1$) and set

$$r = P \tag{4.7}$$

then P and the type of interaction are determined. In Figure 4.3, with r = 0.35, the interaction is Compton. The interaction is photoelectric if $0 \leq r < 0.2$, Compton is $0.2 \leq r < 0.7$, or pair production is $0.7 \leq r \leq 1.0$. Because r is uniformly distributed on the unit interval, each type of interaction is sampled with probability p_1, p_2, or p_3. In a large number of samples the proper number of interactions would be selected. In the example, Compton scattering would be selected in 50% of the collisions. In a small number of samples, statistical fluctuations could be significant, and Compton scattering might be selected in less than or more than 50% of the collisions. The statistical nature of sampling is a fundamental feature of the Monte Carlo method.

Similarly, the continuous probability density p(x) may be defined such that it is the probability of x' being between x and (x + dx). The cumulative probability distribution

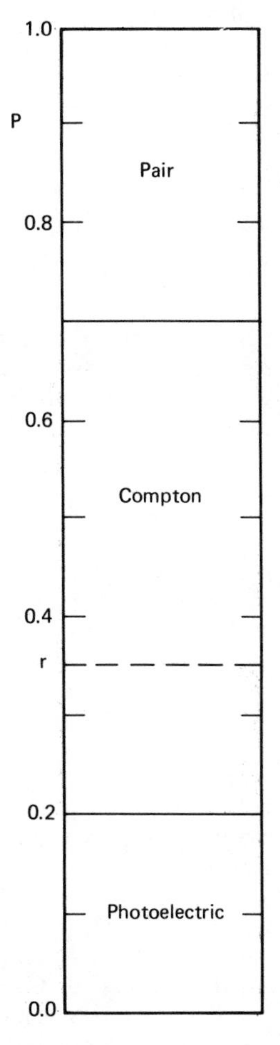

Fig. 4.3. Cumulative-probability distribution and selection of interaction with random number r.

$$P(x) = \int_0^{x'} p(x)\ dx \qquad (4.8)$$

corresponds to the probability that x is less than x'. For example, p(E) plotted in Figure 4.4 for a fission neutron spectrum gives the fraction per MeV that lie between E and (E + dE). The cumulative distribution P(E) gives the fraction that have energies below E. By setting a random number equal to P(E) and solving for E one can select the energy to be assigned to a fission neutron. At r = 0.55 the energy is 1.8 MeV. In the limit of a large number of selections, the spectrum generated will follow the function p(E). One can tabulate a plot of P(E) against E at many energy points and then interpolate to find the E corresponding to r between tabulated values. If P(x) is a simple function, it may be possible to solve for x directly, as illustrated later in the path-selection routine.

Another method of selection is the rejection technique. Consider the problem of selecting from the function f(x) in Figure 4.5 [a semicircle bounded by x = a and x = b and with a maximum f(x) = A]. Generate a random number r_1 and select a trial x_1

$$x_1 = a + r_1(b - a) \qquad (4.9)$$

Note that this distributes x_1 uniformly over the interval (a,b). Generate another random number r_2 and obtain a value between 0 and A by

$$y_1 = r_2 A \qquad (4.10)$$

Now if $y_1 > f(x_1)$, the trial x_1 is rejected and another random number generated. When $y_1 \leq f(x_1)$, the value of x and thus f(x) are accepted. We can see that the pairs (r_1, r_2) distribute points uniformly over (x = a) to (x = b) and (y = 0) to (y = a). The efficiency of acceptance is equal to the ratio of the area under f(x) to the total area A(b - a) or $\pi/4$ in the example. This efficiency is high enough to make the rejection technique worth considering in the example. The technique is useful when the efficiency is high and it is relatively easy to calculate f(x) [or p(x) in general] but

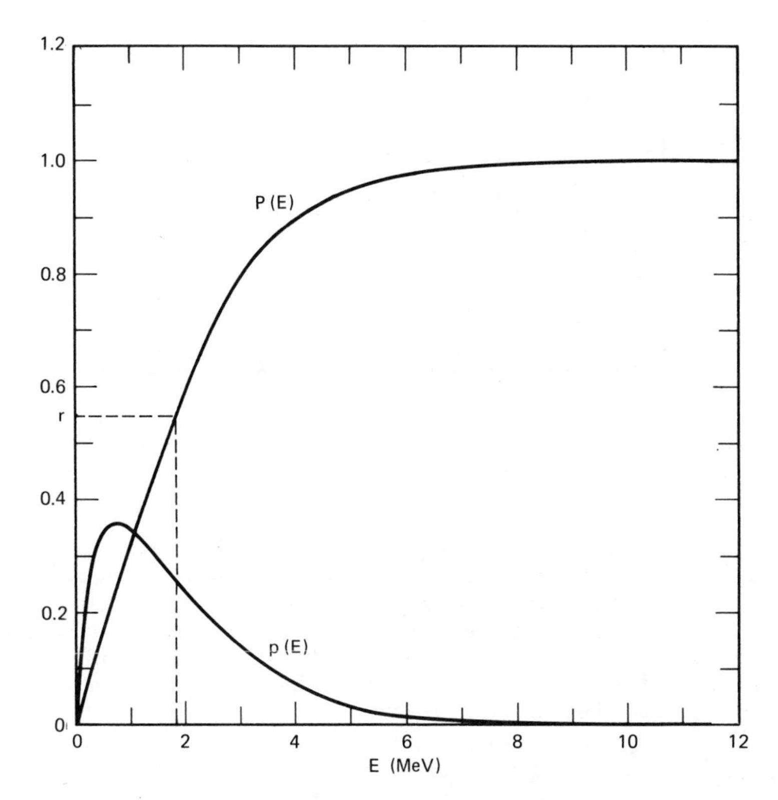

Fig. 4.4. Probability density function p(E) and cumulative-probability distribution P(E) for fission spectrum.

where solution of P(x) = r is difficult. Sampling methods are discussed in the literature [2-4].

Geometry Specification

The first step in specifying the source, material (e.g., shield), and detector geometry is to establish a coordinate system fixed in the laboratory frame. Although special symmetries may exist, it is convenient to use Cartesian coordinates in Monte Carlo because (1) the

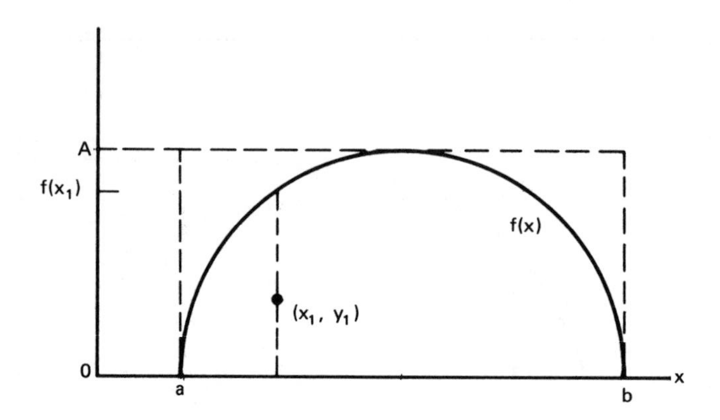

Fig. 4.5. Illustration of rejection technique for sampling function f(x).

path between collisions is a straight line and (2) Monte Carlo is usually used for two-dimensional and three-dimensional calculations. Boundaries between material regions, source regions, and detector regions are specified by mathematical surfaces. The most general surface description available in most Monte Carlo programs is the quadric

$$Ax^2 + By^2 + Cz^2 + Dxy + Eyz + Fzx + Gx + Hy + Jz + K = 0 \tag{4.11}$$

The constants A,B,C,D,E,F,G,H,J,K are chosen to describe planes, spheres, cylinders, cones, or ellipsoids of arbitrary size, position, and orientation. In older codes the programmer was responsible for specifying the constants in the input. Modern codes simplify the task of preparing input (and avoiding mistakes) by allowing certain simple forms such as planes, cylinders, and spheres to be specified by points defining the plane, or axis of the cylinder, or center of the sphere, and the radius of the cylinder or sphere. For example, the cylindrical shield in Figure 4.6 could be specified by two planes, z = 0.0 and z = 10.0, the cylindrical surface with axis passing through (0,0,0) and (0,0,12.0)

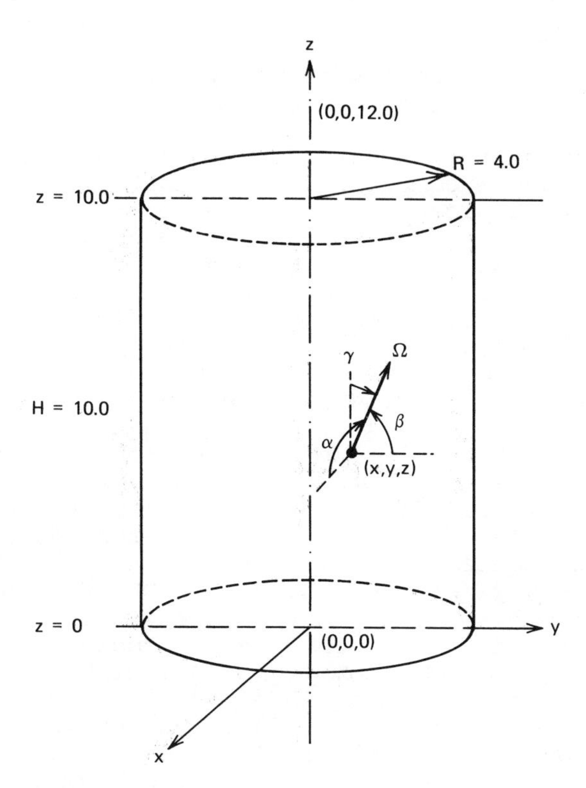

Fig. 4.6. Description of finite-cylinder shield by bounding planes parallel to x-y plane and cylindrical surface whose axis coincides with z-axis. A particle with spatial coordinates (x,y,z) and direction Ω specified by direction cosines $u = \cos \alpha$, $v = \cos \overline{\beta}$, and $w = \cos \gamma$ is shown.

and radius 4.0, and a convention for describing whether the shield region is on the positive or negative side of the surface. For example, the shield region is (+) with respect to the lower bounding plane, (-) with respect to the upper bounding plane, and (-) with respect to the cylindrical surface. The computer solves for the coefficients in the quadric, $z = 0.0$ for the lower plane, $z - 10.0 = 0$ for the upper plane, and $x^2 + y^2 - 16.0 = 0$ for the cylinder. The expression $x^2 + y^2 - 16.0$ is negative for points inside the cylindrical surface, zero for points on the surface, and positive for points outside the surface, hence the sign convention.

Describing a complex geometry accurately and ensuring that all regions are defined with respect to the bounding surfaces is one of the most arduous tasks in preparing input for a Monte Carlo problem. It is helpful to have a computer program that reads the geometry input and plots cross section or perspective views so that any errors may be discovered.

In the version called "combinatorial geometry," every space point is assigned to a zone described by the union or intersection of simple bodies such as spheres, cylinders, and boxes. These bodies are defined as discussed previously. Then a zone may be specified as being outside a given body, inside a given body, or described by the OR operator. For example, in Figure 4.7 zone J is described as inside body 1 (a sphere) and outside body 2 (a finite solid cylinder), corresponding to a sphere with a reentrant cylindrical hole. Zone I, on the other hand, is described as containing all points either inside body 1 or inside body 2 (or both, since the OR operator is inclusive). This describes a sphere with a cylindrical peg sticking out of it. These zones could be described without combinatorial geometry by defining planes that limit the length of the cylinder and by specifying all regions between planes, cylinder, and sphere; however, combinatorial geometry simplifies the input.

Source

It is usually possible to describe the source by separation of variables,

$$S(x,y,z,u,v,w,E,t) = W\, S(x,y,z)\, q(u,v,w)\, p(E)\, f(t) \tag{4.12}$$

Results are normalized to the total number of source particles generated; hence the spatial distribution function need only give the relative number of source particles per cm^3 at x,y,z or the number of source particles emitted from a point. The selection is often made from a uniform distribution within a certain defined volume. If the angular distribution is isotropic in the laboratory system, as is normally the

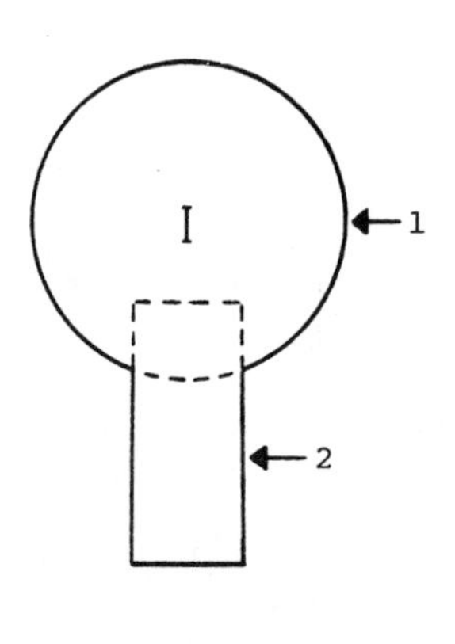

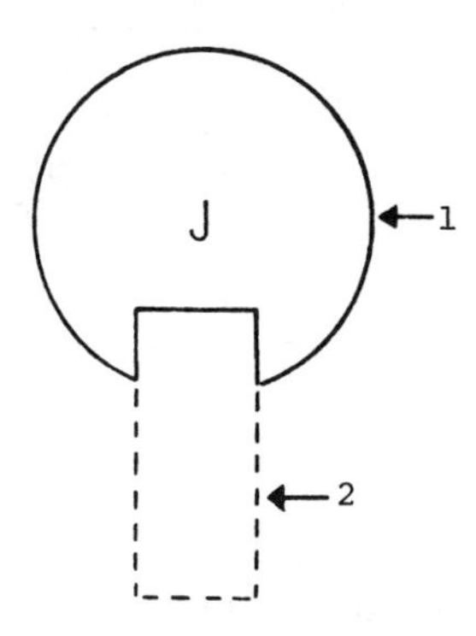

Fig. 4.7. Examples of zones that can be described by combinations of a sphere and a finite cylinder.

case, the cosines with respect to the x,y,z, axis (u,v,w, respectively) may be chosen by selecting a point on the unit sphere $u^2 + v^2 + w^2 = 1$.

Referring to Figure 4.8, we can select an angle θ by

$$w = \cos\theta = 2r_1 - 1 \qquad (4.13a)$$

and projecting the unit direction vector on the u - v plane the projected radius

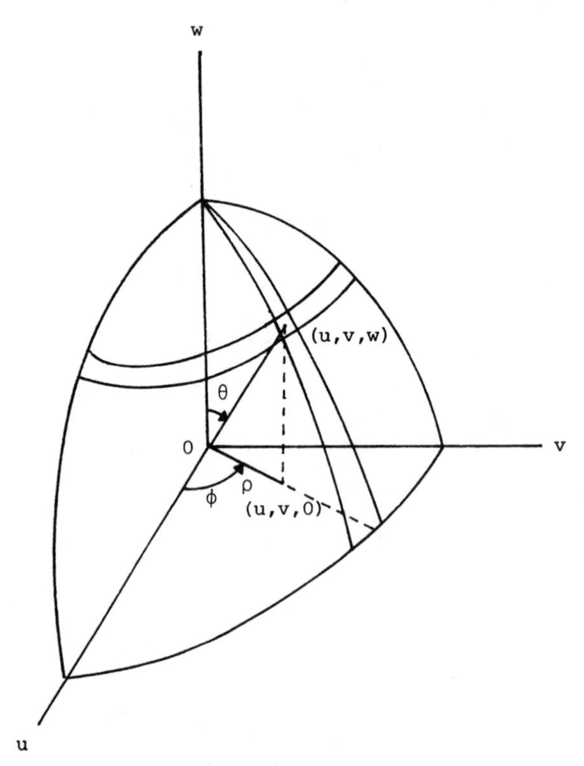

Fig. 4.8. Sampling an isotropic distribution by means of points uniformly distributed on unit sphere $(u^2 + v^2 + w^2) = 1$.

$$\rho = \sqrt{u^2 + v^2} = \sqrt{1 - w^2} \tag{4.13b}$$

The azimuthal angle ϕ is selected uniformly by

$$\phi = \pi(2r_2 - 1) \tag{4.13c}$$

Then direction cosine w is given by Eq. (4.13a) and

$$u = \rho \cos \phi \tag{4.13d}$$

$$v = \rho \sin \phi \tag{4.13e}$$

If the angular distribution is not isotropic, an appropriate reference direction has to be specified and a routine written to select the angles or direction cosines from the appropriate cumulative probability distributions.

In analog or straight Monte Carlo, the initial weight W = 1.0, and the probability density function p(E) is sampled to obtain the energy of the source particle. However, sampling from the fission spectrum such as that plotted in Figure 4.4 can be inefficient in common shielding problems. Fast-neutron cross sections, especially for hydrogen, generally decrease with increasing energy. High-energy neutrons penetrate farther and may contribute most of the dose at deep penetrations, even though their percentage abundance in the source spectrum is quite small. The program should be spending most of the available computer time in tracking "important" neutrons that will make the major contribution to the desired detector response, namely, the dose at a large distance from the source. This can be done through a use of "importance sampling," where sampling is done from a modified probability density function p*(E) designed to enhance selection of important neutrons. The weight is modified to W*(E) such that the mean is unchanged, requiring

$$W^*(E)p^*(E) = W(E)p(E) \tag{4.14}$$

For example, one might sample from a flat distribution such that neutrons are generated uniformly in the interval, say, 0.05 to 12 MeV. Then Eq. (4.14) would be used to modify the weight assigned to each source particle so that the total weight of all source particles is unchanged. The effect of the importance sampling is to generate many more neutrons, each of lower weight, in the high-energy region, which should reduce the variance while leaving the mean score unchanged. One should be careful not to "bias" the distribution too strongly or eliminate all low-energy neutrons, because the results close to the source may be very inaccurate.

Shielding problems seldom involve time dependence. The result can be interpreted in terms of flux density per unit source strength or fluence per source particle. However, it is possible to keep track of the time of flight between collisions and sum up for the chronological "age" of the particle. The time-dependence is of interest in pulsed-source problems.

Path Length

The path length, or distance to the collision point, is measured in units of the mean free path, $\lambda = \Sigma^{-1}$. The "free paths" are exponentially distributed. The probability of a particle traveling a distance L before collision is

$$p(L) = e^{-L} \tag{4.15a}$$

and the physical distance in a homogeneous medium is

$$s = \lambda L \tag{4.15b}$$

The probability of the particle having an interaction at any distance less than L is

$$P(L) = 1 - e^{-L} \tag{4.16}$$

If we set a random number $r = P(L)$ and solve for L, we obtain

$$L = -\ell n(1 - r) = -\ell n\ r \tag{4.17}$$

because r is distributed in the same way as $(1 - r)$.

The sampling of path length may be biased in such a way as to favor transport toward a detector or along a certain direction. The biasing is removed from the mean score by appropriate adjustment of the weight, but the variance should be reduced if the biasing is done properly. In the exponential transform [5] the macroscopic total cross section is modified.

$$\Sigma^* = \Sigma - f(\underline{r},\underline{\Omega},E) \tag{4.18}$$

and L is chosen from Eq. (4.17) with $S = L\lambda^* = (L/\Sigma^*)$. The modified weight is

$$W^* = W \frac{\Sigma e^{-\Sigma s}}{\Sigma^* e^{-\Sigma^* s}} = W \frac{\Sigma}{\Sigma^*} \exp\left[-f(\underline{r},\underline{\Omega},E)\ s\right] \tag{4.19}$$

Various forms of the $f(\underline{r},\underline{\Omega},E)$ function of position $\underline{r}$, direction $\underline{\Omega}$, and energy E have been suggested. Setting

$$f(\underline{r},\underline{\Omega},E) = \underline{\varepsilon} \cdot \underline{\Omega}B \tag{4.20}$$

where $\underline{\varepsilon}$ is a unit vector along the desired direction of propagation and B is the constant determining the strength of the biasing, encourages transport along the preferred direction (as long as Σ^* is positive). The transform

$$f = B\left[(x - x_0)^2 + (y - y_0)^2 + (z - z_0)^2\right]^{1/2} \quad (4.21)$$

encourages transport toward the point x_0, y_0, z_0.

When f is large it is possible for the effective macroscopic cross section Σ^* to become negative. This upsets the scoring and defeats the purpose of stretching path lengths aimed toward the preferred direction, while compressing path lengths in the opposite sense. If $f(\underline{r}, \underline{\Omega}, E)$ is selected for the minimum cross section $\Sigma(E)$ in the medium at position $\underline{r}$, it should be possible to avoid a negative Σ^*. Otherwise, it is desirable to test if $\Sigma^* \leq 0$ and if so, to set it to some arbitrary positive value.

Geometry Tracking

The geometry-tracking routine calculates the coordinates of the collision or intersection with a boundary. In a homogeneous medium the new coordinates are

$$x' = x + us \quad (4.22a)$$

$$y' = y + vs \quad (4.22b)$$

$$z' = z + ws \quad (4.22c)$$

where s is the physical path length. The computer program tests to determine if s is less than d, the distance to any boundary in the direction of travel. If it is, the particle coordinates are advanced to (x',y',z'), and the collision routine is called to determine the parameters after collision. If $s > d$, the particle is assigned the spatial coordinates at the intersection with the boundary and d is used for the track length within the original zone. One could subtract d/λ from L and continue along the original direction until the path length is "used up." Another procedure recognizes that neutral particle transport is a Markov process; that is, it does not depend on the history or how a particle arrived at a certain point with a certain direction, energy, age, and weight. Thus one can start at the intersection with the boundary and compute another path length with the aid of a random number, obtain another distance s for the mean free path in the material of the zone being entered, and so forth. If the crossing occurs at an outer boundary, the particle escapes and another source particle is generated.

Let us consider the finite cylinder shown in Figure 4.6. The flow diagram for a geometry-tracking routine is given in Figure 4.9. A provisional distance to the next collision is calculated. Tests are made to determine whether the trial collision point lies inside or outside of the cylindrical surface and inside or outside of the parallel planes. If the trial point lies within the finite cylinder, the distance to (x',y',z') is s and the program exits to the collision routine. If the trial point lies outside of the cylinder, it is necessary to test if the distance d to the plane boundary is smaller or larger than the distance to the extended cylindrical surface, because the particle could cross the plane boundary first and then intersect the cylindrical surface, or vice versa. The shorter distance decides which surface is crossed first, and the (x',y',z') coordinates are calculated at the corresponding intersection.

Cashwell and Everett [6] discuss a number of geometry-tracking routines. The calculation is simplified if one wants to determine if a particle escapes or not, but not where.

Collision

The collision routine handles absorption and scattering. It computes the scattered energy and the direction cosines in the coordinate system fixed in the source-shield configuration. One of two approaches is used in a Monte Carlo program. In the "microscopic" approach, the nuclide (or atomic species) collided with, and the type of reaction, are selected from discrete probability distributions with the aid of random numbers. The mass A and Q value are obtained as well as the cross sections for that nuclide. The angle of scattering in the CMS is selected from a cumulative distribution function, using another random number. The scattered energy in the laboratory system, and the cosine of the polar angle of scattering, μ_L, are calculated from the kinematics as discussed in

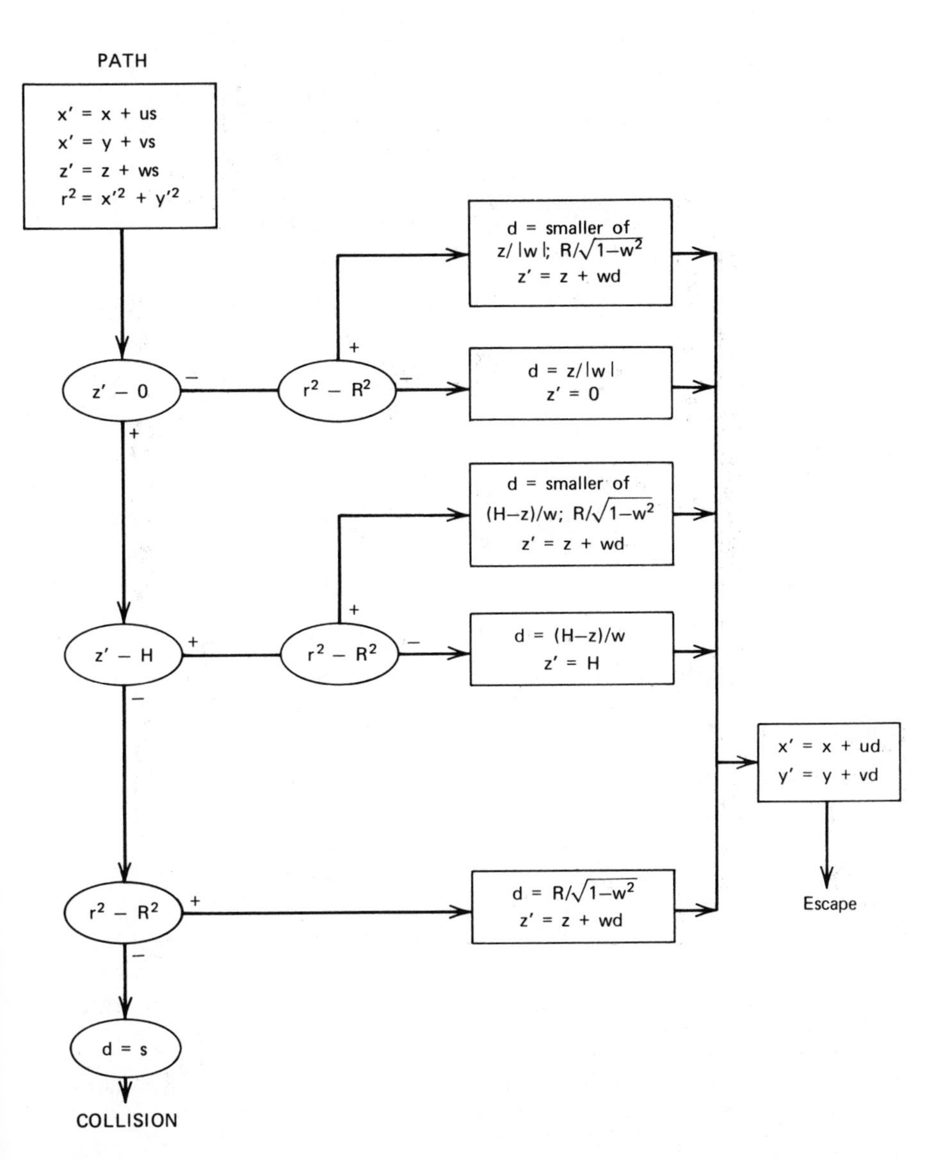

Fig. 4.9. Flow diagram for calculating collision or boundary-crossing coordinates (x',y',z') in a finite cylinder.

Chapter 3 (or from the Klein-Nishina distribution for Compton scattering of photons). The azimuthal angle of scattering ϕ is chosen from an isotropic distribution.

Next it is necessary to transform from the original direction cosines in the fixed system (u,v,w) to the direction cosines (u',v',w') of the scattered particle in the fixed system, by appropriate rotation of the coordinate frames. The equations are [2]

$$u' = u\mu_L + \frac{a}{b}(uw \cos \phi - v \sin \phi) \tag{4.23a}$$

$$v' = v\mu_L + \frac{a}{b}(vw \cos \phi + u \sin \phi) \tag{4.23b}$$

$$w' = w\mu_L - ab \cos \phi \tag{4.23c}$$

where

$$a = \sqrt{1 - \mu_L^2} \tag{4.23d}$$

$$b = \sqrt{1 - w^2} \tag{4.23e}$$

If the particle is traveling along the direction of the negative z-axis, or nearly so, b becomes indeterminate. Hence if $b < 0.0001$, say, use

$$u' = a \cos \phi \tag{4.24a}$$

$$v' = a \sin \phi \tag{4.24b}$$

$$w' = \mu_L \frac{w}{|w|} \tag{4.24c}$$

In the "macroscopic" or multigroup approach, energy-transfer cross sections for all reactions and all nuclides or atoms in a mixture are computed and averaged over the incident and scattered energy bands (groups) as discussed in Chapter 3. The angular distribution is handled by the Legendre polynomial expansion of $\mu_L = \mu$,

$$f(\mu) = \sum_{\ell=0}^{L} \frac{2\ell + 1}{2} f_\ell P_\ell(\mu) \tag{4.25}$$

For each Legendre component there is a table of group-to-group-transfer cross sections. In Monte Carlo, only one particle can be followed at a time, and it is necessary to select a single angle and energy at each collision. The MORSE [7] Monte Carlo program does this in a unique way. The Legendre coefficients for each group-to-group transfer are converted to $n = (L + 1)/2$ angles with discrete cosines μ_i and probability p_i for scattering through that polar angle in the laboratory system. Thus for P_3 cross sections, and 30 groups with downscattering only, there are two angles per transfer, or 60 all together. The cosines and probabilities are derived from a generalized Gaussian quadrature, with the angular distribution as a weight function, in such a way that the first L moments are preserved and the μ values are clustered where the angular distribution is peaked. The procedure for selecting the cosines and probabilities is discussed in Straker, Stevens, et al. [7].

For correlated energy and angle, as in Compton scattering of photons or neutron elastic scattering on a single nuclide, the cosines nearly coincide and the probabilities are equal. The polar angle of scattering is chosen from the discrete probabilities and the scattered energy group is obtained. The azimuthal angle is selected from an isotropic distribution, and the direction cosines in the fixed system are calculated as before.

The discrete angle representation in MORSE can give erroneous results when the source and detector regions are small and single scattering dominates the fluence at the detector, because the discrete angles may not agree with the angle of scattering into the detector. The problem is mitigated by using a higher-order expansion and hence more angles. With multiple scattering, the effect of discrete angles is blurred and is usually negligible. In other Monte Carlo codes the polar angle of scattering is selected from a cumulative probability distribution.

Scoring

Various methods are available [2,8] to score the fluence, current, or response derived from them by integrating over energy with a suitable weighting function (e.g., to calculate dose or reaction rate). The scoring algorithms or "estimators" include:

(1) collision, predicted-collision, and path-length estimators, (2) boundary-crossing estimators, and (3) point-detector or next-event estimator. The number of collisions in volume V at position $\underline{r}$, in energy band E to $(E + \Delta E)$,

$$F = \int_V \phi(\underline{r},E)\ \Sigma(E)\ d\underline{r} \tag{4.26}$$

where Σ is the total cross section. Thus if the weight of the i-th particle making a collision is W_i^{coll}, the fluence is

$$\phi(\underline{r},E) = \frac{F}{\Sigma V} = \underset{i}{S}\ \frac{W_i^{coll}}{\Sigma V} \tag{4.27}$$

This estimator is fast but gives poor results when there are few collisions because V is small or Σ is small. The predicted-collision estimator overcomes this objection by estimating the collisions from the number of particles entering the volume and the probability of collision, $[1 - \exp(-\Sigma d)]$, where d is the path length of the particle through the volume. Then for W_i^{enter} weight of a particle entering V,

$$\phi = \underset{i}{S}\ \frac{W_i^{enter}\ [1 - \exp(-\Sigma d)]}{\Sigma V} \tag{4.28}$$

This method takes more computation time but every particle entering the detector volume contributes to the score.

If Σd is small, the exponential may be approximated by $1 - \Sigma d$ and

$$\phi = \underset{i}{S}\ \frac{W_i^{enter}\ d}{V} \tag{4.29}$$

which is the path-length estimator. It saves taking the exponential. All volume estimators suffer from the fact that the particle must at least enter the "detector" volume, which may be small. The term detector is used in a general sense; it may be simply a geometrical region in which the average fluence is desired. Symmetry should be exploited and the volume made as large as permissible; the more differential the information (fine spatial resolution, small ΔE, etc.), the poorer the statistics in Monte Carlo.

The boundary-crossing estimator may be derived by considering a thin box on the boundary with base area A and small thickness T, so that particles crossing the narrow sides can be neglected. The path length is then approximately $T/\cos\theta$, where θ is the angle between the normal to the surface and the direction of the particle (this can also be derived by considering the definition of fluence). Then if W_i^{hit} is the weight of the i-th particle hitting A, the fluence

$$\phi = \mathcal{S}_i \frac{W_i^{hit}/\cos\theta}{A} \tag{4.30}$$

If $\theta \simeq 90°$, the estimator blows up and $\cos\theta$ should be set to a finite value. One can also set up angular bins and score the fluence of particles hitting A with direction in the bin. If $\cos\theta$ is eliminated and the normal to the surface is taken as the reference direction, the current

$$J = \mathcal{S}_i \frac{W_i^{hit}}{A} \tag{4.31}$$

In scoring fluence or current with a boundary-crossing estimator, the area should be as large as permissible and symmetry should be exploited. Fine angular bins again result in poor statistics.

With the point-detector estimator, all source particles and collisions contribute to the score. The fluence at the detector location $\underline{r}$ for a fixed or scattering source at $\underline{r}'$ is

$$\phi(\underline{r},E) = \frac{S(r',E)}{\rho^2} e^{-\rho/\lambda(E)} d\underline{r}' \tag{4.32}$$

where $\rho = |\underline{r} - \underline{r}'|$ is the distance between source (or collision) point and detector point and $\rho/\lambda(E)$ is the distance in mean free paths. For a scattering collision,

$$S(\underline{r}',E) = \psi(\underline{r},E',\underline{\Omega}') \frac{d\sigma}{d\Omega} (E' \rightarrow E,\underline{\Omega}' \rightarrow \underline{\Omega}) \quad (4.33)$$

In other words, at each collision the code multiplies the weight of the particle by the probability of scattering into the direction of the detector and by the probability of reaching the detector considering geometric and material attenuation. (A separate calculation is made for the actual energy and angle of the scattered particle.) The point-detector estimator takes much more computer time, especially in complex geometry or when there are several detectors, but it can reduce the variance significantly. However, there are a few pitfalls. Only collisions relatively near the detector will make a significant score; hence it is still important to get collisions near the detector and to avoid wasting computation time in following particles in regions where they will not contribute much to the score. Another problem is that a low-order Legendre expansion of the differential scattering cross section often gives negative scattering probabilities over certain angles. An analytical formulation (e.g., the Klein-Nishina formula for photon scattering) may be used, a nonnegativity constraint imposed, or a higher-order P_L approximation applied. The few collisions very near the detector can contribute a disproportionate score, which tends to increase variance. One way to fix this is to replace the point-detector estimator by a volume average,

$$\frac{\int_0^R e^{-\Sigma r}\, dr}{\int_0^R r^2\, dr} = \frac{1 - e^{-\Sigma r}}{\frac{1}{3}(R^3\ \Sigma)}$$

where R is the arbitrary radius of an imaginary sphere around the detector. Usually R is 1/8 to 1/2 mean free path.

The mean score is obtained from the arithmetic average,

$$\overline{F} = \frac{1}{N} \underset{i}{S}\ F_i \quad (4.34)$$

where N is the number of source particles (number of particle histories or tracks) and F_i is the detector response scored from the i-th particle. The mean, $\overline{F}$, may be the fluence per source particle, hence 1/N times the fluence calculated from one of the estimators discussed. It is also useful to calculate the sample variance from the histories,

$$\sigma^2 = \frac{1}{N}\left\{ S_i \; F_i^2 \right\} - \overline{F}^2 = \overline{F^2} - \overline{F}^2 \tag{4.35}$$

from which the estimated variance of the mean can be calculated,

$$\sigma_{\overline{F}}^2 = \frac{\sigma^2}{N} \tag{4.36}$$

It should be realized that the true mean m is only estimated from Eq. (4.34), and the estimate of the variance (or standard deviation, σ) is even more uncertain. Sometimes the estimate of the error is unrealistically low. However, the central limit theorem [2,4]

$$P\left\{ a \le \frac{\overline{F} - m}{\sigma/\sqrt{N}} \le b \right\} \to \frac{1}{(2\pi)^{1/2}} \int_a^b e^{-t^2/2}\, dt \tag{4.37}$$

states that the distribution of the sum of N independent, identically distributed random variables with finite means and variances, normalized to mean 0 and variance 1, approaches a normal distribution as N becomes large. The variance estimated from the sample statistics is then a valid measure of the error. To get some reasonable score while having a basis for the statistical estimation, it is standard procedure to divide the total number of particles into batches and compute the statistics from the results of the batches. For example, 10,000 source particles may be divided into 50 batches of 2000 particles each.

An example shows how difficult it is to achieve good statistical precision in a shielding problem. Consider a source and thick slab shield, and score $F_i = 1$ if a particle is transmitted and $F_i = 0$ if it is not [9]. Then $\overline{F^2} = \overline{F}$ and

$$\sigma^2_{\overline{F}} = \frac{1}{N}\left(\overline{F^2} - \overline{F}^2\right) = \frac{1}{N}\,\overline{F}(1 - \overline{F})$$

and the fractional standard deviation is

$$\frac{\sigma_{\overline{F}}}{\overline{F}} = \sqrt{\frac{1 - \overline{F}}{N\overline{F}}}$$

In the problems of interest $\overline{F} << 1$; hence

$$\frac{\sigma_{\overline{F}}}{\overline{F}} \simeq \frac{1}{\sqrt{N\overline{F}}} \qquad (4.38)$$

If the fractional standard deviation is to be 10% and the average transmission $\overline{F}$ is, say, 10^{-6}, N would have to be 10^8, far beyond the capacity of even the fastest computers. For deep penetration problems where only a small fraction of the particles reach the detector, straight analog Monte Carlo will not do, and variance-reduction methods are essential to keep the computation time within reasonable limits.

Variance Reduction

Since computation time goes as N but the fractional standard deviation goes as $1/\sqrt{N}$. it is inefficient to simply increase N even if the result could be obtained in a reasonable computation time. Techniques for reducing variance are designed to: (1) reduce the variance in events selected in a particle track or (2) make most efficient use of computation time by concentrating on particles in regions of phase space R (meaning coordinates x,y,z, energy E, and direction $\underline{\Omega}$) most likely to contribute significantly to the desired detector response. A technique may increase the computation time per history or track but improve σ^2/N, thus allowing fewer histories to be run.

The variance-reducing techniques commonly applied to radiation transport problems are: (1) analytical equivalence: (a) systematic (quota) sampling, (b) expected value, and (c) next-event estimator; (2) importance sampling: (a) source energy and direction biasing, (b) scattered energy and direction biasing, and (c) energy and age cutoff; (3) Russian roulette and splitting; and (4) exponential transformation.

Analytical values should be used in place of random variables where feasible. For example, the source energy, angle, and position distributions are completely determined. Systematic or quota sampling may be applied by dividing phase space into bins and generating a preassigned number of particles in each bin. If desired, random numbers could be used to select parameters within the bins, and importance sampling can be included as discussed later. Because selection of source parameters takes little time compared to geometry-tracking and collision-parameter generation in the typical shielding problem, systematic sampling of the source may not lead to a large improvement in the results, but it is worth considering. Even better would be to systematically generate the density of first collisions, but this may be difficult in complex geometry.

Nonabsorption weighting in a collision is an example of the use of an expected value, namely, the average fraction of collisions in which scattering rather than absorption occurs. This allows a particle to follow an average track in terms of absorption rather than terminating the track. Nonabsorption weighting can result in wasting computation time on particles of very low weight, which will not contribute much to the score. Therefore, Russian roulette should be applied also to eliminate those particles.

Next-event estimation (sometimes misleadingly termed "statistical estimation") uses analytical expressions in scoring, as in the point-detector and predicted-collision estimators. Next-event estimation may be the only practical way to get a score when the probability of entering a detector or colliding in it is very low. However, the exponential transformation should be used to force particles toward the detector, and importance sampling or Russian roulette and splitting (or both) should be applied to avoid wasting time on unimportant regions. The point-detector estimator significantly increases computation time per source particle since geometry tracking has to be done from each collision point to each detector. Usually only a few point detectors are used in a problem. Furthermore, care should be taken to avoid negative-scattering probabilities, as discussed earlier.

Importance sampling is illustrated by biasing the energy of the source particles to increase the frequency of sampling those particles that penetrate deeply into the shield, while altering their weights to remove the bias in the mean. The source-particle direction might be biased to favor particles initially headed toward the detector. The selection of the energy (and possibly the angle) of scattered particles might be biased to favor energies likely to contribute most to the response (e.g., dose) in a detector. All importance sampling techniques require either *a priori* knowledge of the "important" regions of phase space, or else the relative importance has to be estimated during the course of the calculations. Then a means must be implemented in the program to modify the transport process according to the importance. It should be realized that biasing the transport to favor a particular response in a particular detector may very well worsen the statistics of another response or in a detector at another position. However, if the biasing is not too strong, one can live with this if the variance in the dose, say, of the most distant or best-shielded detector is reduced. The statistics in detectors close to the source will be good anyway, and some increase in variance is acceptable.

The integral form of the transport equation [2] is

$$F(R) = \int K(R;R')\, F(R')\, dR' + S_1(R) \qquad (4.39)$$

where $F(R)$ is the density of particles entering collisions at phase-space point R, $S_1(R)$ is the density of first collisions from the fixed (external) source, and $K(R;R')$ is the kernel giving the next-flight collision density at R due to a collision at R'. Generally one wants not the collision density, but a functional

$$D(\underline{r}) = \int F(R)\, h(R)\, dR \qquad (4.40)$$

where $h(R)$ is the detector response function, giving the contribution to the detector response from a collision at R. If the fluence is desired, $h(R)$ is $1/\Sigma V$. For the dose, the fluence would be weighted by the mass energy-absorption coefficient, for example. Suppose now that we have an importance function $I(R)$. Then the transport can be modified such that K and S_1 are replaced by

$$K^*(R;R') = \frac{K(R;R')\ I(R)}{I(R')} \tag{4.41}$$

$$S_1^*(R) = \frac{S_1(R)\ I(R)}{S_1(R')\ I(R')\ dR'} \tag{4.42}$$

In the unbiased calculation the weight of each particle making a collision (from the source or scattering) is multiplied by h(R) and added to the score D. In the biased calculation the weight is modified by the ratio $K(R;R')/K^*(R;R')$ or $S_1(R)/S_1^*(R)$. The variance is reduced if I(R) peaks in those positions, energies, and possibly angles that contribute most heavily to the score D. If it does not, it is possible that the variance could be increased. It has been shown [4,10] that zero variance is achieved when $I(R) = F^+(R)$, the solution to the adjoint equation

$$F^+(R) = \int K(R';R)\ F^+(R')\ dR' + h(R) \tag{4.43}$$

The adjoint transport equation uses the inverse of the transport kernel K and replaces the source by the detector response function h(R). The adjoint collision density $F^+(R)$ gives the distribution of particle collisions in space, energy, and angle that contribute unit response D_1. Since each track using $F^+(R)$ as the importance function gives the identical contribution D_1, the total response would be just $D = ND_1$, and following one track would be enough! Of course, obtaining $F^+(R)$ is often just as difficult as obtaining F(R). However, one suspects and this has been borne out in practice, that variance is significantly reduced (perhaps by orders of magnitude) if $I(R) \simeq F^+(R)$, that is, if the importance function approximates the adjoint collision density even if it is not reproduced exactly. For example, one might use the adjoint solution from a one-dimensional approximation of the real problem and calculate the adjoint with a discrete-ordinates code, or simply use an estimate based on previous Monte Carlo calculations.

A drastic type of importance sampling is to cut off the tracking of the particle when its energy has fallen so low that it is unlikely to ever reach the detector or would not contribute significantly to a response such as dose. The program may be spending an inordinate amount of time in tracking low-energy particles, especially low-energy neutrons in moderators, where many collisions are made. A similar cutoff is to terminate unusual particles that have a large "age" because of many collisions or long paths between collisions.

Russian roulette and splitting are special forms of importance sampling. The shield is divided into spatial regions, which do not have to coincide with material zones. Generally the boundaries are chosen to be about one mean free path apart for the source spectrum or in such a way that the collision density is maintained approximately constant throughout the shield when Russian roulette and splitting are applied. Each region is assigned an importance I_n. For example, in a slab shield the importance would increase with increasing distance from the source and decreasing distance to the detector outside the shield. To reduce variance, we want to increase the number of tracks followed in important regions and decrease the number followed in unimportant regions (or at uninteresting energies). One way to do this is to split a particle into two particles, each of half the former weight when a region of higher importance is entered from a region of lower importance. One particle is followed while the parameters of the "latent" particle are stored and its track is followed later. Conversely, if a particle enters a low-importance region from a higher-importance region, the game of Russian roulette is played. The particle survives with probability $k < 1$ and is assigned a new weight kW. Another way to implement Russian roulette and splitting is to specify lower and upper weight standards $w_{\ell o}$ and w_{up}, in each energy group and spatial region. Particles entering the region with a weight $W > w_{up}$ for that group are split into two particles, each with half the original weight. If $w_{\ell o} \leq W \leq w_{up}$ nothing is done. If $W < w_{\ell o}$, Russian roulette is applied, with a survival probability W/w_{av}, and is assigned a weight $w_{av} = (w_{\ell o} + w_{hi})/2$. The more important regions are those with lower average weight standard, w_{av}. Still another way to specify Russian roulette and splitting is to give the probability that a particle will be split in two (for each region and group), the probability that a particle will undergo Russian roulette, and the probability that a particle involved in a Russian roulette game will survive (if 1, all survive but if 0, all have their tracks terminated). In scoring from a split particle track, the split particles are treated as one history.

Some feeling for the effectiveness of variance reduction techniques in MORSE-L can be obtained from a γ-ray problem run by Robert Elliott at EG&G, Inc. The source was ^{40}K (single 1.46-MeV γ ray) in soil. The fluence was calculated at an altitude of 152 m in air, with a large boundary-crossing (BC) estimator and a next-event or point-detector (PD) estimator. Each calculation was performed with 10,000 source particles. Table 4.1 summarizes the computation time T, fractional standard deviation (FSD) in the fluence, and the product T • FSD, which we want to minimize. Without variance reduction, the BC estimator requires considerable computation time. The PD estimator requires more time but reduces the FSD. Cutoff of photon histories below 0.1 MeV significantly reduces computation time in this problem but, of course, the information on very low energy photons is lost. Applying Russian roulette to photons below 0.45 MeV, with a probability of 1.0 and survival probability of 0.5, computation time is reduced while the information in the low-energy photons is preserved.

The exponential transform can also be interpreted as a form of importance sampling, where $I \sim \exp(\alpha z)$, for example in a slab shield. The effective number of mean free paths L' to the next collision can be obtained from

$$L' = L(1 - q\mu) \tag{4.44}$$

where μ is the cosine of the angle between the direction of the particle and the line between the collision point and the detector point and q is a number between 0 and 1 that controls the amount of path stretching. If q = 0, there is no stretching. One should be careful not to make q too large (0.5 to 0.7 may be safe). If q is negative, particles will be forced to move away from a point, which could be the source instead of the detector. In Clark's investigation [5], the fractional standard deviation was improved by a factor of about 1/3 as q approached 1, as long as negative values of Σ^* [Eq. (4.18)] are avoided. In thin shields and for large q, it is advantageous to select from a truncated

TABLE 4.1. EFFECT OF ESTIMATOR AND VARIANCE REDUCTION METHOD IN AIR-GROUND γ-RAY PROBLEM

Variance Reduction	Estimator	T(sec)	FSD	T · FSD
None	BC	72	0.16	11.5
Energy cutoff	BC	12	0.21	2.5
Russian roulette	BC	8	0.20	1.6
None	PD	284	0.076	21.6
Energy cutoff	PD	83	0.030	2.5
Russian roulette	PD	73	0.062	4.5

exponential distribution such that all particles collide within the shield.

Adjoint Solution

Some Monte Carlo codes are equipped to carry out an adjoint calculation as well as the usual forward or direct calculation. The inversion of the scattering matrix and replacement of the source by the detector response function are done automatically. One use of an adjoint collision density (or adjoint fluence) has been discussed already, namely, as an importance function for the forward solution of a similar problem. An adjoint solution is also of interest when: (1) an integral response D in a small region is desired as a function of the source distribution (this would require many forward calculations to be performed, e.g., with many monoenergetic sources, but only one adjoint calculation is needed) and (2) the phase-space volume containing nonzero Σ is small, or the detector space is small but the source space is large. The adjoint approach may be more efficient because all histories begin in a small volume and contribute to the adjoint density in a large volume. Adjoint calculations have proved useful in obtaining the count rate in a small detector as a function of the energy and angular distribution of the source in vacuum. They are somewhat more difficult to apply in shielding problems because the variance reduction methods, including importance sampling, are more complicated to implement.

4.2 DISCRETE-ORDINATES METHOD

The discrete-ordinates, or S_n method, has been reviewed by Carlson [11], Lee [12], Mynatt [13], and Lathrop [14]. Here we consider the principles of the method and application to neutral particle transport.

The S_n method is a numerical technique for solving the finite-difference form of the Boltzmann transport equation. The dependent variable is the angular flux density $\psi(\underline{r},E,\underline{\Omega})$, the number of particles per second that cross 1 cm^2 area normal to the direction vector $\underline{\Omega}$, with energy in dE about E, and direction in $d\underline{\Omega}$ about $\underline{\Omega}$, at position $\underline{r}$. The time-dependent transport equation equates losses to gains in the steady state,

$$\nabla \cdot \underline{\Omega}\, \psi(\underline{r},E,\underline{\Omega}) + \Sigma_t(\underline{r},E)\, \psi(\underline{r},E,\underline{\Omega}) = S(\underline{r},E,\underline{\Omega}) + \iint \Sigma_s(\underline{r},E' \to E,\underline{\Omega}' \to \underline{\Omega})\, \psi(\underline{r},E',\underline{\Omega}')\, dE'\, d\underline{\Omega}' \quad (4.45)$$

The first term on the left-hand side is the net convection loss (per $cm^3\ s^{-1}$) because of flow of particles from a differential volume of phase space ($d\underline{r}\ dE\ d\underline{\Omega}$) at $\underline{r}$. The second term on the left-hand side is the loss from the differential volume because of collisions. The first term on the right-hand side is the gain from any external source in the volume. The second term on the right-hand side is the gain from scattering of particles into dE about E, and $d\underline{\Omega}$ about $\underline{\Omega}$, because of collisions of particles with energies E' and directions $\underline{\Omega}'$.

The principles of the discrete ordinates method are discussed first for spherical geometry, with references to other one-dimensional geometries (infinite slab and infinite cylinder), and then extended to two-dimensional geometry. The steady-state Boltzmann transport equation in spherical geometry is

$$\frac{\mu}{r^2}\frac{\partial}{\partial r} r^2 \psi(r,E,\mu) + \frac{1}{r}\frac{\partial}{\partial \mu}\Big[(1-\mu^2)\psi(r,E,\mu)$$

(1) (2)

$$+ \Sigma_t(r,E)\psi(r,E,\mu)\Big] = S(r,E,\mu)$$

(3) (4)

$$+ \int_{-1}^{1}\int_{0}^{\infty} \Sigma_s(r,E' \to E,\mu_0)\,\psi(r,E',\mu)\,dE'\,d\mu' \quad (4.46)$$

(5)

where μ is the cosine of θ the angle to the radius vector as shown in Figure 4.10 and $\mu_0 = (\underline{\Omega}' \cdot \underline{\Omega})$ is the angle of scattering. We now wish to convert Eq. (4.46) into a finite-difference equation.

Finite-difference Equation

In a sphere the differential phase-space cell is

$$dV\,dE\,d\mu = 4\pi\,r^2\,dr\,dE\,d\mu \quad (4.47)$$

The finite-difference cell is

$$V_I\,\Delta\mu_M\,\Delta E_G = \frac{4\pi}{3}\left(r_{i+1}^3 - r_i^3\right)(\mu_{d+1} - \mu_d)$$

$$(E_{g+1} - E_g) \quad (4.48)$$

Thus the volume from 0 to outer radius R is subdivided into intervals bounded by the radii r_i. The number of intervals and the values of r_i are chosen as discussed later. The radii need not be equally spaced. Nested spherical shells or layers of different materials can be handled by specifying a radius at the interface, and specifying the material in each layer or zone. The energy is treated by the multigroup method as discussed in Chapter 3. The G-th group has boundaries E_g, E_{g+1}.

The treatment of the angular variable is peculiar to the S_n method. Discrete angles are chosen, having cosines μ_m with m = 1 to n + 1; for example, for S_8

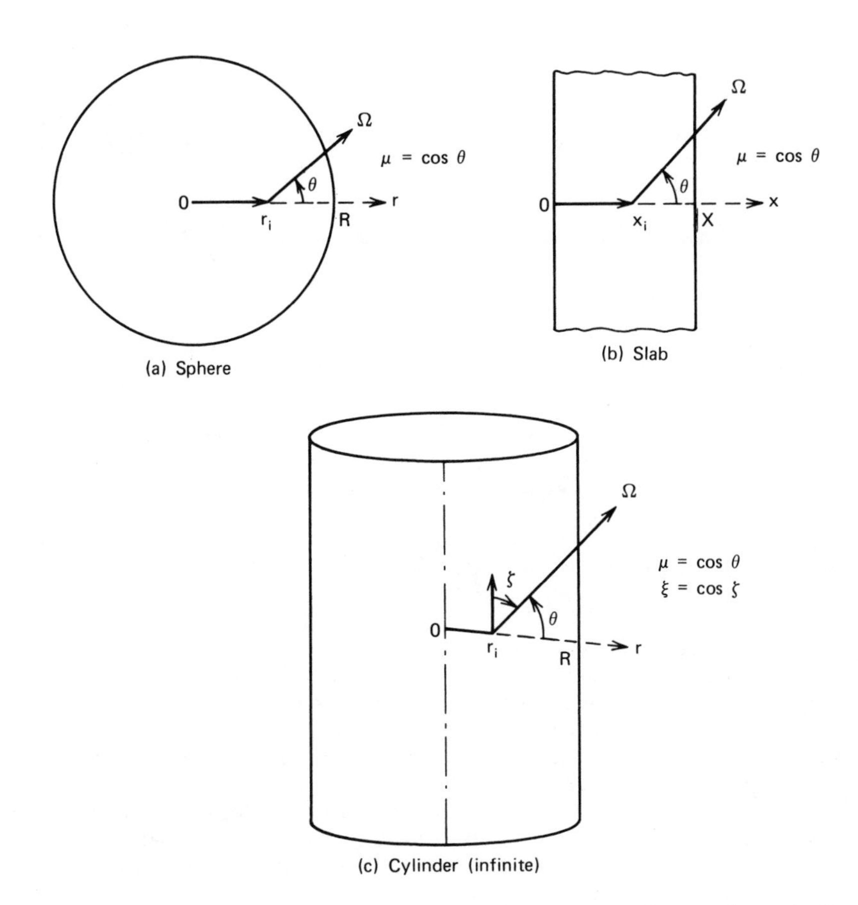

Fig. 4.10. Coordinate systems for one-dimensional S_n calculations.

there are eight discrete cosines plus $\mu = -1$, as shown in Figure 4.11. Unlike the spatial coordinate, these directions define the midpoint or "average" cosine, not the boundaries of the intervals, d. The solid angle associated with each direction is defined by a corresponding weight, w_m. (The sum of the weights is normalized to 1; hence the solid angle in steradians is $w_m/4\pi$. The weight assigned to $\mu = -1$ is zero, but this direction is used in marching back through the cells from the outer boundary.) The number and values of the

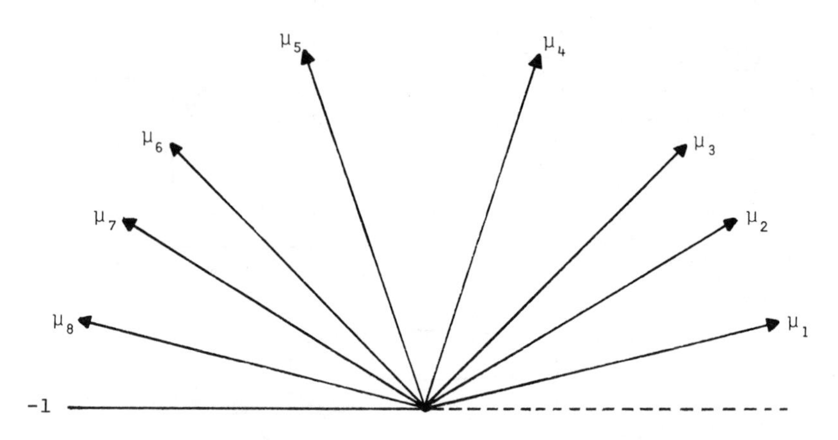

Fig. 4.11. Discrete cosines for an S_8 calculation in plane or spherical geometry.

μ_m and w_m are chosen to meet certain conditions such as accurate representation of the anisotropic angular flux density and accurate integration over the differential scattering cross section as discussed later. The integration is a numerical quadrature with abscissas μ_m and quadrature weights w_m.

The first term on the left-hand side of Eq. (4.46) is converted to finite-difference form by integrating over energy group, spatial cell surface area, and solid angle, giving

$$\textcircled{1} = \mu_m w_m \left[A_{i+1} N_{i+1,m,g} - A_i N_{i,m,g}\right] \tag{4.49}$$

where the area $A_i = 4\pi r_i^2$ and $N_{i,m,g}$ refers to the angular flux density at radius r_i and direction μ_m, integrated over the energy group g [i.e., the group flux found by integrating $\psi(E)$ dE between the group boundaries]. This term accounts for the flow of particles into and out of the cell.

The second term

$$\textcircled{2} = 4\pi\bar{r}_i\ \Delta r_i \left[\left(1 - \mu^2_{d+1}\right) N_{\bar{i},d+1,g} - (1 - \mu^2)\ N_{\bar{i},d,g}\right] \quad (4.50)$$

is evaluated at the average radius of the i-th spatial interval,

$$\bar{r}_i = \frac{r_{i+1} + r_i}{2} \quad (4.51)$$

and gives the loss from the phase-space cell because of angular redistribution. In spherical or cylindrical geometry, the angle changes with flow, even without collision, except along $\mu = \pm 1$. In plane or slab geometry, angular redistribution does not occur.

The third term gives the losses because of collisions,

$$\textcircled{3} = V_i\ w_m\ \Sigma^t_{\bar{i},m,g}\ N_{\bar{i},m,g} \quad (4.52)$$

where the volume $V_i = \frac{4\pi}{3}(r^3_{i+1} - r^3_i)$.

The fourth term is the gain from the source,

$$\textcircled{4} = V_i\ w_m\ S_{\bar{i},m,g} \quad (4.53)$$

The source is usually isotropic. Anisotropic volume sources can be specified by the Q^ℓ coefficients of the Legendre polynomial expansion

$$S_{\bar{i},m,g} = \sum_{\ell=0}^{L} \frac{2\ell + 1}{2} P_\ell(\mu_m)\ Q^\ell_{\bar{i},g} \quad (4.54)$$

Surface sources can be specified at the boundaries.

The fifth term is the inscattering integral. As discussed in Chapter 3, the differential scattering cross section is approximated by a truncated Legendre polynomial expansion

$$\Sigma_s(r,E' \to E,\mu_0) = \sum_{\ell=0}^{L} \frac{2\ell + 1}{2} \Sigma^{\ell}(r,E' \to E)\, P_{\ell}(\mu_0)$$

(4.55)

Then after appropriate discretization, for each angle m and group g,

$$⑤ = V_i \sum_{g'=1}^{G} \sum_{\ell=0}^{L} \Sigma^{\ell}_{\bar{i},g' \to g} \sum_{m'=1}^{n+1}$$

$$w_{m'}\, \mu_{m'}\, P_{\ell}(\mu_{m'})\, N_{\bar{i},m',g'} \quad (4.56)$$

Thus the integrodifferential Eq. (4.45) has been replaced by a series of coupled finite-difference equations in the discrete spatial coordinate r_i, discrete angle $\cos^{-1} \mu_m$, and energy-integrated group g. The discrete ordinates are the $N_{i,m,g}$, that is, the angular flux densities that we want to find. The scalar flux density is obtained from

$$\Phi_{\bar{i},g} = \frac{1}{4\pi} \sum_{m} N_{\bar{i},m,g} \quad (4.57)$$

Boundary Conditions

The finite-difference equations are solved subject to the boundary conditions. The boundary conditions applicable to radiation-transport calculations include:

vacuum (no return):

$$N_{B,m,g} = 0 \quad (\text{all } g,\ \mu_m < 0) \quad (4.58a)$$

reflective:

$$N_{B,m,g} = N_{B,m',g} \quad (\text{all } g,\ \mu_{m'} = -\mu_m) \quad (4.58b)$$

The vacuum condition is usually specified at the outer boundary. Some discrete ordinates codes allow specification of a "white" albedo condition, in which a specified fraction (albedo) of the outgoing current is redistributed equally among the incoming angles. However, the albedo may not be known. It is usually better to apply the vacuum condition with a large outer radius in an "infinite medium" problem, because the (specular, mirror-like) reflective condition can introduce spurious structure in the angular flux density. The reflective condition is required at the "left" boundary ($r = 0$) in the sphere and cylinder because of symmetry.

Method of Solution

Before proceeding with the numerical solution of the finite-difference equation, it is necessary to assume some model for the relationship between the values of N at the edges and midpoint of the angle-and-space cell shown in Figure 4.12. The usual model is diamond difference, which assumes that the angular flux density at the midpoint is the average of the N at the edges

$$N_{\bar{i},m} = \frac{N_{i+1,m} + N_{i,m}}{2} \tag{4.59a}$$

$$N_{\bar{i},m} = \frac{N_{\bar{i},d+1} + N_{\bar{i},d}}{2} \tag{4.59b}$$

These relationships are used to extrapolate from a known value, say, $N_{i,m}$, at a boundary or previous mesh point, to the value $N_{i+1,m}$ at the next mesh point, taking into account the gains and losses from the finite-difference equation. However, if the flux density decreases by a factor of 2 or more from r_i to $\bar{r}_i$, the diamond difference model will predict a negative flux density at the next mesh point (or angular interval), and oscillatory or negative angular flux densities will be calculated. One way to circumvent the breakdown is to reformulate the problem with a finer spatial mesh (smaller Δr) or higher-order S_n approximation, but this requires more computer memory and a longer running time. Another approach is to have the program switch to the step model when negative flux

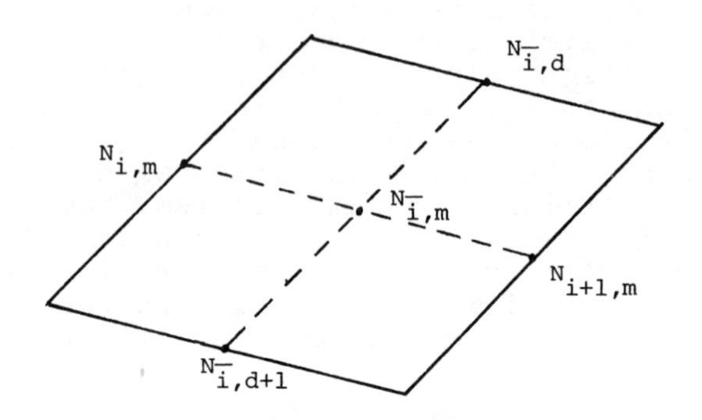

a) Diamond difference model

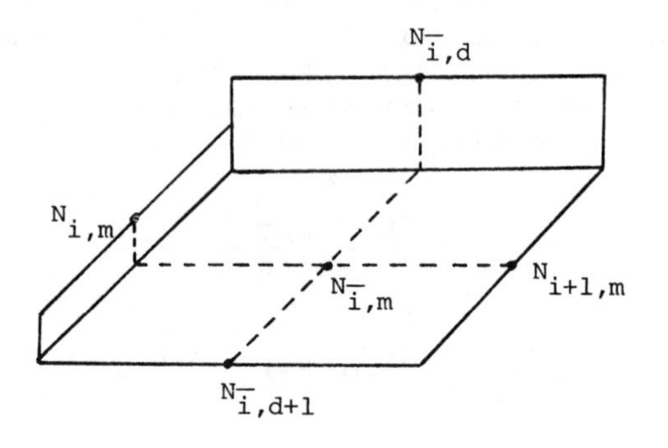

b) Step model

Fig. 4.12. Models for solution of finite-difference equation.

densities are calculated. In the step model, the forward-edge flux density is equated to the midpoint flux density calculated from the finite-difference equation as shown in Figure 4.12,

$$N_{i+1,m} = N_{\bar{i},m} \qquad (\mu > 0) \tag{4.60a}$$

$$N_{i,m} = N_{\bar{i},m} \qquad (\mu < 0) \tag{4.60b}$$

$$N_{\bar{i},d+1} = N_{\bar{i},m} \tag{4.60c}$$

The step model is less accurate than the diamond difference model but cannot give negative fluxes.

The calculation begins with an initial scalar flux density input for every group and spatial interval. The solution from a similar problem may be used, but a guess of all 1.0 values is adequate. The program sweeps through all spatial and angular intervals, starting with the boundary condition at the outer boundary and $\mu = -1$ (or a zero-weight angle near -1 in some angular quadrature sets). The fluxes in each group are calculated, one group at a time, starting with the highest-energy group and working downward. This process is called an inner iteration. In the absence of fission, all scattering is down (including n,2n reactions and pair production), and the scattered particles are picked up in the inscattering integral. If fission occurs, high energy neutrons may be generated by low-energy neutrons, and "outer iterations" are required. In radiation transport calculations it is often possible to precalculate the distribution of fissions in, say, a reflected reactor, and then input the distribution as a fixed source in the shielding calculation.

The solution is obtained by iterating until the boundary conditions are satisfied and the flux density is converged, that is, no longer changes on further iteration, within a specified fractional error, ε. For convergence on the volume integral of the group scalar flux density, ε should be small, say, 10^{-7}. It is better to converge at each spatial interval to, say, 10^{-4} in the group-by-group scalar flux density, which should then give adequate convergence in the angular flux densities as well. To avoid wasting money, it is standard procedure to set a limit on the computing time and number of inner iterations per group, say, around 25 or 50. Output should always be examined for possible lack of convergence and for negative or oscillatory angular flux densities. The computer output will include tables giving the number of iterations, maximum error in each group, and balance between sources and losses.

Radiation transport calculations are performed in the "Q" or source option. Discrete ordinates codes are also used for criticality calculations. They can solve for the k_{eff} of a multiplying assembly, search for the concentration of fissionable material required to achieve criticality, search for the critical size, and also perform an "alpha (α)" calculation for the dieaway time constant.

Selection of Groups and P_L

The multigroup, Legendre polynomial expansion method is used to prepare cross sections, as discussed in Chapter 3. A typical discrete ordinates transport calculation may use 30 to 100 groups and a P_3, or possibly P_5, expansion. One group-to-group-transfer table is required for each term, hence 4 for P_3 (P_0, P_1, P_2, P_3). Computer storage increases rapidly with the number of energy groups and tables. Often the order of P_L is increased in a series of trial calculations until the result no longer changes significantly. Figure 4.13 shows the fast-neutron dose rate from a fission source in water, where P_0 is poor, P_1 is better, and P_2 or higher give essentially the same result in this problem.

Selection of Spatial Mesh

The radii should be chosen such that the linear extrapolation in the diamond difference model is an accurate estimate of the behavior of the angular flux density. Thus Δr should be small in regions of large attenuation or curvature, such as near the origin in spherical and cylindrical geometry. On the other hand, small Δr implies very many space points in a thick shield, and some compromise may be necessary. Typical shielding problems in one dimension may use 100 to 300 radii.

In a slab or at large radii in curved geometries, $(\Delta N/N) \sim (\Delta r\, \Sigma_t/\mu)$. We want

$$\Delta r \le \frac{2\mu_{min}}{\Sigma_{t_{max}}} \tag{4.61}$$

where μ_{min} is the smallest discrete cosine and $\Sigma_{t_{max}}$ is the largest total cross section in the multigroup set. This is a stringent requirement and may be relaxed somewhat with the switch to the step model if negative flux densities or oscillations are encountered with

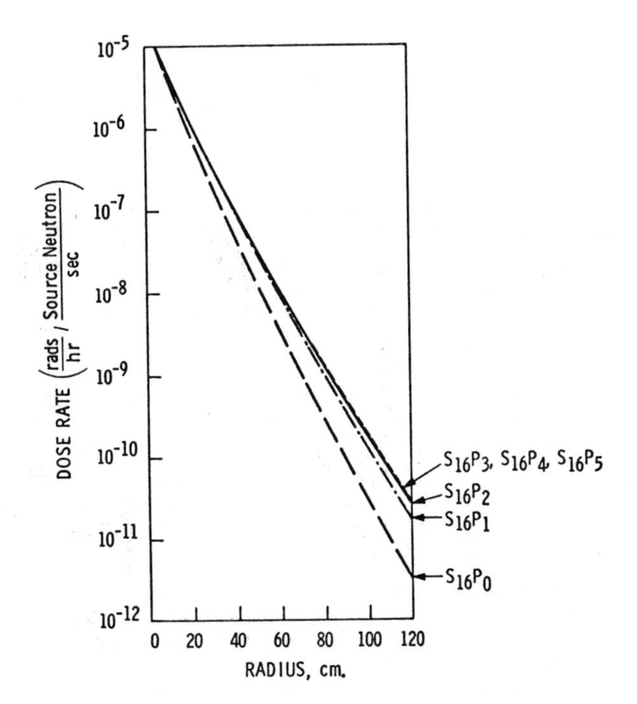

Fig. 4.13. Effect of P_L expansion on fast-neutron dose rate from a point fission source in water.

increasing radius. Near the origin in curved geometry, the angular flux density changes rapidly even without collisions, and a smaller Δr may be required. In a noncentral void one can use Eq. (4.61) with $\Sigma_t = (1/r_{\bar{1}})$. The angular flux density may be nearly isotropic and slowly varying with r within a distributed volume source region, and larger Δr should be permissible. One should not change Δr by more than a factor of 2 from one interval to the next. The usual practice in discrete-ordinates calculations is to make Δr a fraction of the mean free path in each material zone and then decrease Δr until negative and oscillatory fluxes are eliminated and the solution no longer changes significantly on further refinement of the spatial mesh (assuming computer memory permits). If only the scalar flux density or dose is required, negative or oscillatory angular flux densities may be permissible as long as the scalar flux density is correct. Note the magnitude of the angular flux densities at the various μ_m; the code may simply be giving small oscillations around zero as a representation of zero.

Selection of Angular Quadrature

The most accurate quadrature set for integrating the Legendre polynomials for a given n is the Gauss-Legendre quadrature (μ_m at zeros of Legendre polynomials). Lathrop [14] has found, in fact, that in plane geometry with Gauss-Legendre quadrature the S_n method is entirely equivalent to the spherical harmonics method of order P_{n-1}. Thus one may expect that a relatively low-order quadrature, such as S_8 for P_3 cross sections, would be sufficient as far as the inscattering source is concerned. However, in radiation transport, the angular flux density is usually very anisotropic and a higher-order set may be needed to properly represent the angular distribution in the output. Another consideration is the accuracy of calculation of the angular redistribution in curved geometries. We also want symmetric solutions in symmetric geometry, independent of the choice of orientation of the coordinate axis.

Most radiation transport calculations are performed with cosines distributed symmetrically about $\mu = 0$, but $\mu_m \neq 0$, and

$$\sum_m w_m = 1.0 \tag{4.62a}$$

$$\sum_m w_m \mu_m = 0 \tag{4.62b}$$

$$\sum_m w_m \mu_m^2 = 1/3 \tag{4.62c}$$

Thus sets have been developed with the μ_m and w_m somewhat different from Gauss-Legendre quadrature, but still able to integrate Legendre polynomials accurately. Good results have been obtained with the S_8 and S_{16}

sets listed in Table 4.2 for slabs and spheres and the S_8 set in Table 4.3 for cylinders.

In plane and spherical geometry only one angle, $\cos^{-1} \mu$ is needed and there are $(n + 1)$ angles of which n have nonzero weights. In cylindrical geometry, two angles are required, $\cos^{-1} \mu$ measured from the radius vector and $\cos^{-1} \xi$ measured from the z-axis as shown in Figure 4.10. There are then $n(n + 2)/4$ weighted directions and $n/2$ zero-weight starting directions. One can consider the directions to be distributed on an octant of a sphere (the other sectors being symmetric) with different levels of ξ each containing a number of μ values, as illustrated in Figure 4.14 for S_8.

The problem of calculating the uncollided flux (or total flux in vacuum) from a small, spherical source is a severe test of discrete-ordinates approximations. As the distance from the source increases and the angular flux becomes more and more collimated into a narrow forward cone about $\mu = +1$, the angular flux varies rapidly over a very small range of μ and is not integrated accurately by the standard quadrature sets. The problem is of practical significance, not only for the uncollided flux, but for the total neutron flux at energies above several megaelectronvolts, because there is little downscattering into these energies in the typical situation. The essential features of the problem can be found in a one-group model; hence we drop the energy variable. Following an unpublished calculation by Lathrop, we consider a uniform, isotropic source of radius a in vacuum,

$$\begin{aligned} S(r,\mu) &= Q_0 \text{ neutrons cm}^{-3}\text{ s}^{-1}\text{ sr}^{-1}, \quad r \le a \\ &= 0 \qquad\qquad\qquad\qquad\qquad\qquad r > a \end{aligned} \tag{4.63}$$

The analytic solution of this problem is

$$\begin{aligned} \psi(r,\mu) &= Q_0 \left[r\mu + \sqrt{a^2 - r^2(1 - \mu^2)} \right], \quad r \le a \\ &= 2Q_0 \sqrt{a^2 - r^2(1 - \mu^2)}\ , \quad r > a \text{ and} \end{aligned}$$

TABLE 4.2. S_n CONSTANTS FOR PLANE AND SPHERICAL GEOMETRY

	μ_m		w_m
	S_8		
1	-0.9759000	1	0.0
2	-0.9511897	2	0.0604938
3	-0.7867958	3	0.0907407
4	-0.5773503	4	0.1370371
5	-0.2182179	5	0.2117284
6	+0.2182179	6	0.2117284
7	+0.5773503	7	0.1370371
8	+0.7867958	8	0.0907407
9	+0.9511897	9	0.0604938
	S_{16}		
1	-0.9902984	1	0.0
2	-0.9805009	2	0.0244936
3	-0.9092855	3	0.0413296
4	-0.8319966	4	0.0392569
5	-0.7467506	5	0.0400796
6	-0.6504264	6	0.0643754
7	-0.5370966	7	0.0442097
8	-0.3922893	8	0.1090850
9	-0.1389568	9	0.1371702
10	+0.1389568	10	0.1371702
11	+0.3922893	11	0.1090850
12	+0.5370966	12	0.0442097
13	+0.6504264	13	0.0643754
14	+0.7467506	14	0.0400796
15	+0.8319966	15	0.0342569
16	+0.9092855	16	0.0413296
17	+0.9805009	17	0.0244936

TABLE 4.3. S_n CONSTANTS FOR CYLINDRICAL GEOMETRY

	μ_m, ξ_m			w_m
		S_8		
1	-0.9759000		1	0.0
2	-0.9511897		2	0.0604938
3	-0.7867958		3	0.0453704
4	-0.5773503		4	0.0453704
5	-0.2182179	ξ_1	5	0.0604938
6	+0.2182179		6	0.0604938
7	+0.5773503		7	0.0453704
8	+0.7867958		8	0.0453704
9	+0.9511897		9	0.0604938
10	-0.8164965		10	0.0
11	-0.7867958		11	0.0453704
12	-0.5773503		12	0.0462962
13	-0.2182179	ξ_2	13	0.0453704
14	+0.2182179		14	0.0453704
15	+0.5773503		15	0.0462962
16	+0.7867958		16	0.0453704
17	-0.6172134		17	0.0
18	-0.5773503		18	0.0453704
19	-0.2182179	ξ_3	19	0.0453704
20	+0.2182179		20	0.0453704
21	+0.5773503		21	0.0453704
22	-0.3086067		22	0.0
23	-0.2182179	ξ_4	23	0.0604938
24	+0.2182179		24	0.0604938

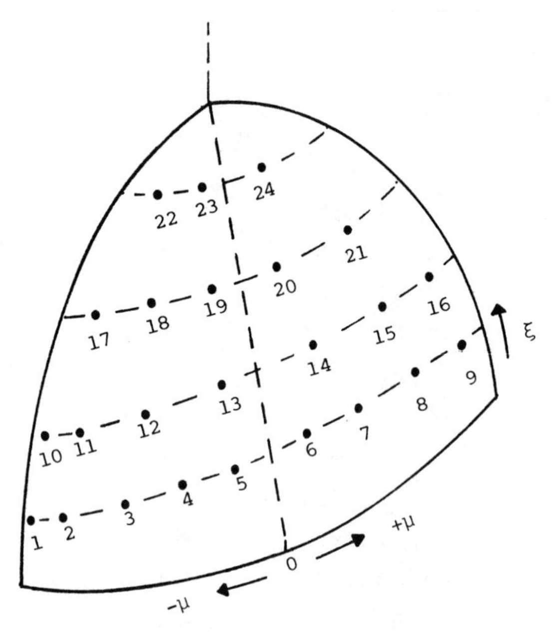

Fig. 4.14. Discrete angles in cylindrical geometry as points on octant of a sphere.

$$\sqrt{1 - \left(\frac{a}{r}\right)^2} < \mu < 1$$

$$= 0 \qquad \text{otherwise} \tag{4.64}$$

For $r >> a$ the angular flux is nonzero only for a small range of μ about +1, but the amplitude of the angular flux at $\mu = +1$ remains a constant. The scalar flux,

$$\phi = \int_{-1}^{1} \psi(r,\mu)\ d\mu \tag{4.65}$$

decreases as $1/r^2$ for $r >> a$, as expected. At $r = 0$ the exact solution is $N(0,-1) = Q_0 a$, whereas the limiting form of the finite-difference approximation gives $N(0,-1) = N_1 + (2/3)\Delta r Q_0$. To make the error small, the first interval Δr in spherical geometry should be especially small. For $r_i > 0$ the S_n scalar flux is close to the exact analytical result for points inside the source, but the angular flux is not.

Numerical calculations were made for $Q_0 = 0.238732$ and $a = 1$ (hence $(4/3)\pi a^3 Q_0 = 1$ neutron/second). Mesh spacings were $\Delta r = 0.025$ for $0 < r < 0.9$, $\Delta r = 0.05$ for $0.9 < r < 1.0$, and $\Delta r = 0.1$ for $r > 1.0$. Gauss-Legendre quadratures S_8, S_{16}, and S_{32} were used. The largest radius was 10 for S_8 and S_{16}, but a maximum radius of 5 was used for the S_{32} calculation because of computer-storage limitations. The analytical solution and the discrete ordinates solutions are plotted in Figure 4.15 for $(r/a) = 1$, 2, and 4. All three quadratures reproduce the cosine angular distribution at the surface of the source, $(r/a) = 1$, but S_8 and S_{16} do poorly at $(r/a) = 2$ and 4. At $(r/a) = 4$, even S_{32} is not very accurate.

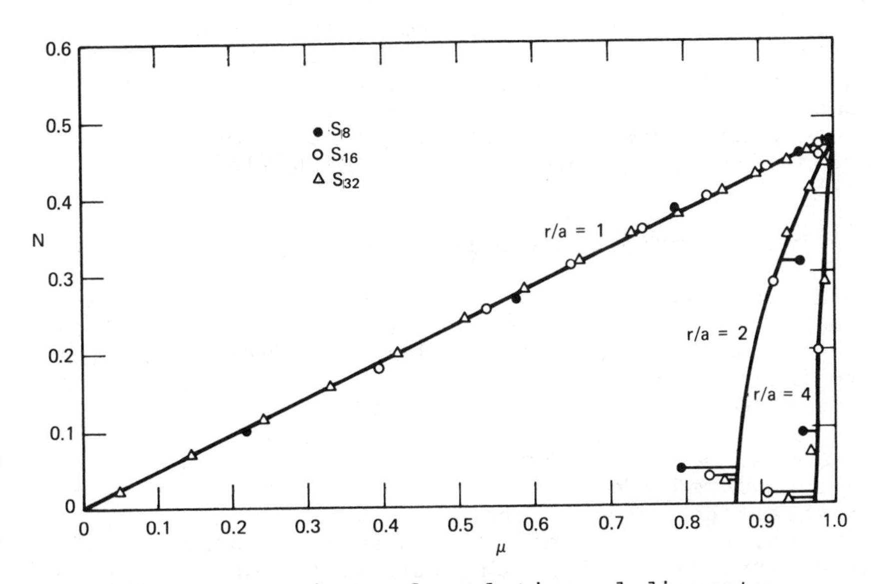

Fig. 4.15. Comparison of analytic and discrete-ordinates calculations (Gauss-Legendre quadrature) for angular flux density in vacuum.

Computer-memory and running-time limitations discourage use of $n > 16$. Special symmetric sets that have cosines spaced closer to +1 (and -1) have been tried, but the improvement is not spectacular. Another approach is to use an asymmetric set when it is known that the angular flux density is anisotropic (e.g., peaked near $\mu = +1$). Cerbone and Lathrop [15] describe a method of generating two Gauss-Legendre sets separately for the interval $-1 \leq \mu \leq 0.95$ and the interval $0.95 < \mu < 1.0$. The asymmetric S_{16} set is listed in Table 4.4. When applied to transport problems involving a small spherical fission neutron source in water, the asymmetric S_{16} quadrature gave good agreement with a symmetric S_{48} quadrature, whereas a symmetric S_{16} quadrature was significantly in error near 0^0.

Slab and Cylinder Calculations

The infinite slab is calculated in much the same way as the sphere, but the angular derivative term in the transport equation disappears (no angular redistribution). The area of the finite-difference cell is 1, and the volume is $x_{i+1} - x_i$. Only one angular variable, μ, is required. Infinite cylinders can be calculated in one space dimension, but two angles are required. The area element is $2\pi r_i$, and the volume element is $\pi(r_{i+1}^2 - r_i^2)$. One-dimensional, time-independent, forward or adjoint calculations may be performed with the computer code ANISN [16] or DTF-IV [17].

Two-Dimensional Calculations

Two dimensional discrete ordinates theory is discussed in detail by Mynatt, Muckenthaler, et al. [18]. Two-dimensional (r,z finite cylinder; x,y bar; r,θ wedge), time-independent, forward, or adjoint calculations may be performed with the DOT code [19].

Spatial intervals have to be specified in two variables; hence the number of mesh points is essentially squared and large computer memory is essential for shielding calculations. Two angle variables are required, as in the infinite cylinder. Energy grouping and P_L expansion may be the same as for one-dimensional calculations, but the smallest number of groups and lowest-order P_L possible should be used. Not only is the computer storage requirement much higher than in

TABLE 4.4. ASYMMETRIC S_{16} QUADRATURE SET

	μ_m	w_m
1	-1.00	0.0
2	-0.974553	0.032502
3	-0.869843	0.072857
4	-0.687415	0.106815
5	-0.447555	0.131264
6	-0.170148	0.144066
7	+0.120148	0.144066
8	0.39755	0.131264
9	0.637415	0.106815
10	0.819843	0.072857
11	0.924553	0.032502
12	0.951685	0.002145
13	0.9558195	0.0045095
14	0.9690345	0.0058488
15	0.980966	0.0058488
16	0.99153	0.0045095
17	0.99831	0.002145

one-dimensional calculations, but the running time is much longer as $N(r_i,z_j,E_g,\Omega_m)$ has to be calculated at every point. Considerable attention is paid in DOT to accelerating convergence in an attempt to minimize computation time.

Two-dimensional (r,z) calculations may exhibit a "ray effect" when the sources and detectors are small and the scattering mean free path is large compared to Δr or Δz or where the medium is highly absorbing. The angular flux densities are found to peak along rays corresponding to the polar angles (direction cosine ξ) of the angular quadrature, centered on the source. In cylindrical geometry the angles of the quadrature, are arranged in levels of the same ξ, as illustrated in Figure 4.14. Transfer from one ξ level to another occurs only through scattering. In regions where little scattering has occurred the particles tend to concentrate along the discrete ξ levels, and ray effects appear if the source is localized. One solution is to increase the number of ξ levels, and hence number of angles, but this soon becomes prohibitive in terms of computation time and storage. Another approach, implemented in DOT, is to use an analytic first-collision source, which distributes the source throughout the medium. The uncollided angular flux density and distribution of particles emerging from their first collisions are calculated analytically in one part of the code and combined later with the collided flux density from the S_n transport calculation.

4.3 OTHER EXACT METHODS

Monte Carlo and discrete ordinates are the most commonly used exact methods, but there are others available, including spherical harmonics, moments, and invariant imbedding. Here "exact" means that the method is capable of including all the physics of interaction and transport correctly. Accuracy is limited only by the fineness of the meshes (space, angle, energy), truncation of the Legendre series expansion, numerical errors such as roundoff, and convergence (S_n) or statistics (Monte Carlo).

Spherical Harmonics

The method of spherical harmonics [20] is another method of solving the Boltzmann transport equation. The angular flux density and angular distribution of the external source are expanded in spherical harmonics, which in one-dimensional slab geometry are Legendre polynomials; for example,

$$\psi(x,E,\mu) = \sum_{\ell=0}^{L} \frac{2\ell + 1}{2} F_\ell(x,E)\, P_\ell(\mu) \tag{4.66}$$

The differential scattering cross section is also expanded in Legendre polynomials. The integropartial differential Boltzmann equation is reduced to coupled differential equations

$$\frac{\ell + 1}{2\ell + 1} \frac{d}{dx} F_{\ell + 1}(x,E) + \frac{\ell + 1}{2\ell + 1} \frac{d}{dx} F_{\ell - 1}(x,E)$$

$$+ \Sigma_t F_\ell(x,E) + \Sigma_\ell(E' \rightarrow E) F_\ell(x,E')$$

$$+ S_\ell(x,E) = 0 \tag{4.67}$$

where $\ell = 0,1,2,\ldots,L$ and Σ_ℓ is the ℓ-th component of the differential scattering cross section. These equations can be written in finite-difference form and solved by the computer. If the expansion of the flux density is truncated at $\ell = 1$ (P_1 approximation), the method is equivalent to diffusion theory. Shielding calculations require at least P_3 approximation. The method is generally restricted to one-dimensional slab geometry. Equivalent results are obtained by discrete ordinates using a Gauss-Legendre quadrature. The discrete ordinates computer programs are maintained more effectively and are more versatile, and spherical harmonics are not used much anymore.

Moments Method

The method of moments was developed for radiation transport by Spencer and Fano [21]. The angular flux density and source are expanded in Legendre polynomials as in the spherical harmonics method. However, instead of solving a difference equation derived from Eq. (4.67), the spatial dependence is represented by the moments [22]

$$M_{n\ell}(E) = \frac{1}{2(n!)} \int_{-\infty}^{\infty} x^n F_\ell(x,E) \, dx \tag{4.68}$$

The limits of integration imply an infinite homogeneous medium and, in fact, the moments method is almost always restricted to an infinite medium. The series of transport equations becomes

$$\frac{-1}{2\ell + 1} \left\{ (\ell + 1)\, M_{n - 1, \ell + 1}(E) + \ell M_{n - 1, \ell - 1}(E) + \Sigma_t(E)\, M_{n\ell}(E) \right\} = \Sigma_\ell(E' \to E)\, M_{n\ell'}(E') + S(E,0) \quad (4.69)$$

assuming a plane, isotropic source at x = 0, in an infinite slab. The equations can be solved by forward substitution, starting from n = 0, ℓ = 0. An advantage of the moments method is that the expansion can be truncated without affecting the accuracy of the lower-order moments. For example, the scalar flux density, corresponding to $\ell = 0$, can be found without solving for higher-order terms in the angular expansion. The spatial distribution is reconstructed from the moments directly or by fitting the moments of the simple functions such as exp $(-\mu x)$ or exp $[-(\mu x)^2]$ to the calculated moments. The practical restriction to an infinite medium as well as the difficulty of the spatial reconstruction severely limit the applicability of the moments method in shield design. However, it has been used to calculate buildup factors for the kernel technique discussed in Section 4.4. Similar results can now be achieved with the more versatile discrete-ordinates codes.

Invariant Imbedding

Invariant imbedding [20] is a fundamentally different approach to radiation transport, rather than another way of solving Boltzmann equation. So far it has been restricted to one-dimensional slab geometry, which severely limits its utility. Transport is described in terms of the reflection function and transmission function of a slab of thickness X; what happens inside is not considered in detail. The reflection function $R(X;\, \mu,E;\, \mu_0,E_0)\, d\mu\, dE$ gives the number of particles reflected from a slab of thickness X in energy dE about E and direction in $d\mu$ about μ, from a particle incident on the slab with energy E_0 and direction μ_0. The transmission function $T(X;\, \mu,E;\, \mu_0,E_0)\, d\mu\, dE$ is the number of particles transmitted through a slab of thickness X and emerging in $d\mu$ about μ with energy in dE about E, for a particle incident on the slab with direction μ_0 and energy E_0. Equations are developed in terms of dR/dx and dT/dX, the changes in R and T from addition of a differential slab of thickness dX to X, resulting from collisions in dX. These equations can be solved with the initial conditions $R(0;\ \mu,E;\ \mu_0,E_0) = 0$ and $T(0;\ \mu,E;\ \mu_0,E_0)\ d\mu\ dE = \delta(\mu - \mu_0)\ \delta(E - E_0)\ d\mu\ dE$. The results pertain to a whole series of slabs (variable X) but do not give absorption rates in the slabs, for example.

4.4 KERNEL METHOD

The point kernel $K(\underline{r};\underline{r}',E')$ is the response of a detector (e.g., flux density, dose, or energy absorption) at point $\underline{r}$ from a unit source of energy E' at point $\underline{r}'$ [23]. The response of an isotropic detector from an isotropic, distributed, polyenergetic source is obtained by integration over the source

$$D(\underline{r}) = \int_V \int_{E'} K(\underline{r};\underline{r}',E')\ S(\underline{r}',E')\ dE'\ d\underline{r}' \tag{4.70}$$

In an infinite, homogeneous medium, K depends only on the distance $R = |\underline{r} - \underline{r}'|$ from source point to detector.

The uncollided flux density from a point isotropic source in an infinite medium

$$\phi_0(R,E') = K_0(R,E')\ S(E') \tag{4.71}$$

where

$$K_0(R,E') = \frac{\exp -\mu(E')R}{4\pi R^2} \tag{4.72}$$

and $\mu(E')$ is the linear attenuation coefficient (or total macroscopic cross section) at source energy E'. In a heterogeneous medium the uncollided flux density is calculated exactly from

$$K_0(R,E') = \frac{1}{4\pi R^2} \exp \left\{ - \mathcal{S}_i\ \mu_i(E')\ d_i \right\} \tag{4.73}$$

where the sum is over the product of the linear attenuation coefficient for material i and the corresponding thickness d_i along the path from source to detector. Computer codes such as QAD [24] perform the ray tracings in complex geometries and calculate the d_i.

Kernels may also be defined for other source distributions, such as plane isotropic and parallel beam. The kernel for the uncollided flux density from a plane isotropic source can be obtained by integrating over point isotropic sources as shown in Figure 4.16. The source strength is uniform, S_A particles/s per cm^2 of plane area. Consider a ring of radius r and width dr, so all points are a distance R from the detector at P. Then

$$\phi_0(z) = S_A \int_0^{\infty} \frac{\exp -\mu r}{4\pi R^2} 2\pi r \, dr \tag{4.74}$$

Let $y = \mu z \sec \theta$ and $R^2 = (r^2 + z^2)$; hence

$$\phi_0(\mu z) = \frac{S_A}{2} E_1 (\mu z) \tag{4.75}$$

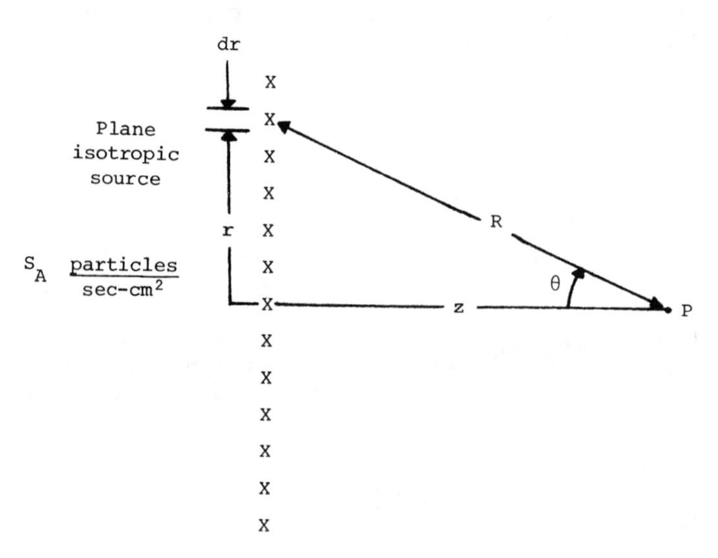

Fig. 4.16. Integration for plane isotropic source.

where E_1 is the exponential integral of the first kind,

$$E_1(x) \equiv \int_x^\infty \frac{e^{-y}}{y}\, dy \tag{4.76}$$

The uncollided flux density from a monodirectional, parallel beam normally incident on a semiinfinite plane is

$$\phi_0(\mu z) = S_B\, e^{-\mu z} \tag{4.77}$$

where S_B is the number of particles per second incident on 1 cm^2. Figure 4.17 plots e^{-x} and $E_1(x)$. Uncollided flux densities for disks, absorbing cylinders, line sources, and many other geometries are given in another work [23]. The kernels for the uncollided flux density follow from Eq. (4.70).

The uncollided particles contribute most of the dose only near the source or in a thin shield. The contribution of the collided (scattered) particles is included by means of a buildup factor

$$B(R,E') = \frac{\text{response from uncollided + collided particles}}{\text{response from uncollided particles alone}}$$

$$= \frac{\int h(E)\ \phi(R,E)\ dE}{h(E')\ \phi_0(R,E')} = \frac{D(R,E')}{D_0(R,E')} \tag{4.78}$$

The buildup factor depends on the source energy E' and angular distribution (e.g., isotropic or monodirectional), the medium, distance R [or mean free paths $\mu(E')R = \mu_0 R$], and the weighting function h(E). Thus there is a different buildup factor for the flux density (h = 1), dose, energy absorption, or other response of interest. Responses are discussed in Chapter 5. The total (collided and uncollided) flux density $\phi(R,E)$ is usually calculated using an exact code such as moments or discrete ordinates. If the total flux density were calculated for the actual configuration of the shield, the response

$$D(R,E') = B(R,E')\ D_0(R,E') \tag{4.79}$$

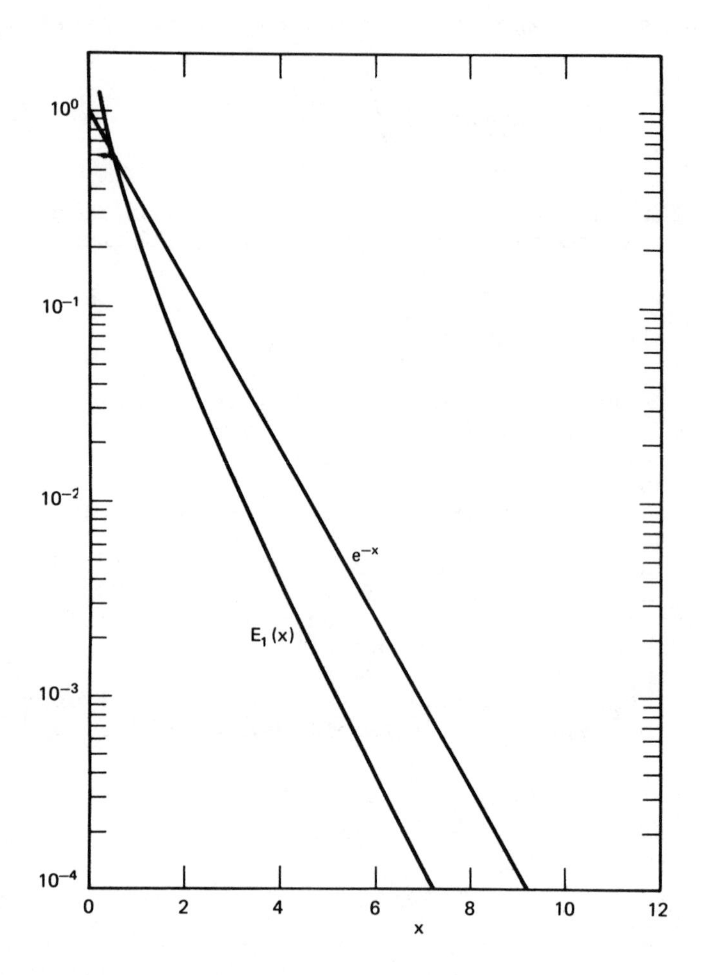

Fig. 4.17. Functions e^{-x} and $E_1(x)$.

would be exact and the buildup factor would be simply an integral parameter obtained from the spectrum of the scalar flux density. In practice, however, the spectrum is nearly always calculated for an infinite medium and the infinite-medium buildup factor is applied in the finite, heterogeneous shield. The approximation may be severe in some situations, as discussed later.

Figure 4.18 plots the γ-ray spectrum for a 1-MeV point isotropic source in infinite water, as calculated

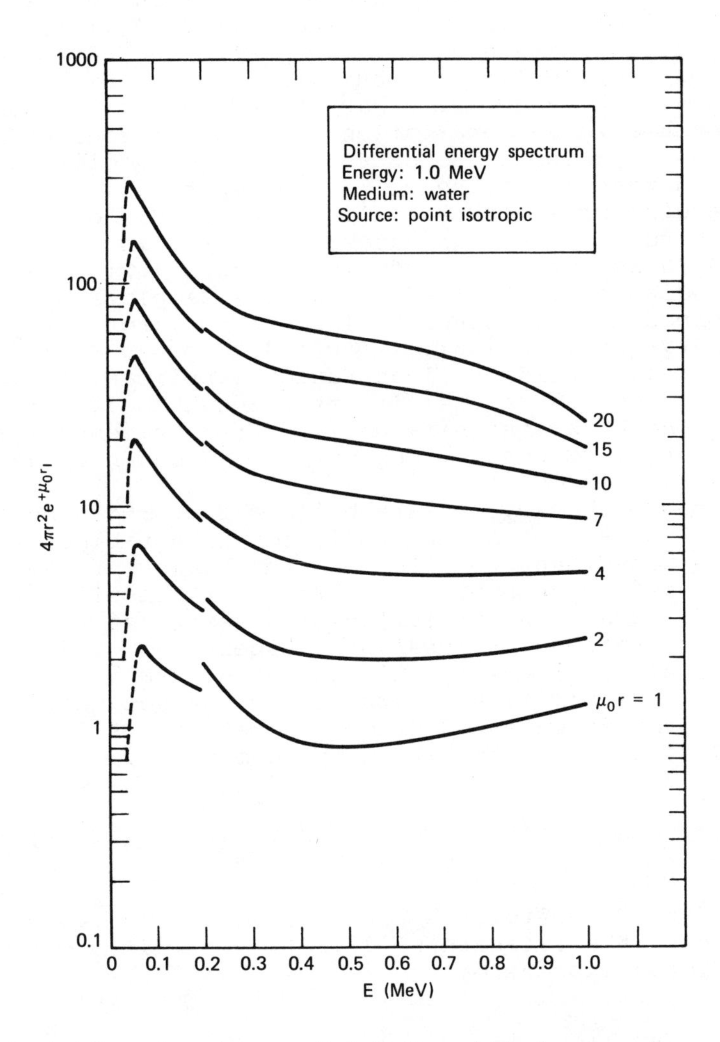

Fig. 4.18. Spectrum of 1-MeV point isotropic γ-ray source in water (after Goldstein and Wilkins [25]).

by the moments method [25]. Figure 4.19 plots for the same source in lead. Parameter $\mu_0 r$ is the radius in mean free paths at the source energy. The ordinate is $4\pi r^2 \exp(+\mu_0 r) I$, the energy flux density $I = E\phi(E)$ for collided and uncollided photons, divided by the uncollided flux density. Note the "buildup" of low-energy photons because of Compton scattering, but the smaller buildup in lead because of photoelectric absorption. The spectrum of scattered photons does not change much after a penetration of three or four mean free paths, but the ratio of scattered to uncollided photons increases continuously with depth.

The dose or exposure buildup factor, B_r for a point isotropic source in water is plotted in Figure 4.20 as a function of $\mu_0 r$. Figure 4.21 is a corresponding plot for lead. Figure 4.22 plots B_a, the absorbed-dose (tissue) buildup factor, for γ rays in ordinary concrete [26]. Figure 4.23 plots B_E, the energy-absorption buildup factor (applicable to heating). Because of the different weighting function $h(E)$, B_a and B_E are significantly different, especially at low energy. Exposure, absorbed dose, and heating are discussed in Chapter 5.

Although the appropriate buildup factor can be obtained from a plot or interpolation in a table, it is sometimes convenient to calculate the buildup factor from the coefficients of an analytical expression fitted to the data. Trubey [27] has investigated the accuracy of various expressions. Linear and quadratic function, $B(\mu R, E') = 1 + A(E')\,\mu R$ and $B(\mu R, E') = 1 + a(E')\,\mu R + b(E')(\mu R)^2$, are simple but not very accurate over an extended range. A better fit is achieved with Capo's four-term polynomial

$$B(\mu R, E') = \sum_{n=0}^{3} \beta_n(E')(\mu R)^n \tag{4.80}$$

Values of the coefficients for polynomial expressions are given in the literature [27,28].

Taylor's expression

$$B(\mu R, E') = A \exp -\alpha_1(E')\,\mu R + (1 - A) \exp -\alpha_2(E')\,\mu R \tag{4.81}$$

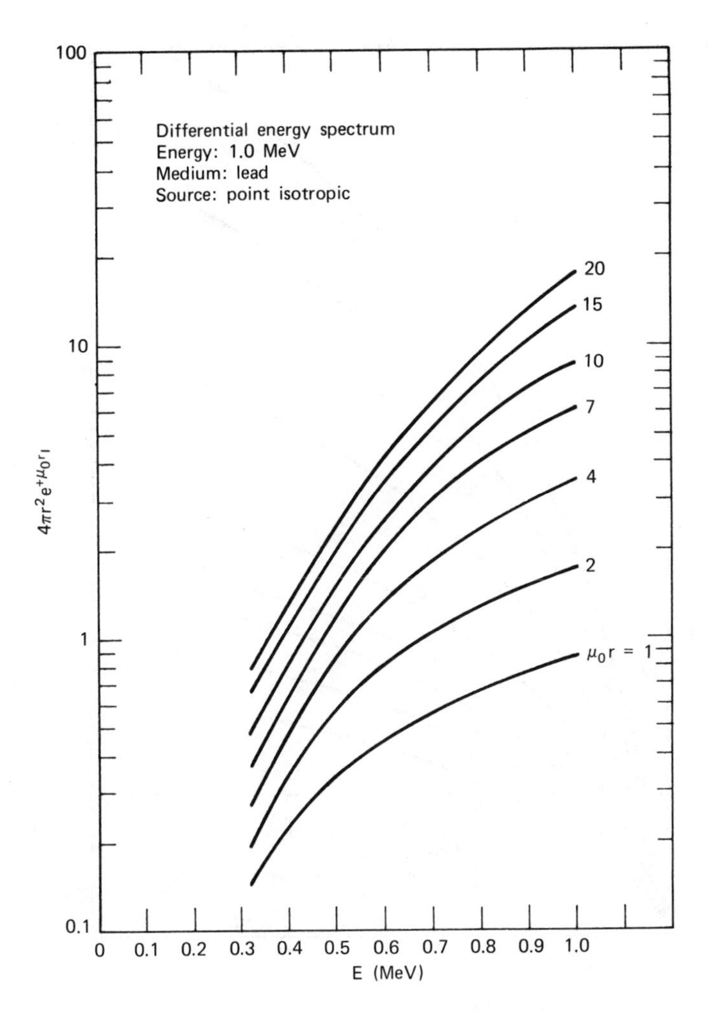

Fig. 4.19. Spectrum of 1-MeV point isotropic γ ray source in lead (after Goldstein and Wilkins [25]).

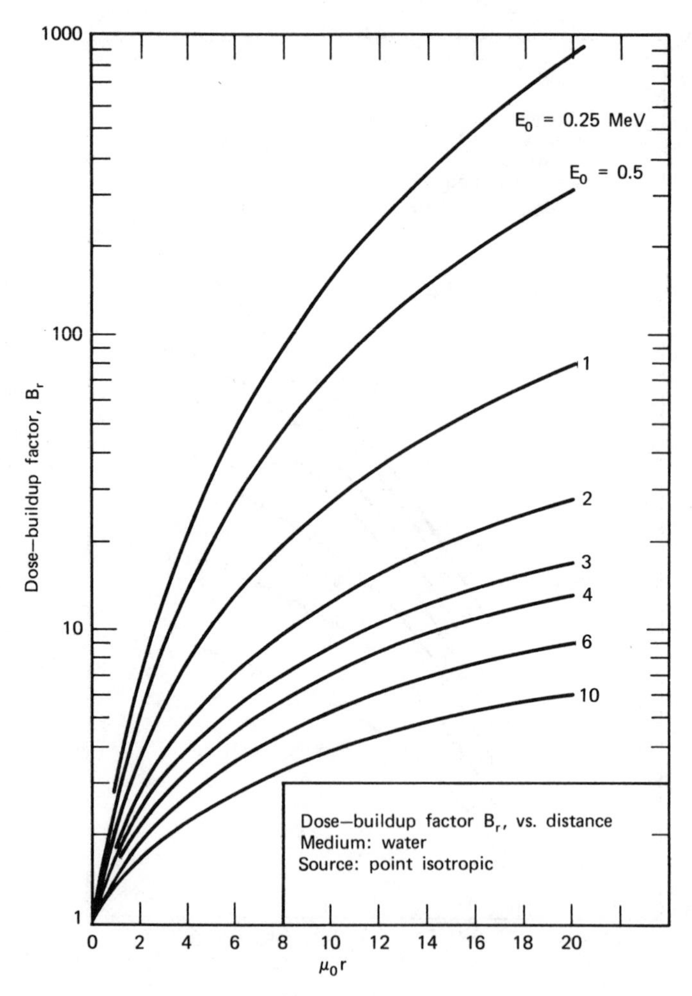

Fig. 4.20. Dose (exposure)-buildup factor for point isotropic γ-ray source in water (after Goldstein and Wilkins [25]).

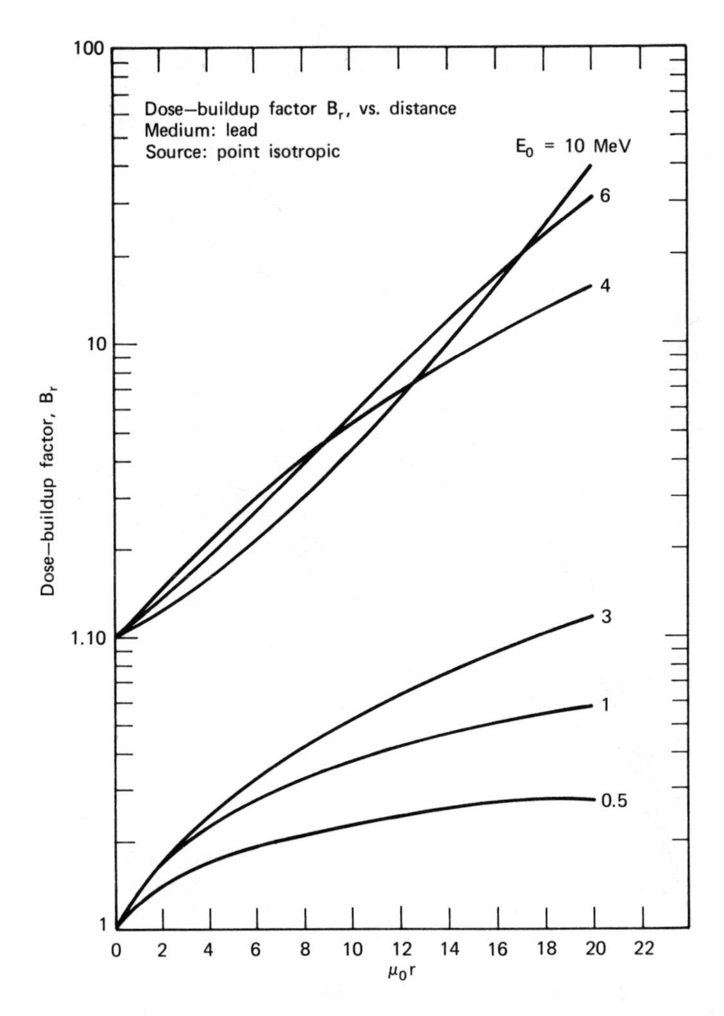

Fig. 4.21. Dose (exposure)-buildup factor for point isotropic γ-ray source in lead (after Goldstein and Wilkins [25]).

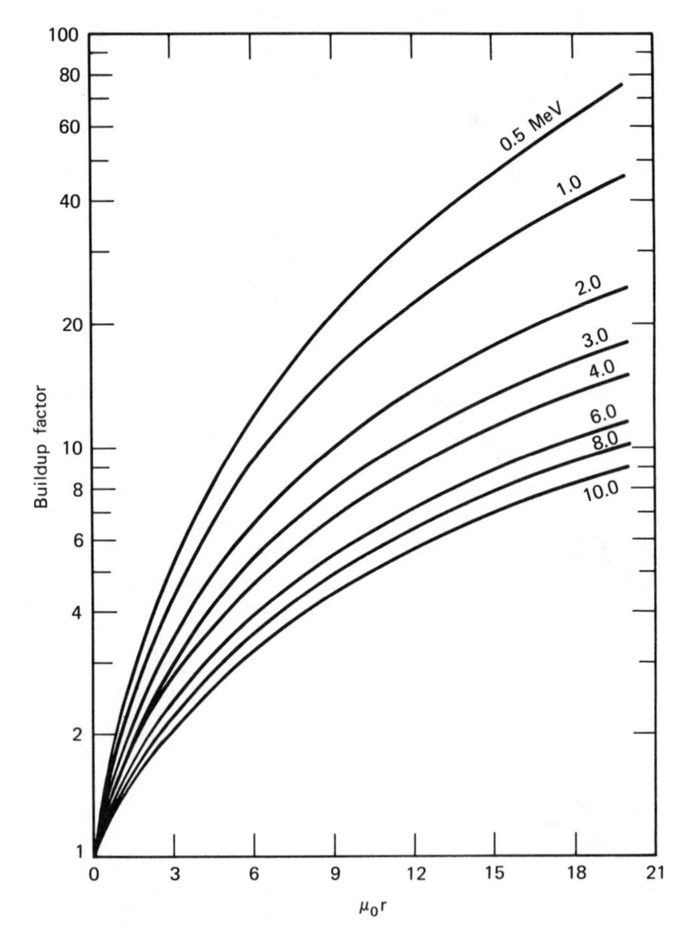

Fig. 4.22. Absorbed-dose buildup factor B_a for point isotropic γ ray source in ordinary concrete [after F. H. Clark and D. K. Trubey, *Nucl. Appl. Technol.*, 4, 37 (1968)].

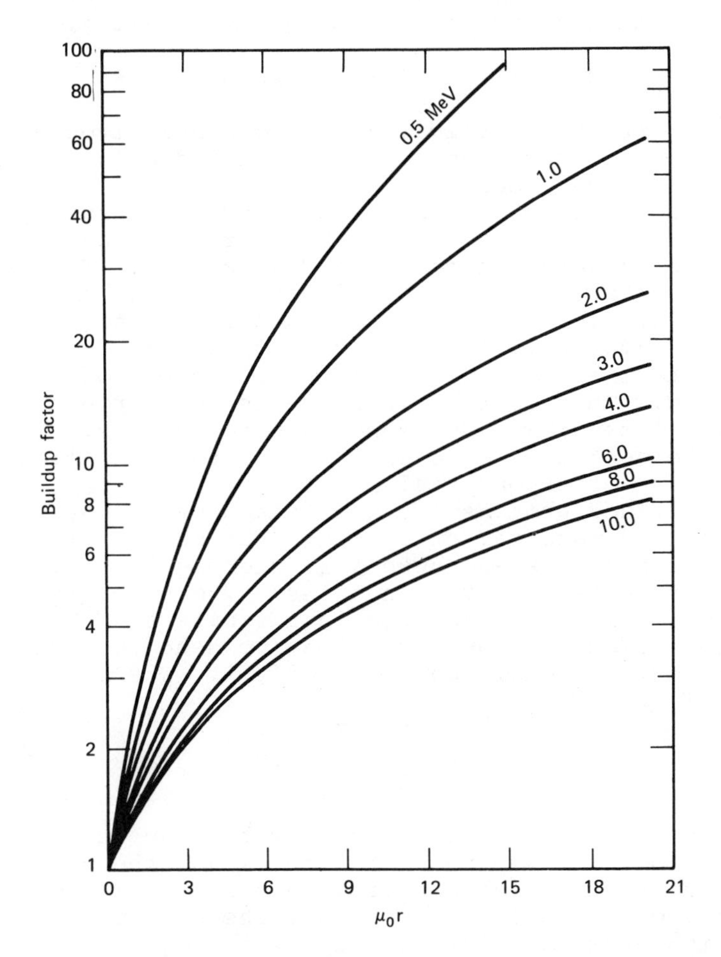

Fig. 4.23. Energy-absorption buildup factor B_E for point isotropic γ-ray source in ordinary concrete [after F. H. Clark and D. K. Trubey, *Nucl. Appl. Technol.*, 4, 37 (1968)].

has the advantage that all of the analytical solutions for the uncollided flux density from various sources can be used with the modified attenuation coefficients $(1 + \alpha_1)\mu$ and $(1 + \alpha_2)\mu$, then multiplying the respective terms by A and (1 - A) and summing. However, the fit is not as good as for the Berger expression,

$$B(\mu R, E') = 1 + C(E')\ \mu R \exp D(E')\ \mu R \qquad (4.82)$$

Plots [29] of the coefficients C and D are given in Figure 4.24 for the dose (exposure)-buildup factor for a point isotropic source.

The behavior of the dose, D, from uncollided photons and from all photons, for a point isotropic 0.26-MeV source in water is shown in Figure 4.25. The dose has been multiplied by $4\pi r^2$ and normalized to unity at zero radius. The dose from the uncollided photons decreases exponentially. However, the total dose increases at first and then decreases, because of the scattering. This initial increase is less pronounced in heavy elements and at higher source enrgies. The total dose decreases more slowly than the uncollided dose and is not exponential.

Uncollided photons are unaffected by the presence or absence of material other than that on the direct line between source and detector, whereas the scattered photons are affected by boundaries. The error involved in approximating the response in a finite slab by the infinite-medium buildup factor may be seen in Table 4.5 for a monodirectional beam source [30]. The ratio

$$k = \frac{B_E(\mu_0 x, \mu_0 x) - 1}{B_E(\mu_0 x, \infty) - 1} \qquad (4.83)$$

where the numerator uses the energy buildup at $\mu_0 x$ in a finite slab $\mu_0 x$ mean free paths thick and the denominator refers to the same position in a semiinfinite medium (half-space). It can be seen that the effect of backscattering from material beyond $\mu_0 x$ is negligible for high-energy γ rays in iron or lead. A correction should be made for low-energy γ rays in low-Z materials such as water. In general, the influence of finite boundaries is small at points more than three or four mean free paths from the boundary. Little information is available for correction of leakage from the lateral surfaces of a slab shield, and exact calculations are needed for accuracies better than 20% or 30%.

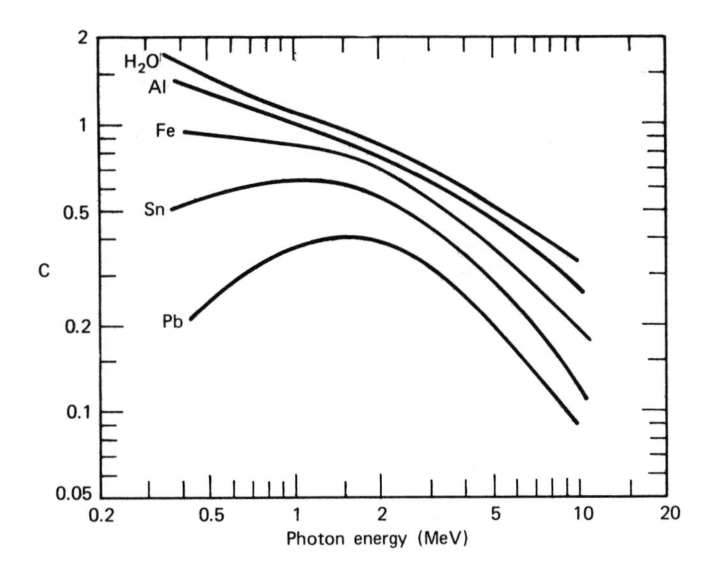

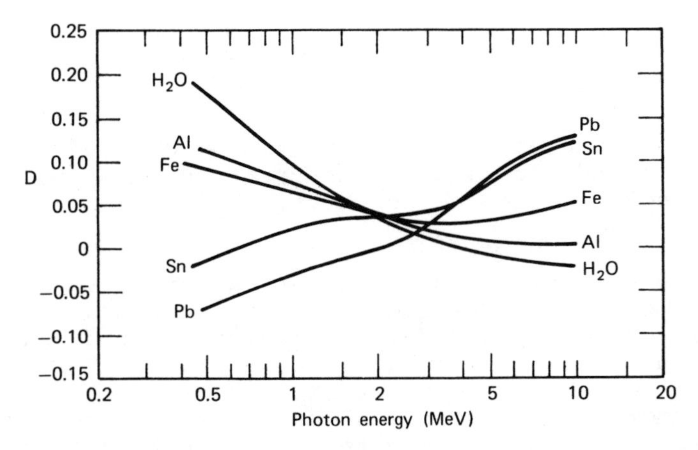

Fig. 4.24. Coefficients C and D for Berger expression for γ-ray dose-buildup factor for a point isotropic source.

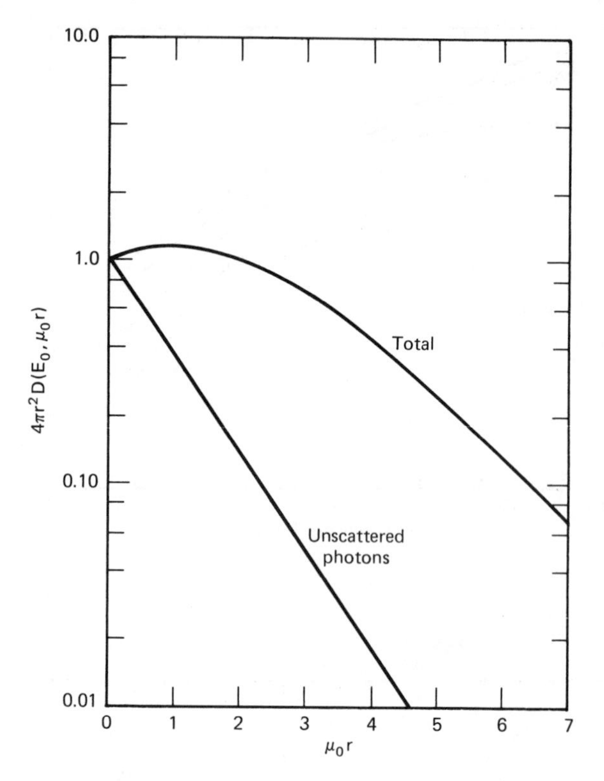

Fig. 4.25. Attenuation of dose from uncollided 0.26-MeV photons and uncollided-plus-scattered photons from a point isotropic source in water.

It is possible though not very accurate to use infinite-medium buildup factors in a laminated shield. Goldstein [31] suggests various recipes. If a low-energy source is surrounded by high-Z material such as lead, followed by low-Z material such as water, the photons emerging from the lead may be considered as a source for buildup in the water. The product of the buildup factors for the lead and the water should be used. When the source energy is well above the minimum in the $\mu(E)$ curve, the photons emerging from the lead will be at the energy of the minimum. The buildup factor for these photons in the water layer should then be used. At intermediate source energies, few

TABLE 4.5. RATIO k FOR BUILDUP IN FINITE SLAB AND HALF-SPACE

Medium E_0 (MeV)	k $\mu_0 x = 1$	2	4	8
		Water		
0.66	0.66	0.71	0.78	0.78
1.0	0.72	0.75	0.82	0.83
4.0	0.88	0.91	0.92	0.93
		Iron		
1.0	0.82	0.85	0.89	0.90
4.0	0.91	0.92	0.94	0.94
		Lead		
1.0	0.95	0.97	0.98	0.98
4.0	0.98	0.98	0.99	0.99

low-energy photons emerge from the lead; hence neglect buildup in lead. The buildup factor for the source-energy photons in the water layer should be used. When the low-Z material is closest to the source, followed by high-Z material, most low-energy photons will be absorbed before emerging from the surface of the shield. The buildup factor in the high-Z material only should be used.

A geometry in which very inaccurate results would be obtained with the kernel method and infinite-medium buildup factors is shown in Figure 4.26. A thick shield wall between source and detector allows very few photons to reach the detector. However, the dose might still be high because of scattering from the roof and air. A calculation considering only the direct path from source to detector seriously underestimates the dose. Any geometry in which any path exists with less attenuation than the direct path may allow scattering or streaming and should be calculated

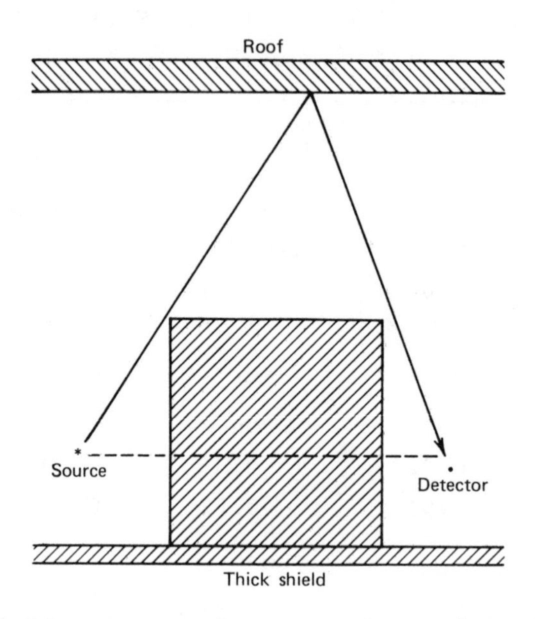

Fig. 4.26. Scattering around a thick shield.

with an exact method such as Monte Carlo, or the dose should be determined by experiment.

The kernel-buildup factor method has been used almost exclusively for γ-ray transport. The attenuation of fast neutrons could be handled, but the buildup factors are not available. Commonly used approximate methods for neutrons are discussed next.

4.5 OTHER APPROXIMATE METHODS

Kernel-Removal Cross-section Method

The attenuation of fast neutrons through hydrogeneous shields may be calculated approximately with the empirical Albert-Welton kernel and removal cross sections. The method applies to mixtures of water and heavy metals, or to laminations, provided the last metal layer is followed by at least 50 cm of water. The method assumes that a collision with hydrogen is equivalent to absorption; hence the cross section for removal from the beam is $\Sigma_H = \Sigma_t$. On the other hand, energy is not reduced so much, and small-angle collisions are more probable in heavy elements. Removal cross sections are determined experimentally and are approximately 2/3 of Σ_t. The fast-neutron flux density goes as

$$\phi(r) \alpha \exp \left(- \underset{i}{S} f_i \Sigma_{R_i} r\right) \int_0^{\infty} S(E') \exp -\Sigma_H(E') r \, dE' \tag{4.84}$$

where f_i is the volume fraction of the i-th heavy element. Removal cross sections Σ_R and mass removal coefficients $\Sigma_{R/\rho}$ are listed in Table 4.6. The dose (rad/hour per neutron/second) from a point fission source [32] in water is

$$D(r) = \frac{2.78 \times 10^{-5}}{4\pi r^2} \left[R^{0.349} \exp -0.422R^{0.58}\right]$$

$$\exp -0.0308R \tag{4.85}$$

where $R = f_w r$, with f_w as the volume fraction of water. This expression should be multiplied by

$$\exp -(1 - f_w) \Sigma_R \tag{4.86}$$

for attenuation in the heavy element to obtain the dose at the shield surface from fast neutrons.

Removal-diffusion Method

The kernel-removal method is restricted to shields containing hydrogen and cannot calculate the thermal neutron flux density (needed for the capture γ-ray source distribution) or intermediate energy neutron dose. The removal-diffusion method couples a removal calculation for fast neutrons to an age-diffusion (actually, multigroup diffusion) calculation for lower energies. Examples of computer programs embodying the method are RASH [33], MAC [34], and NRN [35], available from the Radiation Shielding Information Center. The penetrating component includes uncollided neutrons and those that have made only small-angle elastic

TABLE 4.6. FAST-NEUTRON-REMOVAL CROSS SECTIONS AND MASS-ATTENUATION COEFFICIENTS[a]

Element	Atomic Number	ρ (g/cm^3)	Σ_R/ρ (Calculated) (cm^2/g)	Σ_R (cm^{-1})	Σ_R/ρ (Experimental) (cm^2/g)
Aluminum	13	2.699	0.0293	0.0792	0.0292 ± 0.0012
Antimony	51	6.691	0.0136	0.0907	
Argon	18		0.0244		
Arsenic	33	5.730	0.0173	0.0993	
Barium	56	3.500	0.0129	0.0450	
Beryllium	4	9.013	0.0678	0.1248	0.0717 ± 0.0043
Bismuth	83	9.747	0.0103	0.1003	0.010 ± 0.0010
Boron	5	3.330	0.0575	0.1914	0.0540 ± 0.0054
Bromine	35	3.120	0.0168	0.0523	
Cadmium	48	8.648	0.0140	0.1213	
Calcium	20	1.540	0.0230	0.0354	
Carbon	6	1.670	0.0502	0.0838	0.0407 ± 0.0024
Cerium	58	6.900	0.0126	0.0870	
Cesium	55	1.873	0.0130	0.0243	
Chlorine	17		0.0252		0.020 ± 0.014
Chromium	24	6.920	0.0208	0.1436	
Cobalt	27	8.900	0.0194	0.1728	
Copper	29	8.940	0.0186	0.1667	0.0194 ± 0.0011
Dysprosium	66	8.562	0.0117	0.1003	
Erbium	68	4.770	0.0115	0.0550	
Europium	63	5.166	0.0120	0.0621	
Fluorine	9		0.0361		0.0409 ± 0.0020

Gadolinium	64	7.868	0.0119	0.0938	
Gallium	31	5.903	0.0180	0.1060	
Germanium	32	5.460	0.0176	0.0963	
Gold	79	19.320	0.0106	0.2045	
Hafnium	72	13.300	0.0112	0.1484	
Helium	2		0.1135		
Holmium	67		0.0116		
Indium	49	7.280	0.0139	0.1009	
Iodine	53	4.930	0.0133	0.0654	
Iridium	77	22.420	0.0107	0.2408	
Iron	26	7.865	0.0198	0.1560	0.0214 $\pm$ 0.0009
Krypton	36		0.0165		
Lanthanum	57	6.150	0.0127	0.0783	
Lead	82	11.347	0.0104	0.1176	0.0103 $\pm$ 0.0009
Lithium	3	0.534	0.0840	0.0449	0.094 $\pm$ 0.007
Lutetium	71		0.0112		
Magnesium	12	1.741	0.0307	0.0535	
Manganese	25	7.420	0.0203	0.1505	
Mercury	80	13.546	0.0105	0.1424	
Molybdenum	42	10.200	0.0151	0.1543	
Neodymium	60	6.960	0.0124	0.0861	
Neon	10		0.0340		
Nickel	28	8.900	0.0190	0.1693	0.0190 $\pm$ 0.0010
Niobium	41	8.400	0.0153	0.1288	
Nitrogen	7		0.0448		
Osmium	76	22.480	0.0108	0.2432	
Oxygen	8		0.0405		0.031 $\pm$ 0.002
Palladium	46	12.160	0.0144	0.1747	
Phosphorus	15	1.820	0.0271	0.0493	
Platinum	78	21.370	0.0107	0.2279	
Potassium	19	6.475	0.0237	0.1533	
Praseodymium	59	6.500	0.0125	0.0812	

(Continued)

TABLE 4.6. Continued

Element	Atomic Number	ρ (g/cm^3)	Σ_R/ρ (Calculated) (cm^2/g)	Σ_R (cm^{-1})	Σ_R/ρ (Experimental) (cm^2/g)
Radium	88	5.000	0.0100	0.0498	
Rhenium	75	20.530	0.0109	0.2238	
Rhodium	45	12.440	0.0145	0.1810	
Rubidium	37	1.532	0.0163	0.0249	
Ruthenium	44	12.060	0.0147	0.1777	
Samarium	62	7.750	0.0121	0.0941	
Scandium	21	3.020	0.0224	0.0676	
Selenium	34	4.800	0.0170	0.0818	
Silicon	14	2.420	0.0281	0.0681	
Silver	47	10.503	0.0142	0.1491	
Sodium	11	0.971	0.0322	0.0313	
Stronium	38	2.540	0.0160	0.0407	
Sulfur	16	2.070	0.0261	0.0540	
Tantalum	73	16.600	0.0111	0.1838	
Tellurium	52	6.240	0.0134	0.0837	
Terbium	65		0.0118		
Thallium	81	11.860	0.0104	0.1238	
Thorium	90	11.300	0.0098	0.1111	
Thulium	69		0.0114		
Tin	50	6.550	0.0137	0.0898	
Titanium	22	4.500	0.0218	0.0981	
Tungsten	74	19.300	0.0110	0.2120	0.0082 ± 0.0018

Uranium	92	18.700	0.0097	0.1816	0.0091 $\pm$ 0.0010
Vanadium	23	5.960	0.0213	0.1267	
Xenon	54		0.0131		
Ytterbium	70		0.0113		
Yttrium	39	3.800	0.0158	0.0599	
Zinc	30	7.140	0.0183	0.1306	
Zirconium	40	6.440	0.0156	0.1001	

[a] Table taken from L. K. Zoller, "Fast-Neutron-Removal Cross Sections," *Nucleonics*, 22, 128-129 (1964).

scatterings. Their behavior may be approximated by an exponential attenuation with removal cross sections in some 18 to 30 groups or bands. The removal cross sections may be determined experimentally, or the transport cross section Σ_{tr} may be used. In general, the constants supplied with the codes should be used. Then 16 to 31 diffusion groups are defined, overlapping the removal bands but extending down to thermal energy. Transfer is allowed between removal bands and diffusion groups; hence removed neutrons (corresponding to inelastic scattering or large-angle elastic scattering) serve as a source for the diffusion calculation. Although diffusion theory is not accurate far from the source or near boundaries, the removed neutrons act as a local source, and the diffusion near a boundary is handled well enough for the purpose.

Interpolation Method

Perhaps the easiest method of shield design is to interpolate, extrapolate, scale, or otherwise adapt the results of exact calculations or experiment for a similar situation. For example, the thickness of lead walls for X-ray installations, and shielding for radioactive sources, are often determined in this way. The dose from distributed sources can be obtained by superposition or integration, as in the kernel method, and dose may be scaled to source strength. If results for several source energies and angles of incidence are available, an arbitrary source spectrum and angular distribution can be handled. Adjustment can be made for varying density by measuring material thickness in mean free paths (but using the correct distance for the $1/4\pi r^2$ factor in geometrical attenuation from a point source). The total dose at the shield surface may be used directly, or calculations may be simplified using the coefficients of a curve fit. It is common to work with "attenuation length," the thickness in which the dose falls by a factor of $(1/e) = 0.368$, or "tenth-value layer" for 0.1 the dose, or half-value layer (HVL). However, the relaxation length or similar quantity may vary with depth, especially near the source, where another description should be used.

Figure 4.27 plots the fast-neutron dose (in gray per 10^{10} incident n/cm^2) for 14 MeV neutrons on concrete (0.61% hydrogen, 54% oxygen, 34% silicon, 5.3% aluminum, and 6.5% calcium by weight, $\rho = 2.3$ g cm^{-3}) [36]. From 40 cm to 100 cm the attenuation length is

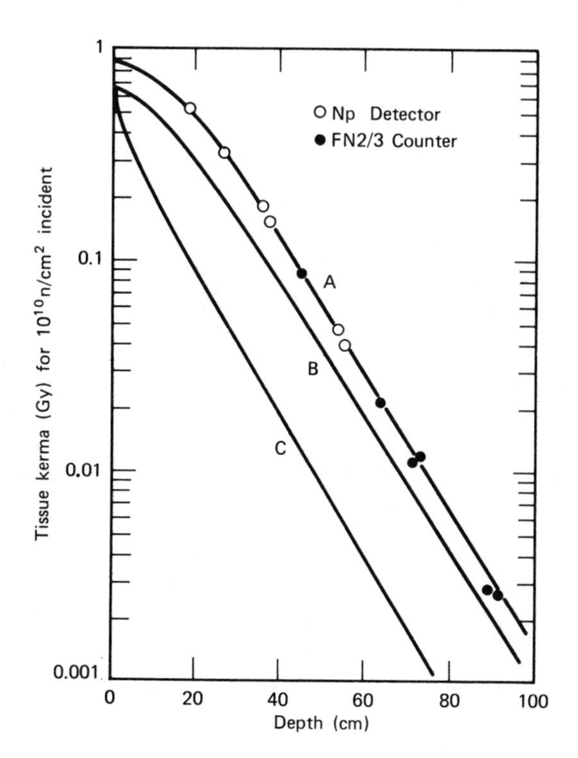

Fig. 4.27. Attenuation of fast-neutron dose in concrete from a normally incident beam of 14-MeV neutrons on thick block (A), slab of given thickness (B), and for isotropic plane source (C).

13.5 cm. Doubling the hydrogen concentration reduces the attenuation length to 12 cm. The shape from 0 to 40 cm depends on the source angular distribution.

Single-scatter and Albedo Methods

Problems such as scattering from the roof in Figure 4.26 may be approximated by a single-scatter method if the roof is thin (much less than Σ_t^{-1}; hence multiple scattering can be neglected), or by an albedo method if the roof is thick. The albedo method is also applied to ducts as discussed in Chapter 7. Consider the single-scatter geometry in Figure 4.28. The uncollided flux density at dV is

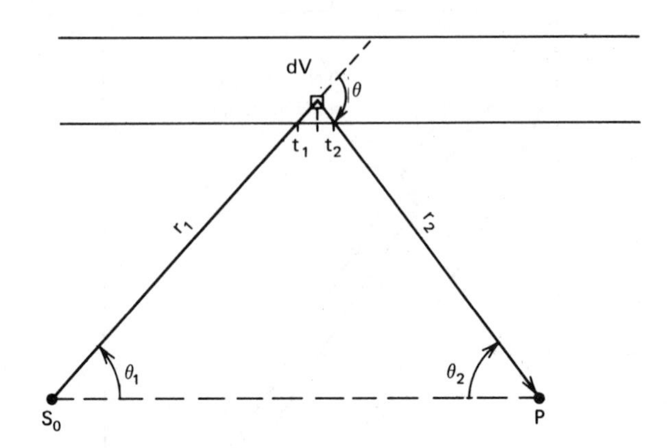

Fig. 4.28. Single-scattering geometry.

$$\phi_0 = S_0 \frac{\exp(-\Sigma_t t_1 \sec \theta_1)}{4\pi r_1^2} \tag{4.87}$$

The number of first scatters in dV is $\Sigma_s \phi_0 dV$, and the flux density at detector point P from these scatters is

$$d\phi_1 = \phi_0 \, dV \frac{\Sigma_s(\theta)}{r_2^2} \exp(-\Sigma_t t_2 \sec \theta_2) \tag{4.88}$$

where $\Sigma_s(\theta)$ is the macroscopic differential scattering cross section. The energy can be obtained from θ also. The contribution from the entire roof is obtained by integrating over the volume. Often the attenuation in the thin slab can be neglected and r_1 and r_2 measured to the midplane.

When the scattering material is thick, one may use an albedo, defined as the fraction of incident particles that emerge. It is assumed that the particles emerge from the same area in which they entered, but their spectrum and angular distribution is different because of the scattering and absorption inside. The albedo is a function of the incident energy and angle and may be specified in terms of a current or dose emerging at another angle. Further discussion of albedos may be found in Chapter 7.

4.6 ATMOSPHERIC TRANSPORT

The diffusion of radioactive gases or aerosols in the atmosphere is a convective rather than radiative transport process. Submersion in a cloud or plume of contaminated air, or inhalation of radionuclides, may be significant modes of exposure to ionizing radiation. Here we are concerned with calculation of the concentration χ (Ci m^{-3}) of activity in the air. Calculation of the dose from the concentration and exposure time is discussed in Chapter 8.

The turbulent diffusion and dilution of contaminated air is governed by the wind and the stability of the atmosphere. For a point emitter of radioactive gas (or particulates if settling is neglected) at height h above plane ground, the concentration given by the Gaussian plume model [37] is

$$\chi = \frac{Q'}{2\pi\sigma_y\sigma_z u} \exp\left(-\frac{y^2}{2\sigma_y^2}\right)\left\{\exp\left[-\frac{(z-h)^2}{2\sigma_z^2}\right] + \exp\left[-\frac{(z+h)^2}{2\sigma_z^2}\right]\right\} \tag{4.89}$$

where

Q' = rate of release of activity (Ci s^{-1});
σ_y = lateral dispersion coefficient (m), at x (m) downwind;
σ_z = vertical dispersion coefficient (m), at x (m) downwind;
u = wind speed (m s^{-1}); and
h = height (m) above ground ($z = 0$).

The rectangular coordinate system is oriented with the x-axis along the direction of the wind, the y-axis crosswind, and the z-axis vertical with the origin at ground level beneath the emitter. The lateral and vertical dispersion coefficients vary with distance x (m) downwind and with the stability. An unstable (turbulent) atmosphere means more mixing hence lower concentration. Usually the Pasquill stability classes A through F (extremely stable to moderately unstable) are used. The lateral dispersion coefficient $\sigma_y(x)$ is plotted in Figure 4.29, and the vertical dispersion coefficient $\sigma_z(x)$ is plotted in Figure 4.30.

Stable atmospheric conditions, hence little vertical mixing, occur when there is a temperature inversion (temperature increasing with altitude near the ground, instead of decreasing). The Pasquill class can be obtained from measurements of the temperature gradient or can be inferred from records of the wind direction as a function of time. In the absence of such information, an estimate of the stability class can be obtained from wind speed and isolation (sunshine) or nightime radiation conditions (affected by cloud cover)

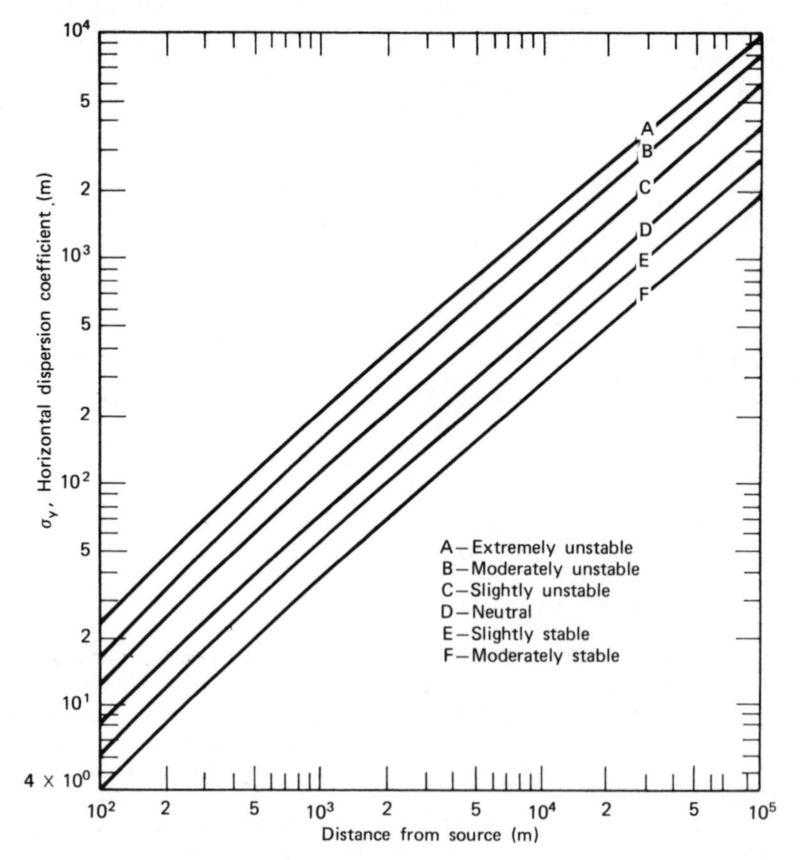

Fig. 4.29. Lateral dispersion coefficient as function of distance downwind and Pasquill stability class (after Slade [37]).

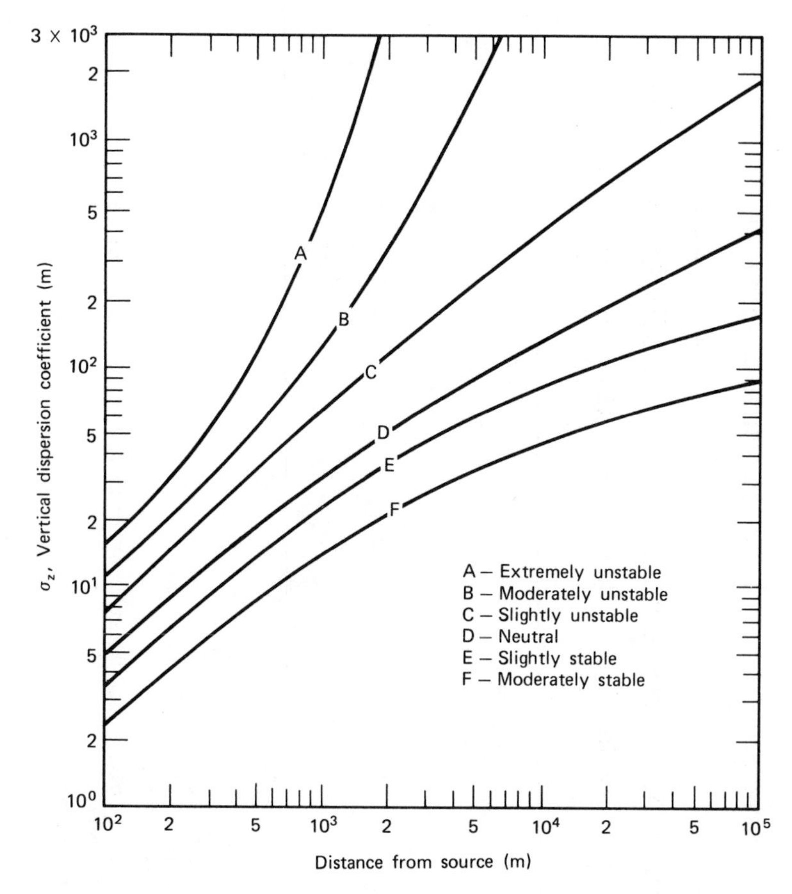

Fig. 4.30. Vertical dispersion coefficient as function of distance downwind and Pasquill stability class (after Slade [37]).

as given in Table 4.7. Regulations governing licensing of nuclear facilities may specify average or pessimistic conditions to be used for a preliminary evaluation, until on-site meteorological data are obtained. Pasquill class F and a wind speed of 1 meter per second are very pessimistic, giving little mixing and transport, hence large concentrations.

TABLE 4.7. PASQUILL STABILITY CLASS[a] AND WEATHER CONDITIONS

Wind Speed ($m\ s^{-1}$)	Daytime (Insolation)			Nightime (Cloudiness)	
	Strong	Moderate	Slight	Thin Overcast or > 1/2 Cloudiness	Cloudiness $\leq$ 3/8
<2	A	A to B	B		
2	A to B	B	C	E	F
4	B	B to C	C	D	E
6	C	C to D	D	D	D
>6	C	D	D	D	D

[a]Classifications:

A = extremely unstable;
B = moderately unstable;
C = slightly unstable;
D = neutral (applicable to heavy overcast, day or night);
E = slightly stable;
F = moderately stable.

If the emitter releases a puff of activity of Q curies, instead of a continuous release, Eq. (4.89) may be used to calculate the average "exposure" ψ(curies-seconds) received at a given point during the passage of the puff, by replacing χ by ψ and Q' by Q.

REFERENCES

1. R. R. Coveyou and R. D. MacPherson, "Fourier Analysis of Uniform Random Number Generators," *J. Assoc. Comp. Mach.*, 14, 100 (1967).
2. L. L. Carter and E. D. Cashwell, "Particle-Transport Simulation with the Monte Carlo Method," TID-26607 (1975). National Technical Information Service, Springfield, Va.
3. E. J. McGrath et al., "Techniques for Efficient Monte Carlo Simulation," ORNL-RSIC-38 (1975).
4. J. Spanier and E. M. Gelbard, *Monte Carlo Principles and Neutron Transport Problems*, Addison-Wesley, Reading, Mass., 1969.
5. F. H. Clark, "The Exponential Transform as an Importance-Sampling Device," ORNL-RSIC-14, Oak Ridge, Tenn. (1966).
6. E. D. Cashwell and C. J. Everett, *A Practical Manual on the Monte Carlo Method for Random Walk Problems*, Pergamon Press, New York, 1959.
7. E. A. Straker, P. N. Stevens, D. C. Irving, and V. R. Cain, "The MORSE Code--A Multigroup Neutron and Gamma-Ray Monte Carlo Transport Code," ORNL-4585 (1970). Available in Radiation Shielding Information Center Code Package CCC-127.
8. T. P. Wilcox, "MORSE-L, A Special Version of the MORSE Program Designed for Neutron, Gamma, and Coupled Neutron-Gamma Penetration Problems," UCID-16680 (1972). Available in RSIC Code Package CCC-261.
9. Gerald Goertzel and Malvin H. Kalos, "Monte Carlo Methods in Transport Problems," in *Progress in Nuclear Energy*, Series I, Vol. 2, Pergamon Press, New York, 1958.
10. R. R. Coveyou, V. R. Cain, and K. J. Yost, "Adjoint and Importance in Monte Carlo Application," *Nucl. Sci. Eng.*, 27, 219-234 (1967).
11. Bengt G. Carlson, "The Numerical Theory of Neutron Transport," in B. Adler, S. Fernback, and M. Rotenberg, Eds., *Methods in Computational Physics*, Vol. 1, Academic Press, New York, 1963, pp. 1-42.
12. Clarence E. Lee, "The Discrete S_n Approximation to Transport Theory," Los Alamos Scientific Lab. Report LA-2595, Los Alamos, N. Mex. (1962).
13. F. R. Mynatt, "The Discrete Ordinates Method in Problems Involving Deep Penetrations," in *A Review of the Discrete Ordinates S_n Method for Radiation*

Transport Calculations, Oak Ridge National Lab. Report ORNL-RSIC-19, Oak Ridge, Tenn. (1968), pp. 25-51.

14. K. D. Lathrop, "Discrete Ordinates Quadratures," ORNL-RSIC-19, pp. 53-63.
15. R. J. Cerbone and K. D. Lathrop, "S_n Calculation of Highly Forward Peaked Neutron Angular Fluxes Using Asymmetrical Quadrature Sets," *Nucl. Sci. Eng.*, 35, 139-141 (1969).
16. W. W. Engle, Jr., "A User's Manual for ANISN, A One-Dimensional Discrete Ordinates Transport Code with Anisotropic Scattering," K-1693 (1967). Included in RSIC Code Package CCC-82.
17. K. D. Lathrop, "DTF-IV, A Fortran IV Program for Solving the Multigroup Transport Equation with Anisotropic Scattering," LA-3373 (1965).
18. F. R. Mynatt, F. J. Muckenthaler, and P. N. Stevens, "Development of Two-Dimensional Discrete Ordinates Transport Theory for Radiation Shielding," CTC-INF-952 (1969). Available in RSIC Code Package CCC-89.
19. DOT Two-Dimensional Discrete Ordinates Transport Code, Radiation Shielding Information Center Code Package CCC-89.
20. "Reactor Shielding for Nuclear Engineers," in N. M. Schaeffer, Ed., TID-25951 (1973), National Technical Information Service, Springfield, Va. 22151, Chapter 4.
21. L. V. Spencer and U. Fano, "Calculation of Spatial Distributions by Polynomial Expansion," *J. Res. Nat. Bur. Stand.*, 46, 446 (1951).
22. R. G. Jaeger, Ed., *Engineering Compendium on Radiation Shielding*, Vol. I, Springer, New York, 1968, pp. 127-135.
23. *Engineering Compendium*, Vol. I, pp. 124-127, 363-416.
24. QAD, Kernel Integration Code System, Radiation Information Center Computer Code Package CCC-48.
25. Herbert Goldstein and J. Ernest Wilkins, Jr., "Calculations of the Penetration of Gamma Rays," NYO-3075 (1954).
26. W. G. Cross and H. Ing, "Spectra and Dosimetry of Neutrons Interacting with Concrete Shielding," IVth Congress International Radiation Protection Association. See also Clark and Trubey, *Nucl. Appl.*, 4, 37-41 (1968).
27. D. K. Trubey, "A Survey of Empirical Functions Used to Fit Gamma Ray Buildup Factors," ORNL-RSIC-10 (1966).

28. *Engineering Compendium*, Vol. I, pp. 210-226.
29. A. B. Chilton, "Two-Parameter Formula for Point Source Buildup Factors," *Nucleonics*, 23 (8), 119 (1965).
30. M. J. Berger and J. Doggett, *J. Res. Nat. Bur. Stand.*, 56, 89C (1956).
31. Herbert Goldstein, *Fundamental Aspects of Reactor Shielding*, Addison-Wesley, Reading, Mass., 1959, pp. 223-226.
32. A. W. Casper, "Modified Fast Neutron Attenuation Functions," General Electric Report XDC-60-2-76 (1960).
33. A. F. Avery et al., "Methods of Calculation for Use in the Design of Shields for Power Reactors," UK Report AERE-R-3216 (1960).
34. E. G. Peterson, "MAC-A Bulk Shielding Code," Hanford Report HW-73381 (1962).
35. L. Hjarne and M. Leimdorfer, "A Method for Predicting the Penetration and Slowing Doses of Neutrons in Reactor Shields," *Nucl. Sci. Eng.*, 24, 165 (1966).
36. F. H. Clark, N. A. Betz, and J. Brown, "Monte Carlo Calculation of the Penetration of Normally Incident Neutron Beams Through Concrete," Report ORNL-3926 (1967).
37. D. H. Slade, "Meteorology and Atomic Energy-1968," TID-24190, NTIS, Springfield, Va. (1968).

PROBLEMS

1. Generate 100 random numbers, normalized to $0 < r < 1$. (You may use a computer or calculator, table of random digits, dice, wheel or spinner, etc.). Check that they are more or less uniformly distributed and that there is no obvious correlation between pairs. Use the random numbers to select fast-neutron interactions from elastic scattering, inelastic scattering, (n,p) reaction, and (n,α) reaction: cross sections are 1.0 b (elastic), 0.5 b (inelastic), 0.1 b (n,p), and 0.1 (n,α). Compare the "Monte Carlo" results with the exact probabilities of each interaction.

2. Write a simple program to generate (x,y,z) coordinates of source particles uniformly distributed within a cube of side dimension A, emit them isotropically, select path lengths with a mean free path λ, test whether the particle escapes or is scattered isotropically, or is absorbed with a probability of 0.1 per collision, and score the fraction of particles absorbed and the fraction escaping. Solve this problem for $A = 3$ and $\lambda = 1$.

3. If a one-dimensional discrete-ordinates program such as ANISN is installed on your computer, and suitable multigroup cross sections are available, calculate the scalar and angular flux density spectrum at the surface of a sphere of shielding material, at least five mean free paths in radius, with a monoenergetic isotropic source imbedded in a small volume at the center. Each student may run a different calculation to investigate the influence of S_n quadrature, P_ℓ truncation, convergence criterion, and other parameters on the precision and time.

4. Use an S_8 quadrature to integrate the product of the laboratory-system differential cross section for neutron elastic scattering in hydrogen and the angular flux density $\psi(\theta) = A(\cos\theta + 1)$ from $\theta = 0°$ to $180°$.

5. Calculate the uncollided and total dose rate at the surface of a 10-cm diameter lead sphere with a 10-mCi ^{60}Co source at the center. Multiply the

uncollided flux density (photons/cm^{2}s) by 6×10^{-10} to obtain the dose rate in rad per second (from Chapter 5).

6. Derive an expression for the uncollided flux density at a certain distance from a line source of radiation emitting S_L particles per second, per meter.

7. Calculate the ground-level concentration under the centerline of a plume, 1000 m downwind from a stack releasing 5 Ci s^{-1}, for slightly stable conditions and a wind speed of 2 m s^{-1}.

5 DOSIMETRY

Biological effects of radiation are correlated with the energy absorbed by ionization in unit mass of tissue and modified by the microscopic spatial distribution of ionization events, the concentration of oxygen, and sometimes the rate of energy absorption because of biological repair. The effects of the energy absorbed are discussed in Chapter 6. Here we are concerned with definition of physical quantities related to energy absorption and methods of calculating and measuring them.

Radiation damage to organic materials and insulators is also related to energy absorbed by ionization. Damage to metals and semiconductors is related to atomic displacements and nuclear transmutations. Usually the spectrum is obtained and folded with an empirical "damage function" for the mechanical, electrical, or other property of interest. For approximate comparison of the effects of neutron irradiation in similar spectra, the fluence above 1 MeV may be adequate.

Heating by nuclear radiation is important in nuclear reactors and the inner regions of the shields.

5.1 DEFINITIONS OF DOSE QUANTITIES

The quantities recommended by the International Commission on Radiological Units and Measurements (ICRU) are summarized in Table 5.1.

TABLE 5.1. QUANTITIES USED IN DOSIMETRY

Quantity	Symbol	h(E)	Units
Flux density	ϕ	1	particles cm^{-2} s^{-1}
Fluence	Φ		particles cm^{-2}
Intensity	I	E	J cm^{-2} s^{-1} (or MeV cm^{-2} s^{-1})
Energy fluence	F		J cm^{-2} (or MeV cm^{-2})
Kerma rate	$\dot{K}$	$E(\mu_K/\rho)$	J kg^{-1} s^{-1}
Kerma	K		J kg^{-1}
Exposure rate	$\dot{X}$	$c\frac{E}{W}(\mu_{en}/\rho)$	R s^{-1}
Exposure	X	(air)	R (roentgen)
Absorbed dose rate	$\dot{D}$		Gy s^{-1} (or rad s^{-1})
Absorbed dose	D		Gy (gray) or rad
Dose-equivalent rate	$\dot{H}$		rem s^{-1}
Dose-equivalent	H		rem
Dose-equivalent index	H_I		rem
Heating rate	$\dot{P}$		W (watt) kg^{-1}
Heating	P		J (joule)
Radiation damage or response	R	$f_D(E)$	depends on property

Radiation transport programs or a suitable spectrometer may be used to obtain $\phi(\underline{r},E)$, the spectrum of the scalar flux density at position $\underline{r}$.

The response of an isotropic, nonperturbing detector or dosimeter at $\underline{r}$ is

$$R(\underline{r}) = \int h(E)\ \phi(\underline{r},E)\ dE \tag{5.1}$$

where h(E) is the response function. If h(E) = 1, the response is the (scalar) flux density, $\phi(\underline{r})$ defined in Chapter 3. The integral of the flux density over time is the fluence

$$\Phi(\underline{r}) = \int \phi(\underline{r})\ dt \tag{5.2}$$

If h(E) = E, we obtain the energy flux density or intensity

$$I(\underline{r}) = \int E\phi(\underline{r},E)\ dE \tag{5.3}$$

The corresponding integral over time is the energy fluence, $F(\underline{r})$.

The kerma (K) refers to kinetic energy released in material. The ICRU [1] defines

$$K = \frac{dE_K}{dm} \tag{5.4}$$

where dE_K is the sum of the initial kinetic energies of all the charged particles liberated by indirectly ionizing particles (photons or neutrons) in a volume element of specified material and mass dm. The d notation signifies a limiting process, but the volume is not allowed to become so small that statistical fluctuations are important. The units of kerma are joule kg^{-1} or in older literature, erg g^{-1}. Kerma may be calculated with

$$h(E) = E\ \frac{\mu_K}{\rho} \tag{5.5}$$

where (μ_K/ρ) is the mass-energy transfer coefficient.

Exposure is defined as

$$X = \frac{dQ}{dm} \tag{5.6}$$

where dQ is the sum of the charge of all ions produced in air by the electrons (including positrons) liberated by photons interacting in the volume element of mass dm. Note that the ions do not have to be generated in dm; energetic electrons liberated in the volume element can travel some distance before losing all their energy. The ionization arising from the absorption of bremsstrahlung emitted by the secondary electrons is not included in dQ. Thus the mass-energy absorption coefficient (μ_{en}/ρ) is slightly smaller than the mass-energy transfer coefficient. Except for this difference, significant only at high energies, exposure is proportional to kerma for air. We have

$$h(E) = c\,\frac{E}{W}\left(\frac{\mu_{en}}{\rho}\right)_{air} \tag{5.7}$$

where W is the energy required to generate one ion-electron pair in air (34 eV). The special unit of exposure is the roentgen,

$$1\ R \equiv 2.58 \times 10^{-4}\ C\ kg^{-1} \tag{5.8}$$

hence $c = 6.209 \times 10^{-13}$ g-R per ion-electron pair which is equivalent to the original definition of the charge in electrostatic units (esu) liberated by photons interacting in 1 cm^3 of dry air at standard temperature and pressure (STP). Note that exposure is defined only for photons. It was the first "dose" quantity defined and is relatively easy to measure with an air-filled ionization chamber, if the energy is below 3 MeV. Exposure is being replaced by absorbed dose and dose-equivalent, which apply to all ionizing radiations.

Absorbed dose is defined as

$$D = \frac{dE_D}{dm} \tag{5.9}$$

where dE_D is the energy imparted to the matter in dm.

The energy imparted may appear as ionization and excitation, or as increase in chemical bond energy. Since the distribution of energy absorption has to be determined as well as the energy transferred, absorbed dose is not simply related to the scalar flux density spectrum alone, and h(E) cannot be defined. However, when charged-particle equilibrium occurs (as discussed later) and bremsstrahlung losses are negligible, absorbed dose is proportional to kerma. The special SI unit of absorbed dose is the gray,

$$1 \text{ Gy} \equiv 1 \text{ J kg}^{-1} \tag{5.10}$$

but a more common unit is the rad,

$$1 \text{ rad} = 0.01 \text{ J kg}^{-1} \tag{5.11}$$

Dose equivalent [2] is a more accurate measure of potential biological effect than absorbed dose. It is defined as

$$H = DQN \tag{5.12}$$

where D is the absorbed dose (rad), Q is the quality factor, and N is the product of any other modifying factors [such as oxygen enhancement, but for radiation protection (N = 1)]. The quality factor takes into account the effect of the microscopic distribution of the energy absorbed. It is equal to 1 for X rays, γ rays, and electrons, but Q increases to 10 for fast neutrons and protons and may be larger for α particles and other densely ionizing ions. The special unit of the dose equivalent is the rem (rad-equivalent-man). It is assumed that dose equivalents from different types of radiation are additive for radiation protection purposes.

Dose equivalent and similar quantities are defined for a small, nonperturbing detector at position $\underline{r}$. For biological effects, however, one would like to know the dose equivalent in specific organs and thus the spatial distribution of the dose equivalent in the body, or at least the maximum dose equivalent. In general, this is not the same as the entrance dose equivalent because of attenuation, backscattering, and absorption of γ rays that might otherwise escape from a small detector. The ICRU has proposed the quantity dose-equivalent index,

H_I, defined as the maximum value of the dose-equivalent in a 30-cm-diameter sphere of unit density and composition equivalent to soft tissue, centered at $\underline{r}$.

Heating is not discussed in the ICRU reports. Except for the small amount of energy that may be retained in chemical bonds, heating is equivalent to absorbed dose because ionization and excitation energy are eventually dissipated as heat.

Radiation damage from neutrons in metals and semiconductors is usually defined in terms of the change in some property (e.g., yield strength or electrical resistance) per unit fluence, measured under defined conditions of irradiation including temperature, stress, and so on. A damage function $f_D(E)$ relates the influence of the neutron energy to the change in the property. The change in property, or damage, in a similar environment but different spectrum may then be predicted by the integral of the product of the damage function and spectrum.

5.2 FLUENCE

The particle or energy fluence may be calculated by the methods discussed in Chapter 4.

Radiation detectors and electronics are discussed in the literature [3-6]; hence only applications to measurements in shielding and radiology are considered here.

Ions

The particle fluence and energy spectrum of α particles and other ions can be measured with a silicon surface barrier detector, such as illustrated in Figure 5.1. Only very pure crystalline semiconductors are suitable for radiation detectors, and only silicon and germanium are produced in large quantities. An advantage of semiconductors is the small W, or energy required to generate a hole-electron pair (3.66 eV for Si, 2.96 eV for Ge). The number of hole-electron pairs is

$$N = \frac{E}{W} \tag{5.13}$$

where E is the energy deposited by ionization in the sensitive volume of the detector. If the ion is stopped within this volume, then E is the ion kinetic

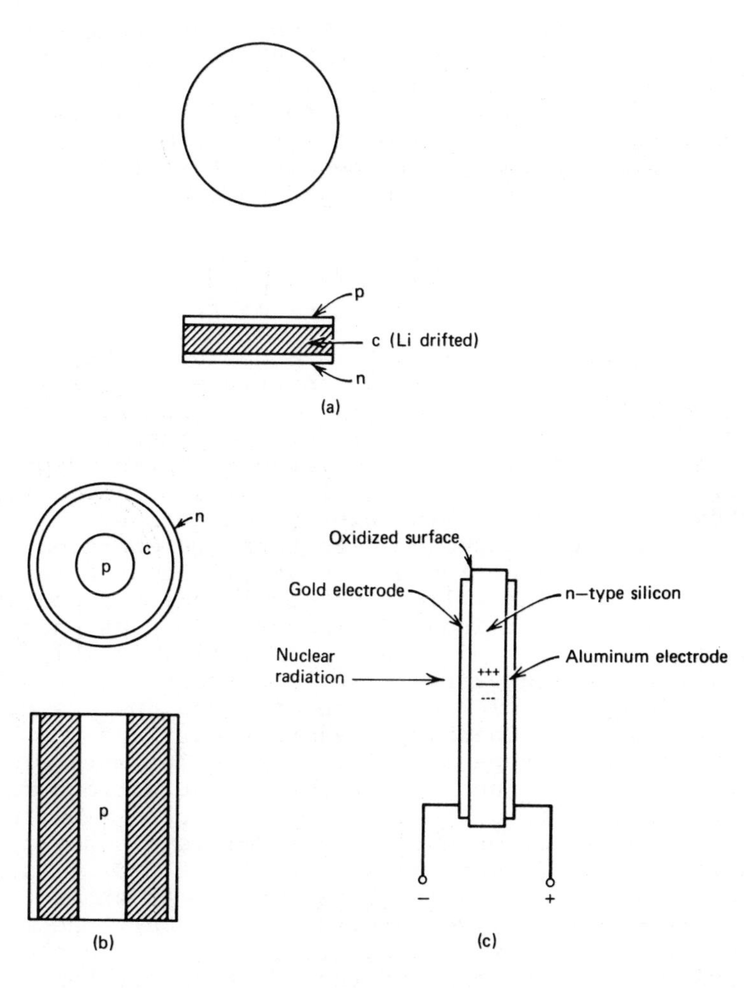

Fig. 5.1. Types of semiconductor detectors: (a) planar lithium-drifted, (b) coaxial cylindrical lithium-drifted, and (c) surface barrier (from A. Edward Profio, *Experimental Reactor Physics*, Wiley, New York, 1976, p. 220).

energy. The charge generated is

$$q = eN \tag{5.14}$$

where $e = 1.60 \times 10^{-19}$ C. Thus if the detector and electronics are arranged such that q can be measured for individual incident ions, their energies can be obtained. The contribution to the energy resolution (expressed as full-width at half-maximum of the counts vs. energy in the peak), because of statistical fluctuations in the number and charge, is

$$\left(\frac{\Delta E}{E}\right)_s = 2.35 \left(\frac{FW}{E}\right)^{1/2} \tag{5.15}$$

where F, the Fano factor, is about 0.3. Typical energy resolution for a 5.48-MeV α particle is 20 keV, of which 5 keV is attributable to the preamplifier noise. The noise increases with input capacitance; hence a detector of small area and large depletion depth, short low-capacitance cable, and low-noise preamplifier should be used for best energy resolution.

Most semiconductor detectors are operated as reverse biased junction diodes as illustrated in Figure 5.2. The silicon or germanium is doped with small amounts of Group V elements, such as phosphorus or arsenic, to produce n-type material in which the majority charge carriers are electrons. When doped with Group III elements such as boron, p-type material is produced in which the majority carriers are holes. When p- and n-type materials are in intimate electrical contact, a p-n junction is formed. Electrons flow from the n region into the p region and holes flow in the opposite direction until the Fermi levels coincide and a contact difference of potential ($\sim$0.5 V) exists. Now if a reverse bias voltage is applied as shown in Figure 5.2 (positive polarity to n-type), the majority carriers are further separated, forming a depletion region on either side of the junction, of depth d, which is effectively an insulator. The current from thermally released charge is small and noise from fluctuations in the reverse bias current is tolerable. Holes and electrons released by ionizing radiation are separated and drift in the applied electric field and generate the desired electrical signal. The <u>polarity must never be changed</u>; the forward-bias current is large and will destroy the detector.

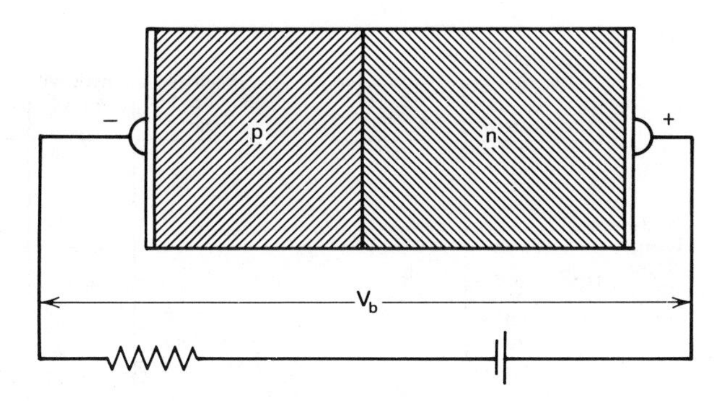

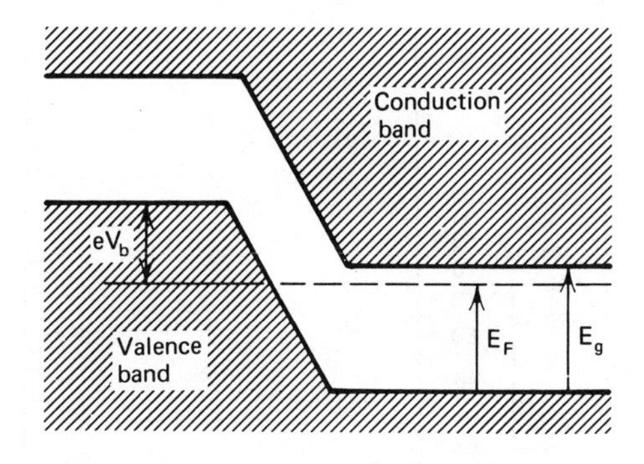

Fig. 5.2. Electrical and energy-band diagram of reverse-biased p-n diode (from A. Edward Profio, *Experimental Reactor Physics*, Wiley, New York, 1976, p. 266).

In the silicon surface-barrier detector, the p-n junction is formed at the surface by etching and controlled oxidation of the n-type material. A gold film, 40 $\mu g/cm^2$ thick, protects the surface and makes electrical contact. Ions traverse this "window" or dead layer and lose a small amount of energy.

The depletion depth at a certain applied voltage V_b is usually specified by the manufacturer. It can be obtained from the nomogram in Figure 5.3 if the resistivity or impurity concentration is known. The

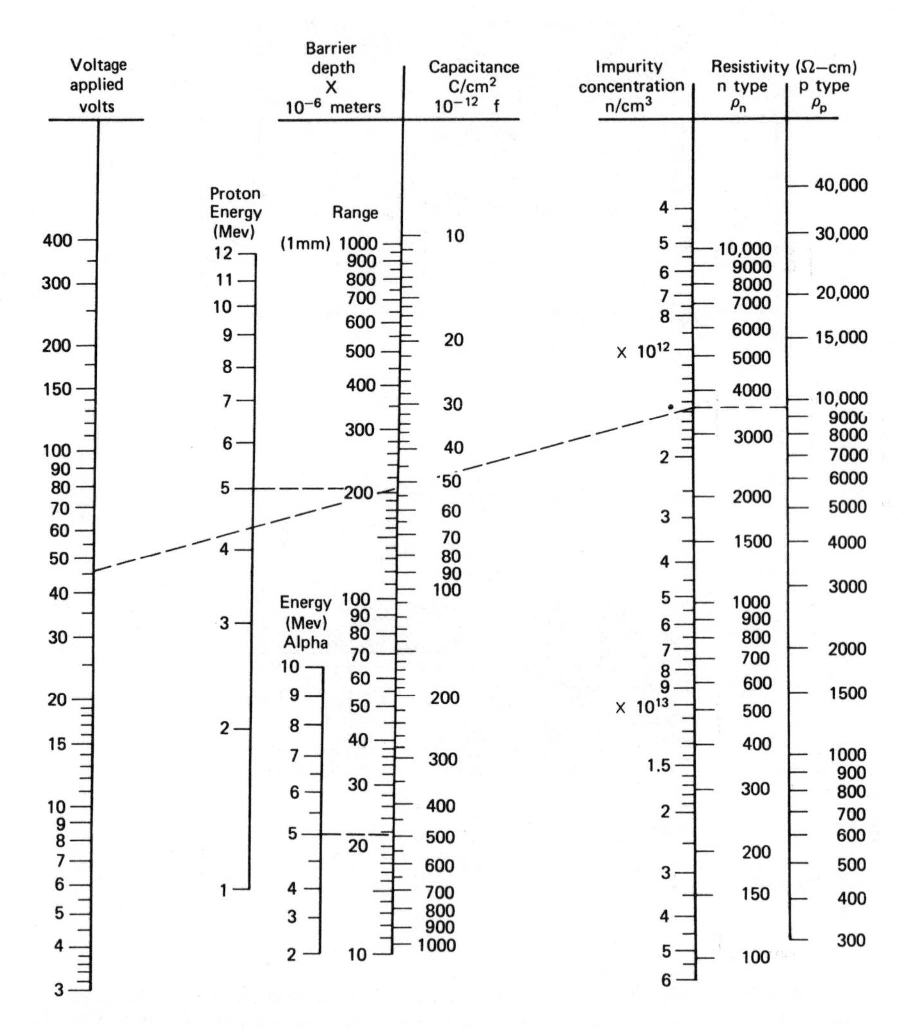

Fig. 5.3. Silicon junction diode nomogram. For p-type silicon with a resistivity of 9500 Ω-cm and 46 V bias, depletion depth is 205 μm and will stop 5-MeV protons (after J. B. Marion and F. C. Young, *Nuclear Reactor Analysis Graphs and Tables*, North-Holland, Amsterdam, and Wiley, New York, 1968) (from A. Edward Profio, *Experimental Reactor Physics*, Wiley, New York, 1976, p. 269).

energies of protons and α particles whose range is equal to the depletion depth is given. The capacitance of the detector is (in picofarads)

$$C_d = \frac{1.05A}{d} \quad \text{pF} \tag{5.16}$$

where A is the area in cm^2 and d is in cm.

The fluence of normally incident ions is

$$\Phi = \frac{C}{\varepsilon A} \tag{5.17}$$

where C is the number of counts and ε is the intrinsic efficiency. The interaction efficiency is essentially 100%, but the counting efficiency may be less if the discriminator is set above noise and some ions deposit less than their total energy (e.g., because the source is thick and energy is lost in the source or in the window). Efficiency is usually measured with a standard source of known activity, emitting into the solid angle defined by the distance and A (or an aperture smaller than A).

Electrons

The fluence of electrons can be measured with a proportional counter or Geiger-Mueller counter. Detection efficiency is 100% for those electrons producing enough ionization to generate a signal above noise and the discriminator threshold (proportional counter) or even a single ionization in the gas (GM counter). For spectrometry of moderate energy electrons, a scintillation counter is preferred to achieve a total absorption of the electron's kinetic energy in a reasonable volume. A low-Z scintillator such as stilbene or scintillating plastic is desirable to minimize backscattering.

The proportional or GM counter consists of a gas-filled space between two electrodes such as that shown in Figure 5.4. The counter is usually cylindrical with a small-diameter tungsten wire anode and metal or graphite-coated glass cathode. A thin cathode or a thin window of mica or plastic (at either the side or end) is needed to admit low-energy electrons. A typical gas filling for the proportional counter is P10 (90% argon, 10% methane), at 1 atm or less. The GM tube may be filled with about 10-cm Hg pressure of argon, plus a small amount of either ethanol or bromine

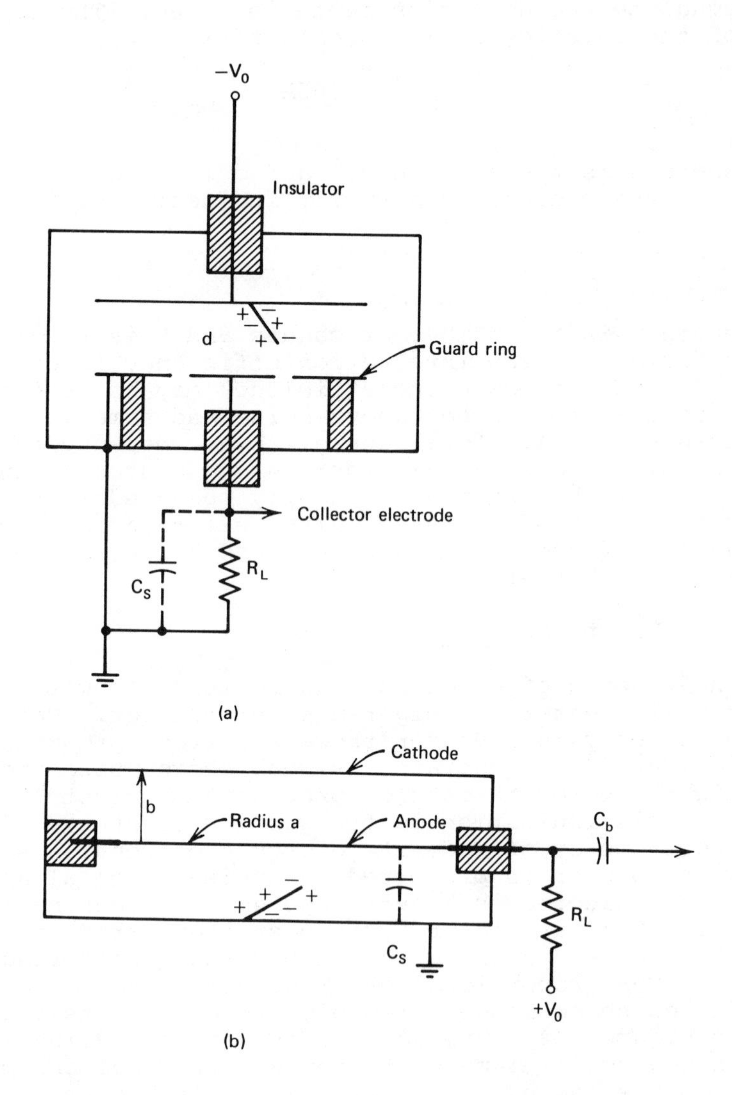

Fig. 5.4. General features of gas-ionization detectors: (a) parallel-plate ionization chamber with guard ring to define field and intercept leakage current and (b) cylindrical proportional or GM counter (from A. Edward Profio, *Experimental Reactor Physics*, Wiley, New York, 1976, p. 244).

to quench the discharge initiated by an ionizing particle. Figure 5.5 shows how the total number of electrons collected at the anode varies as the applied potential V_0 is increased. In region 1 some of the ions and electrons recombine before they can be separated by the electric field. In region 2 the counter performs as an ionization chamber and the number of electrons $N = (E/W)$, where W is the energy required to produce one ion-electron pair, as given in Table 5.2. As the voltage is increased, the proportional region (3) is reached, where

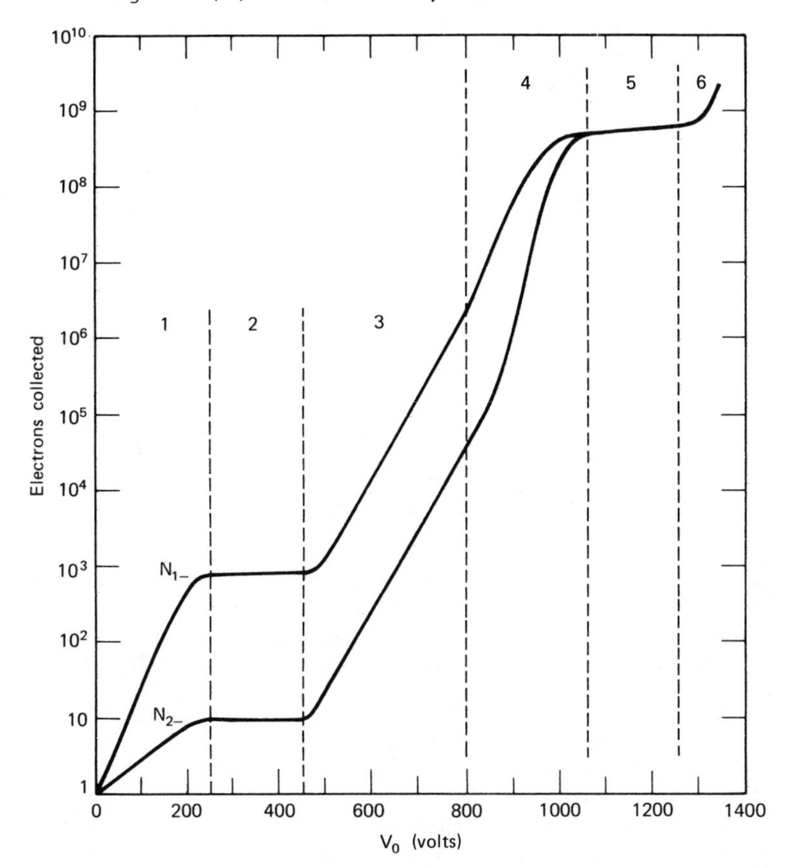

Fig. 5.5. Number of electrons collected as function of applied voltage for energy depositions $E_1 = WN_1$ and $E_2 = WN_2$. Region 2 corresponds to ionization chamber, region 3 to proportional counter, region 5 to GM counter.

TABLE 5.2. W, ENERGY PER ION-ELECTRON PAIR IN GASES

Gas	W (eV)
H_2	37.0
He	46.0 (pure)
	32.0 (normal)
N_2	36.3
O_2	32.2
Ar	26.4
Air	37.7 (pure)
	34.0 (standard)
CH_4	29.4
CO_2	34.3
BF_3	36.0

$$N = M \frac{E}{W} \tag{5.18}$$

and M is the gas multiplication. The total number of electrons collected is proportional to the original number released by the ionizing particle, but greater because the electrons are accelerated in the high electric field near the anode, and produce additional electrons by collisions in the gas. In region 4 the proportionality begins to break down (M no longer independent of the initial ionization); region 5 applies to the Geiger-Mueller counter, where the discharge spreads, the number is independent of the initial ionization, and the discharge is terminated by the loss of electrons to the quenching gas. Finally, in region 6 the counter goes into multiple or continuous discharge and may be destroyed. The magnitude of the voltage in each region depends on the gas pressure and composition and the anode radius.

If the count rate is plotted as a function of applied voltage, at fixed discriminator threshold, a curve such as Figure 5.6 is obtained for the GM counter. The counts begin when the amplitude of the pulses exceeds the discriminator threshold, reaches a plateau where the count rate increases only slightly with voltage, and then increases again in the multiple discharge region. Normally the operating voltage V_0 is set at about 1/3 up on the plateau, for best stability against drift in voltage or discriminator threshold. The counter should never be operated in the multiple-discharge region.

Photons

γ-Ray and X-ray photons, as well as moderate-energy electrons, are usually counted with a scintillation detector. The fluence spectrum can be measured with a scintillation or semiconductor detector.

The design of a scintillation detector is shown in Figure 5.7. The scintillator fluoresces under ionization and excitation by a charged particle, releasing light with time constants from a few nanoseconds to microseconds, depending on the material. The scintillator is mounted in a container to exclude extraneous light and moisture that would attack a sodium iodide crystal. Inside surfaces are coated with a reflector such as titanium-dioxide white paint or magnesium oxide, except for the glass window. The window is optically coupled to the faceplate of a

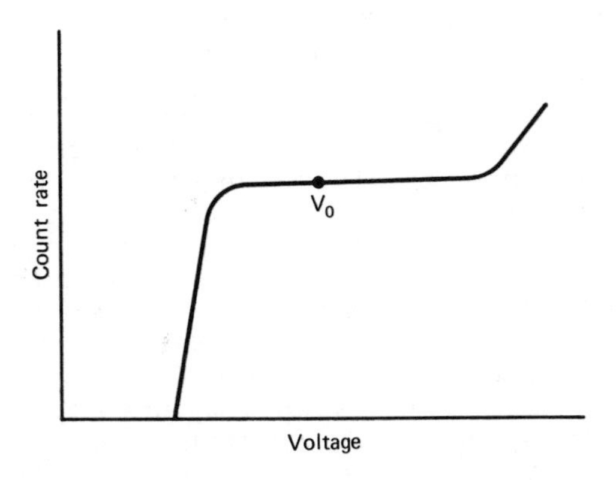

Fig. 5.6. Plateau of GM tube.

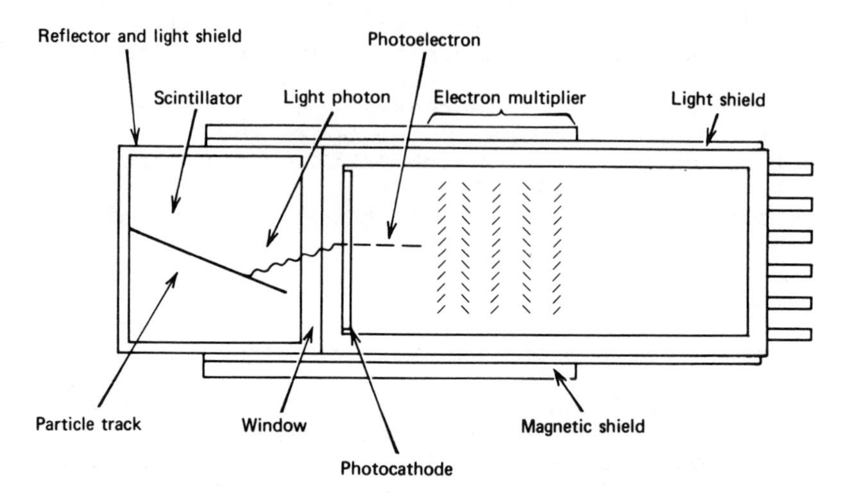

Fig. 5.7. Elements of a scintillation detector (from A. Edward Profio, *Experimental Reactor Physics*, Wiley, New York, 1976, p. 274).

multiplier phototube with refractive-index matching silicone grease. The photocathode is evaporated on the inside of the faceplate in vacuum. The scintillation light releases low-energy electrons from the photocathode. Some of the electrons are accelerated and focussed on the first dynode (secondary-emission electrode) of the electron multiplier. Some two to four electrons are released per incident electron. These secondary electrons are then collected, accelerated, and hit the next dynode, and so on. A multiplier phototube for scintillation counting may have 10 to 14 dynodes, giving an overall charge multiplication of 10^3 to 10^8 depending on the applied interdynode voltage, dynode surface material, and number of dynodes (stages) with very little noise contribution. However, noise does arise from thermally released electrons at the photocathode, which can be cooled if necessary. The phototube must never be exposed to room light with voltage applied, or it will be ruined by excessive current and heating. The electrons can be perturbed by external magnetic fields; hence a magnetic shield of low-permeability alloy encloses the electron multiplier and photocathode-to-dynode space. The required potentials between electrodes is provided by a voltage-divider resistance chain as shown in Figure 5.8. The multiplied charge q is collected at the anode. The corresponding signal-voltage amplitude, developed on the stray capacitance to ground, is $V = (q/C_s)$. The series capacitor (0.001 μF in diagram) blocks DC while transmitting the pulse signal. The risetime of the pulse is ∿10 ns to ∿1 μs, depending on the fluorescence decay time of the scintillator.

Properties of some scintillators are summarized in Table 5.3. The organic crystal and plastic scintillators are often used to detect electrons, and plastic or a liquid organic scintillator (NE 213) are used for detection and spectrometry of fast neutrons. The most commonly used scintillator for γ rays is thallium-activated sodium iodide, NaI(Tℓ), because of its high density, large Z value (iodine), and good light output. Table 5.3 gives a value of 68 eV to produce a scintillation photon in anthracene, for electrons, but as NaI(Tℓ) generates 2.1 times as much light, the energy is 32 eV/light photon. Unfortunately, there are losses in light collection, and the quantum efficiency of the photocathode for blue-violet light is roughly 10%; hence the energy required for one photoelectron at the first dynode is some 300 to 400 eV. Light yield for α particles is 40% of the yield for electrons.

The scintillation light spectrum is broad but peaks at the wavelength given in Table 5.3. It is important to select a multiplier phototube with good sensitivity to the scintillation light. For NaI(Tℓ) the standard is the so-called S-11 response of oxidized cesium antimonide (Cs_3Sb-0) on lime glass. The bialkali cathode (K_2CsSb) on Pyrex glass has a higher sensitivity. In addition to the spectral response, one chooses a multiplier phototube with cathode diameter to match the diameter of the scintillator, a suitable number of stages hence gain at the specified HV, and a low dark current from thermal noise if low-energy particles are to be detected. In some applications, such as at high counting rates or for timing experiments, a small anode rise time (ART) is desired. The linear-focus dynode structure is good for timing. For spectrometry, linearity of anode pulse current with number of scintillation photons on the photocathode is essential. Stability of gain against aging or

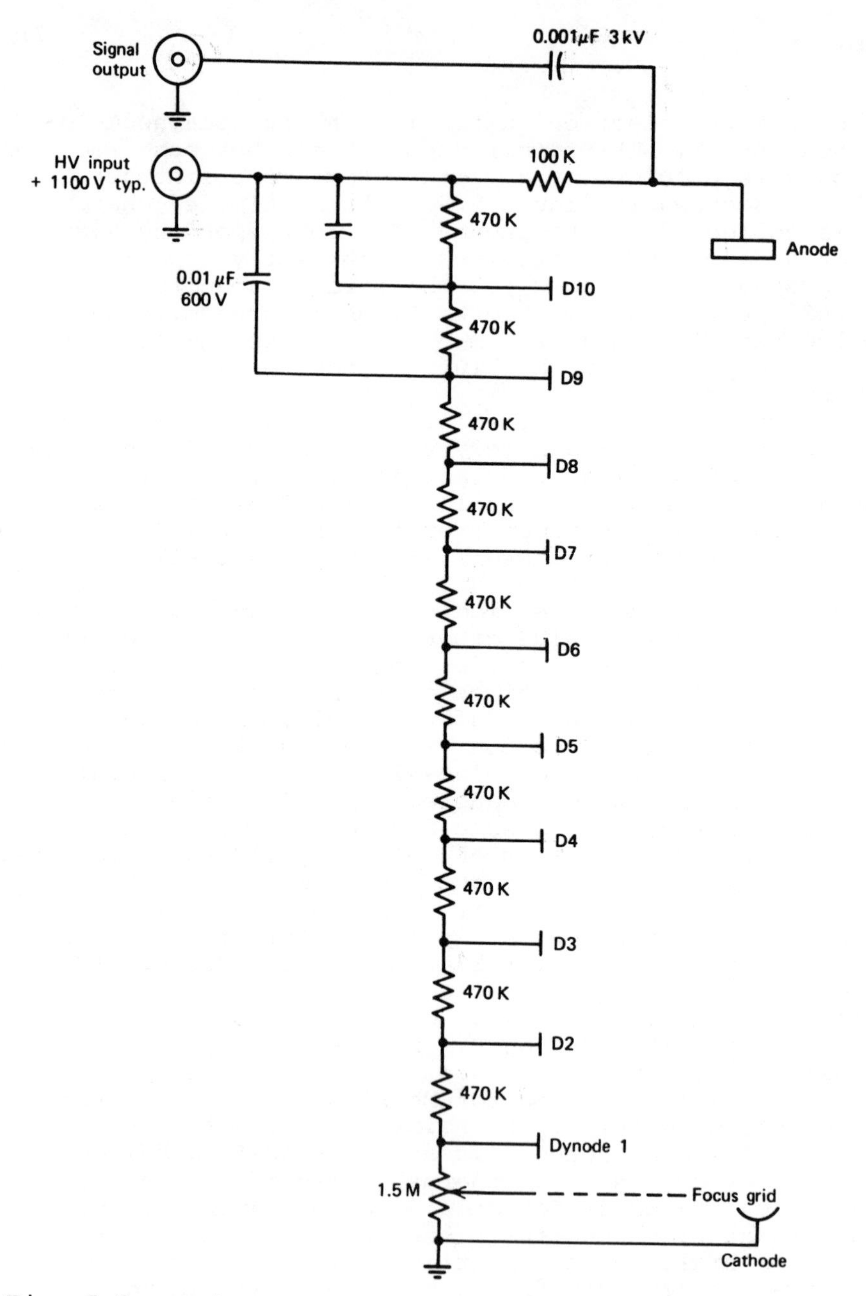

Fig. 5.8. Voltage divider and signal circuit for a multiplier phototube (from A. Edward Profio, *Experimental Reactor Physics*, Wiley, New York, 1976, p. 287).

TABLE 5.3. PROPERTIES OF SCINTILLATORS[a]

Type	Scintillator	Composition	ρ (gm/cm^3)
Organic crystal	anthracene	$C_{14}H_{10}$	1.25
	trans-stilbene	$C_{14}H_{12}$	1.16
Organic liquid	NE213	$CH_{1.213}$	0.88
	toluene + 4 g/liter TP + 0.1 g/liter POPOP	C_7H_8	0.87
Plastic	NE102	$CH_{1.105}$	1.03
	PVT + 36 g/liter TP + 0.2 g/liter TPB	C_9H_{10}	1.03
Inorganic crystal	sodium iodide	NaI + 0.1% Tℓ	3.67
	cesium iodide	Csl + 0.1% Tℓ	4.51
	lithium iodide	LiI + 0.05% Eu	4.06
	zinc sulfide	ZnS + 0.01% Ag	4.09
Glass	NE908	7.8% Li + SiO_2,Al_2O_3	2.67
Gas	xenon	Xe (1 atm)	5.85 g/liter

[a]Data from Birks, Nuclear Enterprises, Ltd., and Harshaw Chemical Co. Abbreviations: PVT, polyvinyl toluene; TP, _p_-terphenyl; TPB, tetraphenyl butadiene; POPOP, 2,2'-_p_-phenylenebis-(5-phenyloxazole).

TABLE 5.3. (Continued)

Refr. Index	Rel. Light Output	λ_{max} (Å)	$\alpha:\beta$	τ Fast (μs)
1.62	1.00	4470	0.1	0.030
1.63	0.50	4100	0.1	0.006
1.50	0.78	4250	0.1	0.004
1.50	0.60	4300	0.1	0.003
1.58	0.65	4250	0.1	0.002
1.58	0.45	4450	0.1	0.004
1.78	2.1	4100	0.4	0.25
1.78	0.95	4200 to 5700	0.5	0.4(α) 1.1(e)
1.95	0.74	4400	1.0	1.4
2.36	3.00	4300	0.5	0.2
1.57	0.20	3990	0.2	0.005?
1.0	1.5	∿3300	1.0	0.002

TABLE 5.3. (Continued)

Scintillator	τ Slow (μs)	Melting Pt. Boiling Pt. (°C)	Remarks
anthracene		217	68 eV/photon (e)
trans-stilbene	0.37	125	PSD
NE213	0.1	(139)	xylene, napthalene, POPOP
toluene base		(110)	
NE102		75	PVT base
PVT base		75	
sodium iodide		650	hygroscopic
cesium iodide		620	resists shock
lithium iodide		445	hygroscopic 96% ^{6}Li
zinc sulfide	∿10	1850	multicrystal
NE908	75	1200	0.1% Ce_2O_3 9.6% ^{6}Li
xenon	impurity independent	(-107)	needs uV PM, or wavelength shifter

temperature fluctuations is also important in critical spectrometry applications.

Low-energy (<150 keV) X-ray photons interact mainly by the photoelectric effect in sodium iodide. The intrinsic efficiency (counts per incident photon) is given by

$$\varepsilon = 1 - \exp(-\mu d) \tag{5.19}$$

where μ (cm^{-1}) is the linear attenuation coefficient and d (cm) the path length (equal to scintillator thickness for normal incidence). The linear attenuation coefficient of sodium iodide, and the photoelectric, Compton, and pair-production components are plotted in Figure 5.9. At 100 keV, $\mu = 6\ cm^{-1}$ and $\varepsilon \geq 0.99$ for $d \leq 0.8$ cm.

At higher energies, Compton scattering and pair production have to be taken into account. The photon transfers only part of its energy to a Compton electron, and the scattered photon may escape without further interaction. One or both of the annihilation quanta from pair production may escape, giving a full-energy peak, a single-escape peak, and a double-escape peak such as seen in Figure 5.10. The amplitude or pulse height is proportional to the secondary-electron energy deposited in the NaI by ionization. Some of the energy may be lost because of escape of electrons or fluorescence X-rays, or generation and escape of bremsstrahlung. The pulse-amplitude distribution is blurred by the effect of statistical fluctuations, irregularities in the scintillator or phototube, and electrical noise. In Figure 5.10 the distribution was calculated by a special Monte Carlo code [7]. An improved version is available [8]. The measured and calculated distributions are in good agreement at the full-energy peak (1.78 MeV) and Compton edge (1.56 MeV), but the measurement is higher than calculation at lower energies because of backscattering from the surroundings, not included in the calculation. The single-escape peak appears at 1.78 - 0.51 = 1.27 MeV, and the double-escape peak at 1.78 - 1.02 = 0.76 MeV, as they should. The calculated intrinsic efficiency ε = 0.6029 and the full-energy peak:total ratio f_p = 0.2821 (also called the photofraction although full-energy absorption here occurs because of pair production and absorption of both annihilation quanta). The efficiency for detecting counts in the full-energy peak is

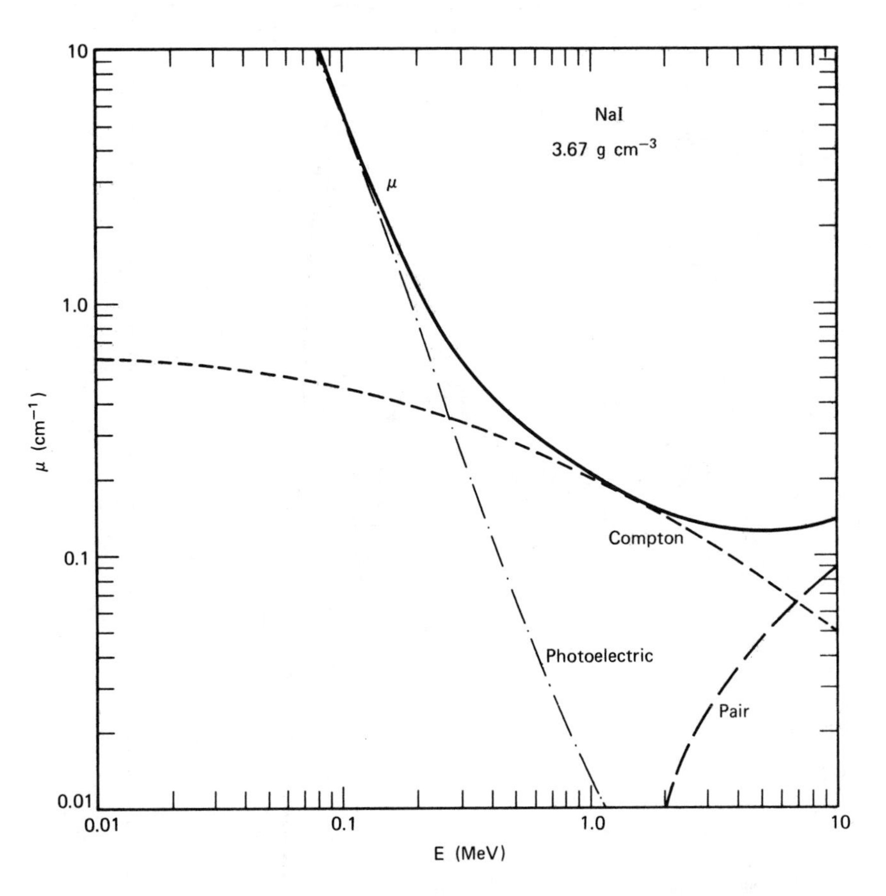

Fig. 5.9. Linear attenuation coefficient of sodium iodide.

$$\eta_p = f_p \varepsilon G \tag{5.20}$$

where the geometry

$$G = \frac{\Omega_d}{4\pi} = \frac{A_d}{4\pi h^2} \tag{5.21}$$

with Ω_d as the solid angle (in sr) subtended by the detector at the point source, and h as the distance

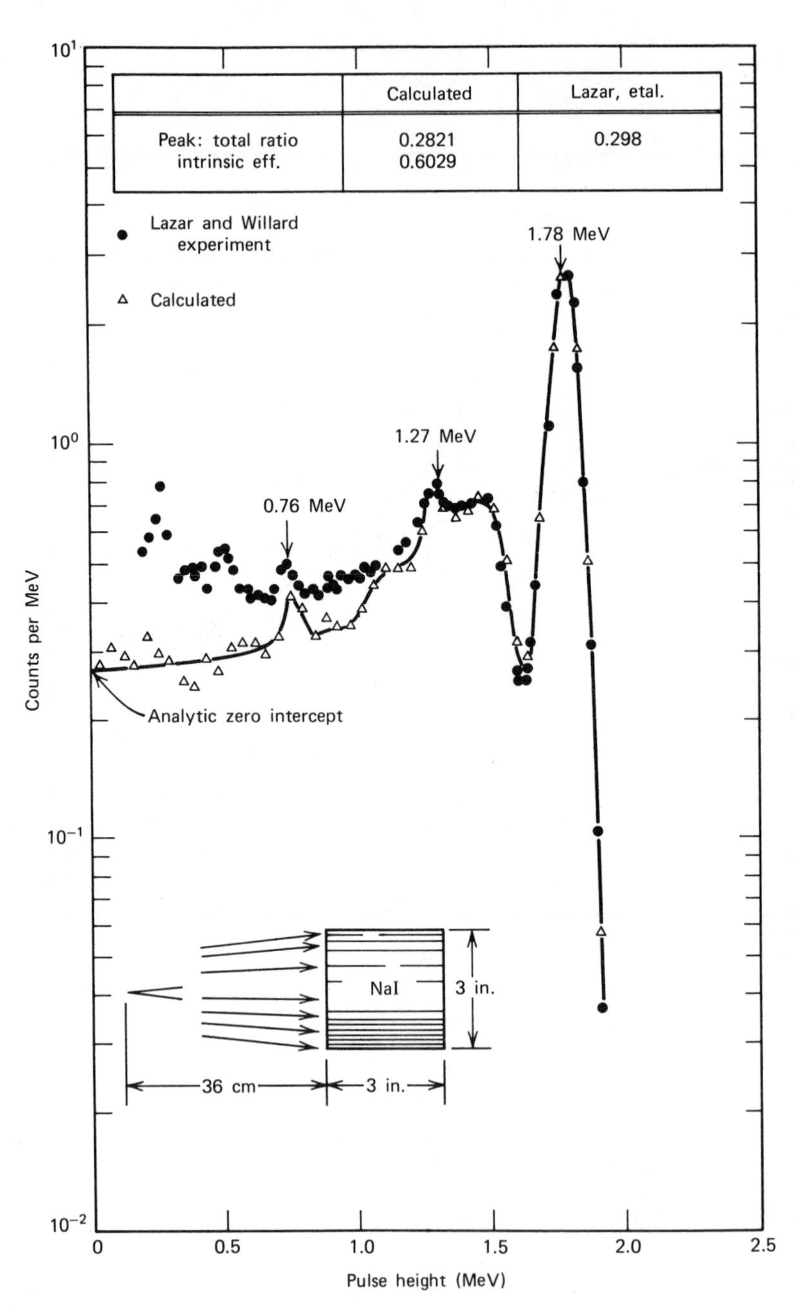

Fig. 5.10. Pulse-amplitude distribution for 1.78-MeV γ rays from a point isotropic source at 36 cm from a 7.6-cm-diameter by 7.6-cm-long NaI(Tℓ) detector, as measured and calculated by Monte Carlo (after Zerby [7]).

from source to base of the detector of projected area A_d. Plots of ε and f_p are given in Figure 5.11 and 5.12, respectively. The photon fluence from a point isotropic source emitting S photon/s and on axis at h cm is

$$\Phi = \frac{C}{4\pi h^2} = \frac{C_p}{\varepsilon f_p A_d} \tag{5.22}$$

where C_p is the observed number of counts under the peak.

If there is more than one γ-ray energy, fluence and spectrum measurements are more complicated. If the energies can be identified from peaks, one can perform "spectrum stripping" by subtracting the pulse-amplitude distribution from each γ ray, starting at the highest energy. Another technique is to express the unknown pulse-amplitude distribution as the sum of distributions from known-energy γ rays, with coefficients to be determined by a fitting process. A more general method is unfolding, knowing the response functions for many closely spaced γ-ray energies. Unfolding is used with neutron spectra and is discussed later under the subheading Neutrons.

Sodium iodide detectors are efficient, but energy resolution is only a few percent. Much better resolution is achieved with semiconductor detectors. Silicon is useful for photon energies less than about 50 keV, but germanium is preferred for higher energies. High-purity germanium is now available but expensive. Most detectors achieve a large sensitive volume by the technique of lithium drifting; thus we have Si(Li) and Ge(Li) detectors. Lithium is mobile and acts as an electron donor. One starts with p-type material, which has an excess of acceptor-atom impurities. Lithium is diffused into the crystal at high temperature. Then a reverse-bias voltage is applied at above room temperature. The Li ions drift in the electric field until they neutralize the charge on the acceptor atoms, thus automatically achieving exact compensation. The Si(Li) detectors are usually planar, whereas the Ge(Li) detectors are often coaxial-cylindrical, as shown in Figure 5.1. Lithium-drifted detectors _must_ be kept refrigerated (at the boiling point of liquid nitrogen, 77K) to avoid migration of the lithium. The semi-

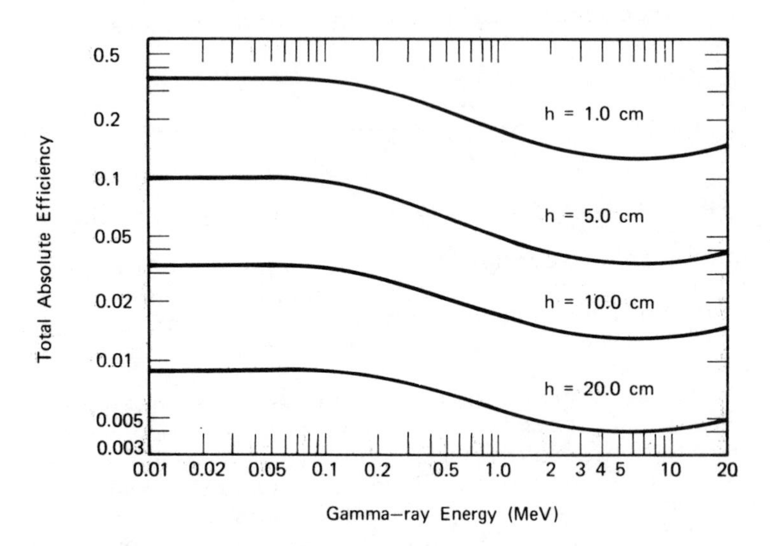

Fig. 5.11. Total efficiency η for γ rays on a 7.6-cm-diameter by 7.6-cm-long NaI(Tℓ) scintillator with on-axis isotropic point source at h cm distance (after J. B. Marion and F. C. Young, *Nuclear Reaction Analysis Graphs and Tables*, Wiley, New York, 1968)(from A. Edward Profio, *Experimental Reactor Physics*, Wiley, New York, 1976, p. 485).

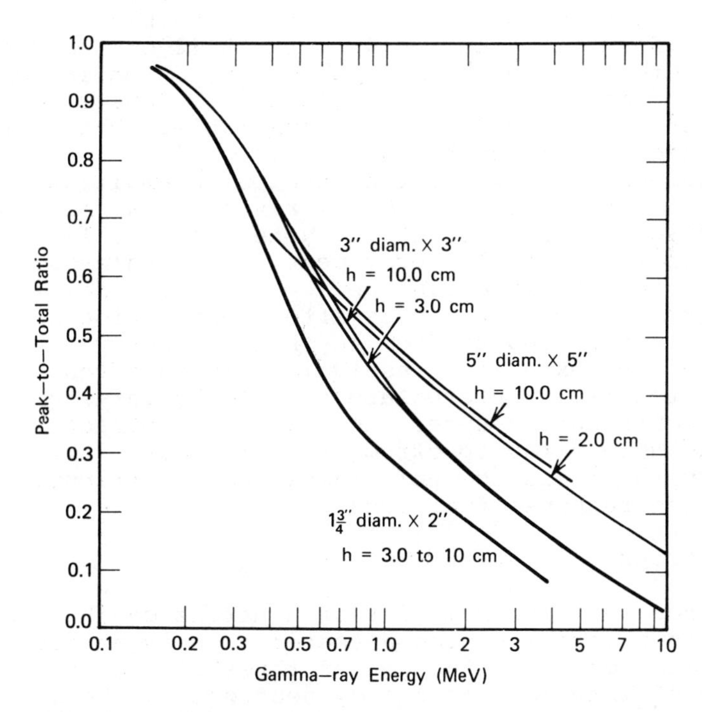

Fig. 5.12. Peak:total ratio f_p for γ rays on sodium iodide (after J. B. Marion and F. C. Young, *Nuclear Reaction Analysis Graphs and Tables*, Wiley, New York, 1968; from A. Edward Profio, *Experimental Reactor Physics*, Wiley, New York, 1976, p. 486).

conductor detectors are operated at 77 K to reduce noise from thermally released electrons and holes.

Figure 5.13 plots the measured peak efficiencies $\eta_p = f_p \varepsilon G$ for two Ge(Li) detectors. The planar detector was 2.5 cm^2 in area and 4 mm thick (1 cm^3) and the coaxial detector was 26 mm in diameter, 30 mm in length, and had a 7-mm-thick Li drift (12 cm^3). Sensitive volumes of more than 30 cm^3 are available. The source was located on axis at h = 3 cm.

Figure 5.14 shows part of the pulse-amplitude distribution for the 2.754-MeV γ ray of ^{24}Na, measured with a 31-cm^3 Ge(Li) detector. Note the excellent resolution in the full-energy peak, the single-escape peak, and the double-escape peak. The peak efficiencies correspond to net counts after subtraction of Compton electron counts from any higher-energy γ rays. Because of the excellent resolution, it is often possible to simply interpolate the slowly varying Compton contribution under the peak and subtract it. If the unknown γ-ray spectrum is continuous, it will be necessary to unfold the pulse-amplitude distribution knowing the response functions.

Neutrons

The fluence and spectrum of fast neutrons can be measured with a recoil-proton detector such as an organic scintillator or by activation of threshold foils. The spectrum of intermediate-energy neutrons can be measured with proportional counters filled with hydrogen or methane. The fluence of slow neutrons is usually measured with a proportional counter filled with ^3He or $^{10}BF_3$, by a fission counter (^{235}U-coated pulse ionization counter), or by foil activation. If the source is pulsed, the neutron energy can be measured by time of flight. A major problem in neutron counting is background from γ rays.

A favorite scintillator is the NE 213 organic liquid (Nuclear Enterprises, Ltd.), encapsulated in a glass cell of 5.08 cm diameter and 5.08 cm length. Its fluorescence dieaway time constant is about 4 ns for electrons but about 100 ns for protons. Thus the risetime of γ-ray and neutron pulses are different, and appropriate electronics allow the γ-ray pulses to be rejected. The response functions (count/light unit vs. light units, per incident neutron) can be calculated with the special Monte Carlo code O5S [9]. In

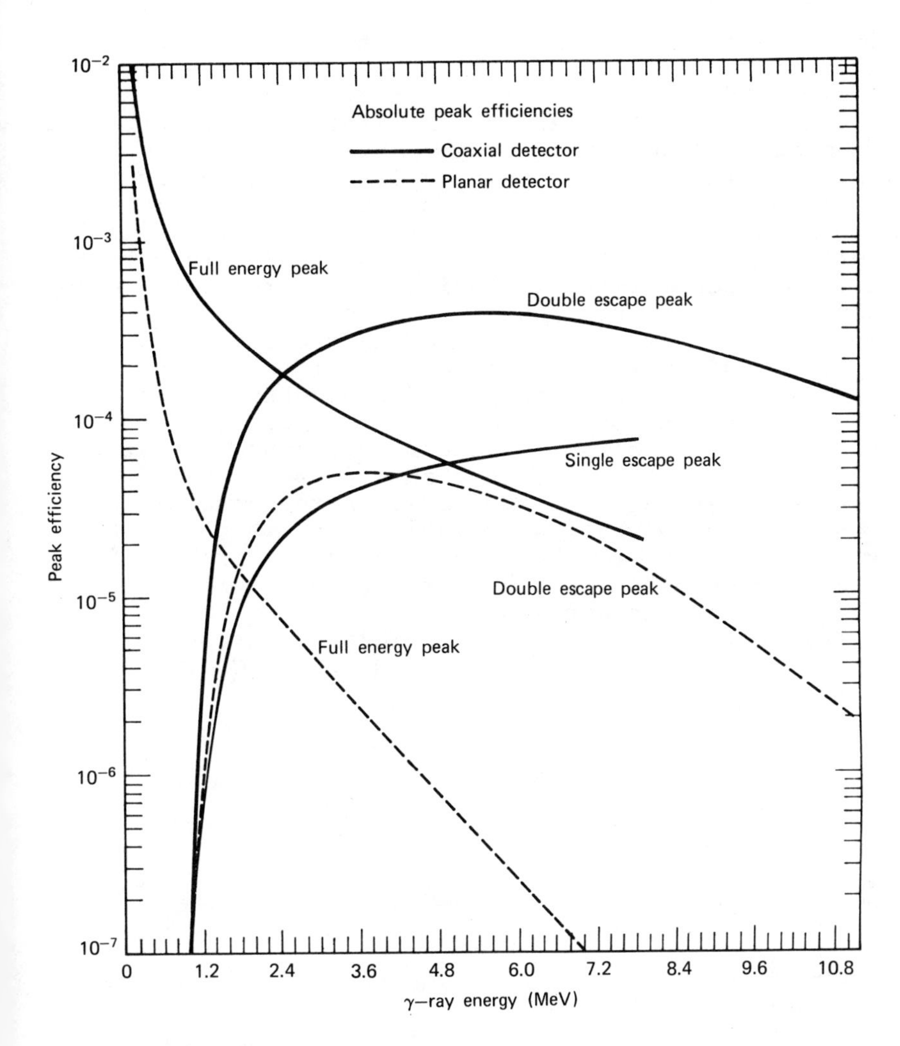

Fig. 5.13. Peak efficiencies for a 1 cm^3 planar and a 12 cm^3 coaxial cylinder Ge(Li) detectors for a point source at 3 cm (after J. B. Marion and F. C. Young, *Nuclear Reaction Analysis Graphs and Tables*, Wiley, New York, 1968).

addition to calculating the energy transferred to recoil protons, recoil carbon ions and carbon reaction products, 05S takes into account the leakage of protons from the scintillator, and the nonlinear light output, which depends on dE/dx as well as E. Resolution blurring is incorporated using a Gaussian smearing function whose standard deviation is an empirical function of light output. The signal pulse amplitude is proportional to scintillation light output. The light unit in 05S was originally defined in terms of the response to ^{60}Co γ rays but is now defined as 13% above the half-height of the Compton edge from the 1.275-MeV γ ray of ^{22}Na. The 0.511-MeV annihilation quantum from the positron decay of ^{22}Na provides another convenient point (half-height of Compton edge) for calibration of gain and any zero-offset of the multichannel analyzer. Figure 5.15 shows the response functions of NE 213 for several incident neutron energies. A response matrix A_{ij} is constructed by interpolating between neutron energies, grouping and binning; A_{ij} is the probability that an incident neutron in energy group j will generate a count in amplitude bin i.

The neutron spectrum N_j (incident neutrons in group j) is related to the observed counts C_i in amplitude bin i by

$$C_i = \sum_j A_{ij} N_j \tag{5.23}$$

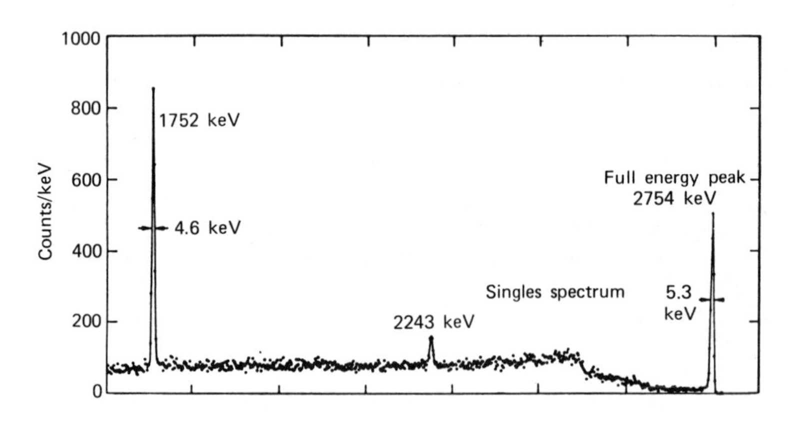

Fig. 5.14. Pulse-amplitude distribution for ^{24}Na 2.754-MeV γ ray on a 31-cm^3 Ge(Li) detector (courtesy V. J. Orphan; from A. Edward Profio, *Experimental Reactor Physics*, Wiley, New York, 1976, p. 481).

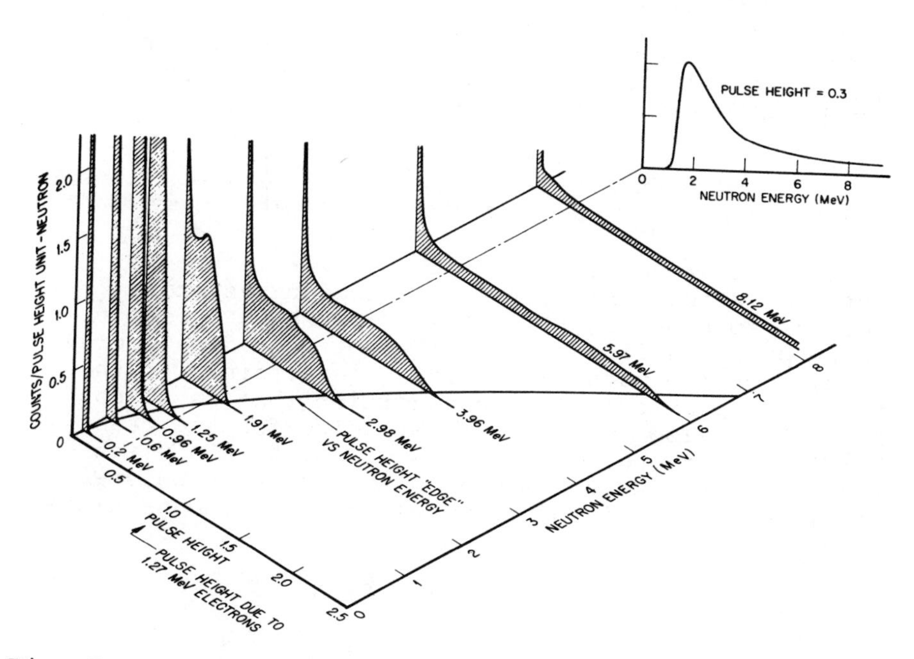

Fig. 5.15. Response functions for a 5.08-cm × 5.08-cm NE 213 scintillation detector (from A. Edward Profio, *Experimental Reactor Physics*, Wiley, New York, 1976, p. 446).

Matrix inversion could be used to solve for the N_j, but poor results are obtained. Instead a constrained least-squares approach is used, such as embodied in the FERDOR/COOLC code [10] and its later version FORIST [11]. In ordinary weighted least-squares technique one minimizes the sum of residuals squared,

$$r^2 = \mathop{S}_{i} \sigma_i^{-2} \left\{ C_i - \mathop{S}_{j} A_{ij} N_j \right\}^2 \tag{5.24}$$

where σ_i^2 is the variance of the counts in the i-th bin. But statistical errors in the data can result in serious spurious oscillations. The FERDOR code tries to suppress oscillations (while not degrading energy resolution too much) by applying constraints to the solution, particularly the condition that is be non-negative. Also FERDOR does not solve for the mean N_j but for upper and lower bounds on p_k, within which are the solutions consistent with the input pulse-amplitude data and the response matrix. The parameters

$$p_k = \int W_k(E)\ N(E)\ dE \tag{5.25}$$

where $W_k(E)$ is an input smoothing or "window" function. For example, it may be a Gaussian with width corresponding to the energy resolution of the detector, or some larger width if spurious oscillations are troublesome. The FORIST code starts with input $W_k(E)$ and iterates to find a solution that gives the best resolution consistent with the statistical errors in the data. Another code, MAZE [12], uses information theory to devise another function to minimize, intended to smooth out structure due to statistical fluctuations but not to smear real Gaussian peaks. Figure 5.16 shows the pulse-amplitude distribution from a Pu-Be source and the neutron spectrum unfolded by FERDOR. A good review of unfolding methods is given in the proceedings of a conference at Oak Ridge [13].

The neutron spectrum from about 10 keV to 1.5 MeV can be measured by unfolding the pulse-amplitude distribution in a proportional counter filled with hydrogen or methane. Several runs are made at various fillings, pressures, and high voltages to cover the entire span. The pressures are a compromise between γ-ray background sensitivity and excessive distortion

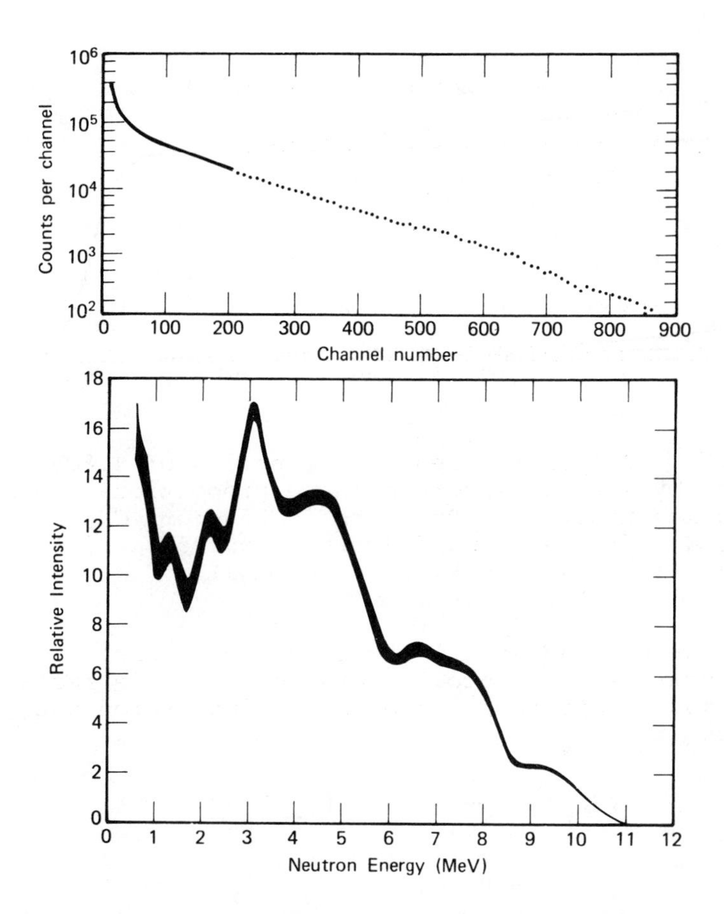

Fig. 5.16. Pulse-amplitude distribution in an NE 213 detector and neutron spectrum of Pu-Be unfolded by FERDOR (from A. Edward Profio, *Experimental Reactor Physics*, Wiley, New York, 1976, p. 456).

of the recoil-proton distribution because of wall and end effects. Figure 5.17 shows the calculated [14] pulse-amplitude distribution or response function of a 4.82-cm ID, 7.62-cm-long cylindrical counter filled with methane (258 torr) and nitrogen (12 torr) and irradiated in an isotropic flux density of 500-keV neutrons. The nitrogen provides a calibration peak at

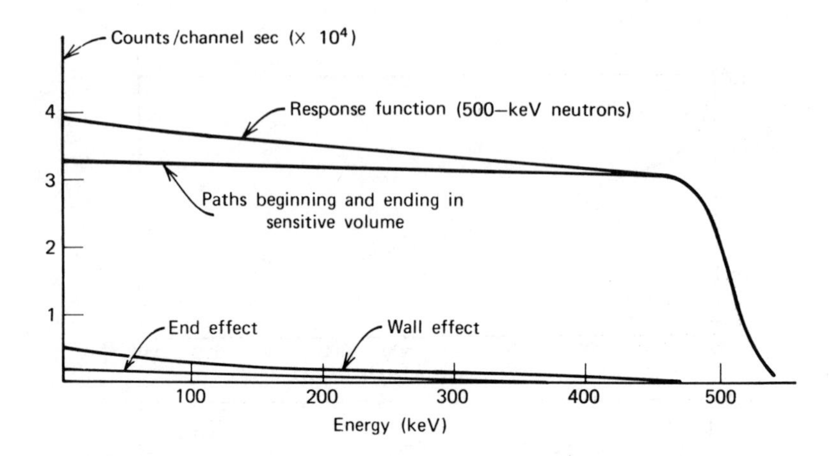

Fig. 5.17. Calculated response function of 4.82-cm-diameter by 7.62-cm-length methane proportional counter for 500-keV neutrons. Pressure is 258-torr CH_4 and 12-torr N_2 (from A. Edward Profio, *Experimental Reactor Physics*, Wiley, New York, 1976, p. 453).

615 keV when irradiated with thermal neutrons. For single scatter, the recoil proton spectrum is rectangular, and the probability of observing proton energy E_p in dE_p is

$$p(E_p) = \frac{1}{E_n} \qquad 0 \le E_p \le E_n$$

$$= 0 \qquad E_p > E_n \tag{5.26}$$

The neutron spectrum could then be obtained by taking the derivative

$$N(E_n)\ \varepsilon(E_n) = \frac{d}{dE_p}\ n(E_p) \tag{5.27}$$

where ε is the efficiency (protons generated per incident neutron). However, the rectangular distribution is distorted when the proton does not deposit all its energy by ionization in the sensitive volume of gas. The end effect refers to loss of energy (hence shifting of the count to a smaller amplitude) because a portion of the track lies in a dead, or low multiplication, volume near the ends of the anode. The wall effect refers to tracks that are truncated by hitting the wall (cathode). Multiple scattering would shift counts toward larger amplitudes. If these distorting effects are small, the observed pulse-amplitude distribution can be corrected for them, and the neutron spectrum obtained by the derivative method [Eq. (5.27)]. A correction is needed for the response to high-energy neutrons that produce truncated tracks and thus obscure the true spectrum from lower-energy neutrons in the span being measured. Another problem is γ-ray background, which can be rejected by pulse risetime (pulse shape) discrimination if it is not too large. The risetime of pulses from a proportional counter is related to the radial projection of the ionization-track length, because electrons from regions near the cathode take longer to drift to the anode than do electrons generated nearer the anode. The secondary electrons from γ rays have large ranges, hence long tracks on the average, and slowly rising signal pulses. Recoil protons from elastic scattering of neutrons have short tracks and rapidly rising signal pulses. Details and pitfalls in measuring neutron spectra with proportional counters are discussed by Bennett and Yule [15].

Table 5.4 lists characteristics of typical proportional counters for neutrons. The boron trifluoride and helium-3 counters are especially sensitive to thermal neutrons but can have a useful sensitivity to over 100 eV at high pressures. The sensitivity

$$S = \frac{C}{\Phi} \tag{5.28}$$

where C is the number of counts and Φ the fluence, assuming the fluence spectrum is the same in the calibration and measurements. The sensitivity is also equal to the count rate per unit flux density. It is more convenient than efficiency when neutrons impinge from many directions, because the effective area is already included in the sensitivity.

If the neutron source is pulsed, the spectrum of the angular fluence can be measured by time of flight, from thermal to over 15 MeV. A collimated beam of neutrons is extracted from a reentrant probe hole in the shield, on the order of 1 to 10 cm diameter, with the end at the position of interest and at the desired angle, as shown in Figure 5.18. The neutrons drift

TABLE 5.4. CHARACTERISTICS OF TYPICAL NEUTRON PROPORTIONAL COUNTERS

	Boron Trifluoride	Helium-3	Methane
Filling gas	BF_3, 96% ^{10}B	^3He + quench	94% CH_4, 6% N^2
Pressure	400 torr	4 atm ^3He	2 atm
Cathode diameter	2.5 cm	2.5 cm	5.1 cm
Anode diameter	0.005 cm	0.005 cm	0.005 cm
Sensitive length	20 cm	20 cm	60 cm
Operating potential	1 to 2 kV	1 to 2 kV	3 to 3.5 kV
Multiplication	20	20	30
Thermal neutron sensitivity	$\frac{4 \text{ count/s}}{n/cm^2 s}$	$\frac{32 \text{ count/s}}{n/cm^2 s}$	

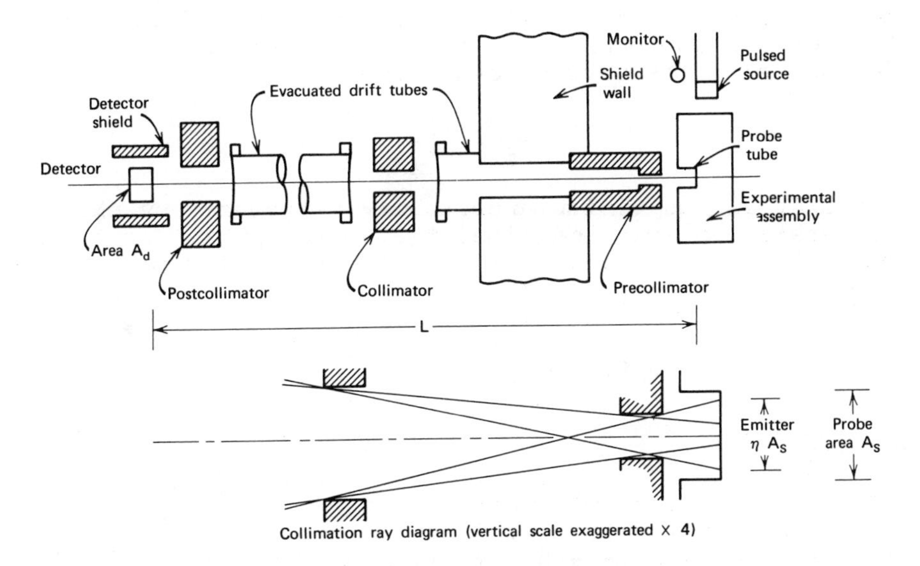

Fig. 5.18. Arrangement for measurement of angular fluence spectrum by pulsed-source, time-of-flight method (from A. Edward Profio, *Experimental Reactor Physics*, Wiley, New York, 1976, p. 434).

down an evacuated tube some 10 to 100 m long, to a detector at distance L. The flight time

$$t_f = \frac{L}{v} = 72.3\ L\ E^{-1/2} \tag{5.29}$$

in microseconds with L in meters and E in eV. The time distribution of counts is analyzed, with time zero as the instant of the source pulse, and

$$t_d = t_f + t_m \tag{5.30}$$

where t_d is the observed time delay from the source pulse and t_m is the mean emission time (for source neutron to slow down and diffuse to the measurement position). The mean emission time is negligible for fast neutrons but can be significant for slow neutrons in moderators. It may be calculated or measured by calibrating the time scale by transmission of neutrons through a slab of material having known-energy resonances. The energy resolution

$$\frac{\Delta E}{E} = \frac{2\Delta t}{t_f} \tag{5.31}$$

where Δt is the timing uncertainty introduced by the dispersion in emission time, the source pulse duration, and jitter in the detector and electronics. The angular fluence spectrum

$$\psi(E) = \frac{C(E)}{\varepsilon(E)\ T(E)\ \Omega_d\ \eta A_s} \tag{5.32}$$

where

$C(E)$ = counts per unit energy;
$\varepsilon(E)$ = detector efficiency;
$T(E)$ = flight-path transmission;
Ω_d = solid angle subtended by detector = (A_d/L^2); and
ηA_s = effective emitting area, a collimator view factor η times probe cross-sectional area A_s.

Counts are accumulated in time bins by repetitively pulsing the source. The pulse duration has to be short for good energy resolution, whereas the time between pulses has to be longer than the flight time of the lowest-energy neutrons detected. Thus the duty cycle is low and an intense pulsed source is required, particularly for shielding experiments.

Foil activation is often used to measure the thermal flux density. Foils are small, insensitive to γ rays, and require no power or signal cables, and some can be irradiated in a high-temperature environment. Table 5.5 gives characteristics of some foil materials, including the activation cross section at 0.025 eV (2200 m/s). The foil is irradiated bare and then the same or similar foil is irradiated inside a cadmium cover about 1 mm thick, to exclude thermal neutrons (<0.5 eV). The saturated activity per gram from thermal-neutron activation

$$A_{th} = A_{bare} - A_{Cd} \tag{5.33}$$

and the thermal flux density

$$\phi_{th} = \frac{A_{th}}{(N/\rho)\sigma} \tag{5.34}$$

where N/ρ is the number of target atoms per gram. Activities measured with thick foils should be corrected for self-shielding, highly absorbing foils in moderators should be corrected for flux depression, and a small correction may be needed for the nonideal cadmium filter, as discussed in Chapter 7 of Profio [3].

The fluence spectrum of fast neutrons can be measured by activation of threshold foils, although the energy resolution is only fair. Table 5.6 lists characteristics of some threshold foils. The effective threshold and cross section are defined for a fission spectrum and are only an approximate guide. The neutron spectrum is unfolded from the saturated activities, where the activation cross section acts as the response function. Because only a few foils are used, an unfolding code such as SAND-II is preferred [16].

The neutron flux density may be measured with a pulse ionization chamber such as illustrated in Figure 5.4, with the cathode covered with a thin deposit of ^{235}U or other fissionable material. The electric field strength between electrodes is $-V_0/d$. The electrons generated by the ionizing fission fragment drift toward the collector (anode) while the ions drift toward the cathode. The drift velocity, w, is a function of the field strength, gas pressure, and mobility. The mobility of electrons in common gases may be 10^3 times that of the ions; hence it is the electrons that are collected in the ionization chamber. (Semiconductor detectors are similar to solid-insulator ionization chambers, but the mobility of holes and electrons are comparable. Electrons are collected in proportional counters where most of the multiplication occurs in a very small volume about the anode wire, where the electric field strength is greatest.) The electron and ion induce a charge on the collector; hence the signal voltage

$$- V_s = \frac{e}{C_s} \left\{\frac{d - x}{d} + \frac{y}{d}\right\} \tag{5.35}$$

TABLE 5.5. THERMAL AND RESONANCE ACTIVATION FOILS[a]

Element (Form)	Melting Pt. (°C)	Reaction	Isotopic Abund. (%)
Au (metal)	1063	$^{197}Au(n,\gamma)^{198}Au$	100.0
Co (metal)	1495	$^{59}Co(n,\gamma)^{60m}Co$	100.0
		$^{59}Co(n,\gamma)^{60}Co$	
Cu (metal)	1083	$^{63}Cu(n,\gamma)^{64}Cu$	69.1
		$^{65}Cu(n,\gamma)^{66}Cu$	30.9
Dy (oxide or Dy-Al)		$^{164}Dy(n,\gamma)^{165m}Dy$	28.18
		$^{164}Dy(n,\gamma)^{165}Dy$	
In (metal)	157	$^{113}In(n,\gamma)^{114m_1}In$	4.23
		$^{113}In(n,\gamma)^{114}In$	
		$^{115}In(n,\gamma)^{116m_1}In$	95.77
		$^{115}In(n,\gamma)^{116}In$	
Mn (alloy or solution)		$^{55}Mn(n,\gamma)^{56}Mn$	100.0

[a]Data from Table of Isotopes, *Handbook of Chemistry and Physics*, (1969) and from K. H. Beckurts and K. Wirtz, *Neutron Physics*, Springer, New York, 1964.

σ (0.025 eV) (b)	Prin. res. (eV)	$T_{1/2}$	Principal Radiations, Energies (MeV)
98.8	4.91	2.7 d	γ 0.412 (95%) β 0.96
16.9	132.0	10.5 min	IT 0.059, 99% to ^{60}Co
20.2		5.26 a	γ 1.173 + 1.332 (100%) β 0.132
4.4	580.0	12.8 h	β 2.63
1.8		5.1 min	$β^+$ 0.656 (19%) $β^-$ 0.573 (38%)
2000.0		1.3 min	IT 0.108
800.0		139.2 min	β 1.29 (83%) 1.19 (15%)
56.0		50.0 d	IT 0.19
2.0		72.0 s	β 2.0
160.0	1.46	54.0 min	β 1.0 (55%), many γ
42.0		14.0 s	β 3.3
13.2	337.0	2.58 h	γ 0.845 (99%)

TABLE 5.6. THRESHOLD ACTIVATION FOILS[a]

Element (Form)	Melting Pt. (°C)	Reaction	Isotopic Abund. (%)
^{237}Np (oxide)		$^{237}Np(n,F)$	
In (metal)	157	$^{115}In(n,n')^{115m}In$	95.8
S (pellet)	120	$^{32}S(n,p)^{32}P$	95.0
Ni (metal)	1453	$^{58}Ni(n,p)^{58}Co$	68.0
Ti (metal)	1675	$^{46}Ti(n,p)^{46}Sc$	8.0
Al (metal)	660	$^{27}Al(n,\alpha)^{24}Na$	100.0
I (NH_4I)		$^{127}I(n2n)^{126}I$	100.0

[a]Data from Table of Isotopes, EG&G bulletins, and *Handbook of Chemistry and Physics* (1969).

E_{th} (eff.) (MeV)	σ_{eff} (mb)	$T_{1/2}$	Principal Radiations, Energies (MeV)
0.6	1650	FP	FP
1.5	365	4.5 h	IT 0.335 (95%)
3.0	300	14.2 d	β 1.71
3.0	550	71.3 d	γ 0.80, β^+, EC
5.5	230	83.9 d	γ 2.01 + 0.89 (100%) β 0.36
7.5	70	15.0 h	γ 2.75 + 1.37 (100%) β 1.39
11.0	980	13.3 d	γ 0.66, 0.39, etc. β^-, β^+, EC also

where x is the distance of the electron from the anode and y the distance of the ion. The amplitude is $V_{max} = (q/C_s)$ when all the electrons and ions reach the corresponding electride, where q = Ne and N is the number of ion-electron pairs generated.

When several neutrons are detected in sequence, the signal waveform at the anode appears as shown in Figure 5.19a. To separate the individual events and to measure the amplitude, the voltage pulses are processed by a low-pass filter or differentiating network, giving output pulses such as those shown in Figure 5.19b. The differentiating circuit is usually located in the main amplifier and should be set for about 4 times the pulse risetime to avoid loss of signal amplitude. If the amplifier is provided with an adjustable integrating circuit, the time constant can be the same as for differentiation.

Figure 5.20 is a block diagram of a pulse-counting system. The preamplifier for scintillation and GM counters is voltage sensitive and has a nominal gain of one but matches the detector impedance to that of the coaxial cable transmitting signals to the main amplifier (usually 93 Ω, sometimes 50 Ω). The preamplifier for semiconductor detectors, pulse ionization chamber, or proportional counter should be charge sensitive with a gain of 10, have low noise, and perform impedance matching. The main amplifier has an adjustable gain of about 10 to 1000 and performs the pulse shaping (differentiation and integration). The discriminator is a trigger circuit that gives no output if the signal amplitude is below the adjustable discriminator threshold (0.1 to 10 V). If the signal exceeds the threshold, a logic output pulse is generated, for example, with an amplitude of 5 V and duration of 0.5 μs. These pulses are accumulated in an electronic register or scaler (controlled by a timer) or a count-rate meter. The discriminator threshold is set above noise and any low-energy background; however, some true signal pulses may also be rejected; hence the overall detection efficiency is reduced. A suitable compromise must be found.

Amplitude (hence energy) selection can be performed with a single-channel or a multichannel analyzer. The principle of the single-channel analyzer is shown in Figure 5.21. Two discriminators are used plus an anticoincidence network. Pulses such as A are rejected because the amplitude is less than the threshold of the lower-level discriminator, LLD. Pulses such as B

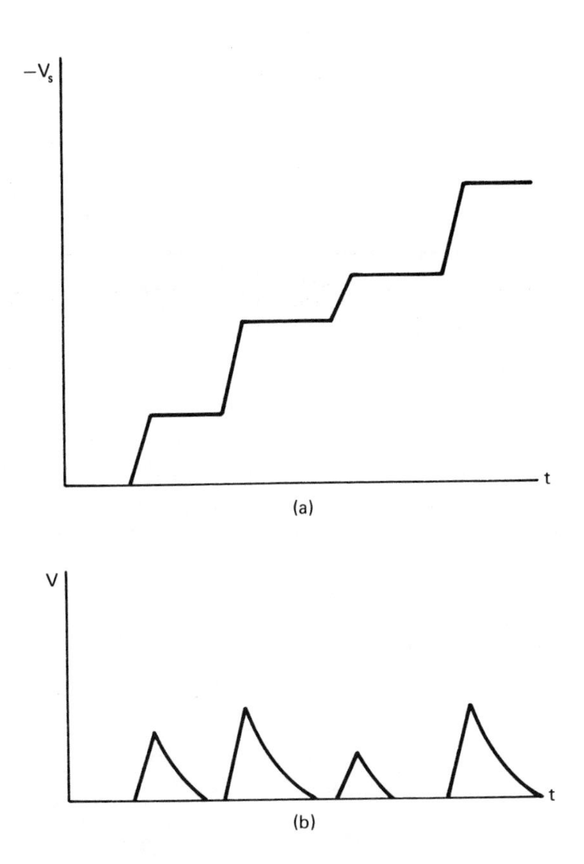

Fig. 5.19. Voltage signal before differentiation (a) and after differentiation (b) to return to baseline.

generate a logic output pulse. Pulses such as C are rejected; the triggering of the upper-level discriminator (ULD) generates an anticoincidence or veto pulse that suppresses the output. Thus pulses with amplitudes and corresponding energies between E and $(E + \Delta E)$ are counted. The ULD and LLD are ganged such that ΔE is fixed. The entire spectrum can be measured sequentially by scanning the LLD or E control. The multichannel analyzer uses an analog-to-digital converter or coder to sort incoming pulses into voltage amplitude bins, usually 256 to 4096, depending on the resolution of the detector. Each pulse is analyzed and one count

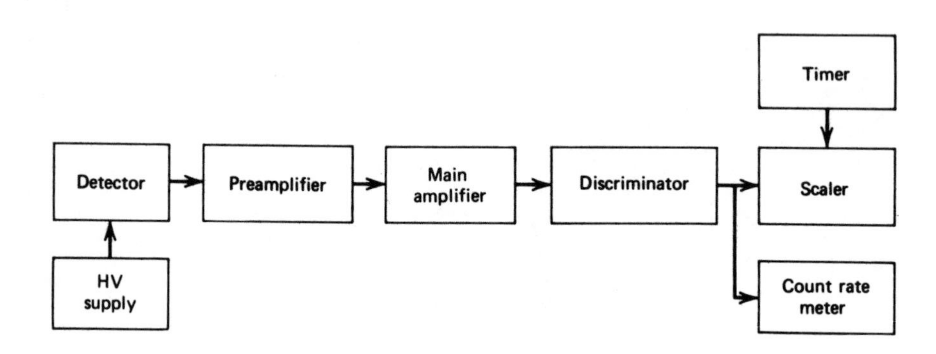

Fig. 5.20. Pulse-counting system (from A. Edward Profio, *Experimental Reactor Physics*, Wiley, New York, 1976, p. 306).

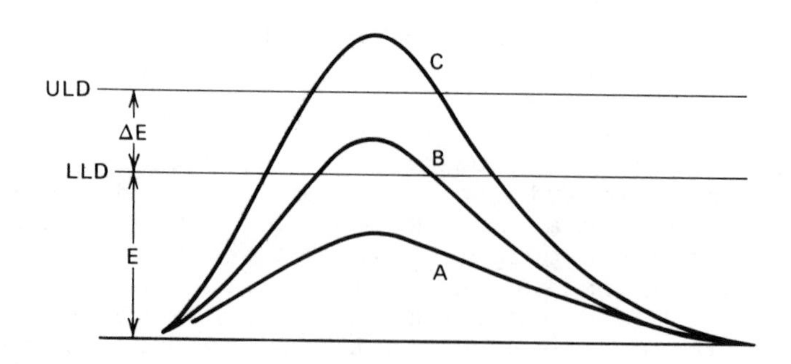

Fig. 5.21. Amplitude selection by a single-channel analyzer (from A. Edward Profio, *Experimental Reactor Physics*, Wiley, New York, 1976, p. 310).

added to the appropriate address or bin in the computer-like memory. The accumulated pulse-amplitude distribution is read out on a scope or printer at the conclusion of the run.

Statistics

Every measurement of fluence (and dose) is subject to an uncertainty because of the randomness of radioactive decay, nuclear reactions, and interactions in a detector [3]. Emission and detection are described by Poisson statistics, where the probability of observing x events (e.g., counts)

$$P(x) = \frac{m^x}{x!} e^{-m} \tag{5.36}$$

where m is the mean number of events. For $m > 20$, a good measure of the uncertainty is the standard deviation, which for Poisson statistics is

$$\sigma = \sqrt{m} \tag{5.37}$$

Thus 10^4 counts corresponds to a standard deviation of 100 counts, and the fractional standard deviation σ/m is 1%. If only 100 counts are accumulated, the fractional standard deviation is 10%.

Background (counts from other than the desired radiation) has to be subtracted in nuclear measurements. Let m_{s+b} be the mean count from source plus background and m_b be the mean count from background. The uncertainties σ_{s+b} and σ_b are given by Eq. (5.37). The net mean count rate from the source is

$$m_s = m_{s+b} - m_b \tag{5.38}$$

and the uncertainty in the net count is

$$\sigma_s = (\sigma_{s+b}^2 + \sigma_b^2)^{1/2} \tag{5.39}$$

according to the theory of statistics and propagation of uncertainties.

5.3 KERMA

Kerma is the initial kinetic energy of electrons and ions generated by interactions of photons and neutrons in unit mass of specified material. It is calculated by multiplying the fluence by the energy-dependent fluence:kerma factor k(E) and integrating over incident particle energy E (MeV). For γ rays the kerma factor

$$k(E) = CE \frac{\mu_K(E)}{\rho} \frac{J/kg}{\gamma/cm^2}$$

$$= C \frac{N}{\rho} (E - E_B)\sigma_{pe} + f_a E\sigma_C + (E - 1.022)\sigma_{pp} \quad (5.40)$$

where

C = 1.602×10^{-10} (J MeV^{-1} × g kg^{-1});
N/ρ = atoms kg^{-1};
E_B = average photoelectron binding energy (MeV);
σ_{pe} = photoelectric cross section (cm^2);
f_a = fraction of energy to Compton electron;
σ_C = Compton interaction cross section (cm^2); and
σ_{pp} = pair-production cross section (cm^2).

The cross sections and f_a are functions of E, as discussed in Chapter 3. The Compton cross section per atom is Z times the cross section per electron. Values of μ_K/ρ for some common materials [16,17] are listed in Table 5.7. The compositions of muscle (striated) and compact bone are given in Table 5.8. Fluence:kerma factors for muscle and bone are plotted in Figure 5.22.

The kerma factor for neutrons is

$$k(E) = \frac{C}{\rho} \sum_i \sum_j \bar{E}_{ij} N_i \sigma_{ij} \quad (5.41)$$

where the sums are over elements i and reactions j. The energy-dependent quantities are

$\bar{E}_{ij}$ = average energy transferred to ions (MeV)
σ_{ij} = reaction cross section (cm^2)

The $\bar{E}_{ij}$ can be derived from the reaction energetics discussed in Chapter 3. For elastic (Q = 0) and inelastic scattering, the average energy of the recoil nucleus is

$$\bar{E}_R = \frac{2AE}{(A + 1)^2} \left[1 - \frac{A + 1}{2AE} Q - \sqrt{1 - \frac{(A + 1)Q}{AE}} \; \overline{\cos \theta_C} \right] \tag{5.42}$$

where A is the ratio of nuclear:neutron mass and $\overline{\cos \theta_C}$ is the average cosine of scattering in the CMS. For continuum nonelastic scattering or (n,2n) and similar reactions, where the secondary neutron energy distribution $p(E_C')$ is given in the CMS, and assuming isotropic scattering.

$$\bar{E}_R = \frac{AE}{(A + 1)^2} + \frac{\bar{E}_C'}{A} \tag{5.43}$$

where

$$\bar{E}_C' = \frac{\int_0^{E^*} E_C' \; N(E_C') \; dE_C'}{\int_0^{E^*} N(E_C') \; dE_C'} \tag{5.44}$$

and the maximum energy E^* available to the scattered neutron corresponds to excitation of the lowest level, Q_{min}; hence

$$E^* = \frac{A}{A + 1} \; \frac{AE}{A + 1} - Q_{min} \tag{5.45}$$

In radiative capture the recoil energy

$$\bar{E}_R = \frac{E}{A + 1} + \frac{1}{2M'c^2} \left(Q_\gamma + \frac{AE}{A + 1} \right)^2 \tag{5.46}$$

where Q_γ is the rest energy given up as photons and $M'c^2$ is the rest energy of the product nucleus. For reactions emitting charged particles but no γ rays

$$\bar{E} = \bar{E}_R + \bar{E}_I = E + Q \tag{5.47}$$

TABLE 5.7. γ-RAY MASS-ENERGY-TRANSFER COEFFICIENTS[a]

			μ_K/ρ (cm^2/g)			
E (MeV)	Water	Air	Iron	Lead	Muscle	Bone
0.01	4.79	4.61	142.0	127.0	4.87	19.2
0.015	1.28	1.27	49.3	91.7	1.32	5.84
0.02	0.512	0.511	22.8	69.2	0.533	2.46
0.03	0.149	0.148	7.28	24.6	0.154	0.720
0.04	0.0678	0.0669	3.17	11.8	0.0701	0.304
0.05	0.0419	0.0406	1.64	6.57	0.0431	0.161
0.06	0.0320	0.0305	0.961	4.11	0.0328	0.0998
0.08	0.0262	0.0243	0.414	1.87	0.0264	0.0537
0.10	0.0256	0.0234	0.219	2.28	0.0256	0.0387
0.15	0.0277	0.0250	0.0814	1.16	0.0275	0.0305
0.20	0.0297	0.0268	0.0495	0.637	0.0294	0.0301
0.30	0.0319	0.0288	0.0335	0.265	0.0317	0.0310
0.40	0.0328	0.0295	0.0308	0.147	0.0325	0.0315
0.50	0.0330	0.0297	0.0295	0.0984	0.0328	0.0317
0.60	0.0329	0.0296	0.0287	0.0737	0.0326	0.0315

0.80	0.0321	0.0289	0.0275	0.0503	0.0318	0.0306
1.00	0.0311	0.0280	0.0264	0.0396	0.0308	0.0297
1.50	0.0285	0.0257	0.0241	0.0288	0.0282	0.0272
2.00	0.0262	0.0236	0.0225	0.0259	0.0259	0.0251
3.00	0.0229	0.0207	0.0212	0.0260	0.0225	0.0221
4.00	0.0209	0.0189	0.0209	0.0281	0.0204	0.0204
5.00	0.0195	0.0178	0.0211	0.0306	0.0189	0.0192
6.00	0.0185	0.0168	0.0215	0.0331	0.0178	0.0184
8.00	0.0170	0.0157	0.0226	0.0378	0.0164	0.0167
10.00	0.0162	0.0151	0.0238	0.0419	0.0155	0.0159

[a] Hubbell, NSRDS-NBS 29.

TABLE 5.8. COMPOSITIONS OF MUSCLE, BONE, STANDARD MAN

Element	Weight Fraction		
	Muscle[a]	Bone[a]	Standard Man[b]
Hydrogen	0.102	0.064	0.100
Carbon	0.278	0.123	0.240
Nitrogen	0.035	0.027	0.029
Oxygen	0.729	0.140	0.600
Sodium	0.0008	--	0.002
Magnesium	0.0002	0.002	0.003
Phosphorus	0.002	0.070	0.011
Sulfur	0.005	0.002	0.0024
Potassium	0.003	--	0.002
Calcium	0.00007	0.147	0.012
Chlorine	--	--	0.002

[a]From ICRU, NBS Handbook 85 (1964).
[b]Ritts et al., ORNL-TM-2991.

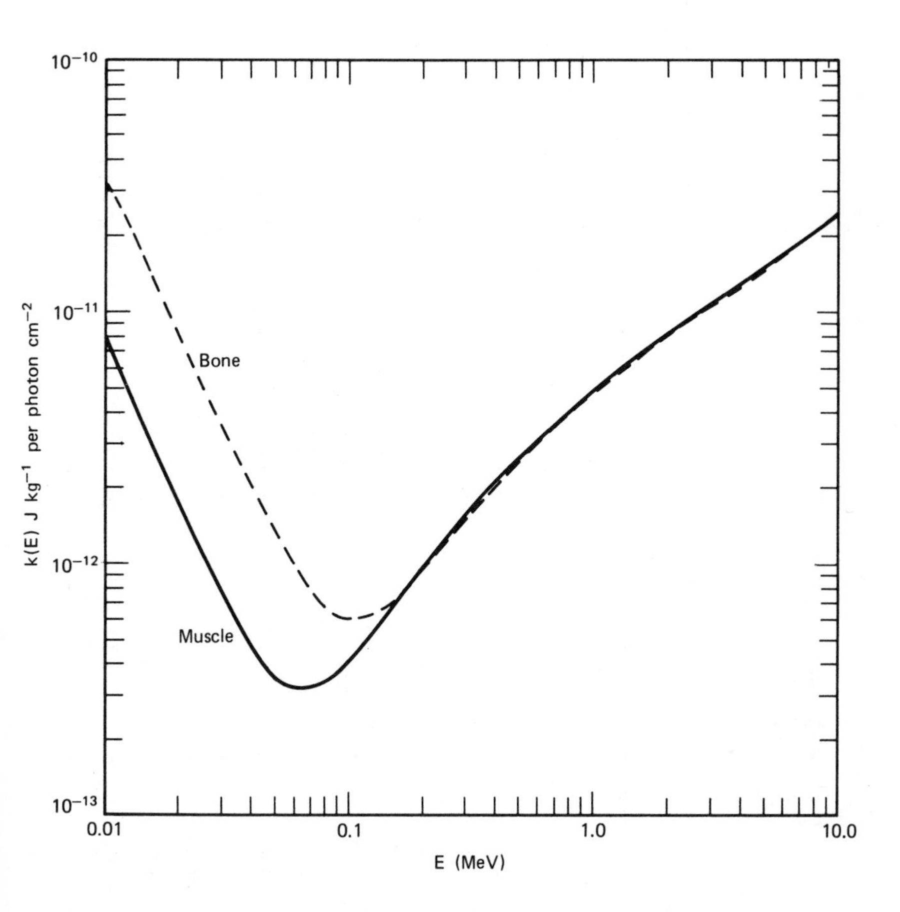

Fig. 5.22. γ-Ray fluence:kerma factor for muscle and bone.

Kerma factors for some materials [18] are listed in Table 5.9. The kerma factor for tissue of "standard man" composition, as given in Table 5.8, is plotted in Figure 5.23 along with the contributions from the different elements. At high energies the major contribution comes from the recoil protons on elastic scattering in hydrogen. At low energies the major contributor is the $^{14}N(n,p)^{14}C$ reaction.

TABLE 5.9. NEUTRON FLUENCE:KERMA FACTORS

E(MeV)	k(E) J kg^{-1}/n cm^{-2}				
	Hydrogen	Carbon	Nitrogen	Oxygen	Calcium
2.23(-8)	4.18(-14)	1.24(-19)	8.34(-12)	6.29(-20)	3.40(-17)
1.07(-7)	2.03(-14)	5.92(-19)	3.83(-12)	2.73(-19)	1.62(-16)
1.00(-6)	7.59(-15)	5.52(-18)	1.26(-12)	2.47(-18)	1.52(-15)
1.26(-5)	1.40(-14)	6.97(-17)	3.59(-13)	3.10(-17)	1.91(-14)
1.01(-4)	9.74(-14)	5.57(-16)	1.29(-13)	2.48(-16)	1.53(-13)
1.03(-3)	9.80(-13)	5.67(-15)	4.82(-14)	2.53(-15)	1.56(-12)
1.02(-2)	9.26(-12)	5.57(-14)	7.91(-14)	2.49(-14)	1.55(-11)
1.01(-1)	6.03(-11)	5.06(-13)	3.71(-13)	2.41(-13)	1.53(-10)
1.00	2.04(-10)	2.79(-12)	2.18(-12)	5.34(-13)	2.78(-9)
2.00	2.77(-10)	3.81(-12)	3.17(-12)	1.88(-12)	4.96(-9)
5.01	3.89(-10)	5.51(-12)	1.38(-11)	4.59(-12)	6.36(-9)
10.00	4.50(-10)	8.82(-12)	1.90(-11)	1.08(-11)	6.82(-9)
15.00	4.63(-10)	3.29(-11)	3.17(-11)	2.51(-11)	7.55(-9)

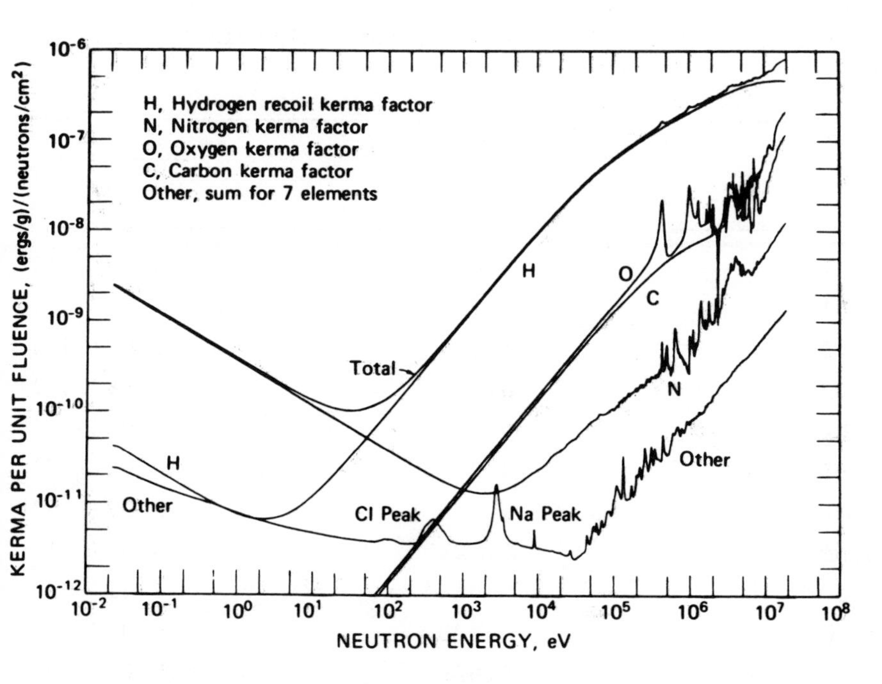

Fig. 5.23. Neutron fluence:kerma factor in standard man and components (multiply erg/g by 10^{-4} for J/kg) (after N. M. Schaeffer, TID-25951).

5.4 EXPOSURE

Exposure is calculated by Eq. (5.7). The mass-energy-absorption coefficient μ_{en}/ρ for air [16] and the fluence:exposure factor are given in Table 5.10, and the factor is plotted in Figure 5.24. Note that with W = 33.7 eV per ion-electron pair, 1 R = 8.69×10^{-3} J kg^{-1}.

Exposure is measured with an air-filled direct current (dc) ionization chamber with air-equivalent walls and electrodes. To be air equivalent, the effective Z value (referred to the photoelectric absorption) has to be the same. A bakelite plastic with graphite filler for electrical conductivity, plus a few percent titanium dioxide, is air equivalent. Design of ionization chambers is discussed under the

TABLE 5.10. MASS-ENERGY-ABSORPTION COEFFICIENT AND EXPOSURE FACTOR FOR AIR

E (MeV)	$\mu_{en}/\rho\,(cm^2/g)$	$\frac{R}{photon/cm^2}$
0.01	4.61	8.49(-10)
0.015	1.27	3.51(-10)
0.02	0.511	1.88(-10)
0.03	0.148	8.18(-11)
0.04	0.0669	4.93(-11)
0.05	0.0406	3.74(-11)
0.06	0.0305	3.37(-11)
0.08	0.0243	3.58(-11)
0.10	0.0234	4.31(-11)
0.15	0.0250	6.91(-11)
0.20	0.0268	9.87(-11)
0.30	0.0288	1.59(-10)
0.40	0.0295	2.17(-10)
0.50	0.0297	2.74(-10)
0.60	0.0296	3.27(-10)
0.80	0.0289	4.26(-10)
1.0	0.0280	5.16(-10)
1.5	0.0257	7.10(-10)
2.0	0.0238	8.77(-10)
3.0	0.0212	1.17(-9)
4.0	0.0194	1.43(-9)
5.0	0.0182	1.68(-9)
6.0	0.0174	1.92(-9)
8.0	0.0162	2.39(-9)
10.0	0.0156	2.87(-9)

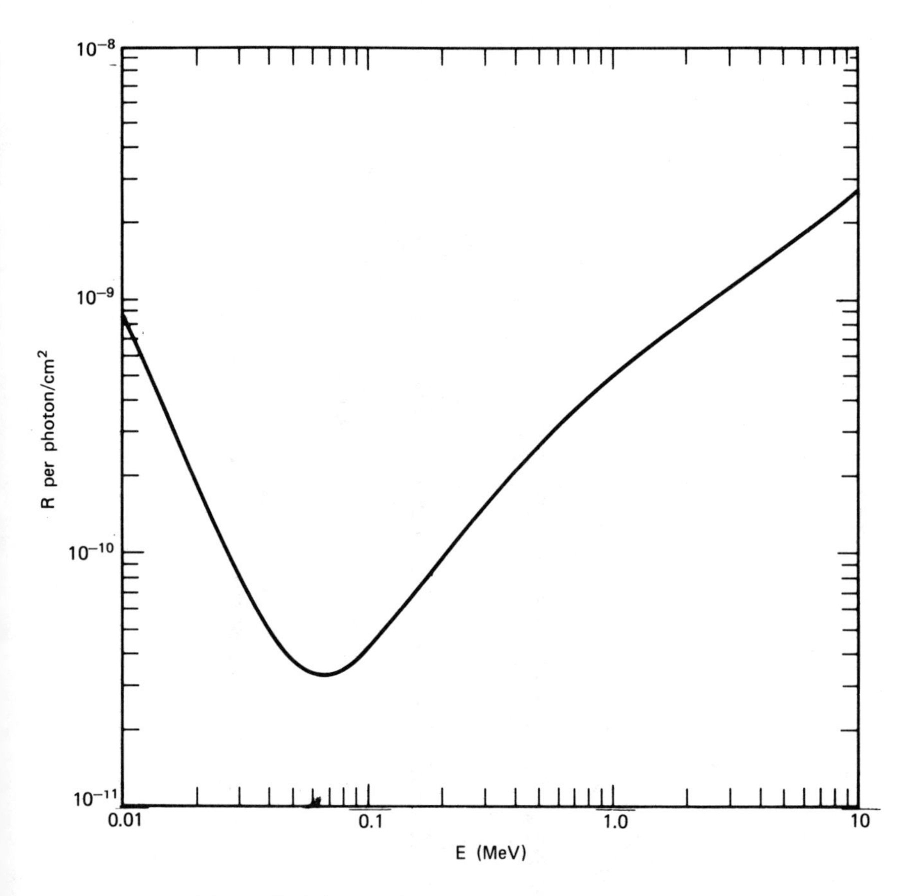

Fig. 5.24. Fluence:exposure factor.

subheading Absorbed Dose. Exposure is difficult to measure above 3 MeV because the range of the secondary electrons becomes comparable to the mean free path of the γ rays, and the ionization produced in a small volume is no longer proportional to the radiation intensity at that point.

Figure 5.25 illustrates a direct-reading condenser (capacitor) type ionization chamber, also known as a "pocket dosimeter". Initially the ion chamber and associated capacitance are charged through the contact in the diaphragm at the left. During irradiation, the ionization current discharges the capacitance. The

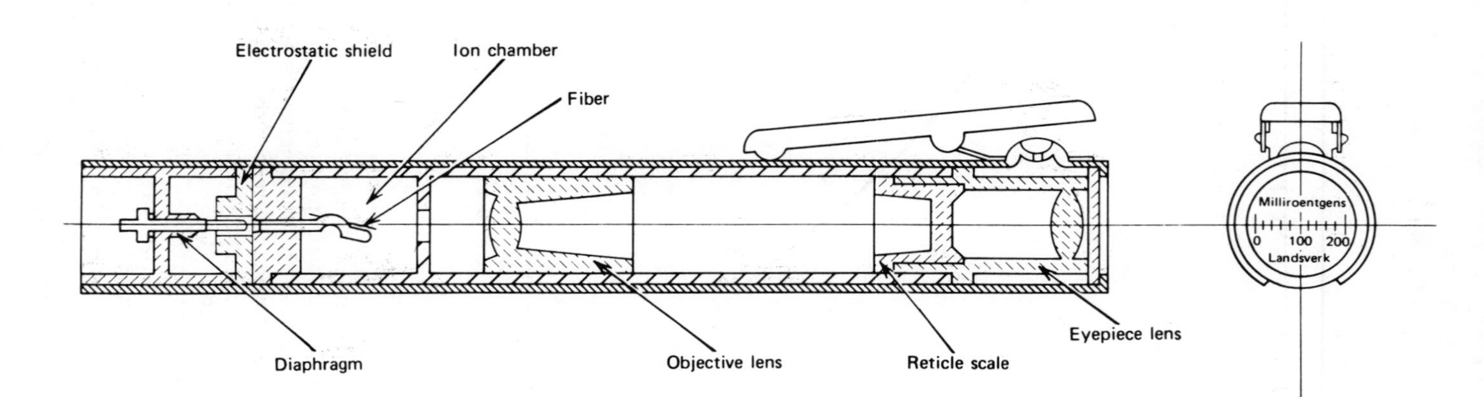

Fig. 5.25. Direct-reading pocket-dosimeter-type ionization chamber for exposure (courtesy Landsverk Electrometer Co.).

residual voltage hence charge is measured by the built-in quartz-fiber electroscope. The position of the fiber is observed through a microscope on a reticle scale calibrated in milliroentgens. Similar chambers are made without the electroscope and microscope. They are read after irradiation on an external electrometer. These condenser-type chambers can be made quite small and are suitable for measurements in a phantom (e.g., a block of plastic or tank of water approximating the body). The voltage at the end of the irradiation cannot be zero but must remain sufficiently high to collect the ions and electrons without excessive loss due to recombination.

Exposure is of interest because the energy absorption per gram of air is not too much different from the energy absorption per gram of soft tissue (e.g., muscle) over an extended energy range [19], as shown in Figure 5.26. The quantity plotted is f, the ratio of absorbed dose in rad, to the exposure in roentgen. The absorbed dose in air is 87 rad/R. If muscle is substituted without perturbing the field in which 1 R is measured, the absorbed dose (muscle) varies from

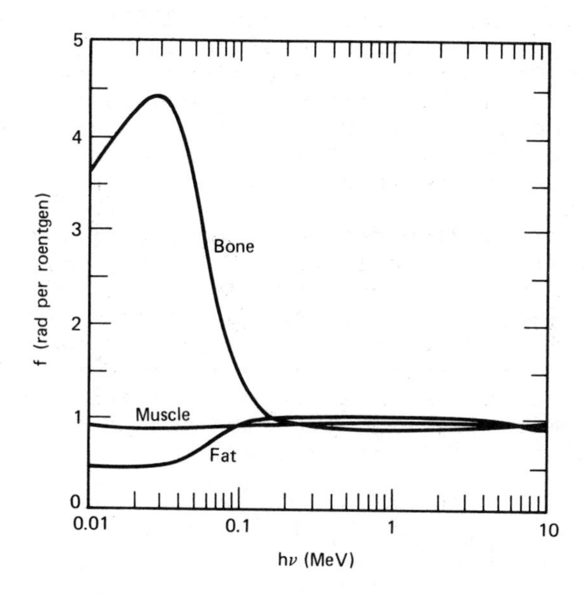

Fig. 5.26. Ratio of absorbed dose (rad) to exposure (R) for muscle, fat, and bone (after Morgan and Turner [19]).

about 0.9 to 1 rad. However, below 0.1 MeV the absorbed dose in fat is only about 0.5 rad, whereas the absorbed dose in bone ranges from about 1 to 4.5 rad. Thus exposure may be used as a measure of the output of an X-ray of γ-ray source, but it is not an accurate measure of the absorbed dose in the body.

5.5 ABSORBED DOSE

The absorbed dose from ions can be calculated from the mass stopping power S/ρ. Consider a monodirectional beam, perpendicularly incident on a differential volume of area dA and thickness $d\ell$, centered at $\underline{r}$, and containing mass dm. An ion loses energy $S(E)\, d\ell$ in traversing $d\ell$, and the number of ions entering dA is $\Phi(E)\, dA\, dE$, where $\Phi(E)$ is the fluence spectrum. Then the absorbed dose at $\underline{r}$ is

$$D(\underline{r}) = \frac{\int S(E)\ \Phi(E)\ dA\ d\ell\ dE}{\rho\ dA\ d\ell} = \frac{S(E)}{\rho}\ \Phi(\underline{r},E)\ dE \qquad (5.48)$$

in appropriate units. The energy and fluence as a function of position may be calculated by the methods discussed in Chapter 3. If a uniform source is distributed in the medium, the energy absorbed per gram can be equated to the energy emitted per gram (except within the surface layer of thickness equal to the range) because of the small range of ions and negligible bremsstrahlung.

The absorbed dose from electrons can be calculated from Eq. (5.48), but it is difficult to obtain $\Phi(\underline{r},E)$ except by transport methods such as moments [20] and Monte Carlo [21]. In practice, the spatial distribution of absorbed dose is often calculated using a calculated or empirical point-kernel for an infinite medium. Loevinger, as discussed by Fitzgerald [22], derived a semiempirical expression for point-source β particles in an infinite medium,

$$D(x) = \frac{k}{(\nu x)^2} \Big\{ c \left[1 - \frac{\nu x}{c} \exp 1-(\nu x/c) \right]$$

$$+\ \nu x \exp 1-\nu x \Big\} \qquad x \leq \frac{c}{\nu}$$

$$= \frac{k}{\nu x} \exp 1-\nu x \qquad x > \frac{e}{\nu} \qquad (5.49)$$

where the apparent absorption coefficient for tissue is

$$\nu = \frac{18.6}{(E_0 - 0.036)^{1.37}} \left(2 - \frac{\bar{E}_\beta}{\bar{E}^*_\beta}\right) \quad \frac{cm^2}{g} \qquad (5.50)$$

with x in g/cm^2, E_0 is the maximum energy of the β spectrum, $\bar{E}^*_\beta$ the average energy, and $\bar{E}_\beta$ the average energy of an allowed spectrum having the same E and Z. The ratio $\bar{E}_\beta/\bar{E}^*_\beta$ is unity for an allowed β spectrum and not much different for forbidden spectra. The parameter

$$c \begin{cases} 2 & 0.17 < E_0 < 0.5 \text{ MeV} \\ 1.5 & 0.50 \leq E_0 < 1.5 \text{ MeV} \\ 1 & 1.50 \leq E_0 < 3.0 \text{ MeV} \end{cases} \qquad (5.51)$$

and

$$k = 1.28 \times 10^{-9} \rho^2 \nu^3 \bar{E}_\beta \alpha \qquad \text{rad/electron} \quad (5.52)$$

where

$$\alpha = [3c^2 - (c^2 - 1)e]^{-1} \qquad (5.53)$$

with energy in MeV and ρ in g/cm^3.

The absorbed dose from γ rays can be calculated if the transport and absorption of the secondary electrons are included as well as the transport of the primary γ rays, fluorescence X rays, Compton scattered photons, annihilation quanta, and bremsstrahlung photons from the electrons. Quite often, however, the approximation is made that the fluorescence X rays and the electrons are stopped where they were generated. The absorbed dose from neutrons can be calculated also, including the generation and absorption of secondary γ rays. The transport of the ions may be treated but it is often a good approximation to assume the ions are stopped where they are generated.

Calculation of absorbed dose is simplified under conditions of radiation equilibrium. Consider a point

P in a homogeneous region V, as shown in Figure 5.27. A small volume ΔV surrounds P, containing mass Δm. The minimum distance d to the boundary is greater than the range of the most penetrating charged particle, or greater than 5 times μ_{en}^{-1}, the energy-absorption mean free path of photons or neutrons. Now if an ion or electron source is uniformly distributed throughout V, on the average as many particles of energy E enter ΔV as leave, and radiation equilibrium exists at P. The absorbed dose is equal to the energy emitted per unit mass of material. The same is true of a distributed photon or neutron source.

If the volume is irradiated by an external photon or neutron source, radiation equilibrium will not exist in general because of attenuation. However, if a charged particle A carrying energy E out of ΔV is replaced, on the average, by another charged particle B of equal energy, then charged-particle equilibrium (CPE) exists. γ Rays of less than 3 MeV or neutrons of less than 20 MeV or so have mean free paths greater than the range of the secondary charged particles, the (collision) source density is almost uniform over the range, and CPE exists to a good approximation. Charged-particle equilibrium is absent at higher energies, near a point source where the flux density changes rapidly because of geometric attenuation, or near boundaries as shown in Figure 5.28. When CPE does exist, the

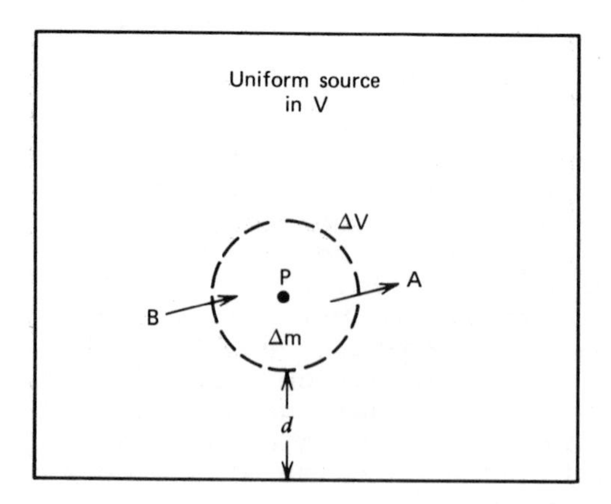

Fig. 5.27. Radiation equilibrium. Energy leaving ΔV (A) is compensated by energy entering (B) when d > range or > $5/\mu_{en}$. Absorbed dose equals energy emitted per unit mass.

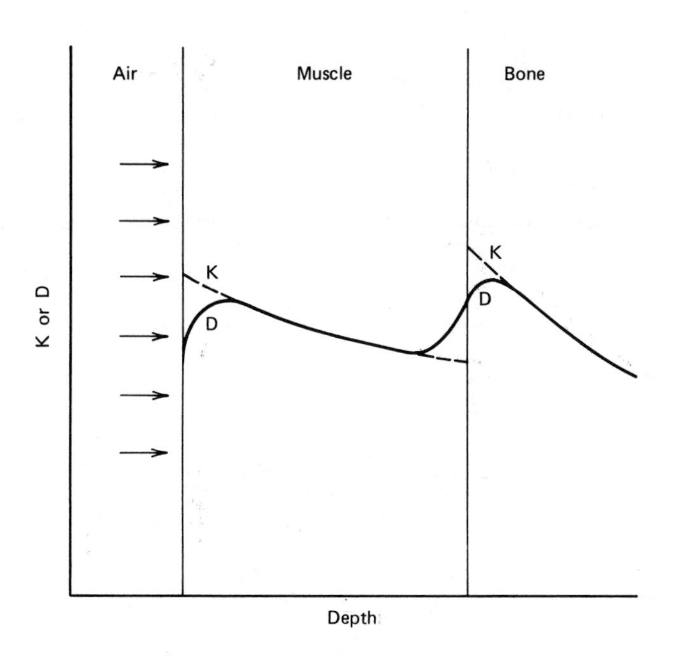

Fig. 5.28. Variation of kerma (K) and absorbed dose (D) with depth for X rays perpendicularly incident on muscle from air. Note the buildup of D as electrons attain equilibrium in muscle, and the larger D in bone and muscle near bone.

absorbed dose is equal to the kerma, calculated as discussed in Section 5.3.

Absorbed dose can be measured with a cavity or extrapolation ionization chamber, thermoluminescent dosimeter, photographic emulsion, or certain chemical systems.

Cavity Ionization Chamber

Cavity theory is discussed in detail by Burlin [23]. The cavity is a material of different density and possibly different composition, introduced in the material for which the absorbed dose is desired. In the ionization chamber, the cavity is a gas-filled space within the material of the walls. The walls are made thick enough for CPE to exist, but not so thick that the primary radiation is appreciably attenuated within them. To measure absorbed dose in soft tissue, the walls are made of a conducting plastic with tissue-equivalent composition ($C_5H_{40}O_{18}N$). The filling gas may be a mixture of methane (64.4 v/o), carbon dioxide (32.4 v/o), and nitrogen (3.2 v/o). Figure 5.29 illustrates a spherical ionization chamber with tissue-equivalent walls 5 mm thick, suitable for measuring the absorbed dose from medium-energy neutrons and γ rays (a thinner wall should be used for low-energy X-rays). The gas volume is 1 cm^3, pressure is 1 atm, and bias voltage is 250 V.

The energy absorbed per gram of gas is

$$D_c = W J_m \tag{5.54}$$

where W is the energy per ion pair and J_m the observed number of ion pairs generated per gram of gas. The problem is to relate D_c to the desired absorbed dose in the wall material, D_w.

According to Bragg-Gray theory,

$$D_w = s_m D_c \tag{5.55}$$

where

$$s_m = \frac{(S/\rho)_w}{(S/\rho)_c} \tag{5.56}$$

the ratio of the mass stopping power in the wall to the mass stopping power in the cavity (gas). Experimentally, it is found that s_m is nearly independent of energy. The Bragg-Gray theory assumes CPE, that the cavity does not perturb the secondary-particle spectrum in the wall, that nearly all secondary particles originate in the wall, and that cavity dimensions are small compared to the range at the gas pressure prevailing. Fano discovered that the size limitation could be relaxed if the atomic composition of wall and gas are matched, and most cavity ion chambers are made this way. More advanced theories relating D_w to D_c take into account the variation of s_m with energy and the influence of the few energetic electrons (δ rays). If cavity dimensions are large compared to the range, theory predicts

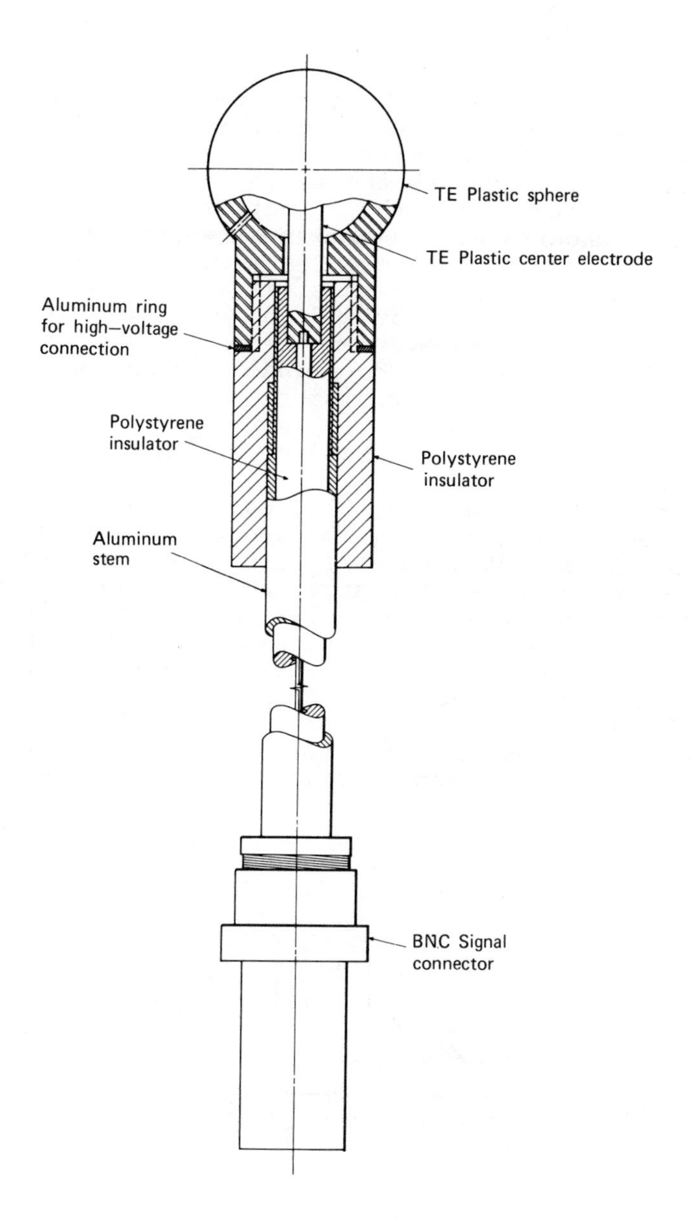

Fig. 5.29. Spherical ionization chamber, 1 cm^3 (courtesy of EG&G, Inc.).

$$D_w = \frac{(\mu_{en}/\rho)_w}{(\mu_{en}/\rho)_c} D_c \tag{5.57}$$

Cavity ionization chambers are made with gas volumes from less than 1 cm^3 to several liters. The larger volumes give larger current per unit absorbed-dose rate, but may perturb the field, and average over a greater distance, so the effective center of detection is more uncertain. It is important to use good insulators to avoid leakage current, and a guard ring may be added to define the sensitive volume and intercept leakage between cathode and anode. The bias voltage is large enough to overcome recombination and saturate the signal current. The signal current is measured with a sensitive electrometer.

Extrapolation Ionization Chambers

Ionization chambers may also be used to measure the absorbed-dose rate from charged particles, but it is difficult to make the wall thin enough and the cavity small compared to the range. The extrapolation chamber, developed by Failla, can overcome these limitations. Figure 5.30 shows the principle of the extrapolation chamber. The incident radiation passes through a thin foil that is separated by a small distance d from the guard ring and collector electrodes. The electrodes are made of the material for which the absorbed dose is desired. The gap is filled with an appropriate gas

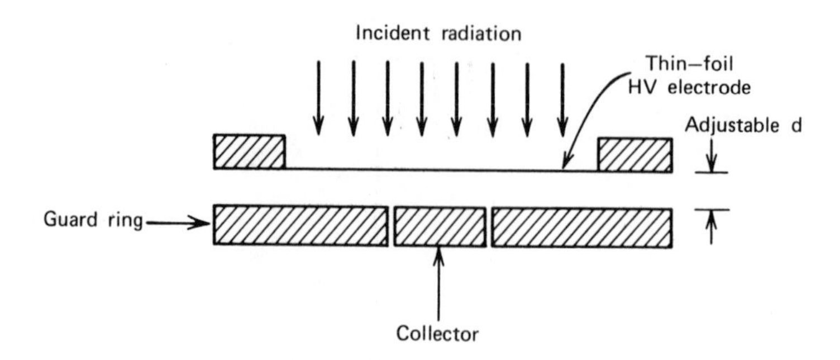

Fig. 5.30. Extrapolation ionization chamber.

and d is adjustable by means of micrometer screws. The active volume $V(cm^3)$ is defined by the collector area and d. The (saturation) ionization current i is measured as a function of V and extrapolated to zero volume; hence a condition in which the Bragg-Gray relation [Eq. (5.55)] holds. As the volume, temperature, and pressure of the gas are measured, the mass is known. The extrapolation chamber may also be used to measure the absorbed dose or exposure at the surface of a material, where the cavity chamber is not suitable.

Thermoluminescent Dosimeter

Thermoluminescence is the property of certain irradiated insulators of emitting light on warming and is a form of thermally accelerated phosphorescence [24]. Ionizing radiation generates hole-electron pairs and excitons. The electrons may return to the valence band quickly, emitting light (fluorescence). However, if the material contains imperfections such as impurities or displacement defects, some of the electrons may be trapped. When the material is warmed to 100° to 300°C, the electrons are released and emit light (phosphorescence). The heating is done in a special oven with a filtered and cooled photomultiplier viewing the thermoluminescent dosimeter (TLD) pellets or capsules. The light output as a function of temperature is characterized by one or more "glow peaks," and the integrated light under defined heating conditions is a measure of the absorbed dose in the TLD. After readout, the TLD can be annealed at relatively high temperature and reused.

Properties of popular thermoluminescent materials are listed in Table 5.11. Calcium sulfate, activated with a small amount of manganese, is the most sensitive to γ rays, but the glow peak is only a little above room temperature and "fading" may occur before readout. Calcium fluoride is more sensitive than lithium fluoride, but because of its higher average atomic number, it is not as good a match (in γ-ray dose-response vs. energy) to air ($\overline{Z} = 7.42$). This is indicated by the rise in response at 50 keV relative to 1 MeV. The response can be improved by surrounding the dosimeter with a low-energy X-ray filter. These materials, especially LiF, are also sensitive to neutrons. However, they will not measure the absorbed dose in tissue because the nuclear properties and kerma are so different. Lithium fluoride is available as

TABLE 5.11. THERMOLUMINESCENT MATERIALS

	LiF	$CaSO_4$:Mn	CaF_2:Mn
Minimum dose (rad)	10^{-2}	20×10^{-6}	10^{-3}
Maximum linear dose (rad)	700	10^4	10^5
Maximum dose (rad)	10^5	10^5	3×10^5
Glow Peak (°C)	210	80 to 100	260
Peak wavelength (Å)	4000	5000	
Fading (25°C)	5%/year	∿50%/10 h	10%/16 h; then 1%/day
Density (g/cm^3)	2.64	2.61	3.18
Average Z (photoelectric)	8.14	15.3	16.5
Response (50 keV/1 MeV)	1.2	1.4	1.6
Dose-rate independence (R/s max)	$>10^8$		120

powder, extruded pins (1 mm diameter by 6 mm length), and plates (3 mm by 3 mm) weighing 25 mg. The CaF_2:Mn material is available as powder, solid forms, and encapsulated in glass under vacuum to reduce spurious luminescence. Thermoluminescent dosimeters are calibrated in a field measured with a cavity ionization chamber.

Photographic Emulsion

Photographic emulsion contains grains of silver bromide dispersed in gelatin, on a plastic or glass backing [25,26]. Nuclear emulsions differ from ordinary photographic emulsions by having a larger fraction of AgBr (80% vs. 40%), smaller grains (typically 0.3 μm compared to 1 μm), and a wider range of emulsion thickness (5 to 600 μm, compared to 10 to 25 μm). Ionizing radiation forms a latent image (few atoms of Ag per grain energized), as does light. These grains are reduced to metallic silver under the action of the developer, a chemical amplification process with a gain of about 10^{10}. The unreduced grains are dissolved in sodium thiosulfate solution ("hypo"), and then the emulsion is washed and dried. In the processing, emulsion thickness is reduced by a factor of 2, which has to be taken into account in examination of particle tracks under the microscope. Track counting (fluence measurement) is performed for ions, including recoil protons from interactions of fast neutrons on the hydrogen in the emulsion. However, for γ rays and electrons, the absorbed dose is obtained from the general blackening as measured by the optical density $D_0 = \log (I_0/I)$, where I_0 is the intensity of the light incident on the blackened film and I is the intensity of the light transmitted. A typical characteristic curve for X-ray film is shown in Figure 5.31. Over a certain range, D_0 is proportional to log exposure. However, the optical density also depends on the batch of film, processing conditions, and other factors. It is necessary to calibrate the optical density as a function of absorbed dose (or exposure), measured by an absolute instrument such as an ionization chamber. As a guide, $D_0 = 0.3$ may be obtained at 10 mR to 1 R, depending on the sensitivity of the film. Measurement of exposure or absorbed dose from photons in tissue is only approximate below 200 keV because of the large photoelectric absorption in AgBr. The fidelity to tissue-dose response may be improved by filters.

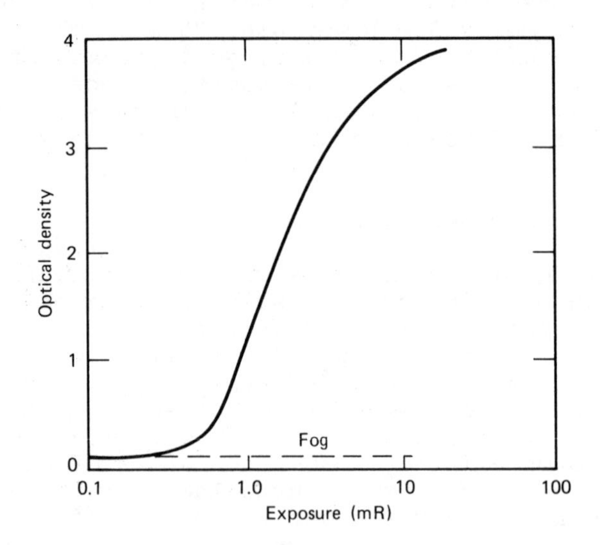

Fig. 5.31. Characteristic curve for X-ray film.

Chemical Dosimeter

Chemical changes induced by ionizing radiation may be used to measure the absorbed dose in the dosimeter material [27]. Various systems have been developed, including dyes in plastic and chlorinated hydrocarbons. The Fricke ferrous sulfate dosimeter is suitable for γ-ray absorbed dose in the 5000 to 50,000 rad range. It consists of a solution of ferrous ammonium sulfate in 0.8N sulfuric acid in a glass ampoule. Purity of reagents and cleanliness of glass are crucial. Ferrous ions are oxidized to ferric ions, with the number proportional to energy absorbed. The concentration of ferrous ions can be measured in a spectrophotometer at the 510-nm absorption peak, or the concentration of ferric ions can be measured at the 304.5-nm absorption peak. Accuracy is about $\pm 5\%$ in the stated dose range.

5.6 DOSE EQUIVALENT

Dose equivalent H from particles of energy E is calculated from the absorbed dose D by multiplying by the quality factor Q. The quality factor expresses the biological damage capability per rad, relative to X-rays, and is a function of L, the linear energy transfer (LET). The ICRU defines

$$L = \frac{dE_L}{d\ell} \tag{5.58}$$

where dE_L is the average energy "locally imparted" to the medium by a charged particle of specified energy in traversing the distance $d\ell$. The statement "locally imparted" is ambiguous because δ rays (energetic electrons generated in close collisions) may have a large range and form their own tracks. Thus L may be defined in terms of energy deposited within a certain specified distance of the primary track or in terms of some maximum energy transfer. The ICRP [28] recommends L_∞, which includes all the imparted energy; hence L_∞ is numerically equal to the collision stopping power. The L_∞ value for protons and for electrons in water is plotted in Figure 3.14, and the linear energy transfer in soft tissue is nearly identical. Average L for a γ ray of given energy can be obtained from the secondary electron spectrum. Average L for a neutron of given energy can be obtained from the spectrum of recoil protons (and other ions).

By definition the quality factor is 1 for X-rays which have an L of $\leq$ 3.5 keV/μm in water or soft tissue. The biological effect chosen for definition of Q in radiation protection is induction of mutations, the genetic effect discussed in Chaptér 6. Radiobiology experiments in mammals have determined Q as a function of L_∞. The values recommended by the ICRP are listed in Table 5.12 and plotted in Figure 5.32. It is assumed that Q = 1 for X-rays, γ rays, and electrons of any LET. The quality factor for neutrons is plotted in Figure 5.33. The quality factor for protons varies from 14 to 1, but a conservative value of Q = 10 can be used if the spectrum is unknown. For α particles and other heavy ions, Q = 20 is conservative.

Maximum Dose Equivalent

The ICRU defines the dose-equivalent index H_I as the maximum dose equivalent in a 30-cm-diameter sphere of tissue, centered at the point of interest. The maximum dose equivalent will depend on the angular distribution of the incident radiation, with normal

TABLE 5.12. RELATIONSHIP BETWEEN L_∞ AND Q

L_∞ in Water (keV/μm)	Q
$\leq$ 3.5	1
7.0	2
23.0	5
53.0	10
$\geq$ 175.0	20

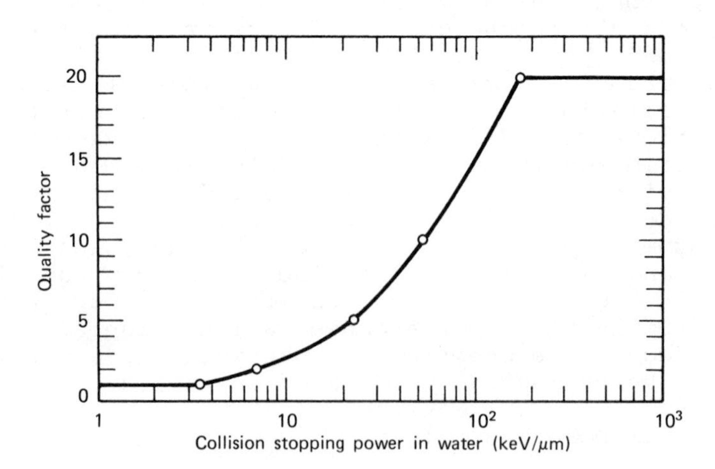

Fig. 5.32. Quality factor as a function of stopping power or linear energy transfer.

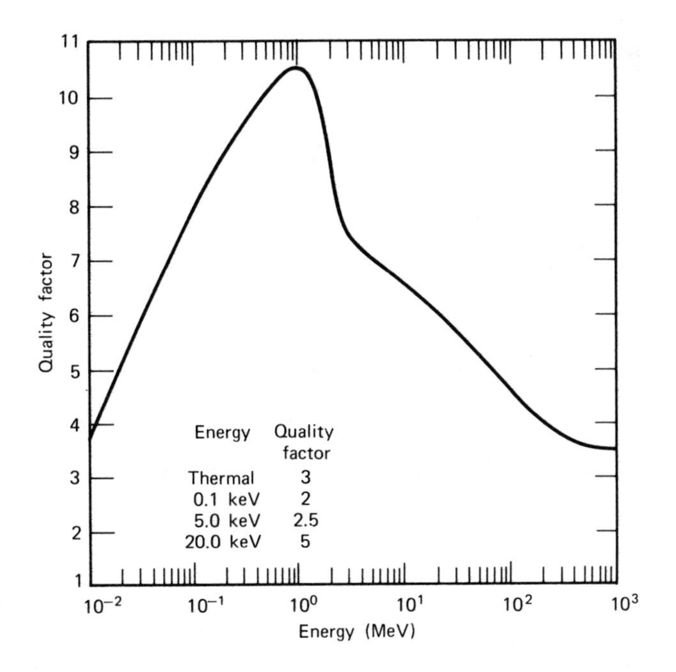

Fig. 5.33. Quality factor for neutrons.

(perpendicular) incidence giving yhe largest H_I. Calculations of the maximum H for neutrons have been performed for normally incident beams on a 30-cm-thick slab of tissue [29] and for normal incidence on a 30-cm-diameter and 60-cm-high cylinder of tissue [30, 31]. The calculations take into account the contribution from multiple scattering and capture γ rays, which do not contribute to the absorbed dose in a small probe or specimen of tissue in a free field (air). The dose equivalent rises to a maximum within a few cm of the surface of incidence and then decreases. Table 5.13 gives the neutron fluence:maximum dose-equivalent factor as a function of energy and the effective Q (maximum H divided by the absorbed dose at the position of maximum H).

The maximum dose equivalent for normally incident γ rays on a slab of tissue 30 cm thick has been calculated [32]. Results are listed in Table 5.14 and

TABLE 5.13. NEUTRON FLUENCE TO MAXIMUM DOSE EQUIVALENT FACTOR

	Slab[a]		Cylinder[b]	
E (MeV)	H/Φ (mrem/n cm^{-2})	Ω_{eff}	H/Φ (mrem/n cm^{-2})	Ω_{eff}
Thermal	1.04(-6)	3.25	1.15(-6)	2.46
10^{-6}			1.34(-6)	2.27
10^{-5}			1.21(-6)	2.34
10^{-4}	1.39(-6)	2.01	1.01(-6)	2.27
0.001			8.85(-6)	2.04
0.005	1.22(-6)	2.14		
0.01			9.92(-6)	2.29
0.02	2.35(-6)	4.12		
0.1	8.30(-6)	7.55	4.86(-6)	6.06
0.5	2.30(-5)	9.61	1.89(-5)	10.40
1.0	3.80(-5)	10.00	3.26(-5)	10.80
2.5	3.41(-5)	7.93	3.50(-5)	8.77
5.0	3.80(-5)	6.55	4.41(-5)	7.70
7.5	4.16(-5)	5.85	4.03(-5)	7.07
10.0	4.16(-5)	5.94	4.31(-5)	5.94
14.0			6.15(-5)	7.40

[a]Normal incidence, 30-cm-thick tissue slab (Snyder-Neufeld).

[b]Normal incidence, 30-cm-diameter by 60-cm-high tissue cylinder (Auxier).

TABLE 5.14. γ-RAY FLUENCE:MAXIMUM DOSE EQUIVALENT FACTOR[a]

E (MeV)	$\frac{\text{rad/h}}{\gamma/\text{cm}^2\,\text{s}}$	H/Φ (rem/photon cm ^{2})
0.15	4.51(-7)	1.25(-10)
0.30	7.21(-7)	2.00(-10)
0.50	1.17(-6)	3.25(-10)
0.75	1.58(-6)	4.39(-10)
1.12	2.14(-6)	5.94(-10)
1.58	2.73(-6)	7.58(-10)
2.00	3.20(-6)	8.89(-10)
2.40	3.60(-6)	1.00(-9)
2.80	3.97(-6)	1.10(-9)
3.25	4.36(-6)	1.21(-9)
3.75	4.78(-6)	1.33(-9)
4.25	5.19(-6)	1.44(-9)
4.75	5.58(-6)	1.55(-9)
5.25	5.96(-6)	1.66(-9)
5.75	6.34(-6)	1.76(-9)
6.25	6.71(-6)	1.86(-9)
6.75	7.07(-6)	1.96(-9)
7.5	7.62(-6)	2.12(-9)
9.0	8.73(-6)	2.43(-9)
11.0	1.02(-5)	2.83(-9)
13.0	1.17(-5)	3.25(-9)
15.0	1.33(-5)	3.69(-9)

[a]Normally incident on 30-cm-thick tissue slab (Claiborne-Trubey).

plotted in Figure 5.34. Since $Q = 1$, rem is equal to rad, but the dose in the slab is larger than the dose in a small sample or probe (dashed line in the figure) because of multiple scattering and absorption of energy that would otherwise escape.

Similar calculations have been performed for electrons and protons, normally incident on tissue-equivalent slabs [28]. Table 5.15 lists the maximum dose equivalent (or absorbed dose) for electrons. For protons from 2 to 60 MeV, the fluence:maximum dose-equivalent factor is 6.94×10^{-4} mrem per proton cm^{-2}, and the effective quality factor is 1.4. Below 2 MeV, protons cannot penetrate the dead layer of the epidermis.

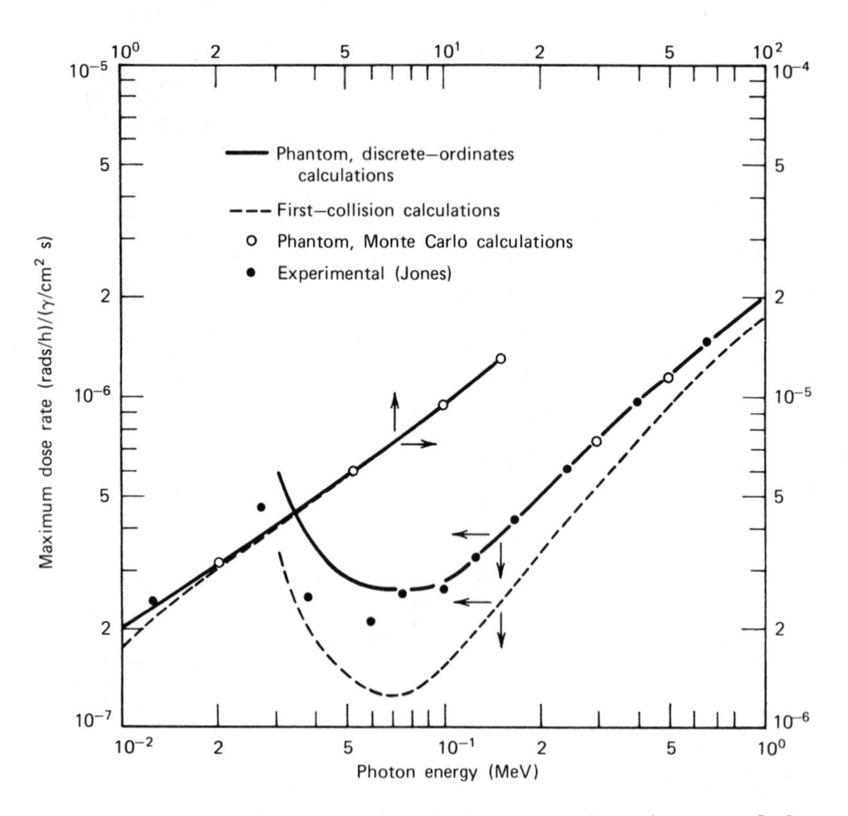

Fig. 5.34. Maximum γ-ray dose rate in tissue slab (after Claiborne and Trubey [32]).

TABLE 5.15. ELECTRON FLUENCE:MAXIMUM DOSE EQUIVALENT FACTOR[a]

E (MeV)	H/Φ (mrem/electron cm^{-2})
0.1	1.74(-4)
0.2	1.07(-4)
0.5	7.12(-5)
1.0	5.79(-5)
2.0	5.05(-5)
5.0	4.48(-5)
10.0	4.15(-5)

[a]Normal incidence on tissue slab 30-cm thick.

Linear Energy Transfer

The dose equivalent in a mixed radiation field

$$H = \overline{Q}D \tag{5.59}$$

where the average quality factor

$$\overline{Q} = \frac{\int_0^{L_{max}} D(L)\ Q(L)\ dL}{\int_0^{L_{max}} D(L)\ dL} \tag{5.60}$$

and D(L) is the absorbed dose per unit interval of L deposited by particles of linear energy transfer L, Q(L) is the corresponding L for these particles, and L_{max} is the maximum value of L. Thus if D(L) can be measured, H can be calculated, knowing the relationship between Q and L from Table 5.12.

The distribution D(L) can be measured with a spherical proportional counter such as shown in Figure 5.35. The tissue-equivalent sphere has a wall thickness of 5 mm and an inner diameter of 5.70 cm. It is filled with tissue-equivalent gas at 200 torr pressure, such that the diameter is equivalent to one micrometer of tissue. The anode wire is 0.127 mm in diameter, and the coaxial helix has an inner diameter of 3.18 mm and a pitch of 3.1 turns per cm. The helix is operated at 20% of the overall voltage (950 V); its purpose is to flatten the electric field distribution near the ends of the anode and thus improve energy resolution. The anode pulses are amplified and accumulated in a multichannel pulse-amplitude analyzer. The amplitude scale is calibrated by an internal ^{241}Am α source, giving 88.84 keV deposited over the 1-μm-diameter cavity. Thus the measured amplitude Y can be expressed in keV/μm. The counts N(Y) are related to D(L) by the theory of Rossi [33]. The

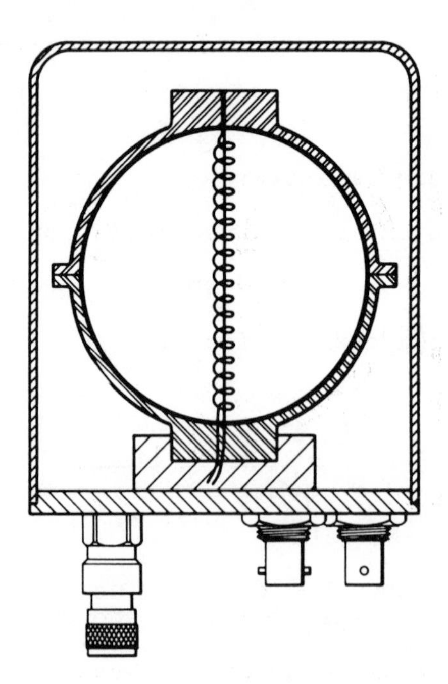

Fig. 5.35. Spherical proportional counter for measurement of absorbed dose as a function of LET (courtesy of EG&G, Inc.).

measured distribution includes tracks along the chords as well as the diameter of the sphere. When this is taken into account, the unnormalized dose distribution

$$A(L) = \frac{2.547 \times 10^{-8}}{r^2} \{YN(Y)(1 - S)\}_{Y = L} \frac{\text{rad}}{\text{keV}/\mu\text{m}} \quad (5.61)$$

where

r = counter inside radius (cm);

$$S = \frac{d \log N(Y)}{d \log Y} = \frac{Y}{N(Y)} \frac{dN(Y)}{dY} .$$

Thus if $N(Y)$ vs Y is plotted on a log/log graph, the slope S can be obtained. Then

$$D(L) = \frac{A(L)}{\int A(L)\, dL} \quad (5.62)$$

Actually there will be some minimum L that can be measured, corresponding to the smallest pulse amplitude. It is convenient to make $L_{min} = 3.5$ keV/μm. Then $\overline{Q}$ for $L > 3.5$ keV/μm can be measured, and the corresponding absorbed dose (integrated from L_{min} to L_{max}) can be subtracted from the total absorbed dose. The remainder is reduced to dose equivalent using $Q = 1$.

Figure 5.36 plots the $D(L)$ spectrum measured [34] near a cyclotron source of neutrons (deuterons on beryllium, broad spectrum from 2 to 14 MeV or greater). The peak from maximum ionizing protons at 100 keV/μm is seen, as well as structure from heavy ion recoils and interactions at higher LET. The contribution below 100 keV/μm comes from protons and from γ rays generated in the surroundings.

5.7 HEATING

Heating can be equated to absorbed dose, since in most instances the energy deposited by ionization is eventually degraded to heat. Methods capable of calculating absorbed dose include moments, discrete ordinates, and Monte Carlo. A comparison [35] of these methods, for a γ-ray source in iron, is shown in Figure 5.37. The source has a fission-γ-ray spectrum and is uniformly distributed in a sphere of 25-cm radius, imbedded in an iron sphere of 70-cm radius. The S_n or discrete

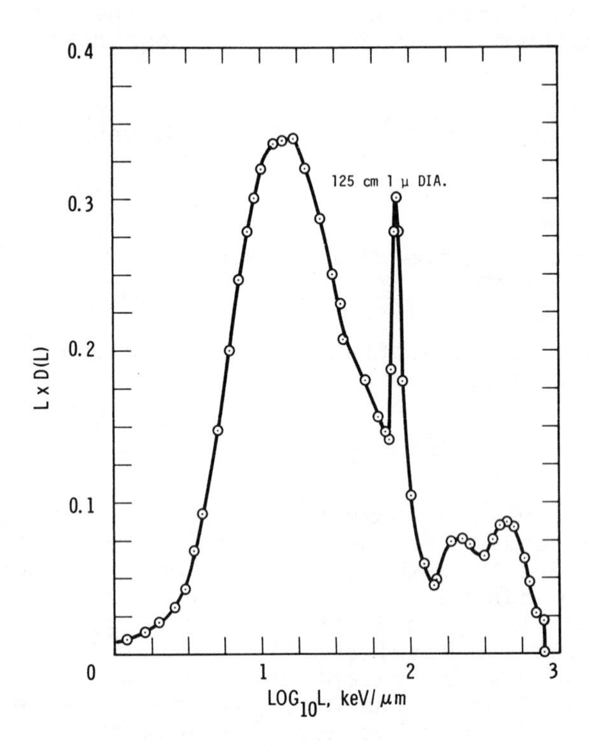

Fig. 5.36. Experimental LET dose spectrum at a cyclotron neutron source (courtesy of EG&G, Inc.).

ordinates calculation used P_3 cross sections and S_{16} angular mesh. Agreement is very good. Generally, heating is significant only in a reactor core and in the inner portions of the shield; hence ability to calculate deep penetrations is not a requirement. However, accuracy may have to be better than for radiation protection purposes, to prevent overheating. Reactors often have a cooled steel "thermal shield" in or just outside the reactor vessel, to absorb and dissipate the radiation heating. It is important to prevent overheating of a concrete shield, too, to avoid thermal stress and loss of water essential for fast-neutron attenuation. Heat generation and temperature distributions are discussed in the *Compendium* [36].

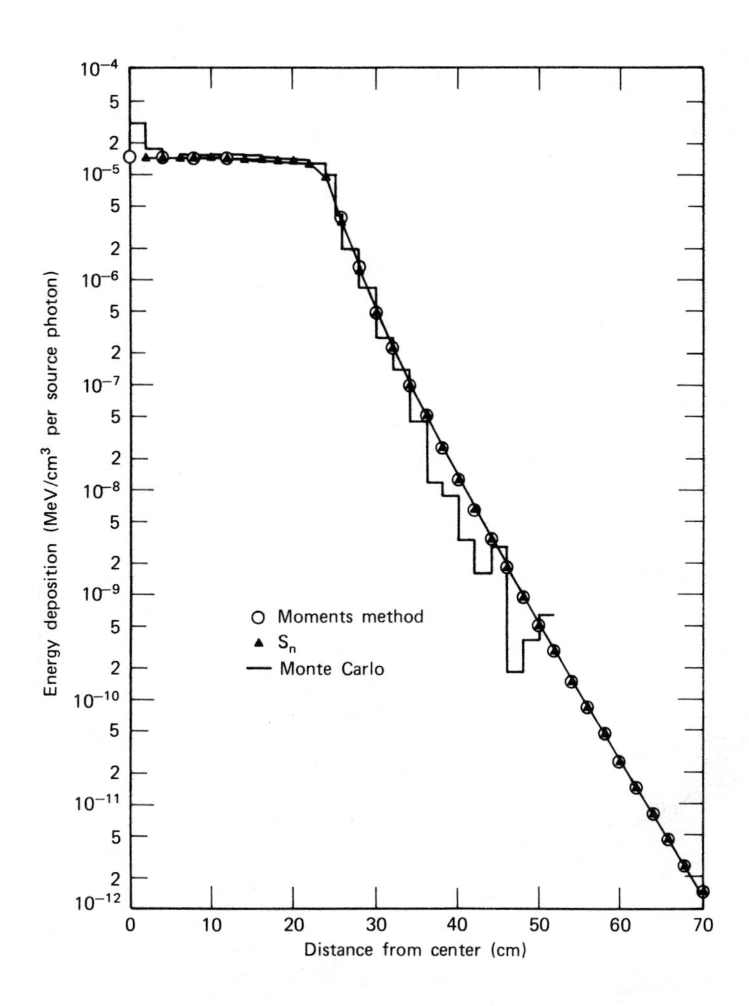

Fig. 5.37. Heating in iron from 25-cm-radius source sphere.

At low flux densities, heating is measured by the absorbed dose in the appropriate material. At high flux densities, heating can be measured directly, in a calorimeter [37]. There are two limiting types of calorimeter: adiabatic and isothermal. In the adiabatic type, the temperature rise is measured in a well-insulated block of the radiation-absorbing material. Negligible heat loss may be achieved only over a short time, even with the best insulation. An alternative is to use a heated jacket whose temperature is made equal to the temperature of the absorber by means of a servo regulator. In an isothermal calorimeter the absorber temperature is held constant by a substance making a phase change, such as melting ice. It is also possible to measure a temperature gradient and deduce the heating rate if the heat-transfer coefficient is known or can be calibrated by comparison with electrical resistance heating, for example.

5.8 DAMAGE

Radiation damage, such as changes in mechanical properties of metals under neutron irradiation, can be

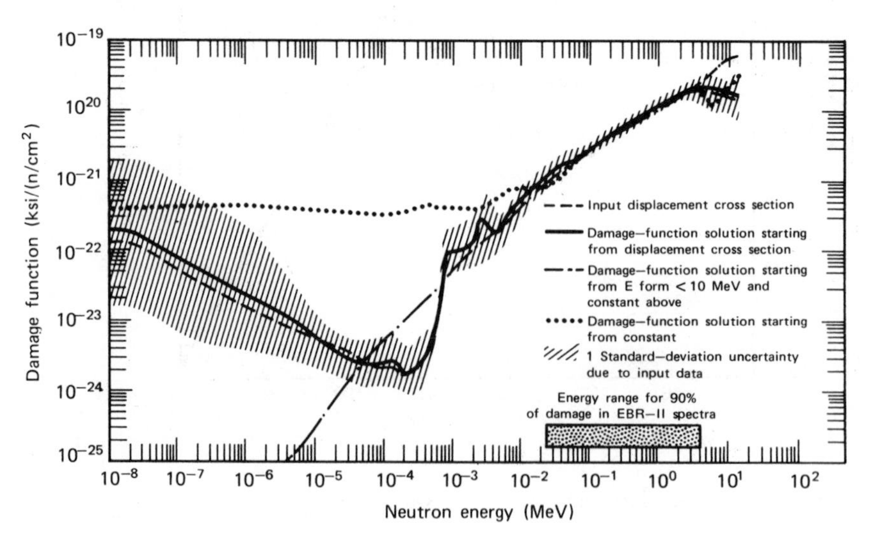

Fig. 5.38. Yield-strength (60 ksi) damage function for Type 304 stainless steel irradiated and tested at 480°C (courtesy of G. R. Odette).

predicted if the energy-dependent damage function is known. Efforts are now underway to calculate theoretical damage functions, including effects of atomic displacements and helium- or hydrogen-filled voids produced by agglomeration of atoms from (n,α) and (n,p) reactions. However, most damage functions are obtained from experiment, unfolding the response function from irradiations in a series of broad but differing spectra. An example of a damage function [38] is shown in Figure 5.38. It gives the fluence of a certain energy required to produce a yield stress of 60,000 psi (pounds per square inch) at 480°C in Type 304 stainless steel. The damage function was obtained by unfolding from the response in differing spectra, starting with a trial function. The best trial function is given by the displacement cross section. The fluence in a certain spectrum, required to reach the same level of the same property (60 kpsi yield stress) can be formed by integrating over energy with the spectrum weighted by the damage function.

REFERENCES

1. "Radiation Quantities and Units," ICRU Report 19 (1971). ICRU Publications, P. O. Box 30165, Washington, D. C. 20014.
2. "Conceptual Basis for the Determination of Dose Equivalent," ICRU Report 25 (1976). ICRU Publications, Washington, D. C.
3. A. Edward Profio, *Experimental Reactor Physics*, Wiley, New York, 1976.
4. William J. Price, *Nuclear Radiation Detection*, 2nd ed., McGraw-Hill, New York, 1964.
5. Kai Siegbahn, Ed., *Alpha-, Beta-, Gamma-Ray Spectroscopy*, North-Holland, Amsterdam, 1965.
6. F. H. Attix and W. C. Roesch, Eds., *Radiation Dosimetry*, 2nd ed., Vol. II, *Instrumentation*, Academic Press, New York, 1966.
7. C. D. Zerby, "A Monte Carlo Calculation of the Response of Gamma Ray Scintillation Counters," in B. Adler, S. Fernbach, and M. Rotenberg, Eds., *Methods in Computational Physics*, Vol. I, Academic Press, New York, 1963, pp. 89-134.
8. "MORN--Calculation of the Response of Sodium Iodide Crystals to Gamma Rays," available as PSR-62 from the Radiation Shielding Information Center (RSIC), Oak Ridge, Tenn.

9. "05S Response Function Generator for Organic Scintillators," available as PSR-14 from RSIC, Oak Ridge, Tenn.
10. "COOLC and FERDOR Spectra Unfolding Codes," available as PSR-17 from RSIC. See also W. R. Burrus and V. V. Verbinski, "Fast Neutron Spectroscopy with Thick Organic Scintillators," *Nucl. Instrum. Methods*, 67, 181 (1969).
11. "FORIST: Neutron Spectrum Unfolding Code-Iterative Smoothing," available as PSR-92 from RSIC.
12. "MAZE II Spectral Unfolding Code," available as PSR-41 from RSIC.
13. "A Review of Radiation Energy Spectra Unfolding," Report ORNL/RSIC-40, Radiation Shielding Information Center (1976).
14. Gene Erwin Bromley, "Neutron Spectroscopy in a Fast Subcritical Reactor with a Proton Recoil Proportional Counter," M.S. thesis, Chemical and Nuclear Engineering, University of California at Santa Barbara (1971).
15. E. F. Bennett and T. J. Yule, "Techniques and Analyses of Fast-Reactor Neutron Spectroscopy with Proton-Recoil Proportional Counters," Report ANL-7763 (1971).
16. J. H. Hubbell, "Photon Cross Sections, Attenuation Coefficients, and Energy Absorption Coefficients from 10 keV to 100 GeV," NSRDS-NBS 29 (1969). Available from Sup. Doc., U.S. GPO, Washington, D. C. 20402.
17. R. D. Evans, "X-Ray and γ-Ray Interactions," in F. H. Attix and W. C. Roesch, Eds., *Radiation Dosimetry*, 2nd ed., Vol. I, Academic Press, New York, 1968, p. 135.
18. J. J. Ritts, E. Solomito, and P. N. Stevens, "Calculation of Neutron Fluence-to-Kerma Factors for the Human Body," Oak Ridge National Laboratory Report ORNL-TM-2079 (1968).
19. K. Z. Morgan and J. E. Turner, *Principles of Radiation Protection*, Wiley, New York, 1967, p. 132.
20. L. V. Spencer, "Energy Dissipation by Fast Electrons," Nat. Bur. Stand., Monograph I (1959).
21. M. J. Berger, "Monte Carlo Calculation of the Penetration and Diffusion of Fast Charged Particles," in B. Adler and S. Fernbach, Eds., *Methods in Computational Physics*, Vol. I, Academic Press, New York, 1963, p. 135.

22. J. J. Fitzgerald, G. L. Brownell, and F. J. Mahoney, *Mathematical Theory of Radiation Dosimetry*, Gordon and Breach Science Publishers, New York, 1967, pp. 540-575.
23. T. E. Burlin, "Cavity Chamber Theory," in Attix and Roesch, *Radiation Dosimetry*, Vol. I, Chapter 8.
24. J. F. Fowler and F. H. Attix, "Solid State Integrating Dosimeters," in Attix and Roesch, *Radiation Dosimetry*, Vol. II, Chapter 13.
25. Robert A. Dudley, "Dosimetry with Photographic Emulsions," in Attix and Roesch, *Radiation Dosimetry*, Vol. II.
26. R. H. Herz, *The Photographic Action of Ionizing Radiation*, Wiley, New York, 1969.
27. Hugo Fricke and Edwin J. Hart, "Chemical Dosimetry," in Attix and Roesch, *Radiation Dosimetry*, Vol. II, Chapter 12.
28. "Protection Against Ionizing Radiation from External Sources," ICRP Publication 15 (1969) and "Data for Protection Against Ionizing Radiation from External Sources: Supplement to ICRP Publication 15," ICRP Publication 21 (1971). Published by Pergamon Press, Oxford, for the International Commission on Radiological Protection.
29. W. S. Snyder and C. Neufeld, "On the Passage of Heavy Particles Through Tissue," *Radiat. Res.*, 6, 67 (1967). Reprinted in NBS Handbook 63.
30. G. A. Auxier, W. S. Snyder, and T. D. Jones, "Neutron Interactions and Penetration in Tissue," in Attix and Roesch, *Radiation Dosimetry*, Vol. I, Chapter 6.
31. "Protection Against Neutron Radiation," NCRP Report No. 38, National Council on Radiation Protection and Measurements, Washington, D. C. 1971.
32. H. C. Claiborne and D. K. Trubey, "Dose Rates in a Slab Phantom from Monoenergetic Gamma-Rays," *Nucl. Appl. Technol.*, 8 (5), 450 (1970).
33. H. H. Rossi, "Microscopic Energy Distribution in Irradiated Matter," in Attix and Roesch, *Radiation Dosimetry*, Vol. I, Chapter 2.
34. A. C. Lucas and W. M. Quam, "Fast-Neutron Source Studies for Radiation Theory," Report S-477-R, EG&G, Inc., Goleta, Calif. (1969).
35. D. K. Trubey, S. K. Penny, and K. D. Lathrop, "A Comparison of Three Methods Used to Calculate

Gamma-Ray Transport in Iron," Report ORNL-RSIC-9 (1965).

36. "Radiation Induced Heat Generation," in R. G. Jaeger, Ed., *Engineering Compendium on Radiation Shielding*, Vol. I, Springer, New York, 1968, Chapter 7.
37. J. S. Laughlin and S. Genna, "Calorimetry," in Attix and Roesch, *Radiation Dosimetry*, Vol. II, Chapter 16.
38. G. R. Odette, R. L. Simons, W. N. McElroy, and D. G. Doran, "Analysis of Reactor Component Fluence Limits Using Damage Function Methods," *Nucl. Technol.*, 32, 125-141 (1977).

PROBLEMS

1. Discuss the relationship between the linear attenuation coefficient μ, the energy absorption coefficient μ_{en}, and the energy-transfer coefficient μ_K for gamma rays interacting in a small mass of tissue.

2. What is the count rate for a 1-cm^2 area by 20-μm depletion depth silicon surface barrier detector located in vacuum, 10 cm from a 1-μCi source of ^{210}Po?

3. What is the flux density of 0.662-MeV γ rays at 10 cm from a point source of ^{137}Cs located on axis of a 7.6-cm-diameter by 7.6-cm-thick NaI(Tℓ) scintillation detector if the count rate in the photopeak is 25 counts per second?

4. A neutron-detection system accumulated 1200 counts in 10 min. With the neutron source turned off, 55 counts were recorded in 20 min. What is the net counting rate and uncertainty?

5. Calculate the kerma and absorbed dose (assuming charged-particle equilibrium) in muscle for a γ-ray fluence of 10^{10} photons/cm^2 if the spectrum is flat from 0.01 to 0.1 MeV, and for monoenergetic 0.1-MeV γ rays.

6. Calculate the dose-equivalent rate and dose-equivalent index rate at 10 m from a 14-MeV

neutron source emitting 10^{11} n s^{-1}, neglecting scattering in the surroundings.

7. Discuss how a damage function can be measured and used to predict the flux density required to produce a certain degree of damage, for an arbitrary spectrum.

6

EFFECTS

6.1 RADIATION BIOLOGY

The sequence of events following interactions of radiation with biological matter is diagrammed [1] in Figure 6.1. In the physical stage, which lasts about 10^{-13} s, two effects or modes of action are distinguished. Direct action involves absorption of energy by ionization and excitation (or nuclear displacement or transmutation) in the biological molecule where the lesion eventually appears. Indirect action refers to absorption of energy in a different molecule, (e.g., water), resulting in diffusible free radicals and other reactive species that can migrate to and damage the biomolecule.

The physicochemical stage lasts about 10^{-10} s. An ionized or displaced atom may produce a primary lesion, such as a break in the molecule, either directly or following intermolecular energy transfer. The indirect action in water results first in ionization of water molecules.

$$H_2O \rightarrow H_2O^+ + e^- \tag{6.1}$$

The positive ion forms a hydroxyl radical by

$$H_2O^+ \rightarrow H^+ + OH^0 \tag{6.2}$$

A hydrogen radical (atom) is formed by

$$H_2O + e^- \rightarrow OH^- + H^0 \tag{6.3}$$

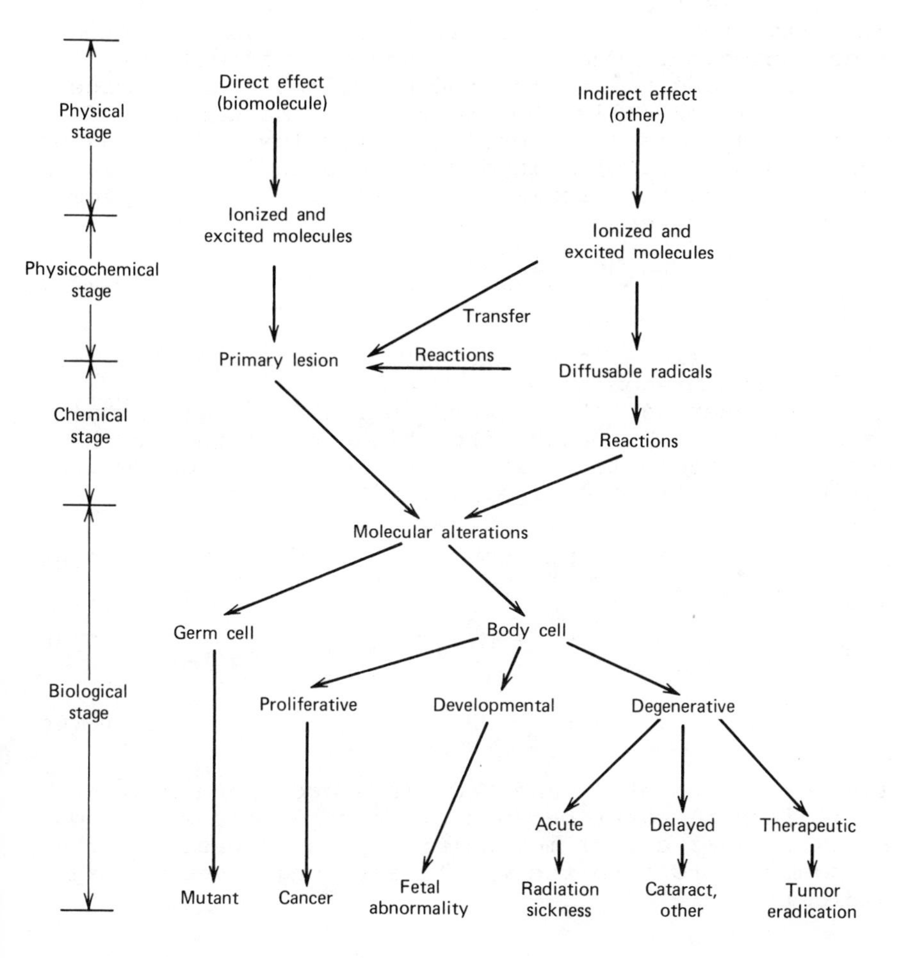

Fig. 6.1. Stages of radiation effects in organisms.

Hydrogen and hydroxyl radicals can also be formed by dissociation of the water molecule following excitation

$$H_2O \rightarrow H_2O^* \rightarrow H^0 + OH^0 \tag{6.4}$$

The free electrons polarize water molecules in the vicinity, forming the relatively long-lived hydrated electron (e^-_{aq}). The e^-_{aq}, H^0, and OH^0 radicals may diffuse to and react with the biomolecule and damage it. In living cells the damage from the direct and indirect actions are about equal for low-LET radiation (electrons, γ rays). In dry systems indirect action can still occur by formation of atomic hydrogen from organic molecules,

$$MH \rightarrow M^0 + H^0 \tag{6.5}$$

At high LET (ions, neutrons) it is usually direct action that predominates.

The chemical stage lasts about 10^{-6} s at normal temperatures. In tracks from high-LET particles the density of radicals is large and recombination occurs, giving nonreactive or less reactive products

$$H^0 + OH^0 \rightarrow H_2O \tag{6.6a}$$

$$H^0 + H^0 \rightarrow H_2 \tag{6.6b}$$

$$OH^0 + OH^0 \rightarrow H_2O_2 \tag{6.6c}$$

The net yields of H_2 and H_2O_2 (hydrogen peroxide) are small in pure water because of back reactions with OH^0 and H^0. However, if molecular oxygen is present, the H^0 combines to form the relatively long-lived peroxyl radical,

$$H^0 + O_2 \rightarrow HO_2^0 \tag{6.7}$$

Peroxyl radicals in turn form hydrogen peroxide,

$$HO_2^0 + HO_2^0 \rightarrow H_2O_2 + O_2 \quad (6.8)$$

If a reducing agent (electron donor) is present as well, more hydrogen peroxide is formed by

$$HO_2^0 + e^- \rightarrow HO_2^- \quad (6.9a)$$

$$HO_2^- + H^+ \rightarrow H_2O_2 \quad (6.9b)$$

The radicals and hydrogen peroxide can reduce or oxidize biological molecules. Typical reactions are

$$MH + H^0 \rightarrow MH_2^0 \quad (6.10a)$$

$$MH + OH^0 \rightarrow MHOH^0 \quad (6.10b)$$

$$MH + H^0 \rightarrow M^0 + H_2 \quad (6.10c)$$

$$MH + OH^0 \rightarrow M^0 + H_2O \quad (6.10d)$$

and similar reactions with (e_{aq}^-), HO_2^0, and H_2O_2.

Certain chemicals or drugs can act as protective or sensitizing agents. Protective agents include the aminothiols, for example, the amino-acid cysteine and 2-mercaptoethylamine (cysteamine). The mechanism of the protective action is uncertain but may involve competitive scavenging of free radicals, restitution of a damaged molecule by donation of a hydrogen atom, or bonding of the protective chemical with the biomolecule in such a way that the protective chemical rather than the biomolecule is attacked. Unfortunately, the known protective chemicals are toxic to the organism at the concentration required for significant protection (maximum protection is equivalent to reducing the absorbed dose by a factor of 1.3 to 1.7).

Restitution of molecules damaged by indirect action can occur even without added protective agents, such as by the addition of a free electron

$$MH^+ + e^- \rightarrow MH \quad (6.11a)$$

or reaction with a hydrogen atom

$$M^0 + H^0 \rightarrow MH \tag{6.11b}$$

Restitution is inhibited by molecular oxygen, which scavenges electrons by

$$O_2 + e^-_{aq} \rightarrow O_2^- \tag{6.12}$$

and removes organic radicals by peroxydation

$$M^0 + O_2 \rightarrow MO_2^0 \tag{6.13}$$

Thus for low-LET radiation, where indirect action predominates, damage is potentiated by the presence of molecular oxygen. The increase in sensitivity is expressed by the oxygen enhancement ratio (OER)

$$\text{OER} = \frac{\text{dose without oxygen for given effect}}{\text{dose with oxygen for same effect}} \tag{6.14}$$

The OER may be as large as 3. Little enhancement is found for high-LET radiation.

Indirect action is increased at higher than normal temperature and decreased at low temperature, as would be expected from chemical kinetics and diffusion.

Inactivation of enzymes may be significant at very high doses. Enzymes are proteins that catalyze metabolic reactions. The primary structure of protein is a string of amino acids. The general formula for an amino acid is

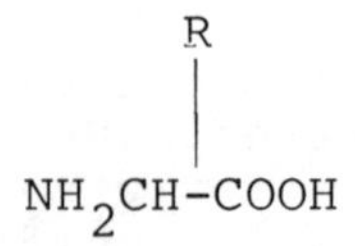

where R, the side chain or group, is characteristic of the particular amino acid. When the protein is formed, the amino ($-NH_2$) group joins with the carboxyl ($-COOH$) group of the next amino acid, with elimination of a molecule of water, to make a polypeptide chain. Proteins also have a secondary structure from coiling and may have a tertiary structure or conformation from folding or twisting, held at various spots by a disulfide bond, hydrogen bond, electrostatic force, or hydrophobic bond. The last refers to the migration of amino acids with nonpolar side chains to the inside of the structure. The biological activity is intimately related to the conformation. Thus ionizing radiation can inactivate an enzyme molecule by disrupting a bond, allowing the tertiary structure to unfold.

Figure 6.2 shows the dose-response curve for inactivation of dry ribonuclease in vacuum and in the presence of oxygen [3]. The ordinate is the percent of enzyme activity remaining after irradiation to the absorbed dose on the abscissa. The exponential function is characteristic of damage in which a single "hit" or interaction is sufficient. The variation with absorbed dose occurs because of the discrete, random nature of the interactions with the molecules (or other units). Let the mean number of hits be vD, where D is the absorbed dose and v may be thought of as an effective volume on the order of 10^{-6} μm^3 according to Lea [4]. The probability of exactly n hits is given by the Poisson distribution

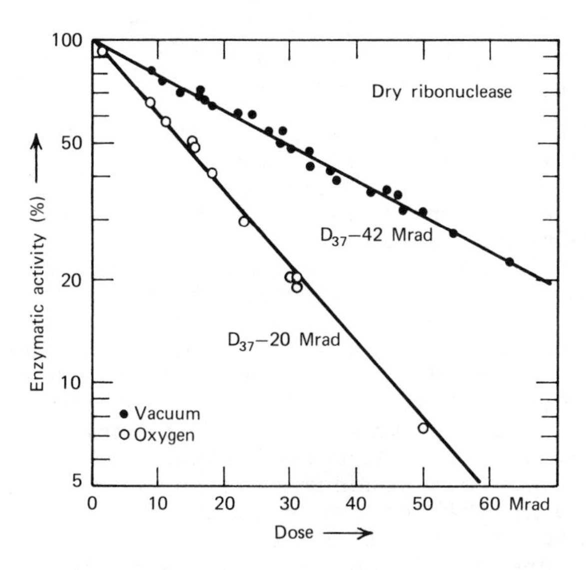

Fig. 6.2. Inactivation of dry ribonuclease by ^{60}Co γ rays in vacuum and under oxygen (after Günther and Jung [3]).

$$P(n) = \frac{(vD)^n e^{-vD}}{n!} \tag{6.15}$$

If n hits are required to destroy or inactivate a unit, then any unit receiving n - 1 or fewer hits will survive. The fraction surviving

$$\frac{N}{N_0} = e^{-vD} \sum_{k=0}^{n-1} \frac{(vD)^k}{k!} \tag{6.16}$$

For a single hit, n = 1,

$$\frac{N}{N_0} = e^{-vD} \tag{6.17}$$

The curve can be described by a single parameter, such as D_{37}, the absorbed dose at which the surviving fraction is 0.37 (hence $vD_{37} = 1$). Under vacuum, D_{37} = 42 Mrad in the dry ribonuclease, whereas in the presence of oxygen, D_{37} = 20 Mrad. Thus the OER is 2.1.

At the biological stage, the effects are mediated by metabolic reactions, and changes may require seconds to years. At the cellular level a distinction is made between alterations in the DNA or chromosomes of a germ cell and alterations in an ordinary body (somatic) cell. Changes in a germ cell may be expressed in offspring if that cell is involved in reproduction (genetic effects). Alterations in other cells are expressed (if at all) in the individual irradiated. As indicated in Figure 6.1, these changes may appear as impairment of proliferative control, possibly resulting in cancer. Damage to development and differentiation may appear as abnormalities in a fetus irradiated *in utero*. Degenerative changes, usually from impaired ability of cells to divide and grow (rather than outright cell death) may appear as acute radiation sickness or a delayed manifestation such as cataract or general life-shortening. Finally, radiation is applied as a therapeutic agent to eradicate tumors. All these biological effects are discussed in the following sections.

6.2 MUTAGENESIS

Production of mutations by X-rays was discovered in 1927 by H. J. Muller in the fruitfly *Drosophila melanogaster*. Since then ample evidence has accumulated to show that ionizing radiation, as well as certain chemicals, cause mutations in man and other mammals. A mutation is an alteration in the deoxyribonucleic acid (DNA) molecule, or in chromsomes. If a sex cell with altered DNA or chromosome is subsequently involved in reproduction, the mutation may be expressed in the offspring or in later generations. Although mutations are essential to evolution, man is presumably in evolutionary equilibrium with his environment, and most mutations are harmful. On the other hand, improved plant strains have been produced by irradiating seeds and selecting the best plants grown from them.

As shown in Figure 6.3, DNA consists of a double-stranded helix, with each strand a long sequence of purine and pyrimidine bases attached to a backbone of sugar (deoxyribose) linked by phosphate groups. The combination of base-sugar-phosphate is called a "nucleotide". The two strands are coupled by weak hydrogen bonds between complementary bases. The purine bases found in DNA are adenine (A) and guanine (G); the pyrimidine bases are cytosine (C) and thymine (T). Adenine is always bonded to thymine in the opposite strand, whereas guanine is bonded to cystosine. A sequence of three bases on a strand is a codon; it controls the synthesis of a specific amino acid. The genetic code is degenerate, with many of the 20 possible amino acids designated by more than one codon, and three of the codons designate the end of a gene. A gene is a sequence of codons that specifies a particular polypeptide chain, often for the synthesis of an enzyme. An average protein with a single polypeptide chain consists of some 300 amino acids and hence requires about 900 nucleotide pairs in the gene. Mammalian cells contain roughly two million genes.

Genetic mutations may be caused by substitution of one base pair for another or several base pairs, deletion of one or more base pairs, or insertion of one or more base pairs, thus altering the sequence of

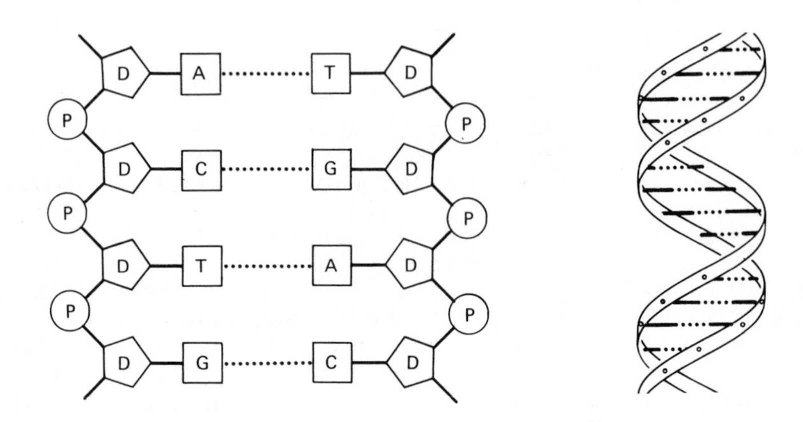

Fig. 6.3. Diagram of DNA structure: P indicates phosphate, D indicates deoxyribose. The bases are A (adenine), C (cytosine), G (guanine), and T (thymine). The opposing chains are twisted into a double helix.

codons [2]. Radiation may change the composition of a base by reaction with OH^0, for example. Radiation may produce a single break in a strand of DNA. Such a break may heal, either by itself or with the aid of an enzyme. However, this restitution or repair is inhibited by oxygen, which may react with the DNA at the break. A doublestrand break may be repaired or the pieces may separate, or if the breaks occur in different molecules, there may be crosslinking between the molecules.

The DNA molecules are complexed with protein (histones) and coiled or folded into chromosomes in the cell nucleus. In germ cells (sperm and egg) of man there are 23 chromosomes, whereas in ordinary somatic (body) cells there are 46. The normal individual has 22 pairs in somatic cells, plus either two X chromosomes (female) or an X and a Y chromosome (male). Division of somatic cells proceeds by mitosis [5], illustrated in Figure 6.4 for one pair. During interphase the chromosomes appear as extended filaments. The DNA is replicated at interphase; hence at early prophase each chromosome is twinned, with two threads (chromatids) joined at the centromere. The threads thicken and at metaphase a "spindle" apparatus forms,

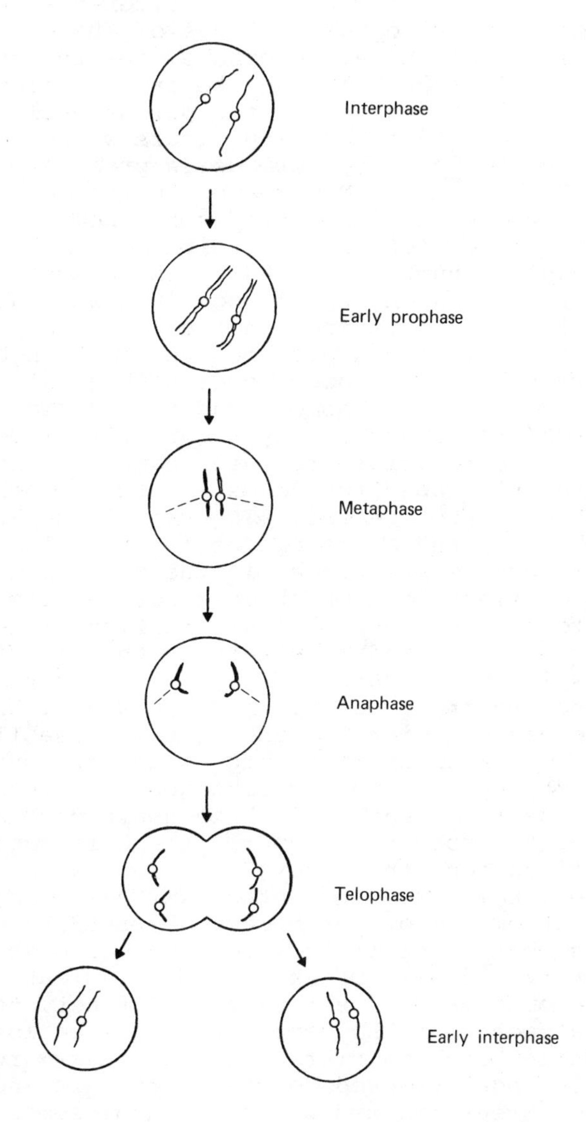

Fig. 6.4. Phases of the mitotic cycle in somatic cell division.

the chromosomes line up on the cell equator, and the centromere divides. At anaphase the spindle apparatus forces the chromatids (now considered as separate chromosomes) to opposite sides of the cell. This process is completed at telophase, and the cell divides, each daughter cell having the same complement of chromosomes and genes as the parent cell, assuming the replication and separation occurs without error.

The division of germ cells proceeds by meiosis (see Figure 6.5). Now there are two cell divisions but only one duplication of chromosomes; hence the number is reduced from the diploid number (2n = 46) to the haploid number (n = 23). The progenitor in the male testis is the spermatogonium, whereas in the female ovary it is the oogonium, both diploid cells. A single pair of chromosomes is shown in Figure 6.5. At prophase I the chromosomes double into two chromatids each, and the homologous pairs align to form a tetrad (crossing over may occur at this point, where parts of chromatids are interchanged). At metaphase I, the spindle apparatus forms and the chromosomes are aligned on the equator. The pair is separated in anaphase I, and at telophase I the cell divides. Two spermatocytes are formed in the male, and an oocyte plus a polar body (which does not develop into an ovum) are formed in the female at the first meiotic division. At prophase II and metaphase II the centromere divides, the spindle apparatus forms, and the chromatids are aligned on the equator. In anaphase II the chromatids (chromosomes) migrate, and at telophase II the cell divides. Each spermatocyte yields two spermatozoa (sperms) or four per spermatogonium. In oogenesis only one ovum (egg) and three nonfunctional polar bodies are formed. A sperm or egg is haploid. At fertilization, the zygote is formed with the diploid number, half from the father and half from the mother.

Chromatid or chromosome aberrations, hence mutations, can occur if radiation causes either a single break or a double break. A single break in a chromatid or chromosome may be repaired. If not, the end that is not attached to the centromere will be lost at anaphase. A double break in sister chromatids may result in union of the ends attached to the centromere and loss of the rest. Breaks in two separate chromosomes that happen to lie near each other may result in joining in different ways, sometimes with deletion of leftover ends. A double break in one arm may rejoin but with an intermediate section omitted. Double breaks in nearby arms

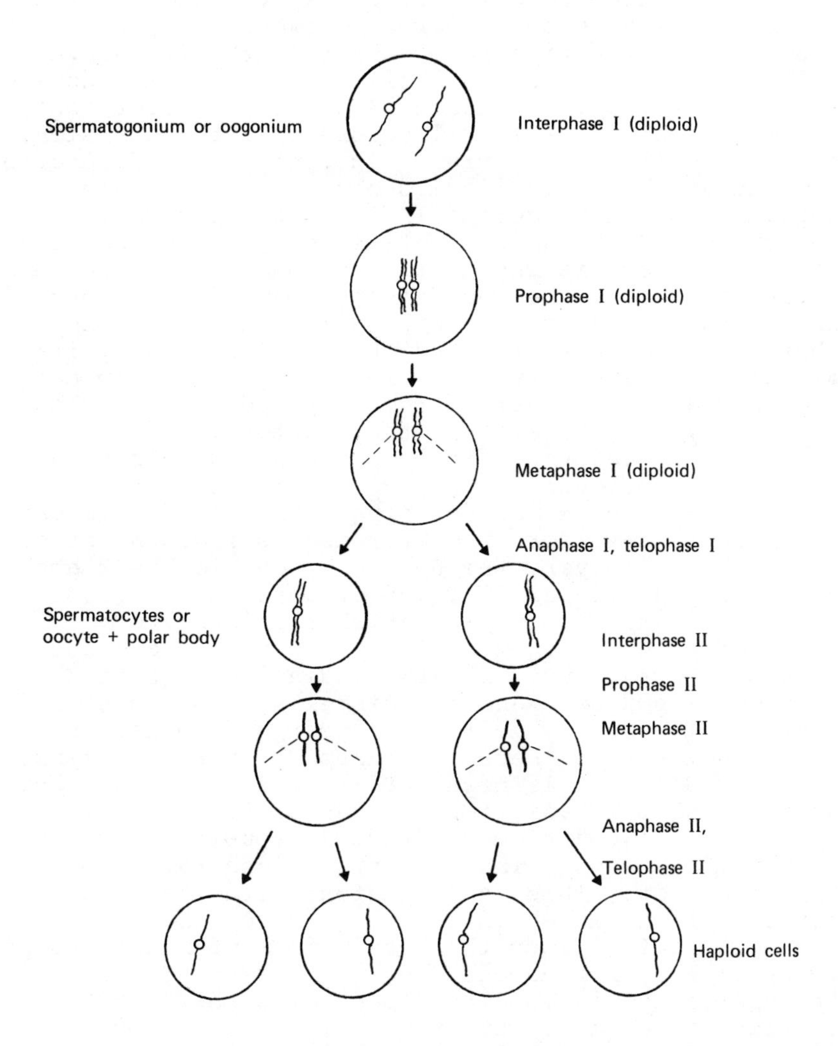

Fig. 6.5. Meiotic cell division and corresponding stages in spermatogenesis or oogenesis.

of the same chromosome may result in joining to form a ring plus two leftover ends or an interchange that leaves the chromosome intact but with a rearrangement of the gene sequence. Whereas damage to a gene may result in malfunctioning or death of the irradiated cell, chromosome and chromatid aberrations may not be evident until cell division. One would expect most of the mutations to be lethal before or after division, but some cells may survive with impairment of function.

Mutations in germ cells occur spontaneously, at a frequency on the order of 10^{-6} to 10^{-5} per gene, per generation. Spontaneous mutations may occur because of inherent errors in replication of DNA and rearrangement of chromosomes or because of environmental factors including natural background radiation. It is known that the frequency of mutations is increased by irradiation. We are interested in the relationship between the increase in mutation frequency and the absorbed dose, LET, dose rate, and other factors.

The frequency of single-break and double-break aberrations produced by X rays [7] is shown in Figure 6.6. The number of single breaks is a linear function of dose (or exposure) and is independent of dose rate. However, the number of double-break aberrations is a nonlinear function of dose and increases with dose rate. Double-break aberrations are not observed at very low dose rates. Oxygen increases the rate of aberrations. Low-LET radiation such as X rays produces a sparse, randomly distributed pattern of breaks. Two breaks must occur near each other in space and in time (because of repair over some minutes to hours) for double-break aberrations to become evident. High-LET radiation such as protons from fast neutrons, or α particles, produces double breaks in proportion to dose, independent of dose rate. Oxygen has no effect. This is consistent with the theory of direct action, in which there is no repair.

Figure 6.7 plots the mean number of mutations per locus (gene), per gamete (sperm), for irradiation of the spermatogonia in mice, together with the 90% confidence levels or error bars [8]. The curves are linear and extrapolate to the spontaneous rate of approximately 10^{-5}. Irradiation at a low dose rate (0.009 R/min) produces only about 1/3 as many mutations per roentgen (or rad) as irradiation at high dose rate (90 R/min), but further reduction in dose rate does not lower the mutation frequency. No dependence on dose

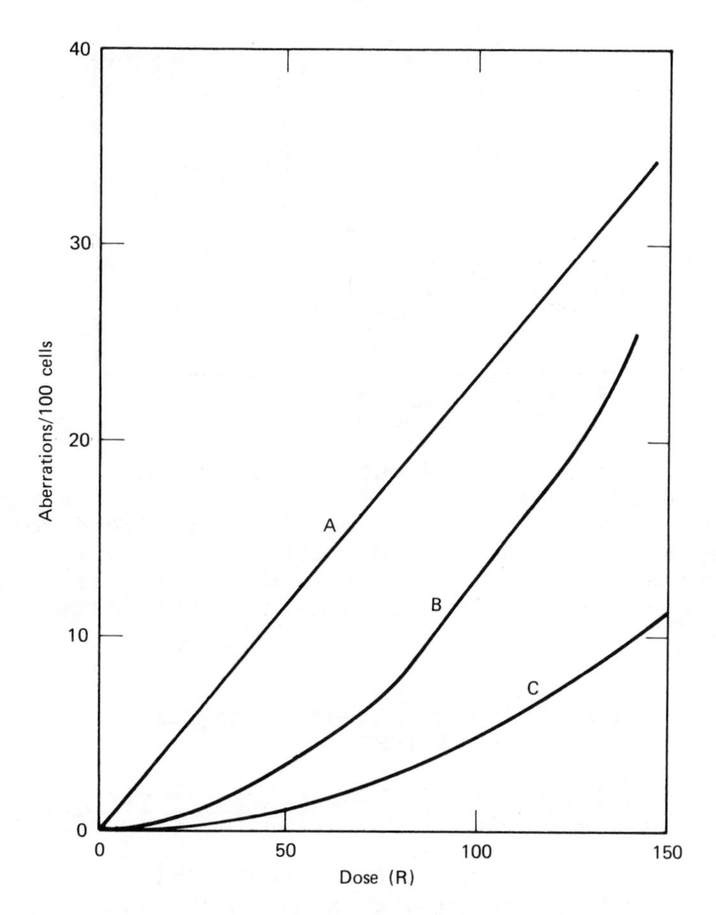

Fig. 6.6. Frequency of chromosome aberrations produced by X rays. Curve A, single breaks; curve B, double breaks at high dose rate; curve C, double breaks at low dose rate.

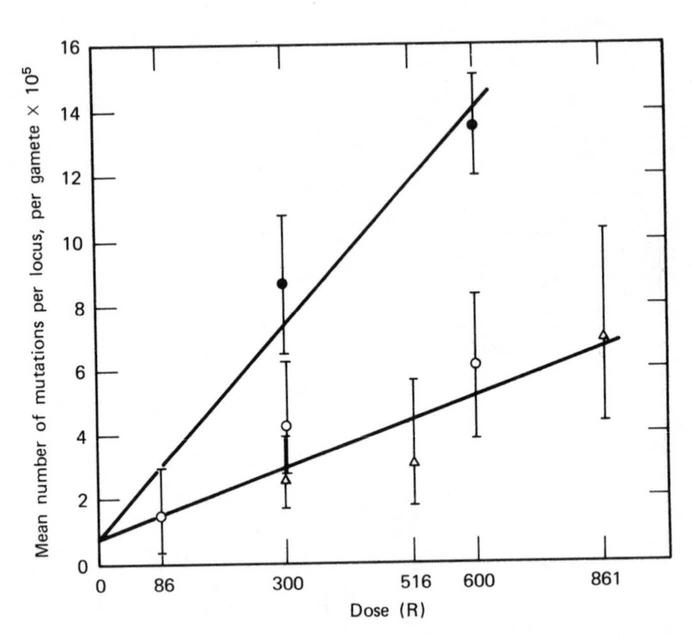

Fig. 6.7. Specific locus mutation rates, with 90% confidence intervals, for various doses and dose rates in spermatogonia of mice. Lower line is fitted to combined 0.001 R/min (open circles) and 0.009 R/min (open triangles) dose-rate data. Upper line is fitted to 90 R/min (solid circles) dose-rate data (after Russell [8]).

rate is found with fast neutrons. The relative biological effectiveness (RBE)

$$\text{RBE} = \frac{\text{absorbed dose of 250 kVp X-rays for given effect}}{\text{absorbed dose of radiation for same effect}}$$

for fast neutrons is about 4 for acute (high-dose-rate) irradiation of spermatogonia in mice, but about 20 for low-dose-rate irradiation. Mutations in the spermatogonia are most significant because some divide to produce new spermatogonia, and spermatogonia continue to produce spermatozoa (which will carry the mutant gene) throughout the reproductive life. Spermatozoa, on the other hand, have a limited life and appear to be relatively resistant to mutagenesis.

Specific locus-mutation frequencies for X-ray irradiation of oocytes in mice are shown in Figure 6.8. The frequency is greatly dependent on dose rate. At high dose rate the frequency is larger than for spermatogonia, whereas the reverse is true at low dose rate. At very low dose rate the frequency cannot be distinguished from the spontaneous mutation frequency. The RBE for acute irradiation with fast neutrons is about 5, whereas the RBE for chronic (low-dose-rate) irradiation is unknown, but probably insignificant. The oogonia disappear soon after birth and population of oocytes is fixed, with one ovum developing and being released in each menstrual cycle from puberty to menopause. Mutations can accumulate in the oocytes (hence ova) during reproductive life.

The date for mice, and observations of the offspring of the Hiroshima and Nagasaki atomic bomb survivors, have been used to estimate genetic effects at low dose rates and doses in man [9]. Low dose-rate irradiation of mouse spermatogonia produces about 0.5×10^{-7} recessive mutations per rem, per gamete (sperm or egg), per generation. It is assumed that the number of mutations increases linearly with dose, and there is no threshold dose below which mutations are absent. Actually, this frequency probably overestimates the risk from radiation because the more important dominant mutations are an order of magnitude less frequent, and the visible genetic effects selected for observation in the mouse experiments may have higher mutability than the average. The mutation frequency can also be expressed in terms of the doubling dose, that is, the dose required to double the rate associated with spontaneous mutations. The spontaneous rate in man is estimated to be between 0.5×10^{-6} and 0.5×10^{-5} per gene, per generation. Thus the doubling dose is between 20 and 200 rem, accumulated in the average generation time of 30 years. The doubling dose for dominant visible mutations is probably about 100 rem. These estimates are consistent with the observed low (almost zero) increase in death rates of children born to parents exposed to the Hiroshima and Nagasaki atomic bomb explosions, compared to unexposed populations. Average dose was about 100 rem, yet the increase in death rate was certainly not doubled. The doubling dose for lethal mutations is probably greater than 50 rem. In fact, no radiation-induced mutations have been observed in human beings (large populations exposed to many rem, together with a very good control

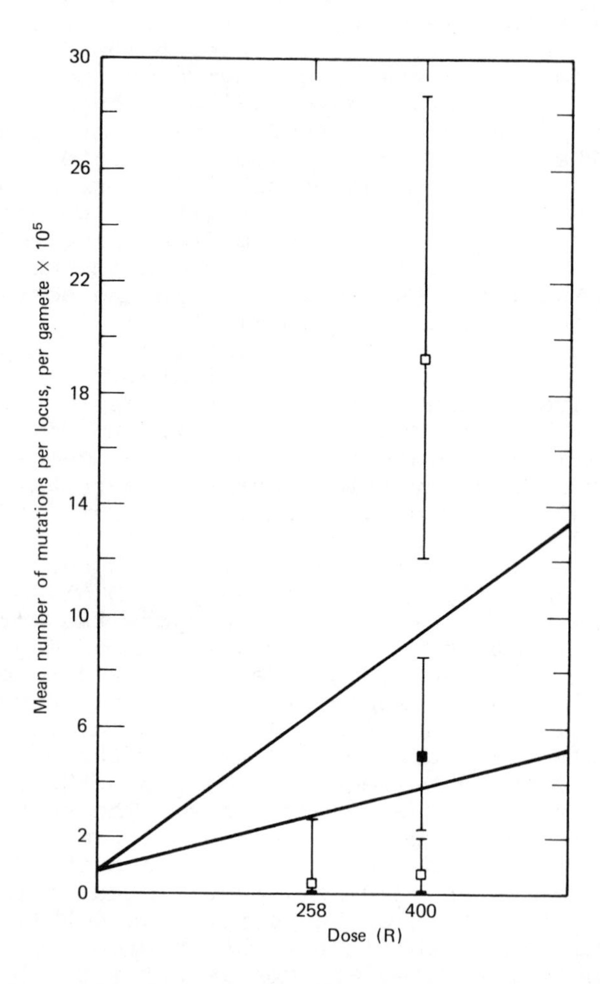

Fig. 6.8. Specific locus mutation rates, with 90% confidence intervals, for various dose rates in oocytes of mice. Dose rates are: open squares, 0.009 R/min; barred square, 0.8 R/min; and solid square, 90 R/min. The control point and straight lines are spermatogonia results shown for comparison (after Russell [8]).

(equivalent unirradiated population), would be necessary before it could be definitely stated that any abnormalities or early deaths were caused by radiation). Nevertheless, experiments in mammals demonstrate radiation mutagenesis, and mutagenesis surely occurs in man.

The risk of radiation mutagenesis and standards for protection against the genetic effects of radiation exposure are discussed in Section 6.6.

6.3 CARCINOGENESIS

Radiation can cause cancer [9,10] as well as cure it. Cancer is the uncontrolled proliferation of cells, in either a solid mass (tumor) or white blood cells derived from bone marrow (leukemia). The reasons why cancer originates and develops are not fully understood. It is possible that a somatic mutation in a single cell can initiate a cancer. Another possibility is that carcinogens somehow trigger or promote the action of a virus, since viruses are associated with some animal tumors. Cancer may be induced by radiation, certain chemicals such as benzpyrene, or prolonged mechanical irritation and regeneration of tissue. Predisposition to some cancers seems to be hereditary. The growth of tumors is influenced by other factors such as hormones. One theory suggests that cancerous changes occur all the time, but aberrant cells are neutralized by the immune-surveillance system. Cancer cells have different antigens on their surfaces compared to normal cells of the host, and lymphocytes can recognize the antigen and destroy the cell. According to the theory, tumor growth is prevented unless the lymphocytes cannot reach the cancerous cells, the production of lymphocytes is impeded for some reason, or the immune-surveillance system is overwhelmed by repeated challenges.

Diagnosis of cancer is usually delayed by several months to several years after exposure to the carcinogen. This "latent period" may occur simply because the tumor grows slowly and is too small to be detected by present methods, or perhaps it takes two independent events to initiate rapid tumor growth. There is evidence that the latent period of some cancers may be decreased by higher doses. Conversely, at very low doses the latent period may extend beyond the time of death from other causes. Radiation-induced cancer cannot be distinguished in kind from spontaneous cancer, only an increase in frequency can be observed under irradiation.

Figure 6.9 shows the incidence of myeloid leukemia in male mice subjected to single (high-dose-rate) or daily (low-dose-rate) doses of γ rays or fast neutrons. The neutron irradiations do not exhibit a dependence on dose rate, but the incidence per rad imparted by γ rays is much smaller at the lower dose rate. The RBE for fast neutrons, since it is referenced to the γ-ray results, varies from about 1 to 5. The drop in the single-exposure curves at high doses is probably the result of deaths from effects other than cancer, rather than a true decrease in the induction of leukemia. These results are consistent with a zero-threshold model; that is, cancer may be induced even at very low dose and dose rate, although the experiments were performed at relatively high total dose and dose rate.

The dose-effect relationship for leukemia in humans has been estimated from the atomic-bomb survivors and from British patients heavily irradiated with X-rays for treatment of ankylosing spondylitis, a rheumatic condition of the spine. Figure 6.10 shows

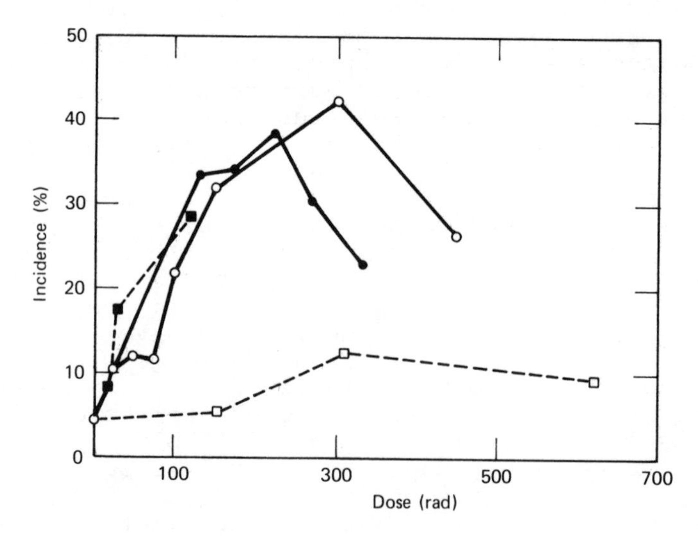

Fig. 6.9. Incidence of myeloid leukemia in mice irradiated with γ rays (open symbols) and neutrons (solid symbols), in single exposures (circles) or lower dose rate daily (squares)(after Ref. 9).

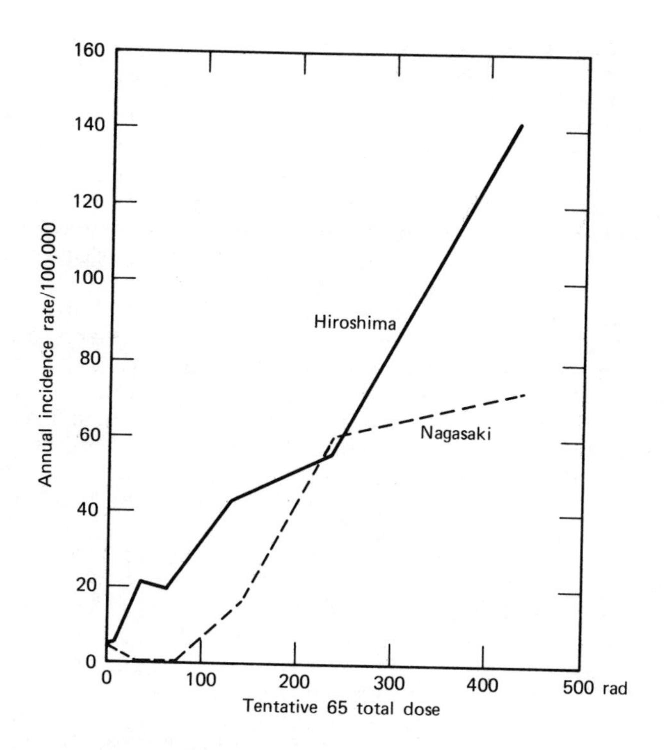

Fig. 6.10. Annual incidence of leukemia in Hiroshima and Nagasaki atomic-bomb survivors as function of absorbed dose (after Ref. 9).

the data for the Hiroshima and Nagasaki bomb survivors in terms of annual number of leukemia cases per 100,000 persons exposed, as a function of absorbed dose in rad. The exposure at Nagasaki was from γ rays, whereas the radiation from the Hiroshima explosion included a sizable fast-neutron component. It is difficult to obtain the RBE from the data, but it is between 1 and 5, and RBE = 5 is often assumed for carcinogenesis. The data are somewhat inaccurate because of the relatively small number of cases and because the dose to each individual had to be estimated from the distance from the explosion. Also, it is difficult to specify a control population for comparison. Nevertheless, it is possible to conclude that the increase in leukemia incidence is about one case per year, per 10^6 persons exposed, per rem, for persons over 10 years of age at the time of exposure. The adult ankylosing spondylitis patients also show an increase of about one case per year, per 10^6 persons exposed, per rem to the bone marrow (patient exposures were partial body but bone marrow, the lungs and other organs in the trunk were irradiated as well as the spine). Although no increase in leukemia incidence can be definitely established for absorbed doses under 50 rad, neither can a threshold dose be established. To be on the safe side in radiation protection, a linear dose-effect model is assumed, with zero cases at zero dose, independent of dose rate and type of radiation. This implies no repair and a cumulative effect of dose no matter how distributed in time.

Examination of excessive deaths from leukemia in children irradiated by diagnostic X-rays *in utero* indicates the rate may be several times that of the adult. Children under age 10 are also more sensitive to induction of leukemia. Conservative values for the annual rate, latent period, and duration of plateau are summarized [9] in Table 6.1. The relative risk is the ratio of the leukemia in the irradiated population to the deaths in the nonirradiated population. The dose for 100% relative risk is the doubling dose. The doubling dose for leukemia deaths in adults is 50 rem, in children it is 20 rem, and for the fetus it is 2 rem.

Breast Cancer

Increased incidence of breast cancer in women has been found in the atomic-bomb survivors, in tuberculosis patients given multiple X-ray fluoroscopies, and in patients given local X-ray treatments for postpartum acute mastitis. Increased incidence has also been observed in irradiated laboratory animals. The absolute risk for the atomic-bomb survivors is six cases/yr per 10^6 women exposed, per rem (if RBE = 1 for the fast neutrons). The latent period is 15 to 20 years. The multiple-fluoroscopic data indicate a slightly higher risk, but the dose is more certain in the mastitis patients, who also exhibit an increase in risk of six cases/year per 10^6 persons per rem. The age-adjusted spontaneous incidence in the United States is 720 cases/year per 10^6 women; hence the relative increase in risk is 0.83% per rem.

The data for human breast cancer are consistent with a zero-threshold, linear model, although they are inadequate to prove such a model. There appears to be little dependence on dose rate in animals, and increased incidence has been observed for doses as low as 10 rem. Thus it is prudent to assume there is no threshold dose for induction of breast cancer. This is important because film mammography and xeroradiography are used to diagnose breast tumors. The dose per breast per examination is about 4 rad, and annual examinations may be needed for early detection. Many women without breast cancer may be irradiated, and depending on the cumulative absorbed dose, it is conceivable that as many tumors would be induced as would be found. The National Cancer Institute has recommended that multiple X-rays of the breast be restricted to the population of women with a higher than average risk of spontaneous breast cancer (over 45 years of age, nulliparous, or family history of breast cancer). Newer techniques are being developed to reduce dose.

Lung Cancer

An increase in lung cancer has been observed in the atomic-bomb survivors, in the ankylosing spondylitis patients, and in underground miners who inhaled radon and daughters in the air for many months. The atomic-bomb survivors received 10 to 600 rad in a single exposure. Their risk is 0.6 cases/year per 10^6 persons exposed, per rem (assuming RBE = 5 for fast neutrons). The bronchial epithelium of the spondylitis patients was irradiated with X-rays to an average of 400 rad over many months. Their risk is 1.2 cases/year-10^6-rem.

The miners were irradiated by α particles (RBE = 10) from radon daughters in the lung. Figure 6.11 shows the number of cases per year per 10^6 uranium miners as a function of the dose equivalent to the bronchial epithelium. For the white uranium miners the risk of lung cancer is 0.63 cases/year-10^6-rem. Canadian fluorspar miners were also exposed to high concentrations of radon and daughters, and the risk is 1.6 cases/year-10^6-rem. An average risk of death is 1.3 deaths/year-10^6 persons, per rem. Radiation and cigarette smoking are synergistic agents for induction

TABLE 6.1. RISKS OF CANCER INDUCTION

Age at Irradiation	Type of Cancer	Duration of Latent Period (years)	Duration Plateau Region (years)
In utero	Leukemia	0	10
	All other cancer	0	10
0 to 9 years	Leukemia	2	25
	All other cancer	15	30
10+ years	Leukemia	2	25
	All other cancer	15	30

[a]Based on:

Type of Cancer	Deaths/10^6/year/rem
Breast	1.5
Lung	1.3
Gastrointestinal, incl. stomach	1.0
Bone	0.2
All other cancer	1.0
Total	5.0

Breast-cancer deaths derived from the value of 6 cases/ 10^6 women/yr/rem, corrected for a 50% cure rate and the inclusion of males as well as females in the population.

Risk Estimate	
Absolute Risk (Deaths/10^6/year/rem)	Relative Risk (% increase in Deaths/rem)
25	50.0
25	50.0
2	5.0
1	2.0
1	2.0
5[a]	0.2

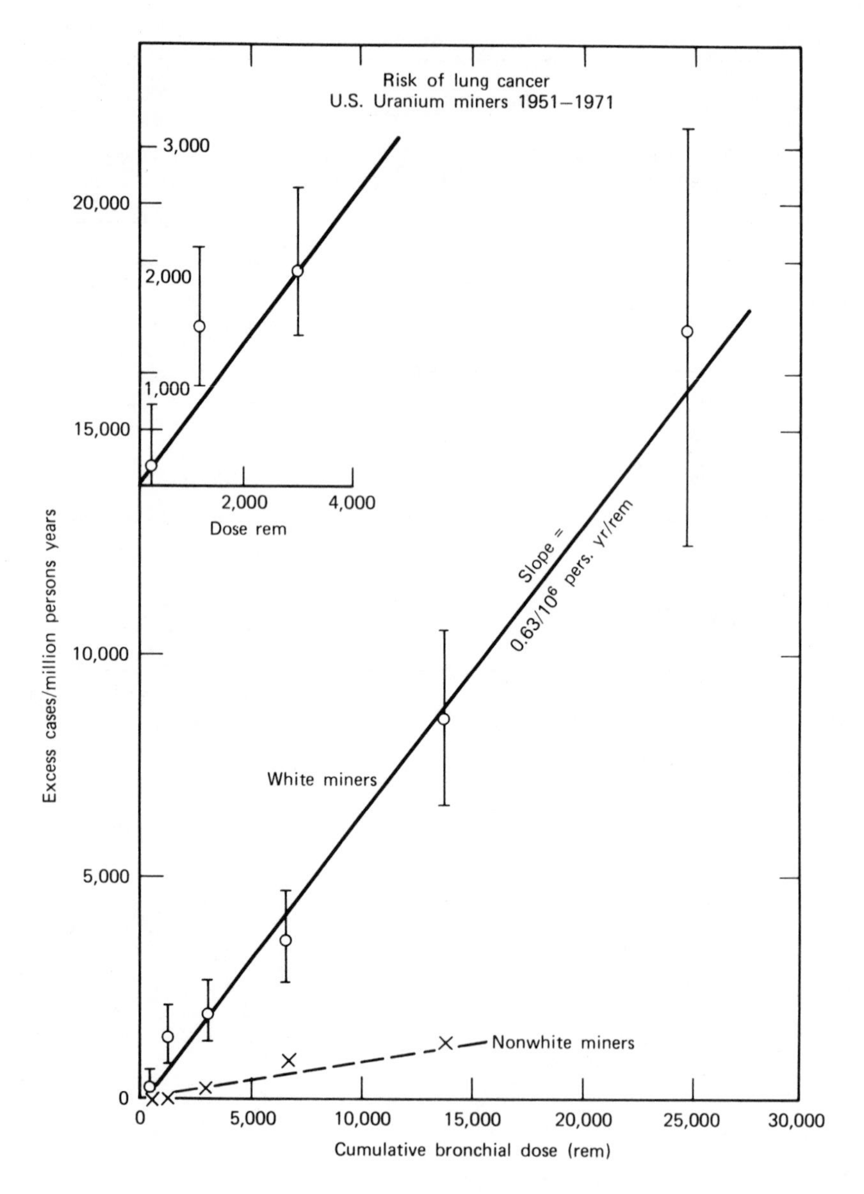

Fig. 6.11. Risk of lung cancer in underground uranium miners (after Ref. 9).

of lung cancer. Miners who smoke are much more likely to develop lung cancer than are nonsmoking miners, and smoking miners are more likely to develop lung cancer than are smoking nonminers. The risks quoted include the nonsmokers as well as smokers, but the percentage of smoking miners is high.

Considerable controversy has arisen as to whether carcinogenesis is enhanced by α-particle emitters in particulate form (especially for particles of $^{239}PuO_2$) rather than uniformly dispersed in lung tissue as is assumed for β- and γ-ray emitters. Because of the short range of α particles, the cells in the immediate vicinity of a radioactive particulate are irradiated to extremely high doses, whereas beyond the range the cells are not irradiated at all. The effective mass for computation of the absorbed dose would be quite small according to the "hot particle" theory. However, the NCRP [11] has concluded that available evidence does not support the contention that the concentration of the dose in a small mass or number of cells increases the absolute risk of carcinogenesis. The energy absorbed may even be less effective because many cells are killed rather than undergoing transformation to a cancerous state. The particles are not fixed but move about in the bronchus. Most significantly, no increase in carcinogenic risk has been found in dogs or in humans who have accidentally inhaled particulate plutonium, compared to nonparticulate plutonium.

Bone Cancer

The spondylitis patients received on the average 400 rad (X-ray) to the bone over an extended period. The increase in incidence of bone cancer is 0.10 cases/year per 10^6 persons exposed, per rem. Bone cancers have also been induced by radioactive substances taken up and held in bone. The most famous example is the group of radium-dial painters, who ingested ^{226}Ra in 1915-1935 and have been closely observed since. The excess (over spontaneous) cases per year, per 10^6 persons exposed, is plotted as a function of mean dose to bone from the α particles (RBE = 10) in Figure 6.12. A straight line can be drawn within the error bars, but no cases have been observed for a dose equivalent less than 5000 rem or for a bone burden of less than 1 μCi. Possibly the latent period is so long at the smaller doses that the persons die from other causes

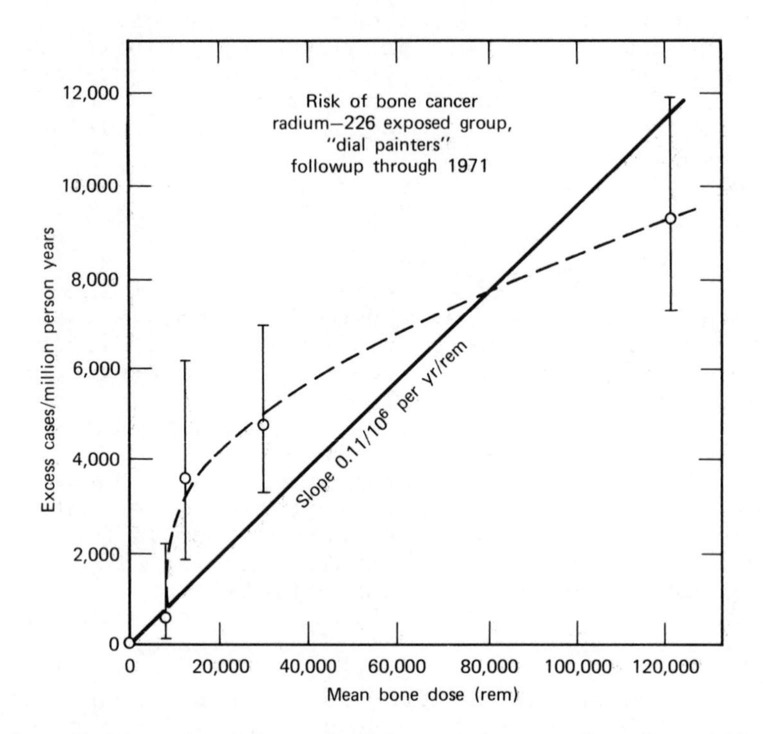

Fig. 6.12. Incidence of bone cancer in the radium-dial painters (after Ref. 9).

before bone cancer is diagnosed. Additional information is accumulating from a series of German patients given therapeutic doses of ^{224}Ra, which also accumulates in bone. So far these data do not indicate a threshold, nonlinear behavior, and the risk appears to be greater (0.55 cases/year-10^6-rem). Children appear to be more sensitive to induction of bone cancer by ^{224}Ra. The BEIR Report [9] suggests a risk of 0.2 deaths/year-10^6-rem for adults and one death/year-10^6-rem for children.

Other Cancers

Radiation carcinogenesis has been observed in the stomach, skin, lymphatic system, and salivary glands of humans, and other organs in animals. X Rays and radioactive iodine, either injected or ingested, produces "nodules" or tumors in the thyroid, at least of children, with an incidence of 2.5 cases/year-10^6-rem. Nearly all are benign and can be removed surgically; risk of death from thyroid cancer is low.

Total Cancer Risk

The risks of death from leukemia and other forms of cancer are summarized in Table 6.1. Since some cancers are not fatal the total incidence may be twice as great. For adults, the risk is six deaths per year, per 10^6 persons, per rem, occurring over 25 to 27 years. Thus the total number of deaths would be about 150 per 10^6 persons exposed, per rem. On the no threshold, linear model assumed for radiation protection purposes, it does not matter how the dose is distributed among the exposed population. A person-rem (or man-rem) is equivalent to one person receiving one rem, or 10 persons each receiving 0.1 rem, and so forth.

The effects on populations exposed to low-level radiation from natural and man-made sources and the establishment of standards for protection against radiation are discussed in Section 6.6.

6.4 RADIATION SICKNESS

Radiation sickness, also known as the "acute radiation syndrome," is the complex of symptoms and signs associated with cell death, observed in an organism subjected to a relatively high dose in a short time [12-14]. The symptoms appear in some minutes to some days.

Cells may be killed immediately, with lysis, at a sufficiently high dose (thousands of rad) because of membrane rupture, or at still higher dose because of inactivation of enzymes. At lower doses (hundreds of rad), the cells may continue to function, but because of damage to chromosomes and other components, the cells do not divide (or if they do, the daughter cells cannot function). Low doses (tens of rad), especially at late prophase, may inhibit mitosis, but most cells eventually divide after a delay of some hours.

Cells differ in their sensitivity to radiation damage, as measured by their ability to survive and continue growth and division. In 1906 the radiobiologists Bergonié and Tribondeau proposed that cells are more radiosensitive if they have a high mitotic rate and normally divide many times and if they are more primitive (i.e., less differentiated). The highly radiosensitive lymphocytes are primitive but do not divide until stimulated by appearance of an antigen. In the adult mammal, including man, the most radiosensitive group of cells includes lymphocytes, erythroblasts (precursors of red blood cells), and spermatogonia. The next most radiosensitive are granulosa cells in the ovary, myelocytes in the bone marrow, crypt cells in the lining of the small intestine, germinal cells in the epidermis, gastric gland cells, and endothelial cells lining small blood vessels. Moderately radiosensitive cells include bone cells, precursors of cartilage, spermatocytes and spermatids, sperm, granulocytes, and cells lining the gastrointestinal tract. Relatively radioresistant cells include those in most glands, fibroblasts (which form the intercellular fibrous matrix), cartilage and connective tissue cells, and erythrocytes. The most resistant to radiation are muscle and nerve cells, which are highly differentiated and normally do not divide in the adult.

Figure 6.13 shows the fraction of cells irradiated *in vitro* that survive a given absorbed dose of low-LET radiation such as X-rays [15,16]. The curve exhibits a shoulder followed by an exponential. If the exponential is extrapolated back to the y-axis, the intercept is a number greater than one, the extrapolation number n. The exponential portion may be described by $\exp(-D/D_0)$, where D_0 is the reciprocal of the slope (for $n = 1$, D_0 is the absorbed dose at which the surviving fraction is $1/e = 0.37$). Values of n and D_0 for X-ray irradiation of various tissues are listed in Table 6.2. The existence of a shoulder, hence $n > 1$, is related to sublethal damage. According to target theory, more than one target or sensitive volume in the cell has to be inactivated for the cell to cease division. Another explanation is biological repair. Repair is facilitated by low-LET and low dose. Assuming repair takes some time, it should also be facilitated by low dose rate or by fractionation.

The effect of splitting the dose into two fractions, separated by an interval allowing for repair, is seen in Figure 6.14. The first dose was 505 rad, resulting in a fraction 0.082 surviving. If a second dose of 487 rad is given immediately, the surviving fraction is 0.00186, the same as for a single dose of

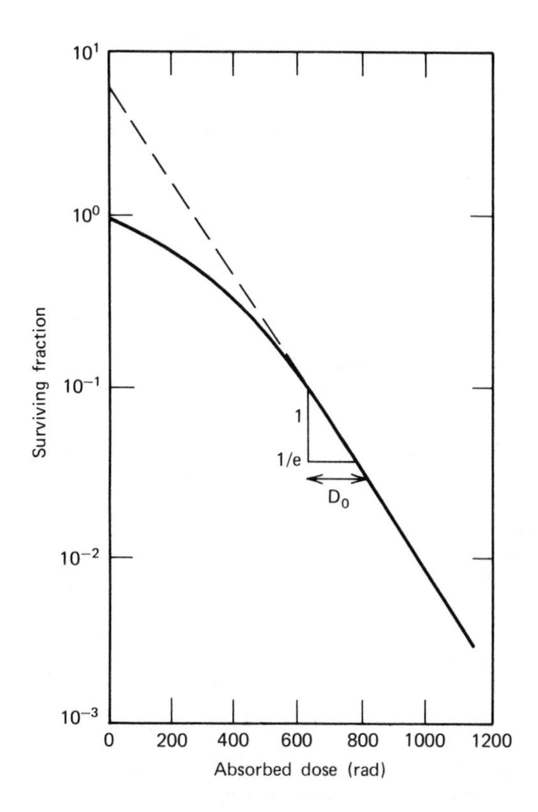

Fig. 6.13. Survival of cells irradiated with X-rays, illustrating the extrapolation number n and the slope D_0 (absorbed dose for factor of 1/e = 0.37 reduction).

(505 + 487) = 992 rad. However, if the interval is longer, some repair occurs. A dose of 487 rad would give a surviving fraction of 0.095, and (0.082 × 0.095) = 0.0078 survive two doses separated by $\geq$ 12 h.

Let us now turn our attention to the effects of cell-reproductive death on organs and organisms. The hematopoietic system is especially sensitive to radiation. Figure 6.15 shows the decrease of blood-cell counts in the rat after X-ray irradiation to hundreds of rad. The decrease in lymphocytes is marked and apparent, even at doses as low as 10 rad. At higher doses the decrease is greater and the recovery slower.

TABLE 6.2. PARAMETERS n AND D_0 FOR CELLS IN TISSUE CULTURE IRRADIATED WITH X-RAYS

Cell Type or Strain	n	D_0 (rad)
HeLa	2.0	128
Human embryonic lung	2.0	221
Various other human tissues	∿1.3	∿80
Hamster	3.2	110 to 155
Mouse	2.0	115
Mouse lymphoma	1.0	114

Lymphocytes are found in the bone marrow, lymph nodes, spleen, and thymus (in the child). They are involved in the immune-surveillance system and fight infection. The granulocytes also fight infection by engulfing and destroying bacteria and foreign particles. Their precursors, the myelocytes, are formed in the bone marrow. Platelets, derived from megakaryocytes in the bone marrow, are needed for coagulation of blood. The erythrocytes are radioresistant, but like other blood cells, they have a limited life. Damage to the erythroblasts in the bone marrow reduces the replacement of erythrocytes; hence their concentration in the blood also decreases after some time. Blood-cell counts eventually increase again if undamaged stem cells remain and multiply.

The gastrointestinal (GI) tract, especially the crypt cells of the small intestine, is sensitive to somewhat larger doses of radiation. The cells lining the stomach and small intestine are continuously sloughed off but are normally replaced by division and migration of stem cells in the gastric pits or crypts. Damage to the stem cells can result in decrease in function, poor nutrition, or even ulceration, hemorrhage, and denudation of the lining of the small intestine with consequent infection, and loss of

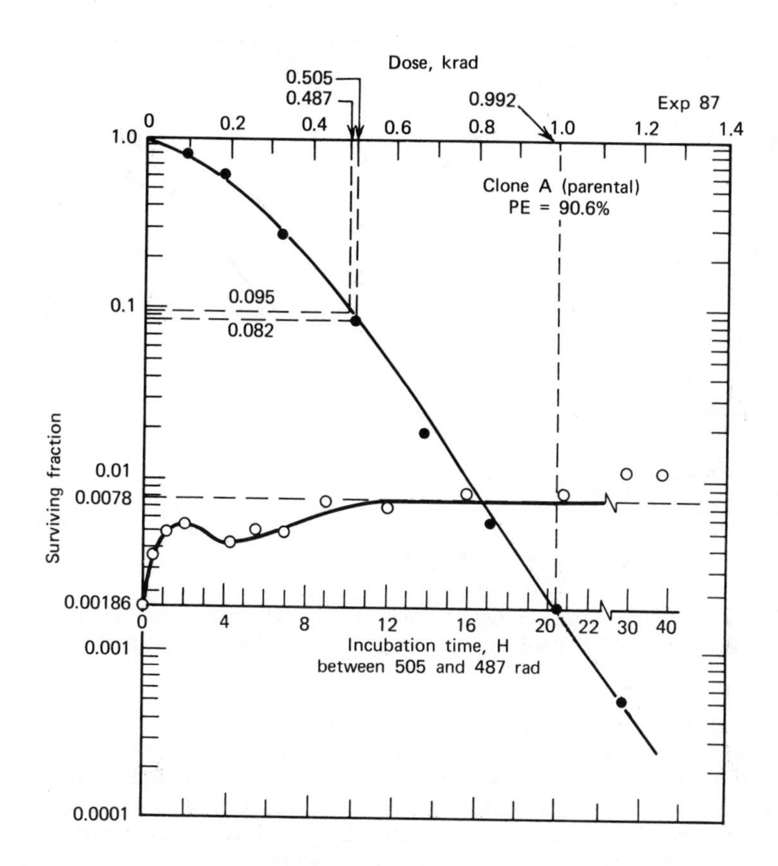

Fig. 6.14. X-Ray irradiation of hamster cells *in vitro*, showing effect of splitting dose into two fractions separated by a variable interval [after M. M. Elkind and H. Sutton, *Nature*, 184, 1293 (1959)].

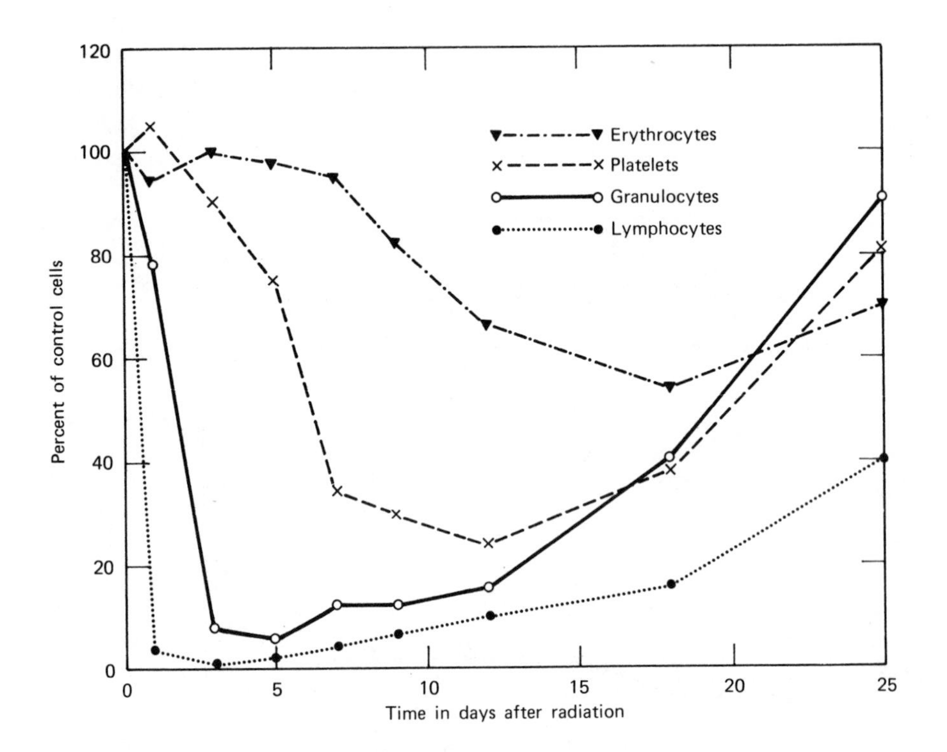

Fig. 6.15. Depression of blood-cell counts in rats irradiated to ∿100 rad with X rays (after Casarett [2]).

fluids and electrolytes. Symptoms of radiation sickness associated with the GI tract include anorexia (loss of appetite), nausea, vomiting, and diarrhea. The hematopoietic system is also damaged; hence the ability to combat infection and hemorrhage is impaired.

Other organs are relatively radioresistant, but blood vessels may be damaged, resulting in rupture or increased permeability of the walls, and glandular secretions may be reduced. At very high doses (5000 to 10,000 rad), death may occur in less than a day because of damage to the central nervous system (CNS) (brain) as evidenced by periods of agitation and apathy, ataxia, and convulsions.

The acute radiation syndrome in man for whole-body irradiation is summarized in Table 6.3. Although depression of lymphocyte and other blood-cell counts can be detected at 10 to 50 rad, clinical symptoms are not evident until about 100 rad. Between 100 and 500 rad, symptoms of nausea and vomiting may be experienced within hours, but these disappear after a couple days and the individual appears to be recovering. However, damage to the bone marrow results in leukopenia (depression of the white blood cells) and anemia in 2 to 3 weeks because the cells are not being replaced. The person becomes susceptible to infection and hemorrhage, which, however, can be treated with antibiotics and transfusions. Thus the threshold for death depends to some extent on the adequacy of the medical treatment. Death, if it occurs, may come between 3 weeks and 2 months for doses above the threshold of about 200 rad (at lower doses, everyone would be expected to recover even without treatment).

Between 500 and 2000 rad, the most serious effects occur because of damage to the small intestine, although the hematopoietic system is also damaged. Death may occur in 2 weeks to 3 days (with the higher dose associated with a shorter survival time), usually by circulatory collapse. Treatment includes replacement of fluids and electrolytes, transfusions, and antibiotics. Bone-marrow transplantation may be considered at the upper end of the range. Death is virtually certain above 2000 rad, characterized by CNS and GI syndromes and occurs in a matter of hours, even with therapy.

Lethality in man is approximated by the sigmoid curve shown in Figure 6.16. There is a threshold at 200 rad, and essentially 100% deaths are expected to occur above 2000 rad. The dose for 50% probability of death within 30 days, LD_{50}^{30}, is listed in Table 6.4 for several species. The LD_{50}^{30} applies to acute (instantaneous or delivered within a minute), whole-body irradiation with X-rays. The same dose protracted over hours or days will have less effect because of repair and repopulation by the surviving stem cells. Figure 6.17 shows the increase in the mean lethal dose for mice [14]. In man, a dose of say 250 rad, delivered continuously at 5 rad/year, would result in no detectable symptoms, not even depression of blood-cell counts. Effects are less severe with partial-body irradiation. For example, if the spleen and bone

TABLE 6.3. ACUTE RADIATION SYNDROME IN MAN (WHOLE-BODY IRRADIATION)

Aspect	Hematopoietic	Gastrointestinal	CNS
Major organ	bone marrow	small intestine	brain
Threshold	100 rad	500 rad	2000 rad
Latency	2 to 3 weeks	3 to 5 days	15 min to 3 hr
Death threshold	200 rad	1000 rad	5000 rad
Death time	3 weeks to 2 months	3 days to 2 weeks	<2 days
Signs, symptoms	leukopenia, anemia, hemorrhage, fever, infection, some nausea and vomiting, fatigue	nausea, vomiting, diarrhea, anorexia, fatigue, fever, dehydration, electrolyte loss, infection	lethargy, tremors, convulsions, ataxia

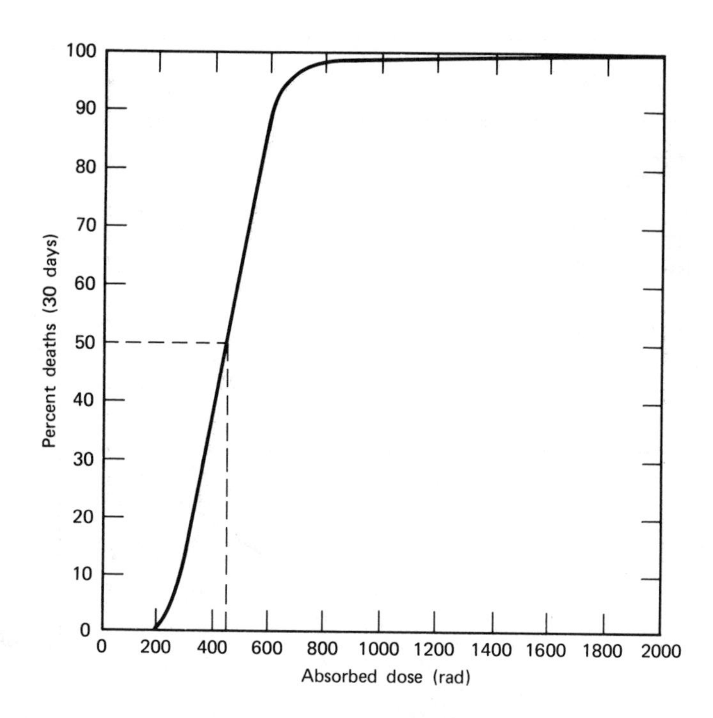

Fig. 6.16. Approximate lethality curve for acute X-ray irradiation in man.

marrow are shielded, the hematopoietic syndrome is alleviated. Partial-body irradiation is used in radiotherapy, as discussed in Section 6.7. The dose to the tumor may be over 3000 rad, and some normal tissue is irradiated, but symptoms of radiation sickness are much less severe than if the whole-body or sensitive organs are irradiated.

Irradiation of cultured cells indicates the RBE for fast neutrons is about 2, as illustrated in Figure 6.13. However, the RBE of fission or 14-MeV neutrons is about 1, for lethality in the dog [17]. As expected, fractionation or low dose rate does not reduce lethality as much as with X-rays.

TABLE 6.4. MEAN LETHAL DOSE (30 DAYS)

Species	LD_{50}^{30}
Man	450
Mouse	550
Monkey	600
Rat	750
Rabbit	800
Dog	350
Guinea pig	400
Goldfish	2300

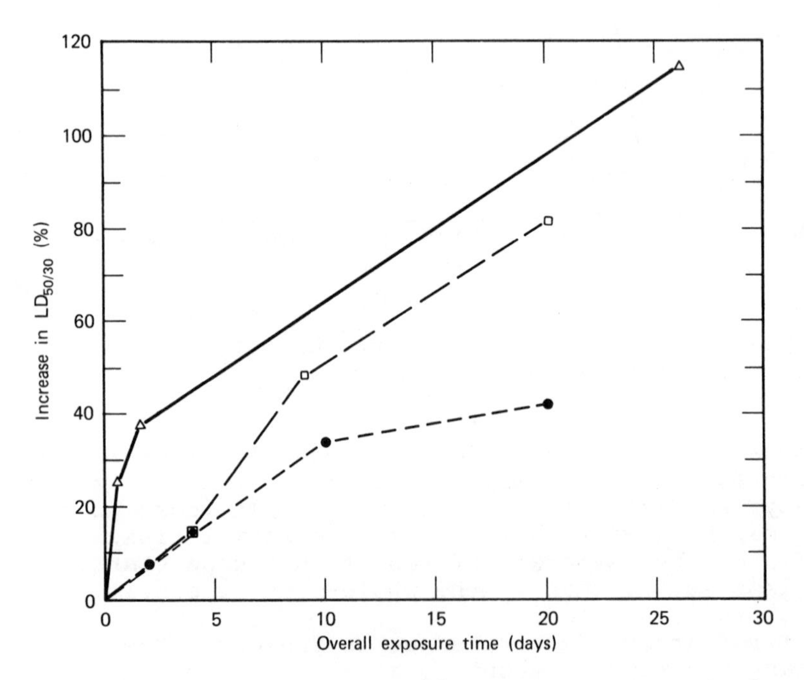

Fig. 6.17. Increase in LD_{50}^{30} for X-ray irradiation of mice for continuous irradiation (triangles), equal to daily doses (squares), or two equal doses separated by the interval (circles) (after Upton [14]).

6.5 OTHER BIOLOGICAL EFFECTS

Fetal Abnormalities

Irradiation of the mouse embryo [12] prior to implantation in the uterine wall increases the prenatal death rate (80% deaths at 200 rad). Irradiation *in utero* during the period of differentiation and organ formation results in congenital abnormalities and neonatal deaths. The sensitivity of the mammalian embryo and fetus is to be expected, either from killing of key cells during development, or from damage to DNA and other essential biomolecules.

Implantation occurs about 11 days after fertilization in the human, and most organogenesis occurs at 11 to 38 days. Radiation can still affect the developing nervous system and germ cells beyond the 40th day, but the crucial period is the first 90 days of pregnancy [9,13]. An increase in miscarriages has been found in women irradiated to over 20 rad. Children irradiated *in utero* to over 25 rad by the atomic bomb exhibited a greater-than-normal incidence of microcephaly (small head and brain) and mental retardation, as well as other abnormalities. Fifty percent of children irradiated to several hundred rad *in utero* during radiotherapy of the mother exhibited microcephaly, retardation, eye defects, stunted growth, or other ill health. Significant increases in abnormalities have not been observed for absorbed doses below 20 rad.

Cataract Formation

The major effect of radiation on the eye is production of opacity of the lens (cataract) because of damage to the epithelial cells, disorganization of lens fibers, and inability to eliminate damaged cells [18]. Threshold for significant impairment of vision in the adult is about 200 rad for acute irradiation with X-rays. Latent period is about 2 to 3 years. Fractionated or low-dose-rate exposure reduces the effect and increases the latent period. The young, growing lens is more sensitive than the lens of an adult. Fast neutrons are more effective in producing cataract than are X-rays. There is considerable disagreement on the RBE, which presumably depends on X-ray dose rate and dose, as well as neutron energy and degree of opacity. The RBE may vary from 5 to 50. Threshold for significant impairment of vision is about 75 rad for fission neutrons. Fast-neutron damage is cumulative; it is not reduced by low dose rate or fractionation.

Reduction in Fertility

Division of spermatogonia is temporarily halted by as little as 50 rem. Fertility is not reduced immediately, however, because the sperm and precursors are relatively radioresistant and may persist for several weeks. The sperm count decreases and then returns to normal after some months. The depression and time for recovery are extended at higher doses. However, the dose for permanent sterility exceeds the lethal dose (acute, whole body). Oocytes are moderately radiosensitive and are not replaced. An acute dose of 300 to 400 rem may result in permanent sterility. Libido, hormone production, and so on are not significantly impaired by radiation at sublethal doses.

Life Shortening

Nonspecific life shortening has been suggested as a long-term effect of radiation. This may occur because of increased susceptibility to a variety of diseases (other than cancer, which is considered separately), or possibly an acceleration of aging. However, the data for humans are inadequate to substantiate a life-shortening effect. If radiation-protection standards are established to minimize risks from the other effects discussed, it is unlikely that nonspecific life shortening will be important.

Effects on Skin and Hair

The hair follicles are sensitive to radiation, especially on the scalp and the beard for men. An acute exposure of 100 to 200 rad (low-LET radiation) is sufficient to delay mitosis in the germinal cells, and temporary (1 to 2 months) loss of hair may occur at 400 to 500 rad. Depilation is permanent above 700 rad. The sebaceous and sweat glands are damaged by 500 to 700 rad. Skin erythema (reddening, like a sunburn) is observed at 200 to 800 rad of acute exposure, depending on the energy of the radiation. Erythema may appear in a few hours, last a day, disappear for a couple of weeks, and then reappear. These changes are related to release of secretions, dilation of capillaries, and obstruction and increased permeability of blood vessels. A single dose of 1000 rad to the skin, or 3000 rad over 3 weeks, causes dry desquamation (sloughing). A single dose of 2000 rad, or 4000 rad over 4 weeks, causes blistering and moist desquamation. The skin recovers, but there may be residual pigmentation and toughening as well as permanent depilation. Ulceration and necrosis follow a dose above 4000 rad.

6.6 RADIATION-PROTECTION STANDARDS

Radiation-protection standards are regulations or guidelines promulgated to reduce the risk to an "acceptable" level while permitting applications of radiation-producing machines and radioactive nuclides. The risk associated with a given dose equivalent can be predicted from the biological effects discussed and the number of persons exposed.

It is easy to set standards for effects with a threshold, such as radiation sickness and cataract. Procedures and equipment are designed to assure that the threshold dose equivalent is not exceeded. The NCRP [20] recommends an emergency, one time only, dose of 25 rem not be exceeded except to save a human life.

Mutagenesis and carcinogenesis are assumed to have no threshold. Thus there is some chance, however small, of adverse genetic effects and induction of cancer, even at very low doses. Setting an acceptable risk from man-made radiation involves a value judgment. Factors influencing acceptability include: the risk relative to unavoidable exposure from natural background and the variability of background, relative risk compared to spontaneous occurrence and other risks (voluntary and involuntary), absolute risk in terms of morbidity and mortality and cost to the individual and society, the technical feasibility and cost of reducing the risk to a lower level, and most of all, the benefit to the individual and society from using the radiation-producing machines and radioactive substances.

Different limits are appropriate for occupational exposure (to workers in a radiation-producing industry or students) and nonoccupational exposure. The radiation worker accepts the risk voluntarily, receives the direct benefit of job and pay, and his exposures can be controlled more easily than for members of the general public. Also, the number of radiation workers constitutes a small fraction of the total population, so genetic effects are diluted. However, the philosophy has been that radiation workers should not be subjected to greater risk of serious illness or death than workers in similar industries are subjected to from other hazards. In fact, the record of occupational health and safety in radiation industry has been very good, except for uranium miners.

Medical applications of radiation for diagnosis and therapy are not controlled by the regulations applying to other man-made sources such as nuclear reactors, accelerators, and radioisotopes for research or industrial and agricultural applications. Much is left to the discretion of the physician and good practice in the field, although standards have been established to reduce unnecessary exposure to the patient, medical personnel, and the public. For example, sources and rooms have to be shielded, and radioisotopes used in nuclear medicine have to be handled and disposed of properly.

Standards and recommendations have been developed by international and national panels of experts, in particular the International Commission on Radiological Protection (ICRP), the National Council on Radiation Protection and Measurements (NCRP) in the United States, and similar bodies in other countries. The reviews and recommendations of special panels of the U.S. National Academy of Sciences-National Research Council, as published in the 1972 BEIR report [9] and the 1956 BEAR report [19], have been most influential, with contributions also from the United Nations Scientific Committee on the Effects of Atomic Radiation (UNSCEAR) and others. Recommendations are given the force of law in federal regulations (Title 10, Code of Federal Regulations) governing the licensing of nuclear reactors and fuel-cycle facilities, radioisotopes produced in reactors, source materials (uranium, thorium), and special nuclear materials (enriched uranium, plutonium). The states control radiation exposures from accelerators and accelerator-produced radioisotopes, radium and other separated natural radioisotopes, X-ray machines, and other sources. The federal regulations are enforced by the Nuclear Regulatory Commission. State regulations are enforced by a Bureau of Radiological Health or similar agency. Certain states also administer the licensing of reactor-produced radioisotopes. State regulations generally follow the federal regulations for allowable dose. Since the regulations are detailed and subject to change, the latest copy of Title 10 and the applicable state code should be consulted.

Occupational

The federal regulations pertaining to occupational exposure are summarized in Table 6.5. All doses are in addition to natural background and medical exposure. The 1.25 rem per calendar quarter (13 consecutive weeks) applies to irradiation of the whole body or critical organs such as the gonads and hematopoietic system. The dose equivalent may exceed 1.25 rem if necessary, but shall not exceed 3 rem/quarter, and the cumulative dose shall not exceed 5(N-18) rem, where N is the age in years. The dose has to be monitored and documented. Minors under the age of 18 are allowed to receive up to 0.5 rem per year. Because of the radiosensitivity of the fetus, pregnant women should receive no more than 0.5 rem in the 9-month gestation period. Higher doses are permitted to the extremities and skin because they are not critical to health, and to allow some flexibility in performing occupational tasks. Doses should be kept as far below these limits as reasonably achievable. Ninety percent of U.S. radiation workers receive less than 0.2 rem per year.

Occupational exposure is based on working 8 h per day, and 40 h per week. Thus the design dose rate is 100 mrem/week or 2.5 rem/h. Limits to the general public are lower; hence access is controlled to the "restricted area." Higher dose rates are permitted in areas that are seldom occupied, or if the source is automatically shut off when anyone approaches it. Any accessible area in which the dose rate may exceed 100 mrem/h is a "high radiation area" and must be provided with appropriate signs, alarms, barriers, and radiation monitor.

The dose limits in Table 6.5 apply mainly to external exposure from γ rays, neutrons, and β particles. α Particles do not penetrate the dead layer of the skin. However, α particles as well as

TABLE 6.5. OCCUPATIONAL DOSE LIMITS (10CFR20) (OVER BACKGROUND AND MEDICAL)

Organ	Rem/quarter	Rem/year
Whole body; or head and trunk, active blood-forming organs, lens of eye, or gonads	1.25[a]	5[b,c]
Hands and forearms; feet, ankles	18.75	75
Skin of whole body	7.50	30

[a]Average; not to exceed 3 rem per quarter.

[b]Accumulated dose 5(N-18) rem. Minors 0.5 rem/year.

[c]Pregnant women should be exposed to less than 0.5 rem in gestation period.

β particles and γ rays from radioactive substances within the body may deliver a significant dose to surrounding tissue. The limit of 5 rem/year applies to radioisotopes such as tritium, which are distributed more or less uniformly throughout the body. Other elements are concentrated in certain organs. For example, iodine is concentrated in the thyroid; therefore, radioactive isotopes such as iodine-131 deliver much of the dose to the thyroid. Certain radioisotopes, including strontium-90, radium-226, and plutonium-239, are taken up in the bone and excreted very slowly. The allowable bone burden (activity) for α- and β-particle emitters is referenced to the equivalent of 0.1 μCi of radium-226 and daughters. (No bone cancers have been observed at bone burdens under 1 μCi.) Determination of organ burden and dose is discussed in Chapter 8. Radioisotopes enter the body by inhalation or ingestion in food and water. Thus it is convenient to express radiation-protection standards in terms of the maximum permissible concentration (MPC), or activity per unit volume of air or water.

The occupational MPC is the concentration that would give the maximum permissible occupational dose to the critical organ, either in equilibrium or at the end of 50 years of continuous breathing or consumption of water at standard specified rates, for the 8-h/day, 5-day/week, 50-weeks/year exposure to the contaminated air or water. In the case of the inert gases such as argon, krypton, and xenon, the MPC is calculated for a whole-body dose of 5 rem/year while submerged in a semispherical volume of contaminated air or water. Examples of MPCs for selected radioisotopes are given in Table 6.6. A distinction is made between chemical forms that are soluble or insoluble in body fluids, because the uptake and distribution within the body are different. The MPCs are calculated for the maximum permissible doses (MPDs) listed in Table 6.7.

Nonoccupational

Radiation exposures to the general public in unrestricted areas are regulated by 10CFR Part 20 (or comparable state codes), except for nuclear power plants that are regulated by 10CFR Part 50 (Licensing of Production and Utilization Facilities) and 10CFR Part 100 (Reactor Site Criteria). Dose limits are summarized in Table 6.8.

The dose limits to the public are basically 0.1 of the dose limits to radiation workers. Part 20 states that an individual in an unrestricted area shall not receive more than 0.5 rem whole body, averaged over the year, from the licensed source. Furthermore the dose rate may not exceed 2 mrem/h or 100 mrem/week. If such dose rates are allowed, an occupancy factor (fraction of time area is occupied by an individual) and usage factor (fraction of time the source is operating) must be established to assure the annual average dose is less than 0.5 rem. The MPC for release to unrestricted areas is equal to the MPC for occupational exposure times 0.1 (40/168) = 0.02, reflecting the lower dose limit and exposure for the full 168 h/week instead of the 40-h work week. It is desirable to keep the exposures as far below the dose limits as reasonably achievable.

Nuclear power plants must follow the "as low as reasonably achievable" (ALARA) objective, and numerical "guidelines" (actually requirements for licensing) have been established in Appendix I of 10CFR50. Most of the dose to an individual off site comes from the

TABLE 6.6. MAXIMUM PERMISSIBLE CONCENTRATIONS (10CFR20) FOR OCCUPATIONAL EXPOSURE (ABOVE BACKGROUND)

Isotope sub = submersion S = soluble I = insoluble		Air (μCi/ml)	Water (μCi/ml)
Argon-41	(sub)	2×10^{-6}	--
Carbon-14	(S)	4×10^{-6}	2×10^{-2}
Cesium-137	(S)	6×10^{-8}	4×10^{-4}
	(I)	1×10^{-8}	1×10^{-3}
Cobalt-60	(S)	3×10^{-7}	1×10^{-3}
	(I)	9×10^{-9}	1×10^{-3}
Fluorine-18	(S)	5×10^{-6}	2×10^{-2}
	(I)	3×10^{-6}	1×10^{-2}
Gold-198	(S)	3×10^{-7}	2×10^{-3}
	(I)	2×10^{-7}	1×10^{-3}
Hydrogen-3	(S or I)	5×10^{-6}	1×10^{-1}
	(sub)	2×10^{-3}	--
Iodine-131	(S)	9×10^{-9}	6×10^{-5}
	(I)	3×10^{-7}	2×10^{-3}
Krypton-85	(sub)	1×10^{-5}	--
Phosphorus-32	(S)	7×10^{-8}	5×10^{-4}
	(I)	8×10^{-8}	7×10^{-4}
Plutonium-239	(S)	2×10^{-12}	1×10^{-4}
	(I)	4×10^{-11}	8×10^{-4}
Radium-226	(S)	3×10^{-11}	4×10^{-7}
	(I)	5×10^{-11}	9×10^{-4}
Strontium-90	(S)	1×10^{-9}	1×10^{-5}
	(I)	5×10^{-9}	1×10^{-3}

TABLE 6.7. MAXIMUM PERMISSIBLE DOSE, INTERNAL EXPOSURE

Organ	Rem/year
Gonads, lens, bone marrow	5
Skin, thyroid	15[a]
Other soft tissues	15
Bone	0.1 μCi ^{226}Ra + daughters equivalent

[a]Previously 30 rem/year.

TABLE 6.8. NONOCCUPATIONAL DOSE LIMITS

Nuclear Power Reactors (10CFR50App.I)	
Whole body[a]	0.005 rem/year
Skin, thyroid[b]	0.015 rem/year
Other sources (10CFR20)	
Whole body	0.500 rem/year
Skin, thyroid	1.500 rem/year
Bone	0.100 of occupational

[a]Dose rate from liquid effluents not to exceed 3 mrem/year to whole body or 10 mrem/year to any organ. Dose rate from gaseous effluents not to exceed 10 mrad/year from γ rays or 20 mrad/year from β particles at site boundary.

[b]Dose rate from radioactive particulates and radioiodine not to exceed 15 mrem/year to any organ.

radioactive effluents released to the atmosphere and to the cooling water in routine operation. The licensee has to show that it is unlikely that the whole-body dose to any individual at the site boundary will exceed 0.005 rem averaged over the year (5 mrem/year), for all pathways of exposure. A dose of 15 mrem/year is allowed to the skin and thyroid. There are other provisions regarding gaseous, particulate, and radioiodine releases to the atmosphere. Furthermore, a cost-benefit analysis has to be performed to see if the doses should be reduced further.

The provisions in 10CFR Part 100 include maximum doses at the site boundary (or exclusion area) and at the boundary of a surrounding "low population zone," in event of a reactor accident releasing radioactive gases and particulates to the atmosphere. It is stated that the reactor shall be located, and containment and other equipment designed, such that the dose at the boundary of the exclusion area does not exceed 25 rem to the whole body, or 300 rem to the thyroid from radioiodine, for an individual remaining at the boundary for 2 h after the accident. The outer boundary of the "low-population zone" (which surrounds the exclusion area) is determined by the whole-body dose of 25 rem or the thyroid dose of 300 rem, which could be received if a person remained there during the entire passage of the radioactive cloud. Certain meteorological conditions are specified in Regulatory Guides, to be used in computing the concentrations and hence the doses. The procedures for predicting concentrations are discussed in Section 4.6 and the doses, in Section 8.4.

Consequences

The most serious risk to the individual is increased chance for death from cancer. In Section 6.3 the risk was expressed as six deaths per year per 10^6 persons exposed, per rem or 150 deaths per 10^6 persons per rem from all cancers, including leukemia. We assume the no threshold, linear model. Thus the risk of an individual dying of radiation-induced cancer is 1.5×10^{-4} chance per rem. For occupational exposure at the maximum of 5 rem/year, over 50 years, the risk is 3.75×10^{-2}. The risk to an individual at the boundary of a nuclear power plant, receiving 0.005 rem/year for 40 years, is 3×10^{-5}. These risks may be compared to the annual risk of accidental death from various causes, or disease, in Table 6.9.

Table 6.10 summarizes the sources of natural and man-made radiation exposure in the United States. The total is about 0.2 rem/year, half from natural background and the other half primarily from diagnostic medical exposures. If the U.S. population of roughly 200×10^6 persons is exposed continuously to 0.2 rem/year, radiation would contribute some 6000 cancer deaths per year. This is 2% of the total cancer death rate.

The most serious risk to society is the genetic effect, expressed as increased incidence of genetically linked disease and neonatal abnormalities. In Section 6.2 the risk was estimated in terms of a doubling dose between 20 rem and 200 rem. Irradiation of the gonads of a large population, at the rate of 0.15 rem/year over the average generation span of 30 years ($\sim$5 rem) would eventually cause an increase of from 2.5% to 25% in the burden of mutation-caused disease. In addition, there may eventually be increase in the rate of neonatal abnormalities. The major contribution to abnormalities is aneuploidy, best known as mongolism. The current incidence in the United States is 4000 cases per million live births. Radiation at 5 rem per generation would add five cases per million live births. The number is less than 2.5% or 25% of 4000, because most cases of aneuploidy do not survive to birth.

The BEIR Committee recommended that steps be taken to reduce unnecessary medical exposures and that other sources of man-made radiation, in particular nuclear power, be kept as low as reasonably achievable and a fraction of background. If the genetically significant dose is kept well below background, we are certain that the consequences of the additional exposure will be less in quantity and no different in kind than we have experienced throughout human history. It is technologically and economically feasible to keep the dose to the public from nuclear power plants to less than 5 mrem/year to any individual (and far less than that averaged over the population).

6.7 RADIOTHERAPY

The objective of cancer radiotherapy is to eradicate the tumor while inflicting only tolerable damage on

TABLE 6.9. RISK OF FATALITY[a]

Accident Type	Probability/year
Motor vehicle	3×10^{-4}
Falls	9×10^{-5}
Fires, hot substances	4×10^{-5}
Drowning	3×10^{-5}
Poison	2×10^{-5}
Firearms	1×10^{-5}
Machinery	1×10^{-5}
Water transport	9×10^{-6}
Air travel	9×10^{-6}
Falling objects	6×10^{-6}
Cancer risk per rem	6×10^{-6}
Electrocution	6×10^{-6}
Railway	4×10^{-6}
Lightning	5×10^{-7}
Tornadoes	4×10^{-7}
Hurricanes	4×10^{-7}
All accidents	6×10^{-4}
All diseases	8×10^{-3}
Cancer	1×10^{-3}

[a]From WASH-1400.

TABLE 6.10. SOURCES OF RADIATION EXPOSURE (U.S. AVERAGE)

Source	mrem/year	
	Whole body	Gonads
Natural radiation		
Cosmic	44.0	
Radionuclides in body	18.0	
Radionuclides in soil, buildings	40.0	
	102.0	90
Man-made radiation		
Medical and dental	73.0	
Fallout	4.0	
Occupational	1.0	
Nuclear power	0.003	
Miscellaneous	2.0	
	80.0	60
Total	182.0	150

normal tissue. It is sufficient to destroy reproductive integrity of tumor cells (the ability to divide indefinitely). The mean lethal dose in this sense is 100 to 200 rad for mammalian cells, but it is necessary to kill every clonogenic malignant cell, or the tumor will regrow. There are about 10^9 cells per gram or cm^3, which would indicate the survival fraction would have to be less than 10^{-9}. On the other hand, not all the cells in a tumor are necessarily clonogenic (able to proliferate), and a survival fraction of only 10^{-4} to 10^{-5} may be sufficient for cure. The absorbed dose required to eradicate a tumor depends on the dose rate and fractionation scheme, oxygen concentration, LET, and other factors. The inherent radiosensitivity of the tumor cell and the normal cells of the affected organ are not much different. However, it is possible to exploit differences in repopulation kinetics, as the normal cells are often replaced more rapidly than the tumor cells. Furthermore, the skin and normal tissues surrounding the tumor are spared as much as possible by directing collimated beams at different angles (intersecting in the tumor) or by taking advantage of the nonuniform spatial distribution of dose around a small radioactive source implanted in the tumor. Design of therapeutic apparatus and dosimetry for radiotherapy are discussed in Chapter 8. Here we are concerned with the radiobiology.

The effect of irradiation on cell reproductive integrity may be measured *in vitro* and *in vivo*. Viability is measured either by counting clones grown from single cells in culture or by using a laboratory animal after transplantation from the irradiated animal [21]. Figure 6.18 is a plot of the survival curve for HeLa cells (originally from a human cervical cancer) irradiated with X rays or 14-MeV neutrons, in air (oxygenated) and in nitrogen.

Both the slope (or D_0) and extrapolation number are affected by the dose rate, at rates low enough for partial repair of sublethal damage during irradiation. Figure 6.19 plots D_0 and n for X-ray irradiation of HeLa cells. The half-life for repair is 1 to 1.5 h. Below 1 rad/min (60 rad/h) there is little dependence on dose rate because essentially all sublethal damage is repaired and the remaining effect is caused by nonreparable damage. Above 100 rad/min (600 rad/h) there is no time for repair, and maximum damage is inflicted.

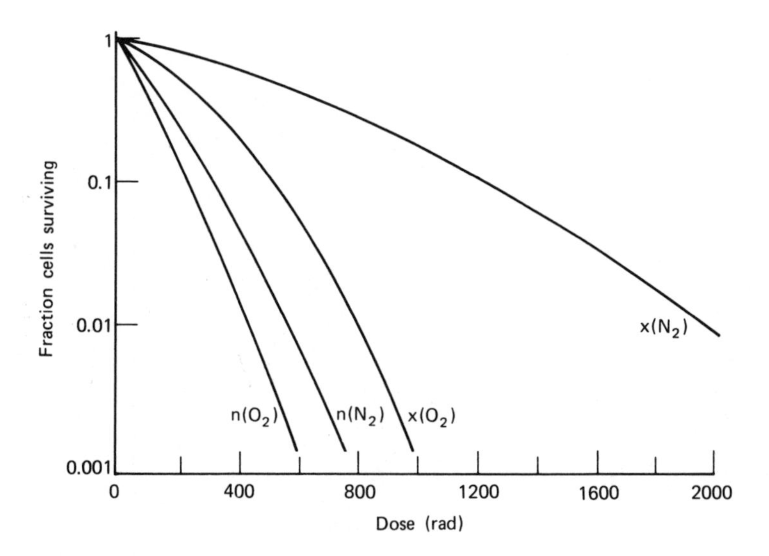

Fig. 6.18. Survival curve for HeLa cells irradiated with X rays or 14-MeV neutrons, oxygenated (in air) and nonoxygenated (in N_2).

The effect of fractionation (dividing the total dose into equal increments spaced by a day or so) was discussed in Section 6.4. Fractionation is required to permit recovery of normal tissues and also helps in reoxygenation of the tumor as discussed later. On the other hand, there is a loss of efficiency because the shoulder portion is repeated in each treatment. A typical schedule in beam teletherapy of squamous-cell carcinoma with a cobalt-60 source is 200 rad per treatment delivered at 100 rad/min, with one treatment per day for 5 days, hence 1000 rad/week, and a total dose of 6000 rad in 6 weeks. More radioresistant cancers such as bone sarcoma may require more than 8000 rad. Very radiosensitive tumors such as seminoma can be cured with 3500 rad [22]. Tumors treated by interstitial or intracavity implants of radium and other radionuclides may be irradiated continuously for 7 days for a total dose of 6000 rad (hence 35.7 rad/h). If the dose rate is raised to 64 rad/h, equivalent biological effect is achieved with 4000 rad in 3 days. Many of the treatment protocols are empirical, not

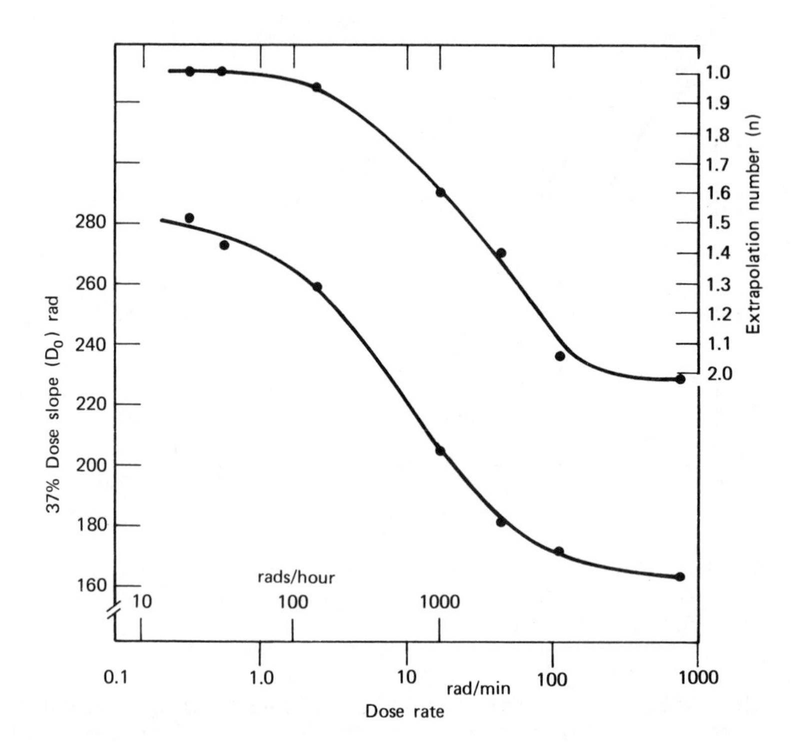

Fig. 6.19. Variation of n and D_0 with γ-ray dose rate for HeLa cells in culture (after Hall [21]).

necessarily optimum, and vary according to the judgment of the radiotherapist in the individual case. The therapeutic ratio (ratio of dose required to eradicate the tumor to the "tolerance" dose for normal tissue) is often small, and the therapist will often have to irradiate to the limit of normal tissue tolerance, accepting the fact that some tumors will not be cured. Failure to cure some tumors has been ascribed to a fraction of cells being hypoxic (deficient in oxygen) because of necrosis in the interior of larger tumors and deficient blood supply. Regression of the tumor under protracted irradiation is thought to allow for reoxygenation of the hypoxic regions.

Oxygen potentiates the effect of low-LET radiation. Typical dependence of radiosensitivity (relative to the anoxic condition) is illustrated in Figure 6.20.

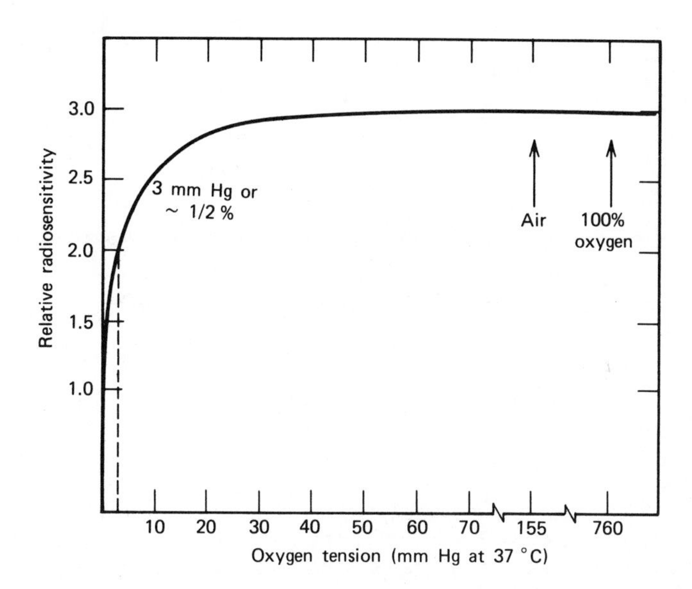

Fig. 6.20. Dependence of radiosensitivity on pressure of oxygen, for low-LET radiation (after Hall [21]).

Half of the enhancement is achieved by 3 mm Hg. Hypoxic tumor cells would thus be less radiosensitive than oxygenated tumor or normal cells. The influence of oxygenation diminishes with an increase in LET, as shown in Figure 6.21. Therapy with fast neutrons is being developed because normal and hypoxic tumor cells are at least equally sensitive to damage from high-LET radiation. The RBE depends on neutron energy and the dose and dose rate for the X rays but is on the order of 3. Fractionation has little effect on the surviving fraction for high-LET radiation.

6.8 MATERIALS DAMAGE

Ionization and excitation, with consequent disruption of chemical bonds, is the primary mode of damage in water and organic compounds. The effect may be expressed in terms of a G value (molecules or product formed per 100 eV of energy absorbed), or the absorbed dose required to produce a certain level or change in

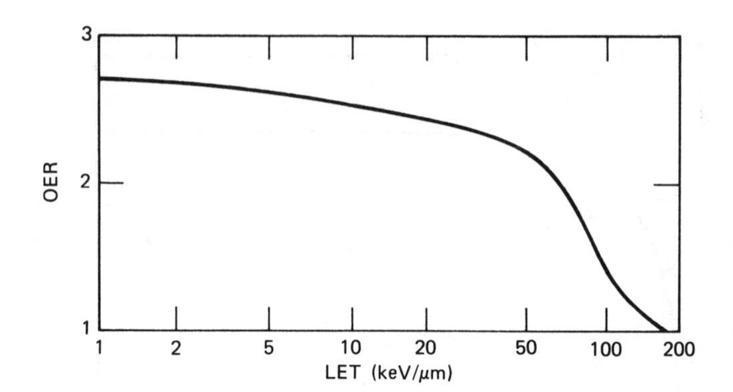

Fig. 6.21. Dependence of oxygen enhancement ratio on linear energy transfer (after Hall [21]).

a physical property. Ionization is also significant in ionic crystals and other insulators. Ionization is not significant in metals and semiconductors, except for a transient increase in conductivity of semiconductors, because electrons are mobile and recombine rapidly with the ions or holes. The main effect of the energy deposited by ionization is heating.

If the energy of a nuclear particle exceeds a certain threshold, enough energy is transferred to an atom to displace it from its original lattice site. The displacement energy is about 25 eV in most solids [23]. The displaced atom will reside between normal lattice sites and constitutes a defect called an "interstitial". The empty lattice site is another defect, the vacancy. At elevated temperature, interstitials and vacancies can migrate and annihilate each other, thus annealing out the radiation damage. On the other hand, vacancies can combine to form a void. When the energy of the primary knock-on atom is much greater than the displacement energy, secondary knock-on atoms can be produced. The result is a cluster, or spike, of disordered atoms. Displacements and displacement spikes contribute to the radiation damage in metals, semiconductors, and insulators.

Fast neutrons can produce (n,p) and (n,α) reactions in many metals. The H and He atoms do not fit in the lattice. They may agglomerate into bubbles, causing swelling. A few reactions with thermal neutrons, notable $^{6}Li(n,\alpha)T$ and $^{10}B(n,\alpha)^{7}Li$, also produce gas. Nuclear transmutations can cause other changes by introducing impurity atoms.

Table 6.11 provides general guidelines for radiation damage in water, organics, graphite, concrete, ceramics, glass, semiconductors, and metals. The composition and processing of many commercial products, such as plastics, are subject to change, and radiation damage is often quite sensitive to the temperature and other conditions of irradiation. Thus Table 6.11 should be used only to aid in selection of materials, rather than for precise design criteria.

Water

Ionizing radiation causes radiolysis of H_2O and D_2O, with evolution of hydrogen and oxygen gas. The fundamental reactions are discussed in Section 6.1. The gases can cause pressure buildup or undesirable bubbles and will recombine explosively if ignited. The G value for formation of H_2 varies from about 1 to 5, depending on the LET [24]. High-LET radiation is most effective. Dissolved impurities tend to enhance decomposition by inhibiting the back reactions, whereas an excess H_2 pressure decreases decomposition. Hydrogen and oxygen can be recombined quietly by either combustion or a catalytic recombiner containing finely divided platinum.

Organics

Solvents and oils are degraded by ionizing radiation, with evolution of hydrogen and an increase in viscosity. The aromatic hydrocarbons are most resistant, and silicones are the least resistant. Radiation also accelerates oxidation if oxygen is present.

Two processes occur in plastics and elastomers [25]; degradation by scission of polymer chains and cross-linking of chains because of formation and reaction of free radicals. Degradation results in evolution of hydrogen and loss of strength. Cross-linking makes the polymer slightly stronger but also more brittle. Polystyrene is most resistant to scission, whereas Teflon (polytetrafluoroethylene) is least resistant. The ratings in Table 6.11 refer to changes in mechanical properties. Electrical resistance is only slightly affected, but insulation

TABLE 6.11. GUIDE TO RADIATION DAMAGE IN MATERIALS

Material	Property
Water	decomposition
Organics	various
Oils	viscosity
Polystyrene	strength
Phenolic, epoxy	strength
Polyethylene	strength
Nylon, Lucite	strength
Teflon	strength
Neoprene	elongation
Wood	strength
Graphite	volume
Concrete	strength
Ceramics	resistivity strength
Glass	transparency
Semiconductors	resistivity
Metals	mechanical
Steel	ductility
Alnico-V	magnetism

Radiation	Dose or Fluence for Significant Damage
γ	10^{10} rad
n or γ	10^{6} to 10^{10} rad
γ	10^{9} rad
γ	10^{10} rad
γ	10^{9} rad
γ	10^{8} rad
γ	10^{7} rad
γ	10^{6} rad
γ	10^{8} rad
γ	10^{7} rad
n (fast)	10^{21} n/cm^2
n (fast) γ	$>10^{21}$ n/cm^2 $>10^{13}$ rad
n (fast) n (fast)	10^{18} n/cm^2 10^{20} n/cm^2
γ	10^{4} to 10^{8} rad
n (fast)	10^{16} n/cm^2
n (fast)	10^{17} to 10^{22} n/cm^2
n (fast)	10^{19} n/cm^2
n (fast)	10^{17} n/cm^2

may crack and fail. Elastomers become too brittle to perform their function.

Wood and wood products such as Masonite are easily damaged by radiation, becoming brittle and losing strength at moderate doses [26].

Graphite

Bombardment by fast neutrons produces displacements in graphite [25]. Interstitial atoms between the layers cause anisotropic distortions, usually swelling but sometimes contraction. The amount of distortion depends on the irradiation temperature and type of graphite. Neutron irradiation also results in an increase in stored energy and hardness and a decrease in thermal conductivity. Much of the damage can be annealed out at high temperatures.

Concrete

The major effect of irradiation of concrete shields is nuclear heating, with the possibility of driving off water and thus decreasing neutron attenuation at temperatures over 93°C, or of excessive thermal stress (hence cracking) at temperature gradients over 0.8°C per cm [27]. Concrete is resistant to radiation damage. No damage has been reported for fast neutron fluences below 10^{21} n/cm^2 or absorbed doses below 10^{13} rad.

Ceramics

Ceramics such as aluminum oxide are quite resistant to radiation damage [25]. There is a temporary decrease in electrical resistance and thermal conductivity at fluences on the order of 10^{18} n/cm^2, but the material recovers after irradiation. Mechanical properties are only slightly affected at fast-neutron fluences below 10^{20} n/cm^2. Boron carbide is sensitive to thermal neutron irradiation because of the $^{10}B(n,\alpha)^{7}Li$ reaction.

Glass

The major effect of irradiation on glass is browning or darkening [28]. Unstabilized glasses are darkened even at 10^4 rad. Cerium-stabilized glasses can withstand 10^8 rad with only moderate loss of transparency.

Semiconductors

Germanium and silicon crystals are easily damaged by neutron irradiation. Figure 6.22 plots the damage function for silicon (as well as carbon and iron), based on the cross section for production of displacements [23]. Fast-neutron irradiation of either n- or p-type silicon produces a rapid decrease in conductivity at fluences of about 10^{16} n/cm^2, followed by a gradual decrease [29]. A similar decrease in conductivity is seen in irradiated p-type germanium. However, defect-acceptor states are generated in n-type germanium. The conductivity decreases, reaches a minimum, and then increases again as the n-type is converted to p-type. Ionizing radiation also produces a transient increase in conductivity because of the release of hole-electron pairs.

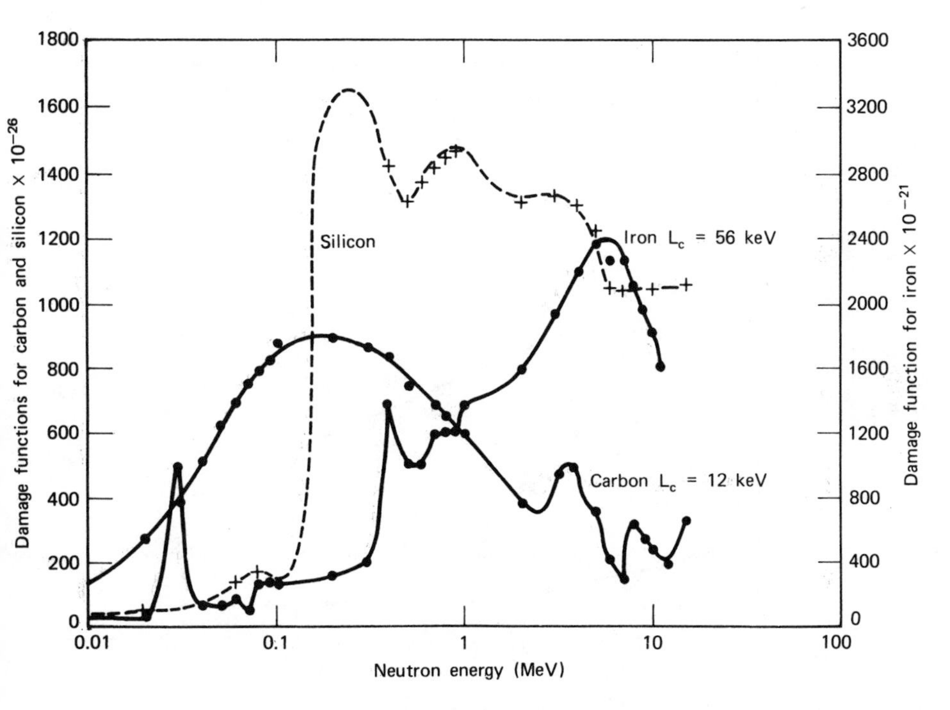

Fig. 6.22. Theoretical neutron damage functions for silicon, carbon, and iron (after Kelly [23]).

Metals

Structural metals such as steel are relatively resistant to radiation. However, displaced atoms can pin dislocations, resulting in an increase in Young's modulus, hardness, yield strength, and tensile strength, and a decrease in ductility and creep [29]. The transition temperature from ductile to brittle fracture is increased. Figure 6.23 shows the variation of these properties for a low-carbon steel [25]. Stainless steels are affected similarly. Permanent magnet alloys such as Alnico-V and soft magnetic alloys such as Permalloy are damaged by fast-neutron fluences as low as 10^{17} n/cm^2.

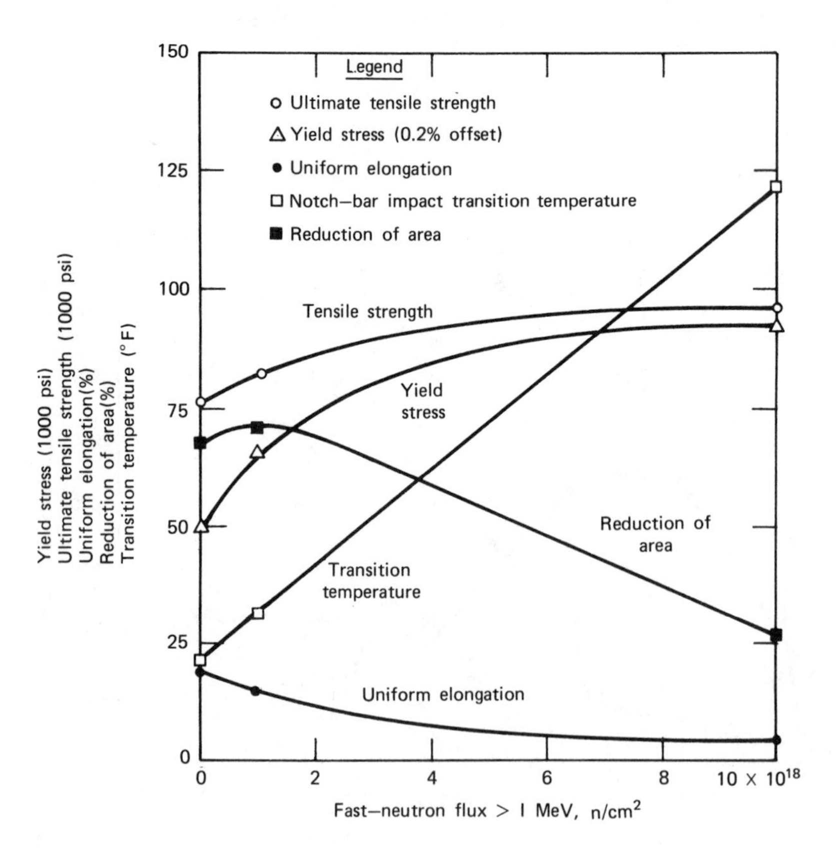

Fig. 6.23. Effects of fast-neutron irradiation on carbon steel (after Kircher and Bowman [25]).

REFERENCES

1. H. Dertinger and H. Jung, *Molecular Radiation Biology*, Springer, New York, 1970.
2. Alison P. Casarett, *Radiation Biology*, Prentice-Hall, Englewood Cliffs, N. J., 1968, Chapter 4.
3. W. Günther and H. Jung, *Z. Naturforsch.*, 22b, 313 (1967).
4. D. E. Lea, *Actions of Radiations on Living Cells*, 2nd ed., Cambridge University Press, 1955 (reprinted 1962).
5. Mitosis and meiosis are discussed in biology texts, for example J. J. W. Baker and G. E. Allen, *The Study of Biology*, Addison-Wesley, Reading, Mass., 1967.
6. Casarett, *Radiation Biology*, Chapters 5 and 6.
7. S. Wolf, in M. Errera and A. Forssberg, Eds., *Mechanisms in Radiobiology*, Academic Press, New York, 1961.
8. W. L. Russell, in F. H. Sobel, Ed., *Repair from Genetic Radiation Damage and Differential Sensitivity on Germ Cells*, Macmillan, New York, 1963.
9. "The Effects on Populations of Exposure to Low Levels of Ionizing Radiation," Report of the Advisory Committee on the Biological Effects of Ionizing Radiations (BEIR Report), National Academy of Sciences-National Research Council, Washington, D. C. 20006, 1972. Reviews data and gives responses to original papers.
10. K. N. Prasad, *Human Radiation Biology*, Harper and Row, Hagerstown, Md., 1974, Chapter 17.
11. "Alpha-Emitting Particles in Lungs," NCRP Report No. 46, National Council on Radiation Protection and Measurements, Washington, D. C., 1975.
12. Casarett, *Radiation Biology*, Chapters 5, 7-10.
13. Prasad, *Human Radiation Biology*, Chapters 5, 8-15.
14. A. C. Upton, *Radiation Injury*, Univ. Chicago Press, 1969, Chapter 1.
15. D. O. Schneider and G. F. Whitmore, "Comparative effects of neutrons and x-rays on mammalian cells," *Radiat. Res.*, 18, 286-306 (1963).
16. M. M. Elkind and G. E. Whitmore, *The Radiobiology of Cultured Mammalian Cells*, Gordon and Breach, New York, 1966.
17. E. L. Alpen, "Effects of Neutrons on Large Animals," in *Symposium on Neutrons in Radiobiology*,

CONF-691106, NTIS, Springfield, Va. (1969). Also see other papers in this report.
18. Prasad, *Human Radiation Biology*, Chapter 19.
19. "The Biological Effects of Atomic Radiation," BEAR Committee, National Academy of Sciences-National Research Council, Washington, D. C., 1956.
20. "Basic Radiation Protection Criteria," NCRP Report No. 39, National Council on Radiation Protection and Measurements, Washington, D. C., 1971.
21. Eric J. Hall, *Radiobiology and the Radiologist*, Harper and Row, Hagerstown, Md., 1973.
22. Milton Friedman, Ed., *The Biological and Clinical Basis of Radiosensitivity*, Charles C. Thomas Publishers, Springfield, Ill., 1974.
23. B. T. Kelly, *Irradiation Damage to Solids*, Pergamon Press, Oxford, 1966.
24. C. R. Tipton, Ed., *Reactor Handbook*, 2nd ed., Vol. I, *Materials*, Interscience, New York, 1960, pp. 878-881.
25. J. F. Kircher and R. E. Bowman, *Effects of Radiation on Materials and Components*, Reinhold, New York, 1964.
26. O. Sisman and W. W. Parkinson, in R. G. Jaeger, Ed., *Engineering Compendium on Radiation Shielding*, Vol. II, Springer-Verlag, New York, 1975, p. 318.
27. H. E. Hungerford, in *Engineering Compendium on Radiation Shielding*, Vol. II, pp. 78-92.
28. Ibid., p. 64.
29. D. S. Billington and J. H. Crawford, Jr., *Radiation Damage in Solids*, Princeton University Press, 1961.

PROBLEMS

1. Estimate the genetic effects in a population of 25 million exposed to an average dose rate of 0.1 mrem per year from release of radioactive wastes.

2. Discuss the trade-offs and costs involved in reducing the occupational dose limit from 5 rem/year to 0.5 rem/year.

3. Determine the dose for a surviving fraction of 10^{-5} if the extrapolation number is 2 and the 37% dose D_0 is 165 rads.

4. Explain the difference between quality factor Q and RBE, and their application in radiation protection and radiotherapy.

5. Estimate the increased risk of cancer for an adult exposed to 5 rem per year for 30 years. What is the chance of radiation sickness from such an exposure?

6. Discuss the difference between biological effects with a threshold and those without a threshold, and the implications in setting radiation protection standards.

7. Calculate the maximum permissible concentration of iodine-131 in air, for a person living near a nuclear power plant (neglect other radionuclides).

7

ENGINEERING

7.1 SHIELD DESIGN

The steps involved in solving radiation shielding and dosimetry problems are outlined in Chapter 1. Here we are concerned with engineering design of economical and effective shields. Table 7.1 summarizes the usual quantities given, the quantities to be calculated or measured, and the items to be specified by the shield designer.

It is assumed that the source parameters are given, along with the maximum dose equivalent at specified locations outside the shield or other design criteria. A practical shield has to allow for penetrations, such as cooling pipes, electrical power and signal cables, rotating shafts, removable plugs or covers, and the like. These represent potential weak spots and special arrangements must be made to assure shield integrity.

The shield may be called on to perform as a structural member and must withstand some mechanical stress from its own weight. In addition, heat generated by absorption of radiation or transmitted from the source or cooling pipes can induce thermal stresses. Heating may also be limited by dehydration (as in concrete) or materials damage. A "thermal" shield (designed to absorb most of the nuclear heating where it can be handled easily) and insulation may be required to protect the biological shield. Shield materials may also have to be protected from corrosion, chemical reactions, and radiation damage for endurance throughout the service life.

The shield is normally optimized on cost, subject to constraints on mass or thickness. Shields for

TABLE 7.1. ENGINEERING DESIGN OF RADIATION SHIELDS

Given

Source strength, spectrum, angular distribution, spatial distribution, and time dependence for each type of radiation

Maximum allowable dose, heating, activation, or damage at specified locations

Requirements for penetrations, including cooling pipes, electrical cables, mechanical drives, access ports, and so on

Structural requirements

Maximum allowable mass, thickness, or cost, and special constraints such as materials capability

Service life

Calculate or measure

Attenuation in materials

Streaming through penetrations

Heating, activation, and radiation damage in shield

Dose, heating, activation, or materials damage outside shield

Iterate until design criteria are met

Specify

Materials, thicknesses, and geometry

Locations and design of penetrations

Shield cooling and thermal insulation

Reinforcing and cladding

Shield erection, inspection, and test procedures

transport of radioactive materials or for mobile radiation-producing machines should not be too heavy, and low mass is essential in space applications. A thin shield may be desirable to reduce the length of penetrations, the size and cost of equipment and building, or to maximize the flux density in a beam at the outside of the shield.

As discussed in Chapter 3, it is easy to shield ion and electron sources. Almost any material, thicker than the range, may be used to stop all ions. Electrons are easily stopped by a few millimeters of any solid or liquid, but a low-Z material is preferred to minimize generation of penetrating bremsstrahlung. Any bremsstrahlung generated can be attenuated by lead or iron, or other γ-ray shielding material.

Shielding X- and γ-ray sources is relatively straightforward. The common shielding materials are lead and iron, but tungsten or depleted uranium may be used (at higher cost) if a minimum thickness shield is required. Concrete and even water have been used where thickness and mass are not of concern. Properties of shielding materials and methods of fabrication are discussed in Section 7.2.

Thermal neutrons are easily absorbed in a few millimeters of high cross-section material such as boron. The boron may be in the form of boron carbide or a soluble boron compound. Boron carbide dispersed in aluminum, and clad in aluminum, is commercially available as Boral. One or two millimeters of cadmium also absorbs nearly all thermal neutrons, but cadmium has the disadvantage of a low melting point and emits penetrating capture γ rays.

Fast-neutron shielding is more difficult. Fast neutrons are often accompanied by γ rays, either from the source or generated by neutron capture and inelastic scattering in the shield itself. Thus the shield has to include elements suitable for attenuating γ rays as well as neutrons. Inelastic scattering (e.g., in iron) helps to moderate the neutrons, but the best neutron-attenuating element is hydrogen. Common neutron-shielding materials such as water, polyethylene, and concrete contain a fairly high concentration of hydrogen. Lithium hydride, though very expensive, has been used in space where low mass is critical. Shields for high-temperature reactors cooled by liquid sodium have been made of stainless steel, graphite, and a high-temperature material containing hydrogen such as serpentine (compact asbestos ore).

The neutron- and γ-ray attenuation elements may be mixed for a homogeneous shield. For example, iron ore and iron scrap may be used as aggregate for heavy or high-density concrete, with better γ-ray attenuation than ordinary concrete. However, thinner or less massive shields can be designed using layers of a dense γ-ray attenuator and a less dense neutron attenuator. Examples are lead-polyethylene, iron-water, iron-concrete, and tungsten (or depleted uranium), followed by lithium hydride. The mass is reduced by locating the dense material close to the source. On the other hand, secondary γ-ray production in the dense material can be reduced by placing some neutron-attenuating material between it and the source. Another consideration is that extra material is needed to attenuate any γ rays generated in the neutron shield, although thermal-neutron capture γ rays can be reduced by adding boron. Once the materials are selected, an iterative process of calculating fast-neutron dose, primary γ-ray dose, and secondary γ-ray dose is required to optimize the lamination thicknesses, for minimum mass or overall thickness.

It is helpful to have the results of previous attenuation calculations and measurements, in homogeneous and laminated shields. Selected results are given in Section 7.3. The design of penetrations is treated in Section 7.4. Heating and stresses are discussed in Section 7.5. Test procedures are given in Section 7.6. Examples of shield designs are given in Section 7.7.

7.2 MATERIALS

Lead

Properties of lead [1,2] are summarized in Table 7.2. The main advantages of lead as a γ-ray shielding material are its high density and atomic number. Disadvantages are the relatively high cost compared to iron or steel, the low melting point, softness, low strength and tendency to creep so that thick or large sections have to be reinforced or bonded to steel supports. Lead shields are lighter than steel or concrete shields of equal γ-ray attenuation and are more economic than steel, if the lead is wrapped closely around the source. Lead is used for shielding radionuclide sources, X-ray machines and rooms, and detectors, and is frequently used in γ-ray collimators.

TABLE 7.2. PROPERTIES OF LEAD

Property	Value
Density (g cm^{-3})	11.35
Atomic number	82
Atomic weight	207.19
Melting point (°C)	327
Specific heat (J g^{-1} $°C^{-1}$)	0.13
Thermal conductivity (J cm^{-1} s^{-1} $°C^{-1}$)	0.339
Coefficient of thermal expansion ($°C^{-1}$ ℓin)	2.95×10^{-6}
Hardness (Brinell)	4
Yield strength (psi)	860
Tensile strength (psi)	1900
Creep (psi, 1% in 10 years at 100°C)	127
Elongation (%)	45
Young's modulus (psi)	2×10^{6}
Radiation damage	nil
Corrosion resistance	good
Thermal-neutron activation	impurities
Approximate cost ($ per kg)	2

Lead is also used in shields of ship and submarine reactors, along with polyethylene or water for neutron attenuation. Methods of installation are illustrated in Figure 7.1.

Cooling coils may be needed to remove nuclear heat because of the low melting point. Lead has good resistance to radiation damage and corrosion and does not activate appreciably if pure. Virgin lead should be used because scrap may contain impurities.

Lead may be cast, extruded, rolled, machined, and "welded" or "burned." It is available in extruded bricks and as rolled sheets or plates, as well as shot and pigs. The main problem in fabrication is to avoid cracks or voids when joining sheets, or in casting. Lead shrinks about 4% in volume on solidification. Arrangements must be made to accommodate the shrinkage and to allow gases to escape. The container should be preheated, the casting completed in one continuous pour if possible, and the temperature

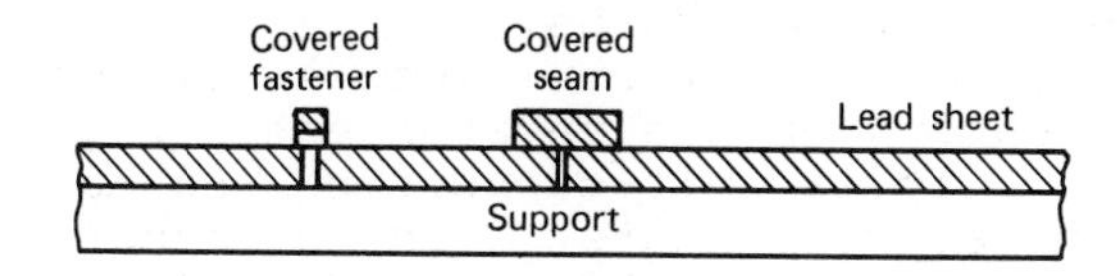

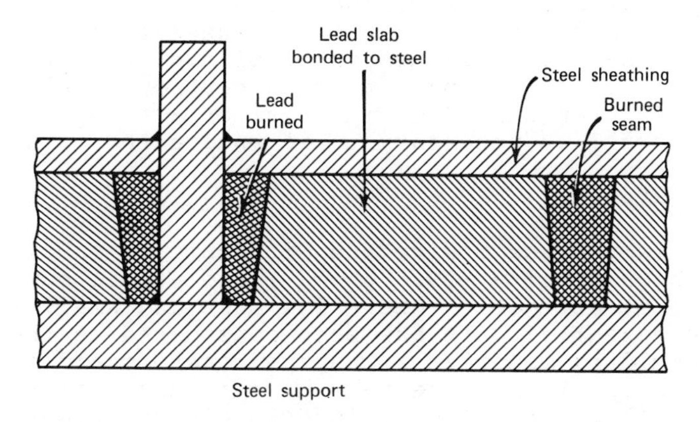

Fig. 7.1. Installation of lead shielding.

controlled to prevent overheating. Slabs and castings should be inspected by γ-ray radiography or scanning with a scintillation counter to detect any hot spots.

Iron

Properties of iron [1,2] are summarized in Table 7.3. The properties of low-carbon steel are similar. The density and atomic number of iron are relatively low, but iron and steel are strong and can be exposed to relatively high temperatures if protected from oxidation. Steel is also relatively economic, is resistant to radiation damage, and activation is moderate unless there are alloying elements or impurities such as manganese, nickel, chromium, or cobalt. Steel is used in pressure vessels, thermal shields, and some reactor shields in combination with water or concrete for neutron attenuation. Thermal shields have to be cooled. Iron ores (magnetite, hematite, limonite) and iron punchings or shot are mixed with Portland cement to make heavy concretes, with improved γ-ray attenuation over ordinary concretes.

Iron may be clad in stainless steel, plated, or painted (depending on the environment) to protect it from rusting. It is self-supporting and may be used to support other components. Iron or steel may be rolled, forged, cast, machined, and welded. It is usually applied as rolled plates or forged and machined shapes in shielding. Cast iron is seldom used. Castings and the seams of welded plates should be inspected by γ-ray radiography or scanned with a scintillation counter to detect any hot spots because of flaws.

Iron is a significant source of capture γ rays and inelastic scattering γ rays in neutron shields. Additional γ-ray shielding material or a thicker shield overall is necessary to attenuate the γ rays.

Tungsten

Tungsten is very expensive and difficult to fabricate, but its high density (19.3 g cm^{-3}) and atomic number (74) make it attractive where low mass and small thickness are essential. It has been considered as the γ-ray shielding material for space reactors and is used in collimators. The very high melting point (3410°C) allows it to be applied in a high-temperature environment but also makes it almost impossible to

TABLE 7.3. PROPERTIES OF IRON

Property	Value
Density ($g\ cm^{-3}$)	7.87
Atomic number	26
Atomic weight	55.85
Melting Point (°C)	1535
Specific heat ($J\ g^{-1}\ °C^{-1}$)	0.46
Thermal conductivity ($J\ cm^{-1}\ s^{-1}\ °C^{-1}$)	0.67
Coefficient thermal expansion ($°C^{-1}$)	11×10^{-6}
Hardness (Brinell)	100
Yield strength (psi)	25,000
Tensile strength (psi)	40,000
Creep (psi)	15,000
Elongation (%)	25
Young's modulus (psi)	30×10^{6}
Radiation damage	small
Corrosion resistance	fair
Thermal neutron activation	moderate
Approximate cost ($ per kg)	0.2

cast; hence powder-metallurgy techniques are used to fabricate tungsten shapes. The shapes are strong but brittle. Tungsten is resistant to radiation damage and corrosion, but thermal neutron activation is high, and it is a prolific source of capture and inelastic scattering γ rays.

Depleted Uranium

Uranium depleted in the ^{235}U isotope is available and not too expensive. Its advantages as a γ-ray shielding material are high density (18.9 g cm^{-3}) and atomic number (92). It is strong, can be fabricated like steel, and has a relatively high melting point (1132°C). It should be protected from corrosion and oxidation in air. It is mildly radioactive and hence not suitable for detector shields, but it has been used in shields for transportation of radioactive materials. Radiation damage and heating, as well as activation and production of secondary γ rays from capture and fast-neutron fission of the ^{238}U (or fission of the residual 0.2% of ^{235}U), make it less useful in shielding neutron sources.

Concrete

Properties of ordinary and heavy concretes [1] are listed in Table 7.4. The relatively low density and atomic number of ordinary concrete are drawbacks for γ-ray shielding. Thinner shields can be achieved with dense aggregates such as magnetite or limonite iron ore, barytes (barium ore), and iron punchings. However, dense-aggregate concrete is much more expensive, both in cost of materials and in the installation. The costs listed in Table 7.4 include forms and reinforcing.

Neutron attenuation is governed mainly by the hydrogen concentration, which should be at least 0.5 weight percent or 0.8×10^{22} atoms per cm^3. The corresponding water content is about 5 weight percent. Higher concentrations can be achieved, but there may be some loss in strength. A very important consideration is rapid loss of water, hence hydrogen, at temperatures above the 93°C heat-generation rate of 0.8 mW cm^{-3}, or temperature rise of 6°C mentioned in Table 7.4. Even at lower temperatures, there is a gradual loss of water, and the shield thickness should be increased by about 12% to allow for this.

TABLE 7.4. PROPERTIES OF ORDINARY AND HEAVY CONCRETES

Property	Ordinary	Heavy
Density (g cm^{-3})	2.2 to 2.4	3.7 to 4.8
Effective atomic number	11	∿26
Hydrogen concentration (atoms cm^{-3})	0.8 to 2.4×10^{22}	
Specific heat (J g^{-1} $°C^{-1}$)	0.8	
Thermal conductivity (J cm^{-1} s^{-1} $°C^{-1}$)	0.01	0.1
Coefficient thermal expansion ($°C^{-1}$)	8×10^{-6}	
Compressive strength (psi)	3500 to 6000	
Tensile strength (psi)	<500	
Elastic modulus (psi)	2×10^{5}	
Maximum temperature (°C)	93.0	
Maximum heating rate (mW cm^{-3})	1.0	
Maximum temperature rise (°C)	6.0	
Maximum temperature gradient (°C cm^{-1})	0.8	
Radiation damage	negligible	
Thermal-neutron activation	small	large
Approximate cost ($ per m^3)	200	1200
($ per kg)	0.09	0.52

Ordinary concrete is made from a mixture of Portland cement, coarse aggregate (crushed rock or gravel 2 to 5 cm across), sand, and water. The standard proportions are 1 part cement, 2 parts sand, and 4 parts coarse aggregate by volume, with 5.5 to 6 gal of water added per 94-lb bag of cement (0.53 liters per kg cement). Local aggregate is used, and the composition and density are variable. A typical ordinary concrete will have a density of 2.3 g cm^{-3} and a composition in weight percent of: oxygen (52.9), silicon (33.7), calcium (4.4), aluminum (3.4), sodium (1.6), iron (1.4), potassium (1.3), hydrogen (1.0), magnesium (0.2), and carbon (0.1). Heavy concrete made of iron coarse aggregate and limonite sand will have a typical density of 4.2 g cm^{-3} and composition (weight percent) of: iron (72.1), oxygen (18.0), calcium (6.1), manganese (1.6), silicon (1.4), aluminum (0.5), magnesium (0.2), sulfur (0.1), and hydrogen (0.05). Boron may be added to suppress production of capture γ rays in the form of boron carbide, boron frits or glass, soluble boron compounds in the water or cement, or a boron-containing aggregate such as colemanite. About 1% boron is adequate; too much boron retards setting.

The concrete is usually mixed and poured into forms, with the reinforcing bars and any penetrations already in place. Permanent steel forms (oiled) or temporary plywood forms with proper shoring may be used. It is desirable that the concrete be cast in one continuous pour. If part must be poured and allowed to set, the joint should be rebated to prevent streaming, as illustrated in Figure 7.2. Care must be taken to fill the entire volume, including spaces under tubes and the like, because voids or cracks will reduce attenuation. The concrete mix should be sufficiently fluid to flow into constricted spaces, but separation of water or segregation of aggregate must be avoided. Heavy concrete has to be mixed in small batches to avoid overloading equipment. Another method, especially good for heavy concrete, is to prepack the coarse aggregate in the forms and then force grout (a thin mortar of cement, water, and sand) into the interstices.

The strength of concrete increases as it cures, reaching about 80% of the final strength in 28 days. For maximum strength and water retention, concrete should be moist cured for at least 7 days, and preferably 28 days. This requires keeping the surfaces wet,

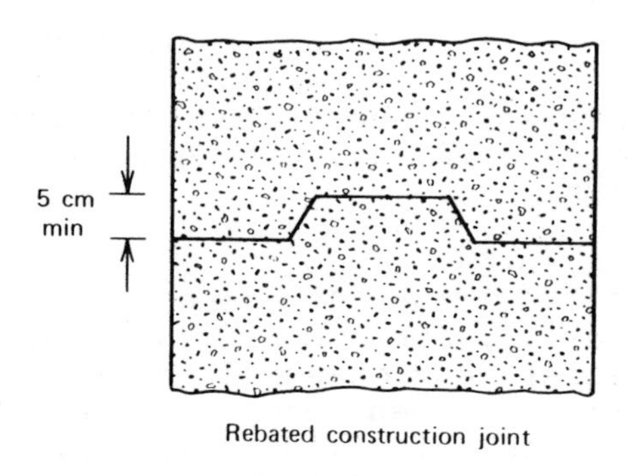

Rebated construction joint

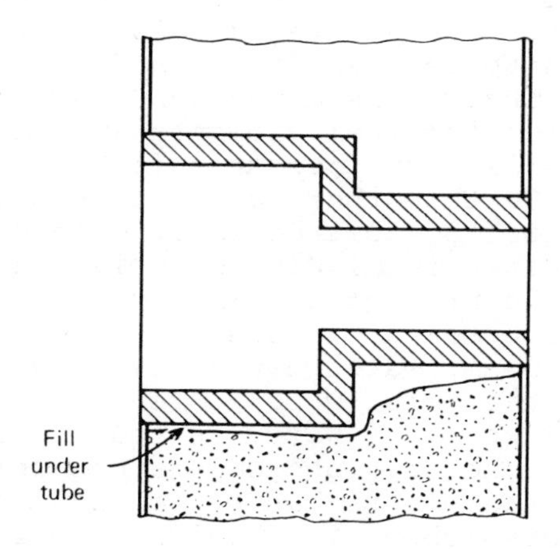

Fig. 7.2. Installation of concrete shielding.

either flooded or covered with damp burlap or sawdust. Concrete must be protected from freezing until it sets.

Heating and thermal stresses are discussed in Section 7.5. Thermal stresses may cause cracking, which can affect shield integrity. The limit of 0.8°C per cm for the temperature gradient is conservative and should prevent damage from thermal stresses. A separate thermal shield or cooling pipes near the source side of the shield may be needed to keep the temperatures and temperature gradient within the limits. The heat-generation rate in ordinary concrete will be within the limit if the incident-energy flux density of neutrons plus γ rays is less than 4×10^{10} MeV cm^{-2} s^{-1}.

Allowance has to be made for thermal expansion in large shields. The concrete is cast in sections with joints. The joints should be filled with lead or other good shielding material, with enough plasticity to permit expansion.

Water

Properties of water [1] are summarized in Table 7.5. Water is an excellent neutron-shielding material because of its high hydrogen concentration. Because of the low density and effective atomic number, water shields have to be thick to attenuate γ rays. However, the number of electrons per gram is relatively high, and water is an effective γ-ray shield on a mass basis.

Water is transparent and fluid and a good heat-transfer agent. Thus it is often used for storage and manipulation of reactor fuel elements and radionuclide sources. The fluidity ensures that it will fill any voids and gaps. On the other hand, water must be contained, and most of the cost of water shielding is in the container. Water is mildly corrosive, and the container should be protected. Shielding would be lost if a leak occurred or if the water evaporated. To avoid this problem, water can be mixed with a polyester resin that gels (water-extended polyester).

The major drawback to water is the low boiling point, precluding use at high temperature unless it is pressurized. Radiolysis has to be considered but is usually a minor problem in shielding applications. Pure water does not activate, but impurities may, and deionized water is preferred for shielding. The thermal-neutron cross section for the $^{1}H(n,\gamma)^{2}H$ reaction is 0.33 barn per hydrogen nucleus, and a 2.23-MeV γ ray is emitted in 100% of the captures. Borax or boric acid may be dissolved in the water to reduce capture-γ-ray production. Boron emits a 0.478-MeV γ ray but is less penetrating than the hydrogen capture γ ray.

Polyethylene

Properties of polyethylene $(CH_2)_n$ are listed [1,2] in Table 7.6. Polyethylene is the most popular of the organic neutron-shielding materials and has a higher hydrogen concentration than water. Its major drawback

TABLE 7.5. PROPERTIES OF WATER

Property	Value
Density (g cm^{-3}, 20°C)	1
Effective atomic number	7.4
Electron number (per gram)	3.34×10^{23}
Hydrogen concentration (atoms cm^{-3})	6.7×10^{22}
Melting point (°C)	0
Boiling point (°)	100
Specific heat (J g^{-1} °C)	4.18
Thermal conductivity (J cm^{-1} s^{-1} $°C^{-1}$)	5.9
Volume coefficient of thermal expansion ($°C^{-1}$)	4.1×10^{-4}
Radiation damage	small
Thermal-neutron activation	impurities
Approximate cost	container

is limited resistance to heat and radiation. Polyethylene softens and creeps at low temperatures and should be supported or contained. It will support combustion. Its resistance to radiation damage is poor, as discussed in Section 6.8. Polyethylene has a rather large coefficient of thermal expansion and a small thermal conductivity. Strength is moderate. Polyethylene is available as sheets and slabs, bars, and pellets. It is a thermoplastic and can be extruded or molded, but it is usually applied as slabs with staggered joints, or machined shapes, in neutron shields. Boron compounds may be added to suppress capture γ-ray production.

TABLE 7.6. PROPERTIES OF POLYETHYLENE

Property	Value
Density (g cm^{-3}, 20°C)	0.95
Effective atomic number	5.5
Hydrogen concentration (atoms cm^{-3})	8.0×10^{22}
Softening temperature (°C)	40
Maximum temperature (°C)	100
Specific heat (J g^{-1})	2.3
Thermal conductivity (J cm^{-1} s^{-1} °C^{-1})	3×10^{-3}
Linear coefficient of thermal expansion (°C^{-1})	1.2×10^{-4}
Compressive strength (psi)	3200
Tensile strength (psi)	2800
Flammability	moderate
Radiation damage	significant
Thermal-neutron activation	nil
Approximate cost ($ per kg)	2

Lithium Hydride

Although rather expensive (more than $30 per kg), lithium hydride (LiH) is used where low mass and fairly good temperature and radiation damage resistance are required, as in shields for space reactors. The density is only 0.78 g cm^{-3}, and the hydrogen concentration of 5.9×10^{22} atoms per cm^3 is less than that of water, but there are 7.6×10^{22} atoms H per gram of lithium hydride, compared to 6.7×10^{22} atoms per gram of water. The melting point is 686°C, but LiH dissociates with evolution of hydrogen at somewhat lower temperatures. Strength is moderate, but LiH is brittle. It reacts with water and must be clad for protection and support. Radiation damage resistance is good, and neutron activation and secondary γ-ray production are negligible.

Graphite

Although not as good a fast neutron shield as hydrogenous materials, graphite has excellent high-temperature properties and is often used as a moderator, reflector, or shield component in fission and fusion devices. Density is somewhat variable but averages 1.65 g cm^{-3}. The sublimitation temperature is 3650°C at 1 atm. Specific heat is 0.71 J g^{-1} $°C^{-1}$. The linear coefficient of thermal expansion is 3.8×10^{-1} $°C^{-1}$. Thermal conductivity is 1.2 J cm^{-1} s^{-1} $°C^{-1}$. Graphite is brittle and the tensile strength is only 1400 psi at room temperature, but the compressive strength is 6000 psi. Strength increases with temperature. Graphite is moderately resistant to radiation damage, and stored energy and damage can be annealed out as discussed in Section 6.8. Thermal-neutron activation and capture-γ-ray production are nil in pure graphite, and graphite can be boronated to attenuate thermal neutrons. Graphite is available as bars, usually 10 cm across, at about $2 per kilogram.

Other Materials

Air is not a material of shield construction, but does contribute to the attenuation of radiation. Density at 20°C and 1 atm pressure (sea level) is 1.29×10^{-3} g cm^{-3}, decreasing with increasing temperature and altitude. Composition is 75.5 weight percent N, 23.2 weight percent O, and 0.013 weight percent Ar, with traces of water and carbon dioxide.

Soil contributes to backscatter and attenuation of radiation from radionuclides deposited on or in the ground and is sometimes used as an inexpensive shield. Density and composition are variable.

7.3 ATTENUATION RESULTS

The dose at radius r from a point, isotropic, monoenergetic source of energy E_0 in vacuum is

$$D_0(E_0,r) = \frac{K(E_0)S(E_0)t}{4\pi r^2} \tag{7.1}$$

where S is the source strength, t the duration of irradiation, and K the fluence:dose factor (for exposure, kerma, etc.). If the point source is imbedded in an infinite medium (Figure 7.3a), the dose may be expressed as

$$D(E_0,r) = D_0(E_0,r)\ T_\infty(E_0,r) \tag{7.2}$$

where T_∞ is the infinite-medium dose-attenuation factor or dose-transmission factor. This factor takes into account the absorption and scattering in the attenuating medium. It may be measured or calculated with an exact transport code such as discrete ordinates, moments, or Monte Carlo, as discussed in Chapter 4. For a spherical shell (Figure 7.3b), the dose is given to a good approximation by

$$D(E_0,r,x) = D_0(E_0,r)\ T(E_0,x) \tag{7.3}$$

where the shell thickness $x = (r_0 - r_i)$. This expression will slightly overestimate the dose because T is evaluated for an infinite medium, and near the boundary of a finite shield the fluence and dose are depressed because there is no material beyond r_0 to reflect or backscatter radiation.

For a parallel beam, perpendicularly incident on a half-space (Figure 7.3c), we have

$$D_0(E_0,x) = K(E_0)\ \phi(E_0)t \tag{7.4}$$

where ϕ is the flux density in the beam and

$$D(E_0,x) = D_0(E_0,x)\ T_s(E_0,x) \tag{7.5}$$

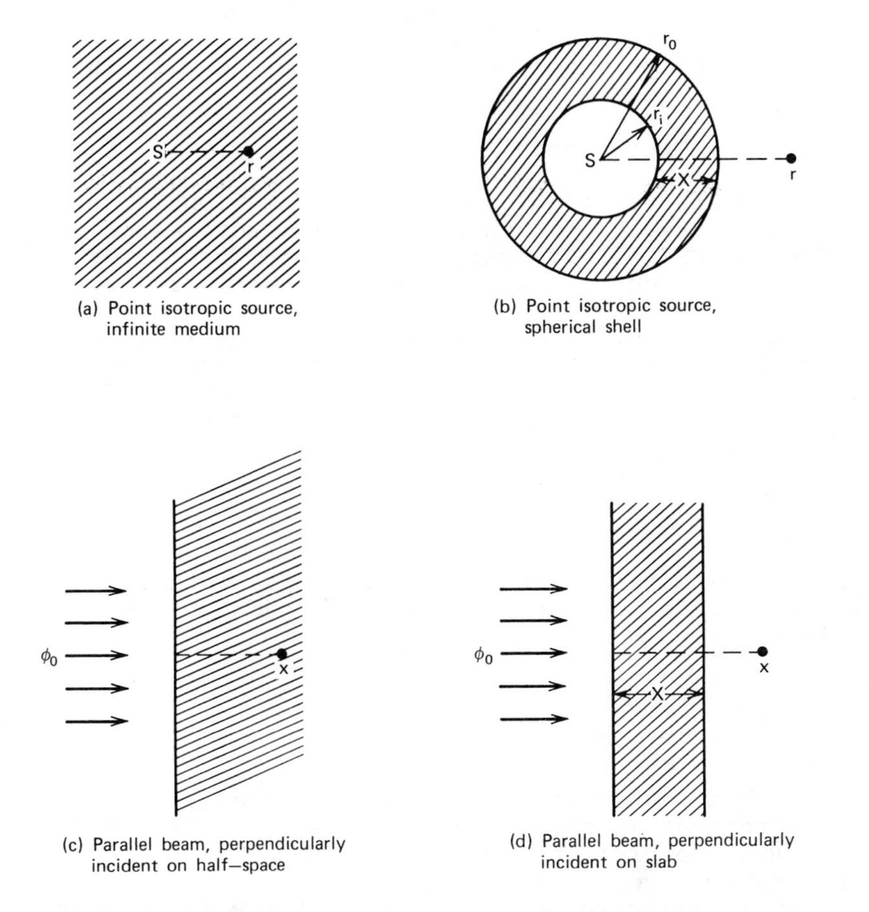

Fig. 7.3. Geometries for definition of dose-transmission factors.

where T_s is the dose attenuation or transmission factor calculated or measured in this geometry. If the parallel beam is incident on a slab of finite thickness X, as shown in Figure 7.3d, we obtain, to a good approximation,

$$D(E_0, x, X) = D_0(E_0, x)\ T_s(E_0, X) \tag{7.6}$$

Again this is a slight overestimate because backscattering from material beyond X does not occur in the finite slab but does contribute to the dose in the half-space.

γ-Ray Attenuation

Some results for the (exposure) attenuation factor for point, isotropic, monoenergetic γ-ray sources in infinite lead, iron, ordinary concrete, and water are plotted in Figures 7.4 through 7.7. Data are from Honig [3].

Results for other geometries and materials can be constructed from the linear attenuation coefficient μ_0 (evaluated at E_0) and buildup factor discussed in Chapters 3 and 4. For a point isotropic source,

$$T_\infty(E_0, r) = B_p(E_0, r)\, e^{-\mu_0 r} \qquad (7.7)$$

where B_p applies to the point isotropic source in the infinite medium. For the parallel beam incident on a half-space,

$$T_s(E_0, x) = B_s(E_0, x)\, e^{-\mu_0 x} \qquad (7.8)$$

If the beam makes an angle θ to the normal, Eq. (7.8) is approximately correct if $x = X \sec\theta$. If the source is not monoenergetic, the dose is found by summation or integration over source energy.

Data may also be presented in terms of the dose per unit source output, integrated over a certain source spectrum, as shown in Figure 7.8 for a constant potential X-ray tube with tungsten target. The exposure (R) is given per milliampere-minute of target current multiplied by time, at 1 m distance. It may be scaled by multiplying by the target current (mA), duration of irradiation (min), and by $1/r^2$ (m^{-2}). The exposure is given as a function of the thickness (cm) of a lead shell for various potentials (hence maximum energy of the bremsstrahlung spectrum) and 2-mm aluminum beam filtration.

The spectrum of γ rays from a unit (1 photon s^{-1}), point isotropic source of 0.662-MeV γ rays (^{137}Cs) at

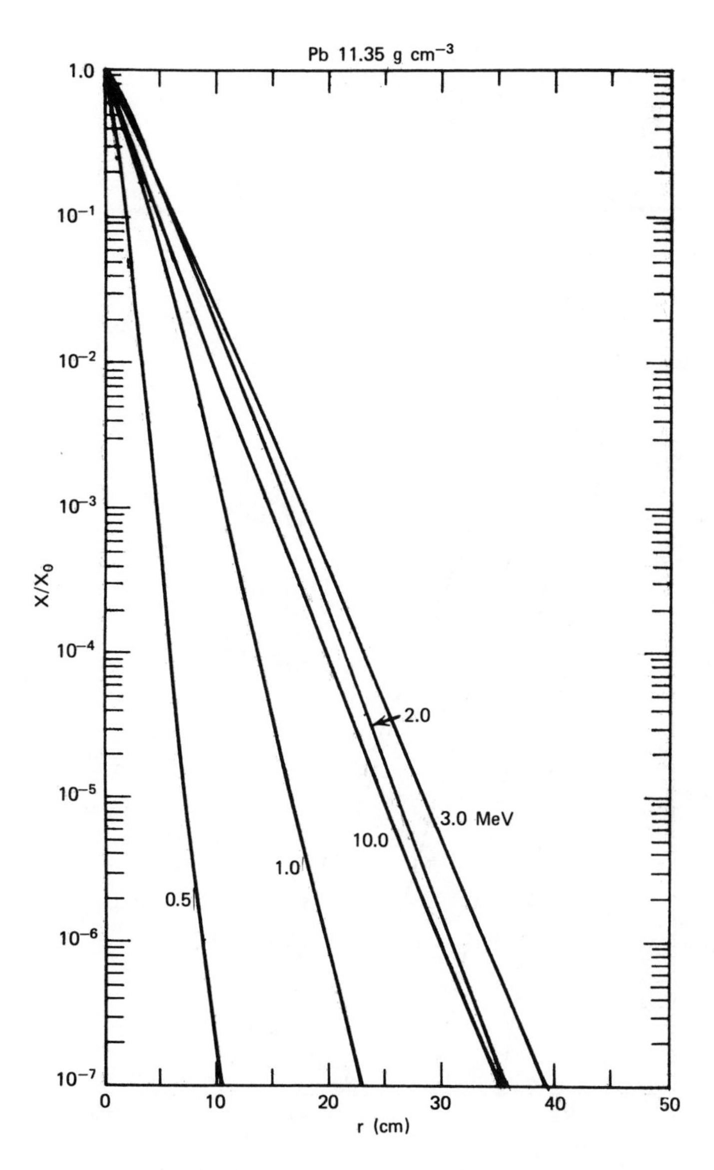

Fig. 7.4. γ-Ray exposure attenuation, point isotropic source in lead, 11.34 g cm^{-3}.

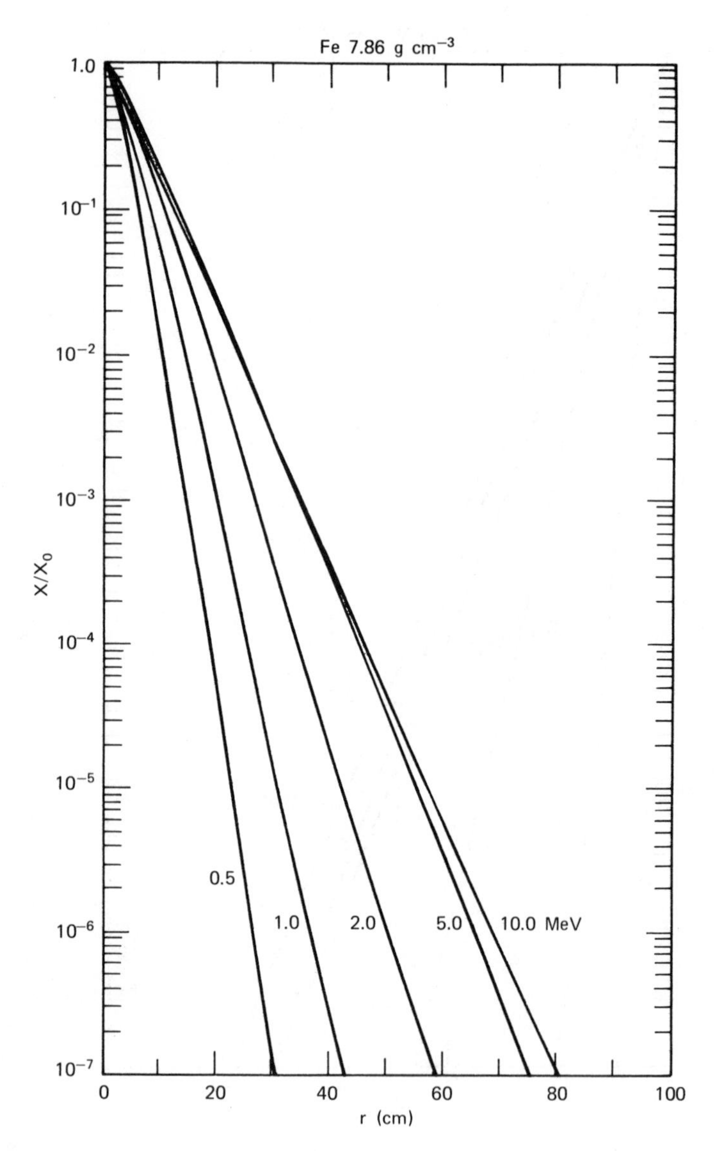

Fig. 7.5. γ-Ray exposure attenuation, point isotropic source in iron, 7.86 g/cm^3.

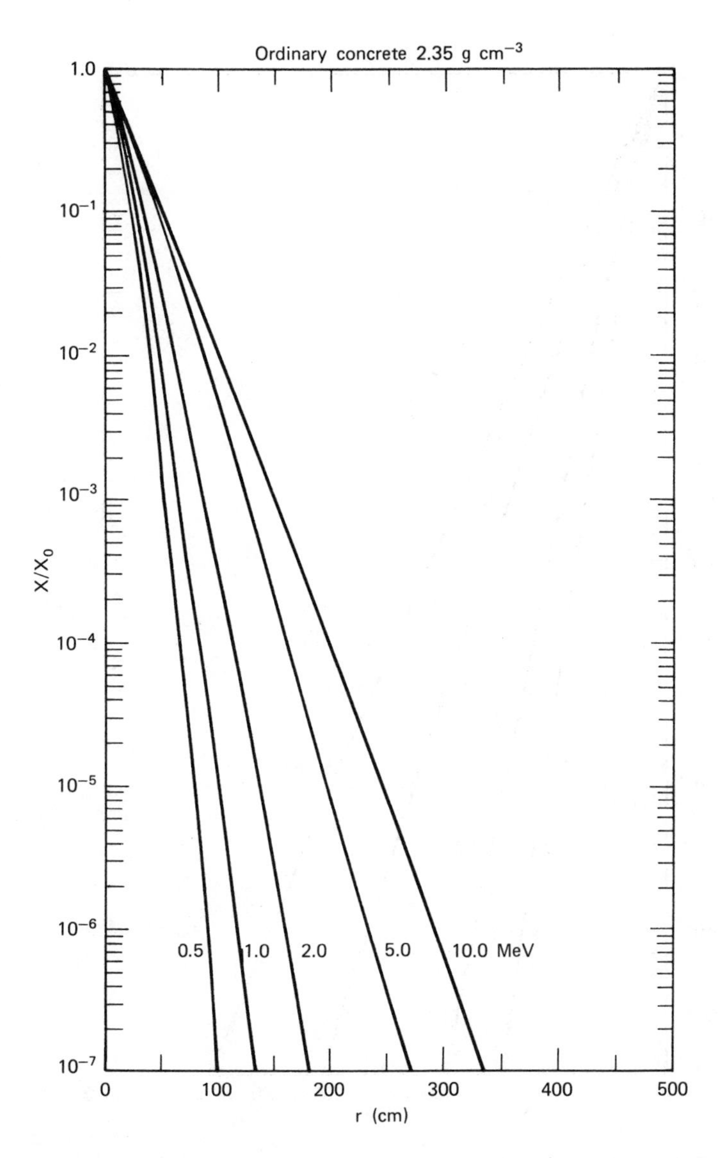

Fig. 7.6. γ-Ray exposure attenuation, point isotropic source in ordinary concrete, 2.35 g/cm^3.

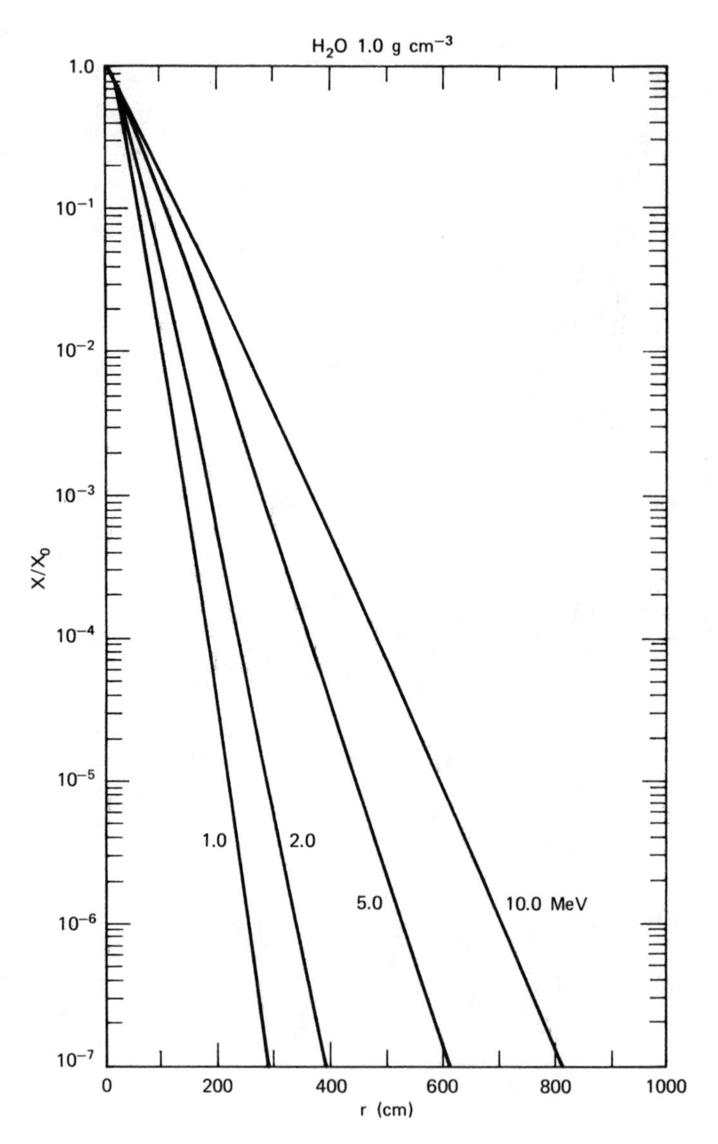

Fig. 7.7. γ-Ray exposure attenuation, point isotropic source in water, 1 g/cm³.

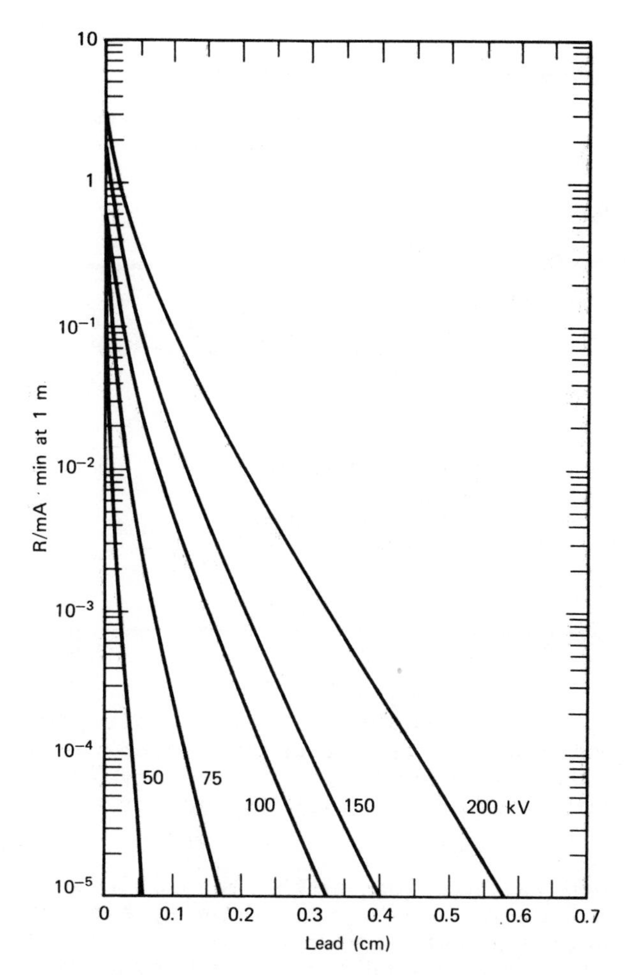

Fig. 7.8. Transmission of exposure from constant-potential X-ray tube (2-mm Al filtration) through lead, per mA-min, at 1 m (after Ref. 4).

3 mean free paths in water [5] is shown in Figure 7.9. The quantity plotted is the energy flux density, $E\phi$ (MeV cm^{-1} s^{-1} per MeV) times $4\pi r^2$ exp $(+\mu_0 r)$, where $\mu_0 r = 3$ and $r = 35.1$ cm. The spectrum measured with a NaI scintillation counter is compared with a multigroup spherical-harmonics transport calculation and a

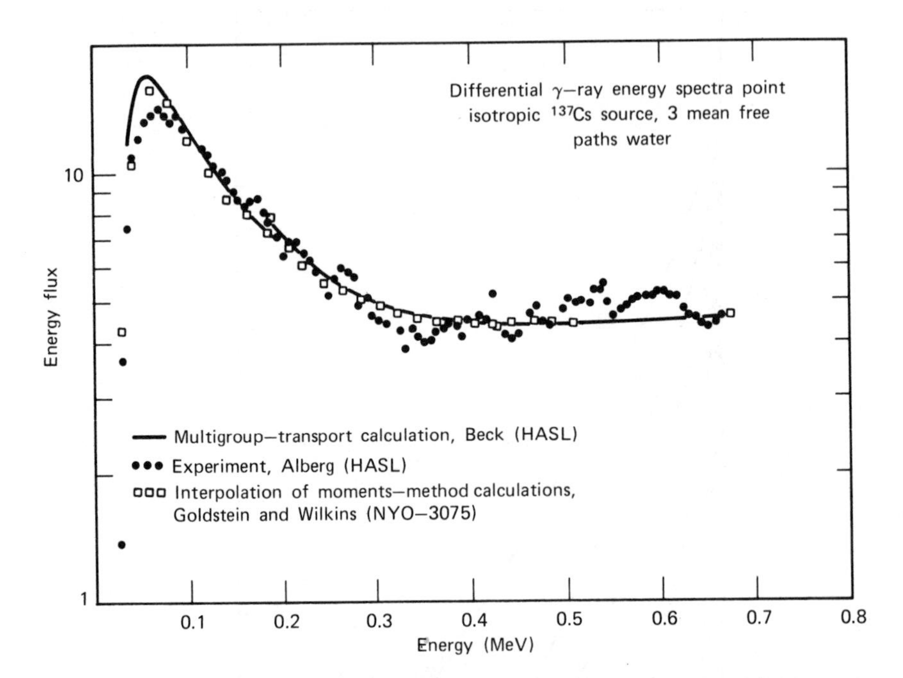

Fig. 7.9. γ-Ray spectrum for a point isotropic source in water. The source emits 1 photon (0.662 MeV) per second. The scattered energy flux density $4\pi r^2 \exp(\mu_0 r) E\phi$ is plotted at $r = 35.1$ cm and $\mu_0 r = 3$.

moments-method calculation, with reasonable agreement. Measurements of the angular flux-density spectrum have also been made, and compared with Monte Carlo calculations [6]. This is an example of a "benchmark" problem, in which exact calculations are compared with experiment or where calculations by at least two different methods are compared for a well-documented situation. The purpose is to provide a standard for testing computational techniques or cross sections.

Another benchmark problem [7] involves the kerma and flux density in air, 3 ft above an infinite plane air/ground interface, uniformly contaminated with ^{60}Co. The γ-ray source is an infinite plane emitting, isotropically, one photon per second per cm^2, with average energy 1.25 MeV. Atomic compositions are for air, 4.19×10^{19} atoms cm^{-3} of nitrogen, 1.13×10^{19} atoms cm^{-3} of oxygen, and 2.53×10^{17} atoms cm^{-3} of argon. The number of atoms cm^{-3} in the soil are: hydrogen (8.55×10^{21}), oxygen (2.27×10^{22}), aluminum (2.01×10^{21}), and silicon (9.53×10^{21}). The scattered flux density and rates are given as a function of energy group in Table 7.7, calculated by the L05 Monte Carlo program. The uncollided component of the kerma rate at 3 ft is 1.185×10^{-7} erg g^{-1} s^{-1}; hence the total kerma rate is 1.43×10^{-7} erg g^{-1} s^{-1}.

The spectrum for a point isotropic source of 1.0-MeV γ rays in infinite lead is shown in Figure 4.19. Spectra for other sources, materials, and geometries may be found in the literature [8,9].

Neutron Attenuation

There have been relatively few calculations or measurements made with monoenergetic neutron sources, except for 14-MeV neutrons from the $T(d,n)^4He$ reaction. Most results are for fission-spectrum sources.

Figure 7.10 displays the results of Monte Carlo calculations [10] of the fast-neutron dose transmission from monoenergetic, parallel beams, incident at various angles to the normal of slabs of polyethylene. They may also be used for water, ordinary concrete, wet soil, or dry soil, using the approximate equivalent thicknesses given in Figure 7.10.

Neutron attenuation in concrete has been reviewed by Schmidt [11], who also performed discrete-ordinates calculations. Figure 7.11 plots the dose equivalent due to neutrons for monoenergetic parallel beams normally incident on slabs of ordinary concrete (density 2.3 g cm^{-3}), with atomic composition (10^{21} atoms cm^{-3}) of hydrogen (8.5), oxygen (35.5), carbon (20.2), magnesium (1.86), aluminum (1.70), calcium (11.1), iron (0.193) and sodium plus potassium (0.057). The doses from fission and 14-MeV neutron source are of special interest. Figure 7.12 shows the calculated dose equivalent from neutrons and the total dose equivalent, including capture γ rays, for a fission-spectrum and the source energy group averaging 14-MeV, as a function of concrete slab thickness.

Figure 7.13 plots the absorbed dose [12] from a point, isotropic source of fission-spectrum neutrons, as a function of radius in normal or ordinary concrete, a heavy concrete, water, boronated polyethylene, and

TABLE 7.7. SCATTERED PHOTON FLUX DENSITY AND KERMA RATE 3 FT ABOVE AN INFINITE ^{60}CO-CONTAMINATED AIR/GROUND PLANE

Energy Interval (MeV)	Flux Density (photons cm^{-1} s^{-1})	Kerma Rate (erg g^{-1} s^{-1})
0.02 to 0.03	1.47(-3)[a]	1.56(-11)
0.03 to 0.04	1.55(-2)	8.18(-11)
0.04 to 0.06	9.32(-2)	2.85(-10)
0.06 to 0.10	2.10(-1)	6.43(-10)
0.10 to 0.18	2.99(-1)	1.66(-9)
0.18 to 0.30	2.94(-1)	3.18(-9)
0.03 to 0.50	1.93(-1)	3.65(-9)
0.50 to 0.75	1.08(-1)	3.15(-9)
0.75 to 1.00	8.31(-2)	3.30(-9)
1.00 to 1.25	1.69(-1)	8.45(-9)
Total	1.47(0)	2.44(-8)

[a] Read as 1.47×10^{-3}.

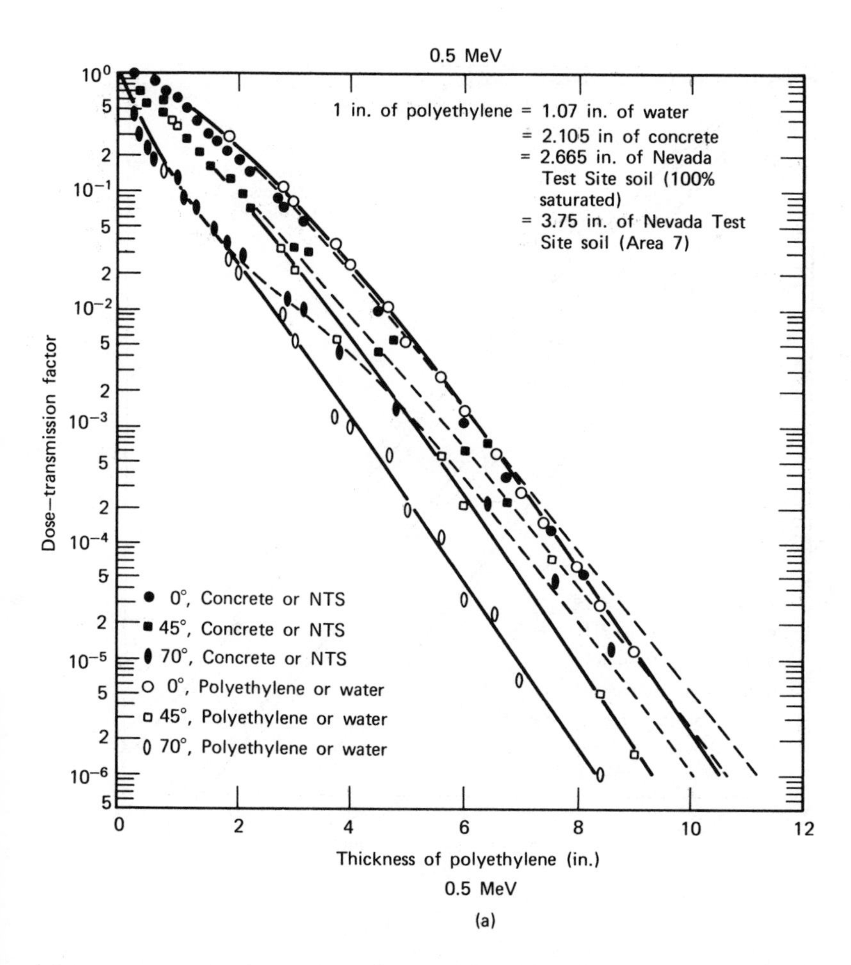

Fig. 7.10a. Neutron dose transmission in polyethylene, 0.5 MeV neutrons (after Allen and Futterer [10]).

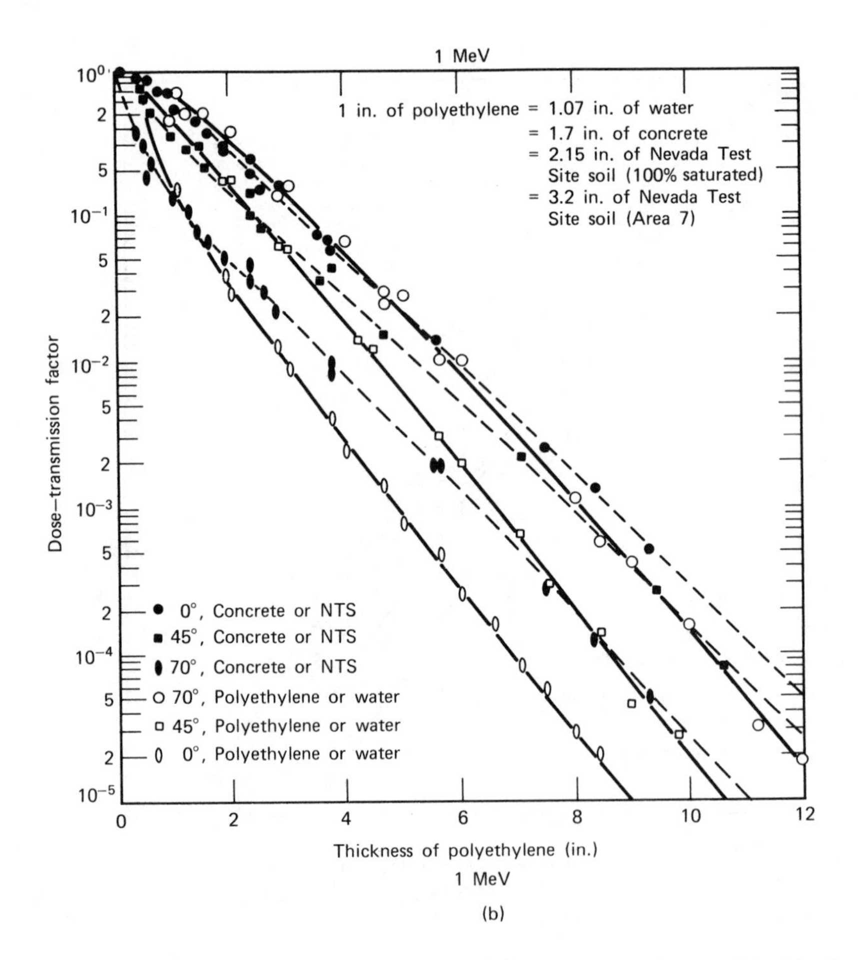

Fig. 7.10b. Neutron dose transmission in polyethylene, 1 MeV neutrons (after Allen and Futterer [10]).

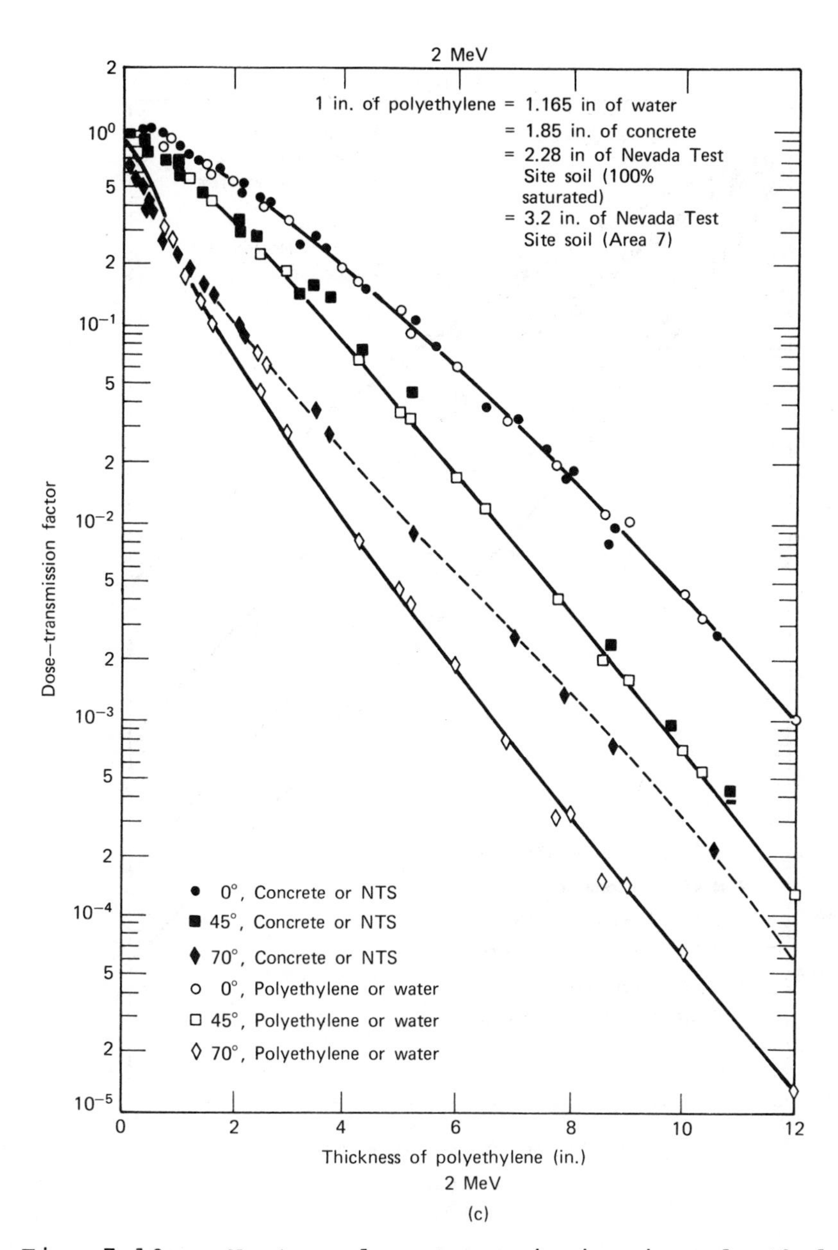

Fig. 7.10c. Neutron dose transmission in polyethylene, 2 MeV neutrons (after Allen and Futterer [10]).

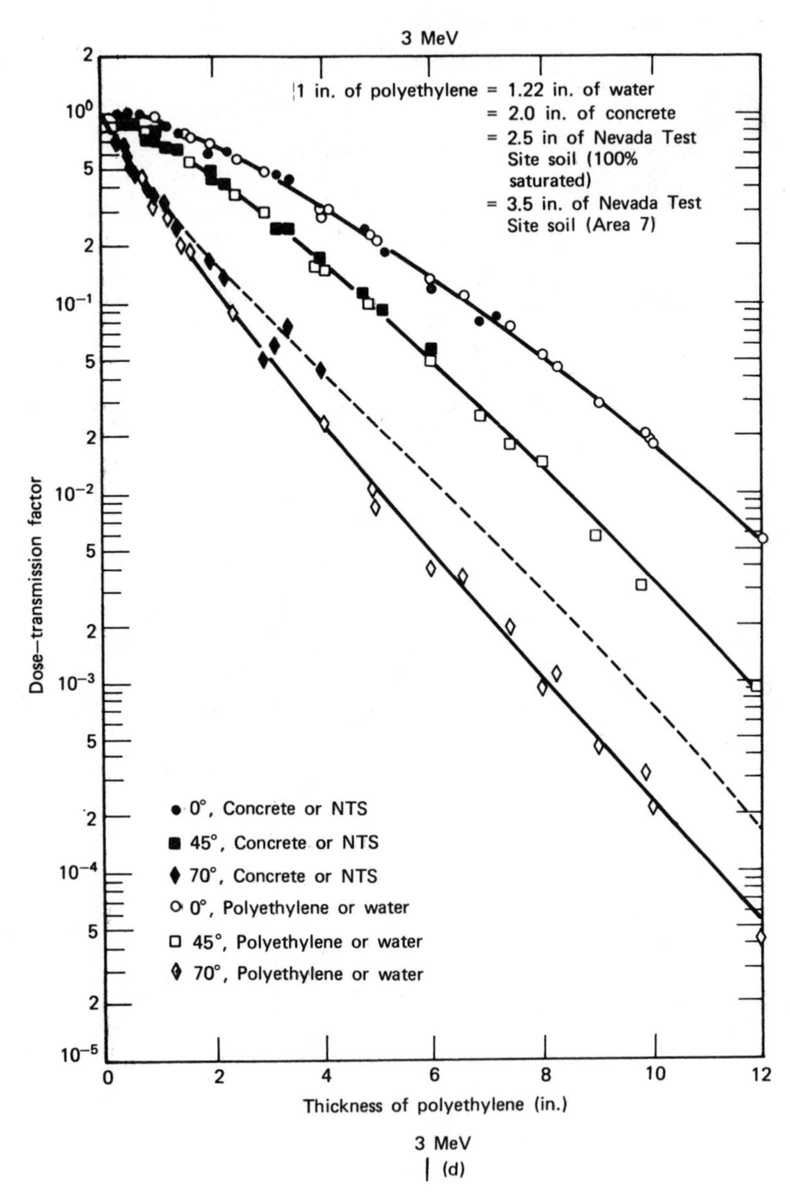

Fig. 7.10d. Neutron dose transmission in polyethylene, 3 MeV neutrons (after Allen and Futterer [10]).

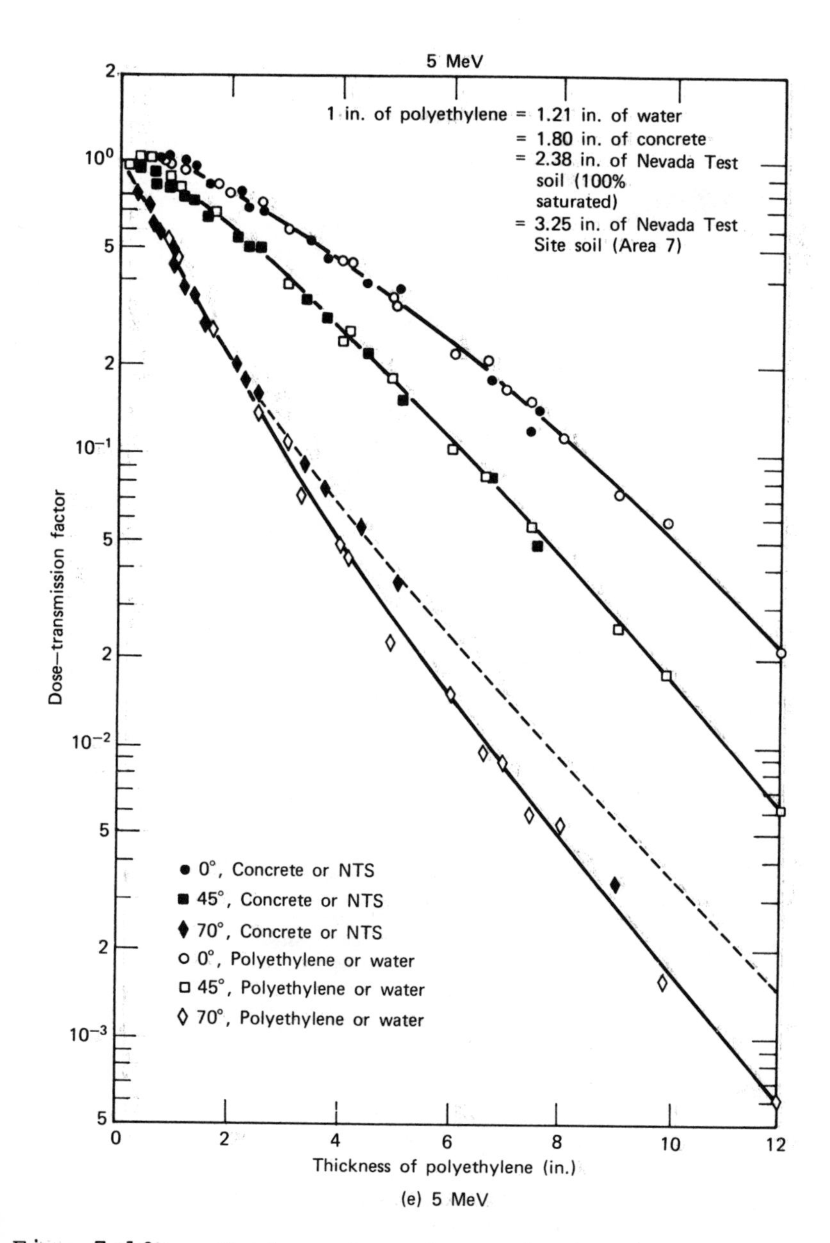

Fig. 7.10e. Neutron dose transmission in polyethylene, 5 MeV neutrons (after Allen and Futterer [10]).

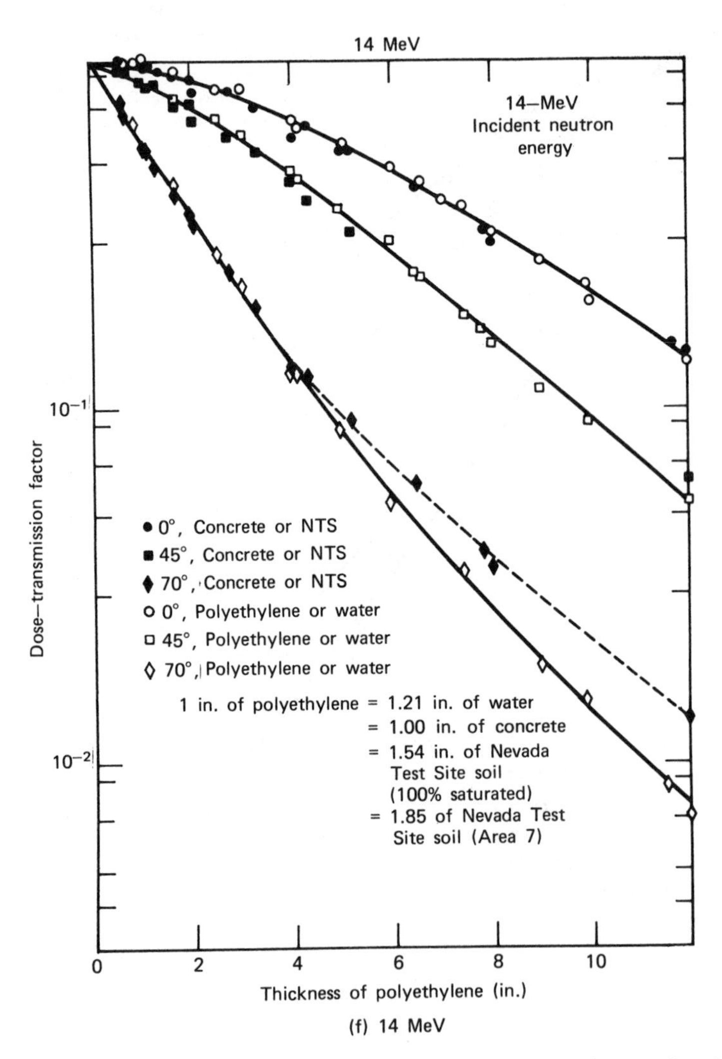

Fig. 7.10f. Neutron dose transmission in polyethylene, 14 MeV neutrons (after Allen and Futterer [10]).

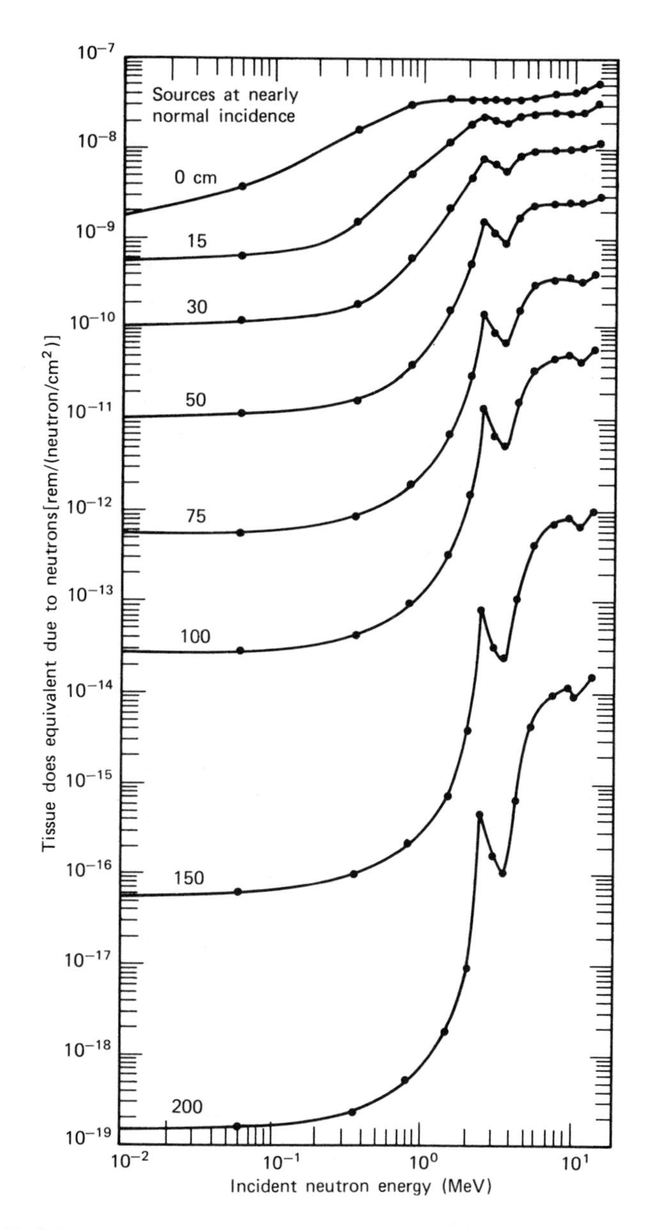

Fig. 7.11. Dose equivalent from neutrons normally incident on concrete slabs (after Schmidt [11]).

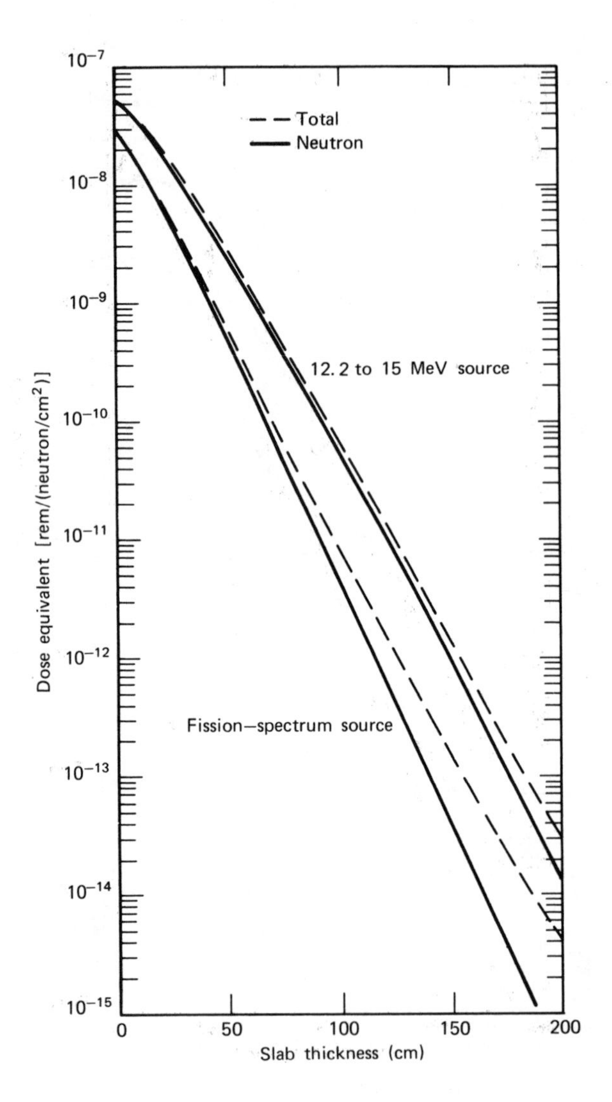

Fig. 7.12. Dose equivalent from neutrons, and neutrons plus capture gamma rays, for fission and quasi-14-MeV neutron beams perpendicularly incident on slabs of concrete (after Schmidt [11]).

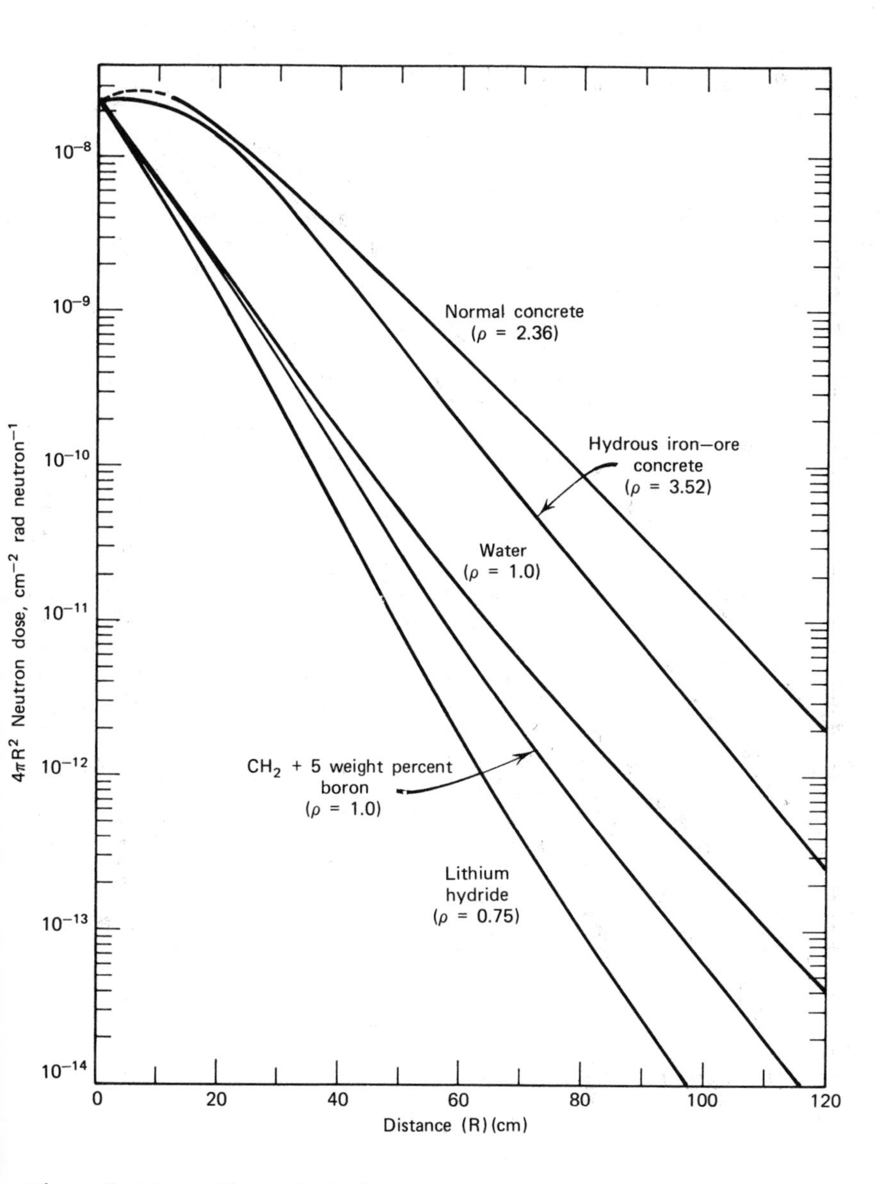

Fig. 7.13. Absorbed dose (neutron) from a point isotropic fission spectrum source in various materials (after Schaeffer [12]).

lithium hydride. Figure 7.14 plots the absorbed dose from capture γ rays generated in the water, normal concrete, and boronated polyethylene. Capture-γ-ray production is not available for the heavy concrete but would be large unless boronated; capture-γ-ray production in lithium hydride is negligible. The density of the normal concrete is 2.357 g cm^{-3} and the composition in weight percent is hydrogen (0.85), oxygen (47.3), silicon (20.8), calcium (26.0), and carbon (5.0). The absorbed dose is normalized to one neutron emitted by the source and has been multiplied by $4\pi r^2$.

Figure 7.15 plots the measured fast-neutron kerma, the γ-ray absorbed dose, and the thermal neutron flux density per watt of reactor power [13] in the water shield of the bulk shielding reactor (BSR), or "swimming-pool"-type research reactor. The change in slope in the thermal neutron flux density curve at about 200 cm is thought to be caused by photoneutrons generated by the reactor γ rays in the small amount of deuterium in water.

Measurements of the fast-neutron spectrum in the water of a similar facility have been made by Harris [14], using a collimated proton-recoil spectrometer. Measurements of the angular flux spectrum in water for a near-fission-spectrum source have been made by Profio, Carbone, et al. [15] using the pulsed-source, time-of-flight technique. Calculations of the scalar flux density spectrum for ^{235}U thermal neutron fission, and ^{252}Cf spontaneous fission sources in water have been performed by Ing and Cross [16].

γ Rays may dominate the dose in thick neutron shields. Figure 7.16 plots $4\pi r^2$ times the dose-equivalent rate from a point source of californium-252 in polyethylene, normalized to one neutron per second emitted by the source [17]. Polyethylene is a relatively poor attenuator of the primary (fission) γ rays emitted by ^{252}Cf, as well as the secondary γ rays generated by capture of neutrons in hydrogen. The dose rate from the primary γ rays can be attenuated by an order of magnitude by adding about 4 cm of lead or 8 cm of iron around the source. The capture-γ-ray contribution can be reduced by an order of magnitude by adding 1 weight percent of B_4C.

Measurements of the fast-neutron angular flux-density spectrum in paraffin wax (similar to polyethylene) have been made [18]. Monte Carlo and two-dimensional discrete-ordinates calculations of the

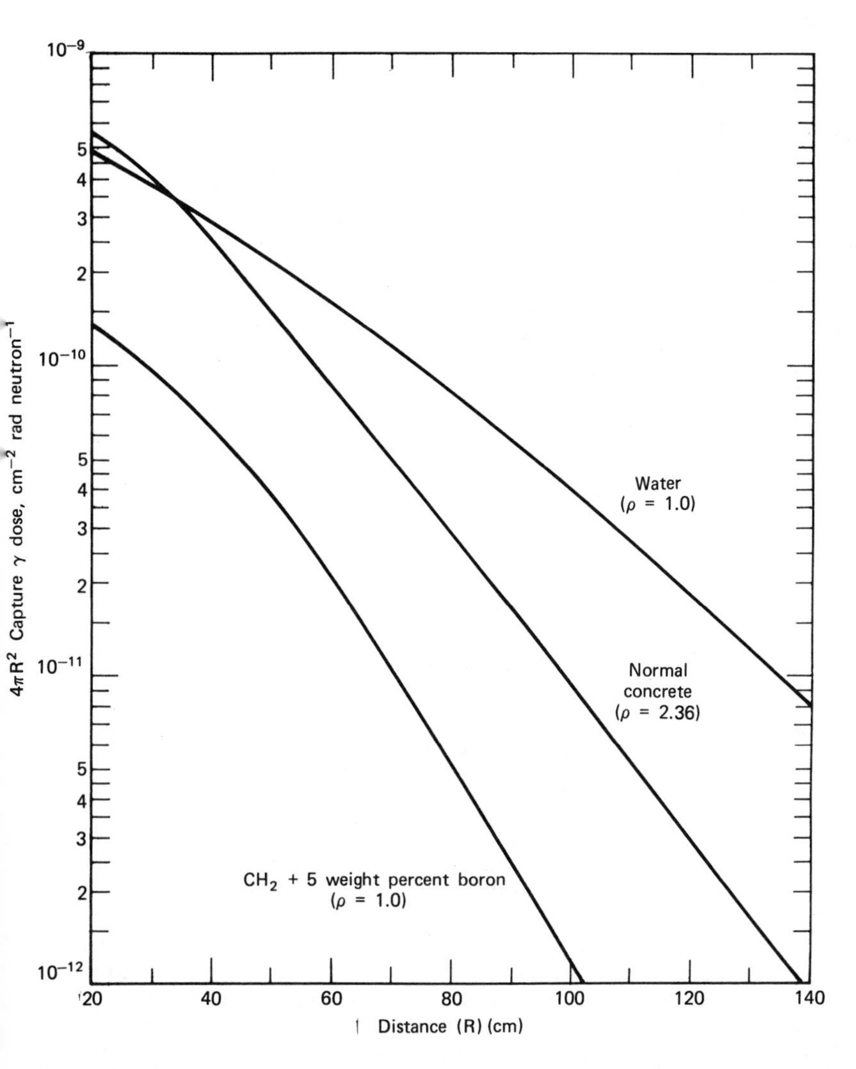

Fig. 7.14. Capture γ-ray dose from a point isotropic fission neutron source in various materials (after Schaeffer [12]).

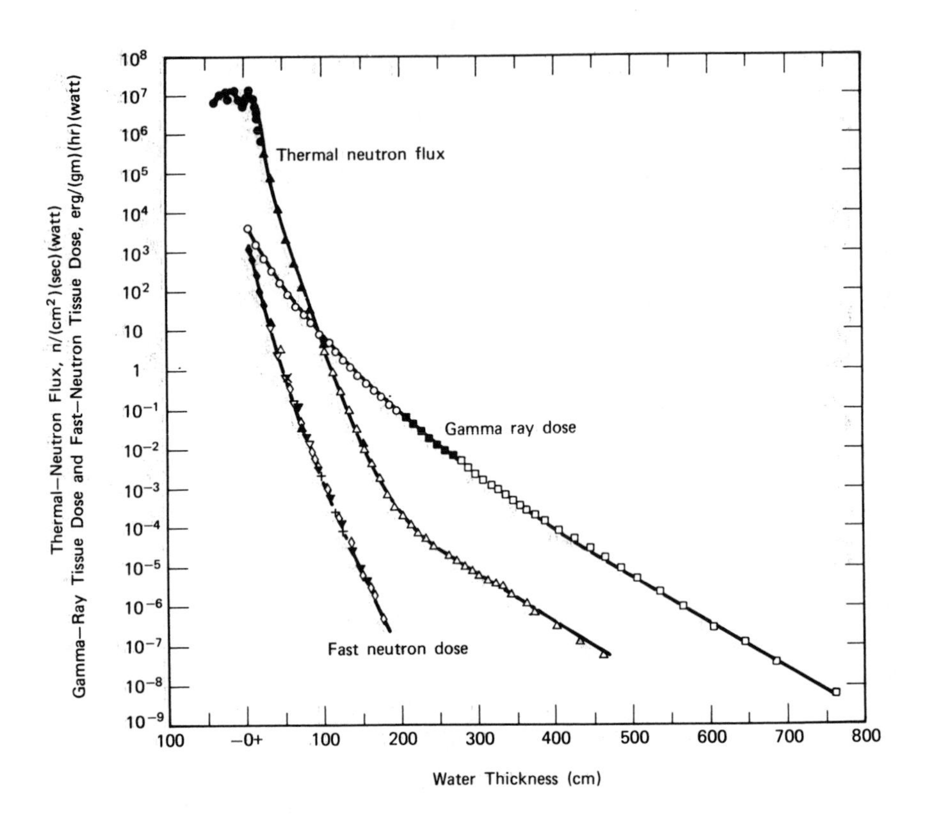

Fig. 7.15. Measured fast-neutron dose, γ-ray dose, and thermal-neutron flux density in water around the bulk shielding reactor (after Ref. 13).

angular flux-density spectrum from a slab of boronated polyethylene have been published by Burgart as a benchmark problem [19].

Laminated shields of tungsten and lithium hydride are of interest for minimum-mass shielding of compact reactors. Figure 7.17 shows the results of discrete-ordinates calculations [20] of neutron dose. Inelastic scattering in the tungsten contributes to reduction of the neutron dose, although its main function is to attenuate the primary γ rays from the reactor core.

The fast-neutron spectrum has been measured in a lithium hydride shield [18] and in a spherical assembly

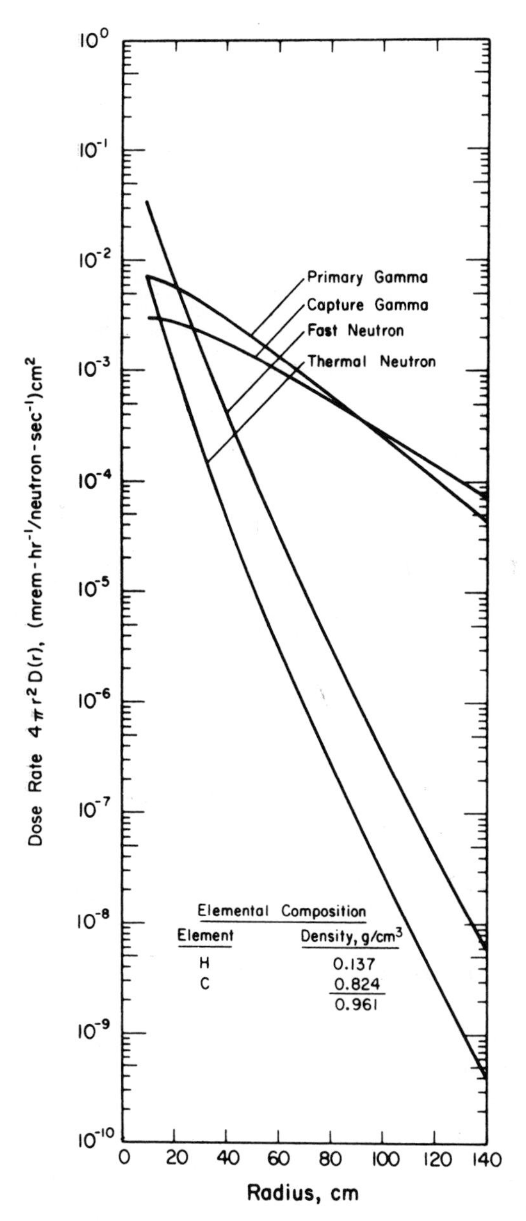

Fig. 7.16. Calculated fast-neutron, thermal-neutron, primary γ-ray, and capture γ-ray dose in polyethylene from ^{252}Cf (after Stoddard and Hootman [17]) (from A. Edward Profio, *Experimental Reactor Physics*, Wiley, New York, 1976, p. 497).

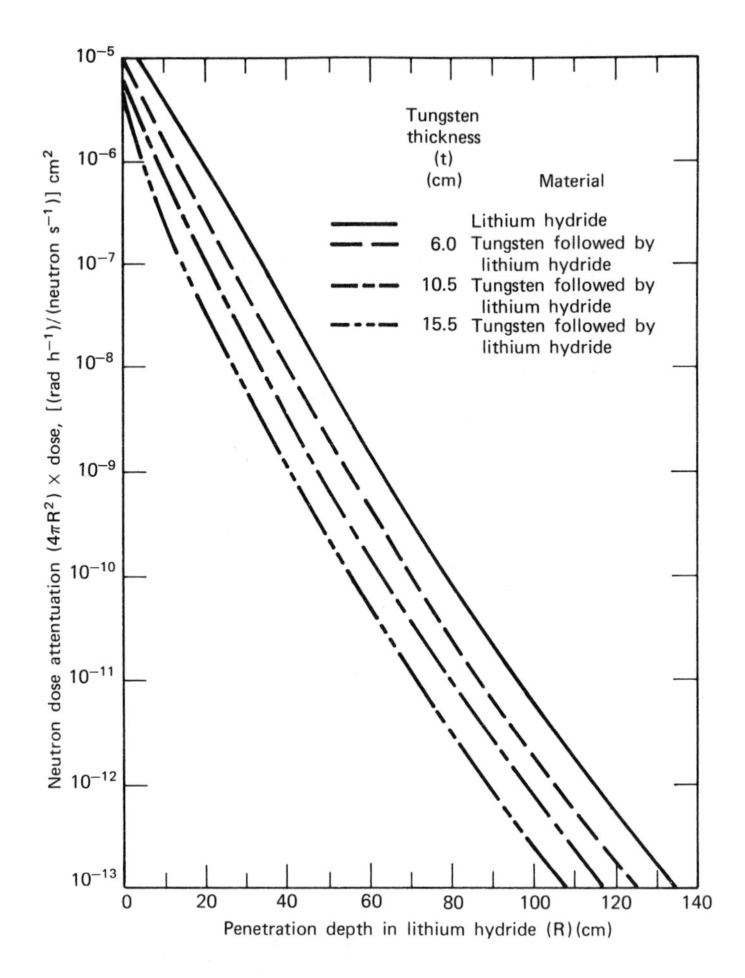

Fig. 7.17. Neutron dose rate from a fission source surrounded by a layer of tungsten, followed by lithium hydride (after Lahti [20]).

of tungsten and lithium hydride [21]. Secondary γ-ray spectra were also measured in the W-LiH assembly. It is difficult to calculate the capture-γ-ray spectrum and dose because of the many closely spaced resonances in the radiative capture cross section for tungsten.

Measurements and calculations of the angular flux-density spectrum from a fission neutron source in graphite have been carried out [14,22]. The calculated scalar flux-density spectrum for fast neutrons in graphite [23] is shown in Figure 7.18. Agreement among the 05R Monte Carlo, 1DF discrete ordinates, and moments calculations is good.

Penetration of neutrons from fission and 14-MeV neutrons in infinite air has also been proposed as a benchmark problem [24]. Neutron and γ-ray transport in air has been reviewed in the proceedings of a seminar at Oak Ridge [25].

7.4 PENETRATIONS

Shield penetrations include conduits for electrical cables, piping, mechanical drives, holes with removable plugs, tunnels for personnel access, and collimators or beam ports. Shield effectiveness is diminished if the material filling such a duct is a poorer attenuator (or generates more secondary γ rays) than the bulk of the shield. The poorest attenuators are air and other low-density materials. Because transmission through the shield may be 10^{-8} or even less, it does not take a very sizable hole to destroy the effectiveness of the shield, at least in the vicinity of the penetration. The shield designer copes with penetrations by minimizing the volume of poor attenuator, by introducing steps or bends to avoid direct streaming of radiation down the duct, and by reinforcing the shield with thicker or more effective material around the penetration. In the case of a collimator or beam port, the objective is usually to minimize the thickness of the shield (hence collimator length) and to control the size, intensity, and quality of the beam.

A collimator is usually a straight cylindrical duct such as that shown in Figure 7.19. The straight duct is also a limiting case for shield-penetration design. With an extended (plane) source, there are three components of the flux density and dose at the detector: (1) line-of-sight component for radiation passing entirely through the duct (usually air with attenuation neglected), (2) a component from radiation penetrating part of the shield before entering the duct, and (3) neutrons or primary γ rays scattered from the duct wall or secondary γ rays from neutrons incident on the wall.

The line-of-sight component can be calculated analytically, at least for simple forms of the angular

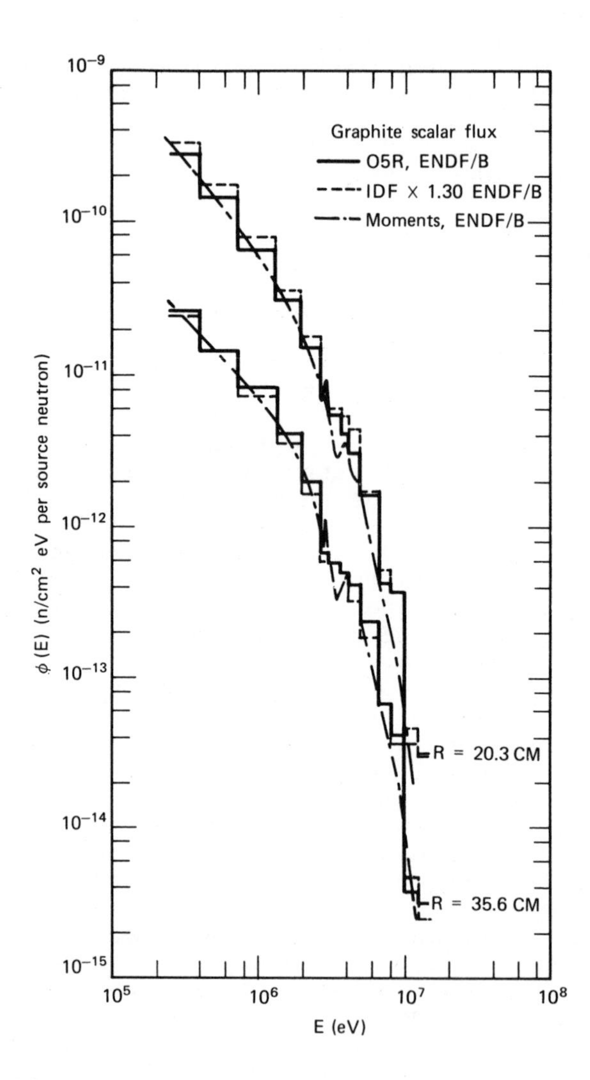

Fig. 7.18. Calculated scalar flux-density spectra at two distances from a point fission source of neutrons in graphite.

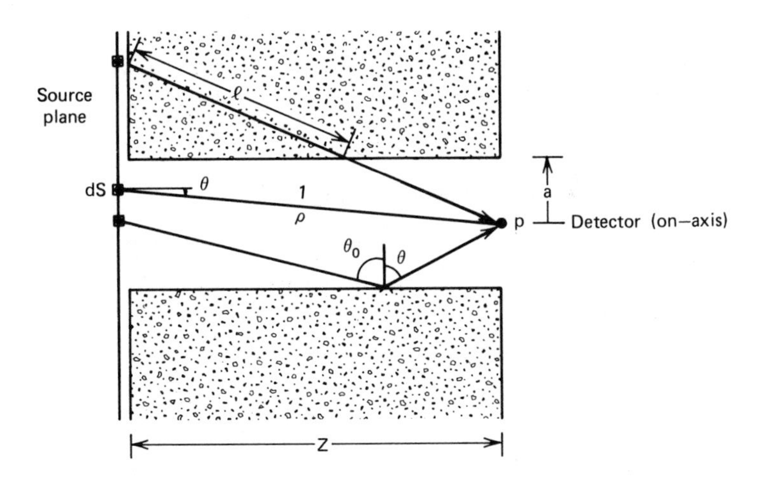

Fig. 7.19. Radiation components transmitted by a straight cylindrical duct: (1) line of sight, (2) transmitted through shield, (3) scattered from wall.

distribution of the source emission [26]. The source may be expressed in terms of the angular flux density $\psi_0(\theta)$ at z = 0, where θ is the angle to the cylinder or z-axis as shown in Figure 7.19, or in terms of the particle emission rate N_0 particles per second, per cm^2 of area in the source plane. For a plane source emitting isotropically into the forward hemisphere only, the angular current density is

$$J(\theta) = \frac{N_0}{2\pi} \quad \text{particles s}^{-1}\ \text{sr}^{-1}\ \text{cm}^{-2} \qquad (7.9)$$

Another source distribution often encountered is cosine, where the intensity is greatest at zero degrees (i.e., parallel to the z-axis) and diminishes as the cosine (but with no emission in the back hemisphere),

$$J(\theta) = \frac{(N_0 \cos\theta)}{\pi} \quad \text{particles s}^{-1}\ \text{sr}^{-1}\ \text{cm}^{-2} \qquad (7.10)$$

The angular flux density and current density are related by

$$J(\theta) = \psi(\theta) \cos \theta \tag{7.11}$$

In general, if $g(\cos \theta)$ is the normalized angular distribution of the source current density, such that

$$\int_0^1 g(\cos \theta)\ 2\pi\ d(\cos \theta) = 1 \tag{7.12}$$

the scalar flux density at slant distance ρ from a differential source area dS is

$$d\phi(\rho) = N_0\ g(\cos \theta)\ \frac{dS}{\rho^2} \tag{7.13}$$

and the total scalar flux density is found by integration over the source area.

For a cylindrical duct with radius a and an isotropic emitter the flux density on axis at a distance z is given by

$$\phi(Z) = \frac{N_{0i}}{2}\ \ell n \left[1 + \left(\frac{a}{Z}\right)^2\right] \tag{7.14}$$

For a cosine source,

$$\phi(Z) = 2\ N_{0c} \left[1 - \frac{1}{\sqrt{1 + (a/Z)^2}}\right] \tag{7.15}$$

For $Z/a > 5$ the isotropic emitter gives

$$\phi \simeq N_{0i}\ \frac{\pi a^2}{2\pi Z^2} \tag{7.16}$$

or approximately the same as for a point source of strength $N_{0i}(\pi a)^2$. With a cosine emitter into the forward hemisphere,

$$\phi \simeq N_{0c}\ \frac{a^2}{Z^2} \tag{7.17}$$

at large distances.

For a long,cylindrical, annular duct (such as the gap between an unstepped port plug and hole), inner radius a_1, outer radius a_2, the scalar flux density at a point at radius r (between a_1 and a_2) and axial distance A, is

$$\phi(r,Z) = \frac{M N_0 a_2^2}{\pi Z^2} \left\{ 1 - \left(\frac{a_1}{a_2}\right)^2 \cos^{-1}\frac{a_1}{r} + \cos^{-1}\frac{a_1}{a_2} - \frac{a_1}{a_2}\left[1 - \left(\frac{a_1}{a_2}\right)^2\right]^{1/2} \right\} \qquad (7.18)$$

where M = 1 for an isotropic emitter and M = 2 for a cosine emitter. Equation (7.18) is equivalent to

$$\phi(r,z) = \frac{M N_0 S}{2\pi Z^2} \qquad (7.19)$$

where S is the area of the source plane viewed from point (r,Z).

The component of radiation penetrating the wall and entering the duct can be estimated by point-kernel integration. The uncollided flux density from a differential source area dS is given exactly by

$$d\phi = N_0 \, g(\cos\theta) \frac{dS}{\rho^2} e^{-\mu \ell} \qquad (7.20)$$

where ρ is the distance between source-area element and the detector, ℓ is the thickness of shield material encountered along the ray (Figure 7.22), and μ is the attenuation coefficient. For a cylindrical straight duct of radius a, in a slab Z thick, the uncollided flux density at P obtained by numerical integration [27] is given in Table 7.8. The ratio of the line-of-sight component to the total uncollided flux density (line-of-sight plus wall penetration) is given in Table 7.9. As expected, the line-of-sight component dominates at large (a/Z) and μZ. Scattered radiation could be included approximately by multiplying by a

TABLE 7.8. WALL-PENETRATION COMPONENT OF THE FLUX DENSITY FOR A CYLINDRICAL DUCT

$\frac{a}{Z}$	Flux density per unit surface source intensity for shield thickness of					
	$\mu Z = 0.1$	$\mu Z = 0.2$	$\mu Z = 0.5$	$\mu Z = 1.0$	$\mu Z = 2.0$	$\mu Z = 5.0$
			Isotropic Plane Source			
0.001	1.823	1.223	0.5602	0.2198	0.04912	1.166×10^{-3}
0.002	1.823	1.223	0.5607	0.2207	0.04935	1.184×10^{-3}
0.005	1.824	1.224	0.5620	0.2215	0.05002	1.242×10^{-3}
0.010	1.825	1.226	0.5643	0.2235	0.05119	1.349×10^{-3}
0.020	1.828	1.229	0.5688	0.2278	0.05363	1.614×10^{-3}
0.050	1.834	1.239	0.5820	0.2406	0.06171	2.977×10^{-3}
0.100	1.842	1.254	0.6024	0.2613	0.07621	7.042×10^{-3}
0.200	1.851	1.272	0.6360	0.2982	0.10490	1.795×10^{-2}
0.500	1.825	1.277	0.6841	0.3659	0.16870	5.223×10^{-2}
0.750	1.765	1.240	0.6804	0.3819	0.19230	6.990×10^{-2}
1.000	1.689	1.185	0.6560	0.3767	0.19800	7.753×10^{-2}

Cosine Plane Source

0.001	1.445	1.1490	0.6539	0.2976	0.07544	0.002026
0.002	1.446	1.1490	0.6545	0.2983	0.07582	0.002059
0.005	1.446	1.1500	0.6564	0.3002	0.07698	0.002166
0.010	1.447	1.1520	0.6595	0.3034	0.07896	0.002365
0.020	1.450	1.1560	0.6657	0.3100	0.08314	0.002862
0.050	1.455	1.1660	0.6836	0.3298	0.09703	0.005458
0.100	1.459	1.1790	0.7095	0.3610	0.12170	0.013230
0.200	1.453	1.1880	0.7462	0.4120	0.16820	0.033430
0.500	1.339	1.1160	0.7510	0.4714	0.24940	0.087280
0.750	1.190	0.9949	0.6827	0.4465	0.25530	0.103600
1.000	1.036	0.8643	0.5953	0.3956	0.23400	0.101400

TABLE 7.9. RATIO OF LINE-OF-SIGHT COMPONENT TO TOTAL UNCOLLIDED FLUX DENSITY FOR CYLINDRICAL DUCT

$\frac{a}{Z}$	Fraction of total flux density for shield thickness of					
	$\mu Z = 0.1$	$\mu Z = 0.2$	$\mu Z = 0.5$	$\mu Z = 1.0$	$\mu Z = 2.0$	$\mu Z = 5.0$
	Isotropic Infinite-Plane Source					
0.001	$<10^{-6}$	$<10^{-6}$	$<10^{-6}$	$<10^{-5}$	$<10^{-4}$	$<10^{-3}$
0.002	$<10^{-5}$	$<10^{-5}$	$<10^{-5}$	$<10^{-5}$	$<10^{-4}$	0.002
0.005	$<10^{-5}$	$<10^{-4}$	$<10^{-4}$	$<10^{-4}$	$<10^{-3}$	0.010
0.010	$<10^{-4}$	$<10^{-4}$	$<10^{-4}$	10^{-3}	0.001	0.038
0.020	$<10^{-3}$	$<10^{-3}$	10^{-3}	10^{-3}	0.004	0.110
0.050	$<10^{-3}$	0.001	0.002	0.005	0.020	0.290
0.100	0.003	0.004	0.008	0.019	0.062	0.420
0.200	0.011	0.015	0.031	0.062	0.160	0.520
0.500	0.095	0.083	0.140	0.230	0.400	0.680
0.750	0.110	0.150	0.250	0.370	0.530	0.760
1.000	0.170	0.230	0.340	0.480	0.640	0.820

Cosine Infinite-Plane Source

0.001	$<10^{-6}$	$<10^{-6}$	$<10^{-5}$	$<10^{-5}$	$<10^{-4}$	$<10^{-3}$
0.002	$<10^{-5}$	$<10^{-5}$	$<10^{-5}$	$<10^{-4}$	$<10^{-4}$	0.002
0.005	$<10^{-4}$	$<10^{-4}$	$<10^{-4}$	$<10^{-4}$	$<10^{-3}$	0.011
0.010	$<10^{-4}$	$<10^{-4}$	$<10^{-3}$	$<10^{-3}$	0.001	0.040
0.020	$<10^{-3}$	$<10^{-3}$	$<10^{-3}$	0.002	0.005	0.120
0.050	0.002	0.002	0.004	0.008	0.025	0.270
0.100	0.007	0.008	0.014	0.027	0.076	0.430
0.200	0.026	0.031	0.050	0.120	0.190	0.540
0.500	0.140	0.160	0.220	0.310	0.450	0.710
0.750	0.250	0.280	0.370	0.480	0.610	0.790
1.000	0.360	0.400	0.500	0.600	0.710	0.850

buildup factor before integrating, or using an "effective" attenuation coefficient, corresponding to the inverse of the relaxation length for the scattered plus uncollided radiation. For more exact results, the penetration may be treated in a Monte Carlo calculation.

The component of radiation scattered or "reflected" from the wall might also be treated by Monte Carlo. However, very good importance sampling is necessary, or most of the computation time will be wasted in following particles that never reenter the duct or contribute to the dose at the detector point, P. Another approach, especially useful for bent or stepped ducts, is to use the concept of albedo.

The term "albedo" refers to the ratio of the current density (or flux density) reflected or backscattered from a surface, to the current density (or flux density) of radiation incident on that surface. The approximation is made that the radiation emerges from the same area on which the radiation was incident, as shown in Figure 7.20. This is valid when surface dimensions are large compared to the effective mean free path, μ^{-1}, in the shield. Let us consider a pencil-beam of particles of energy E_0, incident at

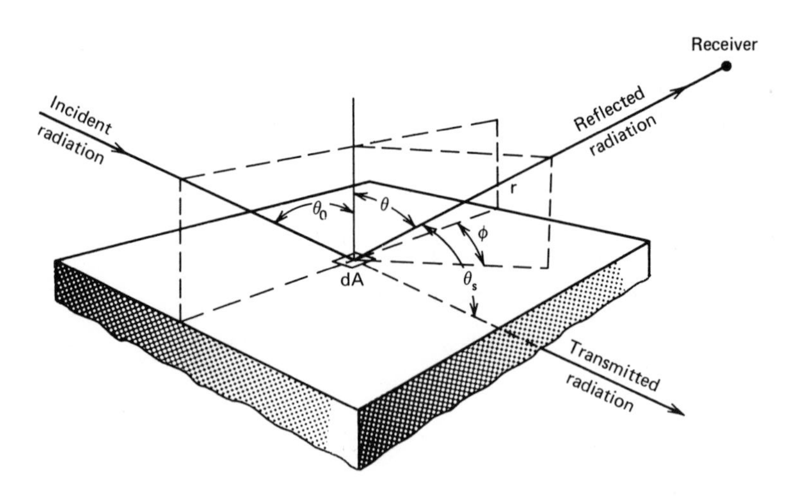

Fig. 7.20. Geometry for definition of reflection and albedo (after Schaeffer [12]).

polar angle θ_0. The reflected current density of particles of energy E and emerging at polar angle θ and azimuthal angle ϕ, per unit energy and unit solid angle, is

$$J(E,\theta,\phi) = J(E_0,\theta_0)\ \alpha(E_0,\theta_0,\theta,\phi) \tag{7.21}$$

where α is the albedo. The albedo thus defined is doubly differential, in reflected energy E and reflected direction described by θ and ϕ. The dose albedo, α_D, is obtained by weighting by the dose conversion factor and integrating over E. If integrated over both reflected energy and direction, a total albedo, A, is obtained, but then it is necessary to make some assumption about the angular distribution if it is to be used in calculations. Albedo may also be defined in terms of an angular flux density, related to the current density by

$$\psi(E,\theta,\phi) = \frac{J(E,\theta,\phi)}{\cos\theta} \tag{7.22}$$

Three kinds of albedo are in common use: α_1 for current density reflected per unit flux density incident, α_2 for the current density reflected per unit current density incident, and α_3 for the flux density reflected per unit flux density incident. They are related by

$$\alpha_1 = \alpha_2 \cos\theta_0 = \alpha_3 \cos\theta \tag{7.23}$$

Most of the experimental and computational results are for concrete. Data are voluminous, even for dose albedos, because they are functions of incident energy and angle as well as reflected polar and azimuthal angles. However, the results can be represented by formulas with empirical coefficients fitted to the data. Monte Carlo calculations, tested against experiment, were made by Maerker and Muckenthaler [28] for the fast-neutron albedo of concrete. The differential dose albedo is fit to within 10% by the expression

$$\alpha_{D_2}(\Delta E_0, \theta_0, \theta, \phi) = \frac{\cos\theta}{\cos\theta + K_1(\Delta E_0)\cos\theta_0}$$

$$\times \sum_{m=0}^{M} G_m(\Delta E_0)\, P_m(\cos\theta_s)$$

$$+ \frac{\cos\theta}{\cos\theta + K_2(\Delta E_0, \theta_0, \theta)}$$

$$\times \sum_{k=0}^{K} B_k(\Delta E_0)\, P_k(\cos\theta_s) \qquad (7.24)$$

where $\cos\theta_s = \sin\theta_0 \sin\theta \cos\phi - \cos\theta_0 \cos\theta$, P_m and P_k are Legendre polynomials of order m and k,

$$K_2(\Delta E_0, \theta_0, \theta) = \sum_{i=0}^{I} (\cos\theta)^i \sum_{j=0}^{J} a_{ij}(\Delta E_0) \cos^j \theta_0 \qquad (7.25)$$

and the coefficients for incident energy groups ΔE_0 are listed in Table 7.10.

Monte Carlo calculations of fast-neutron albedos of concrete, soil, polyethylene, water, and steel have been performed by Allen, Futterer, et al. [29] and by French and Wells [30]. Figure 7.21 illustrates the angular dependence of fast neutrons reflected by a concrete slab for 1.0-MeV neutrons incident. Figure 7.22 shows the dependence of the total dose albedo on the angle of incidence of 14-MeV neutrons for materials of differing hydrogen content. For fission neutrons, the total dose albedo A_{D_1} is given approximately by $1 - 0.430(\Sigma_H/\Sigma_T)$, where Σ_H is the scattering cross section for hydrogen in the material and Σ_T is the total cross section for the material.

Albedos for intermediate-energy neutrons on steel-reinforced concrete have been reported by

Coleman, Maerker, et al. [31]. These include thermal neutrons emerging after moderation of the intermediate-energy neutrons. The total dose albedo A_2 for thermal neutrons incident and emerging from concrete [32] is about 0.8. Mockel [33] has investigated thermal-neutron albedos in strongly absorbing media. Although thermal neutrons are easily absorbed, they can be a serious source of capture γ rays. It may be necessary to add boron or lithium to the shield or at least to the lining of a duct and to avoid materials with large cross sections for production of secondary γ rays.

Backscattering of γ rays occurs because of Compton scattering, and there is a strong dependence on the scattering angle. Annihilation photons from pair production by high-energy γ rays also contribute to the dose. Monte Carlo calculations of γ ray albedos for concrete have been made by Wells [34]. Figure 7.23 plots the differential dose albedo as a function of reflected angle for ^{60}Co γ rays (1.25 MeV) incident at 60°. The results are compared with experiment and other Monte Carlo calculations. Chilton and Huddleston [35] developed a semiempirical expression for the differential-dose albedo.

$$\alpha_{D_2}(E_0,\theta_0,\theta,\phi) = \frac{C\,K(\theta_s) \times 10^{26} + C'}{1 + [(\cos\theta_0)/(\cos\theta)]} \tag{7.26}$$

where $K(\theta_s)$ is the Klein-Nishina differential-energy scattering coefficient for the scattering angle θ_s,

$$K(\theta_s) = 0.392\left(\frac{E}{E_0}\right)^2\left(\frac{E_0}{E} + \frac{E}{E_0} - \sin^2\theta_s\right)\ \text{barn sr}^{-1} \tag{7.27}$$

per electron, and the coefficients C and C' are listed in Table 7.11.

One way to apply the albedo is to trace the rays incident on the duct wall, as shown in Figure 7.19, knowing the albedo corresponding to the incident and reflected angles and integrating over the duct wall as well as the source area at the opening of the duct. However, this soon becomes tedious, especially if multiple reflections have to be taken into account. For straight cylindrical ducts where the line-of-sight component predominates, an approximate treatment with

TABLE 7.10. COEFFICIENTS FOR DIFFERENTIAL-DOSE ALBEDO OF FAST NEUTRONS ON CONCRETE [28]

Constant	Value of constant for ΔE_0 of					
	0.2 to 0.75 MeV	0.75 to 1.5 MeV	1.5 to 3 MeV	3 to 4 MeV	4 to 6 MeV	6 to 8 MeV
G_0	6.585(-2)	7.045(-2)	7.211(-2)	7.024(-2)	6.856(-2)	5.899(-2)
G_1	5.048(-2)	4.393(-2)	5.845(-2)	7.452(-2)	6.294(-2)	6.039(-2)
G_2	3.710(-2)	7.088(-2)	5.968(-2)	1.000(-1)	9.517(-2)	7.524(-2)
G_3	1.544(-2)	1.898(-2)	2.729(-2)	5.591(-2)	7.761(-2)	8.140(-2)
G_4	7.837(-3)	2.408(-3)	1.190(-2)	2.646(-2)	4.292(-2)	6.622(-2)
G_5	0	-3.589(-3)	1.000(-3)	-6.908(-4)	1.824(-2)	3.056(-2)
G_6	0	0	4.637(-3)	-8.087(-4)	5.599(-3)	1.595(-2)
G_7	0	0	6.490(-3)	-1.459(-3)	5.288(-3)	1.277(-2)
G_8	0	0	0	-1.809(-3)	1.046(-2)	9.380(-3)
B_0	6.27(-2)	9.00(-2)	8.80(-2)	9.05(-2)	8.744(-2)	6.374(-2)
B_1	1.50(-2)	8.5(-3)	1.30(-2)	2.15(-2)	2.817(-2)	1.382(-2)
B_2	5.3(-3)	9.7(-3)	6.0(-3)	2.30(-2)	2.344(-2)	1.178(-2)
B_3	0	0	0	0	1.779(-2)	1.084(-2)

B_4	0	0	0	0	8.517(-3)	6.801(-3)
K_1	1.0	1.0	1.1	0.9	1.1	1.06
a_{00}	0.36	0.51	0.56	0.60	0.43	0.35
a_{01}	1.29	0.32	0.18	0.15	2.02	0.95
a_{02}	0	1.00	1.32	0.48	-0.38	0
a_{10}	0.06	-0.04	-0.14	-0.61	0.05	0.10
a_{11}	-3.06	-2.46	-2.76	-1.08	-9.13	-2.28
a_{12}	0	0	0	0	5.93	1.11
a_{20}	-0.20	0.05	0.05	0.32	0.04	0
a_{21}	1.68	0.95	1.14	0.30	5.97	0
a_{22}	0	0	0	0	-4.39	0

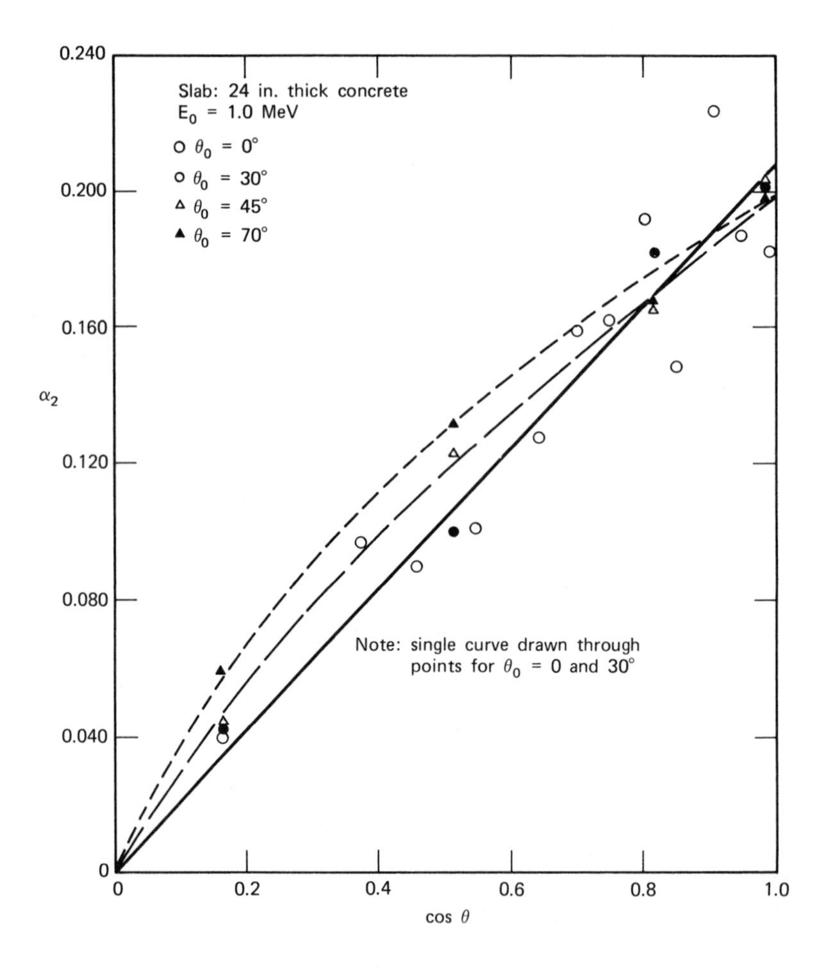

Fig. 7.21. Angular distribution of fast neutrons reflected from concrete, for 1-MeV neutrons incident (after Allen, Futterer, et al. [29]).

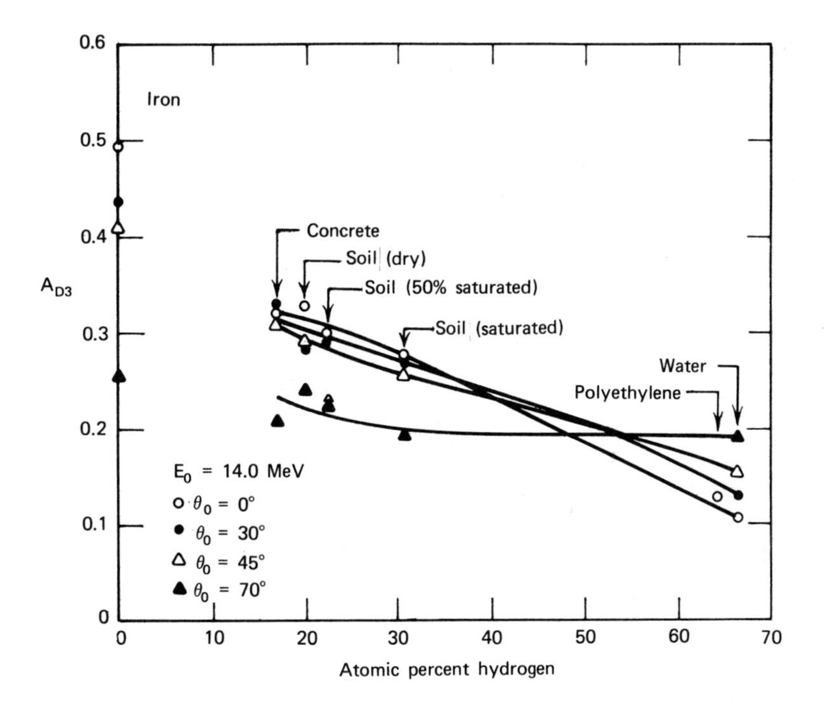

Fig. 7.22. Dependence of total dose albedo on hydrogen content for 14-MeV neutrons incident at various angles (after Allen, Futterer, et al. [29]).

an assumed analytical form for α may be sufficient. Simon and Clifford [36] obtain

$$\phi(Z) = \frac{N_0}{2}\left(\frac{a}{Z}\right)^2 \left[1 + \left(\frac{A_2}{1 - A_2}\right)\left(\beta + \frac{4\gamma a}{Z}\right)\right] \quad (7.28)$$

where the albedo A_2 is the fraction of neutrons scattered from the wall and the angular distribution is assumed to be part isotropic (fraction β of all reflected neutrons) and part cosine (fraction γ). The first term is the line-of-sight component, and the second term is the wall-scattered component. For fast

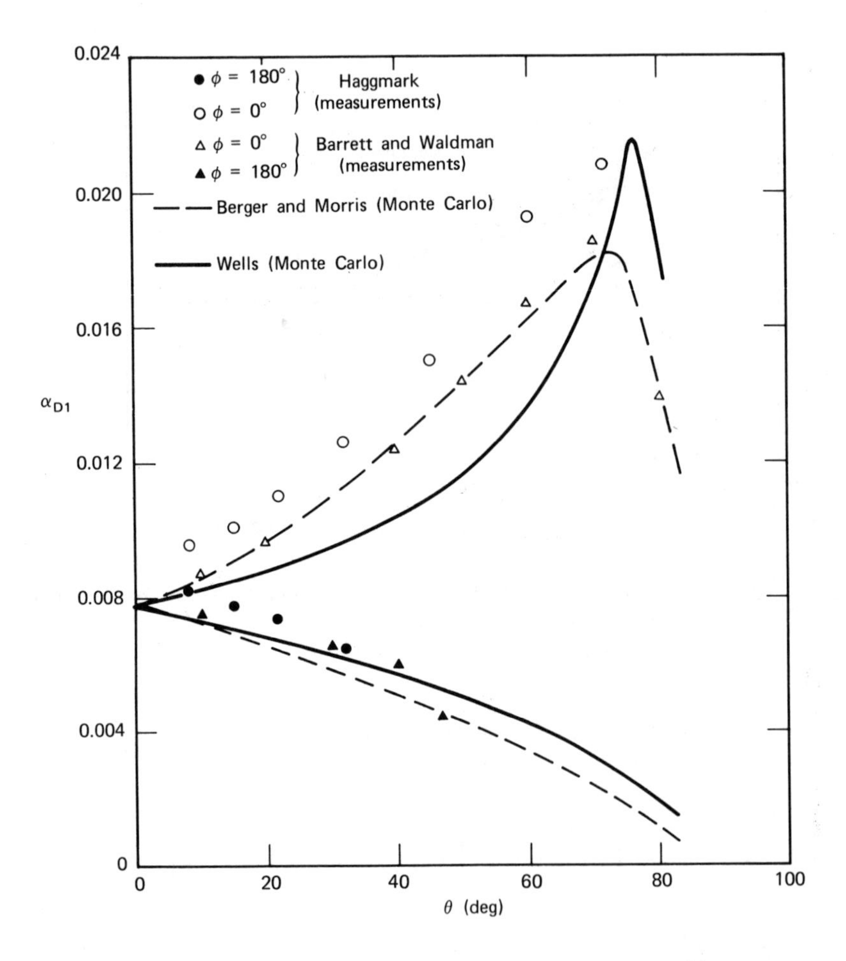

Fig. 7.23. Differential-dose albedo for ^{60}Co (1.25 MeV) γ rays incident on concrete at 60° (after Wells, [34]).

TABLE 7.11. CHILTON-HUDDLESTON COEFFICIENTS FOR γ-RAY DIFFERENTIAL DOSE ALBEDO[a]

Material	E_0 (MeV)	C	C'
Water	0.2	-0.0187	0.1327
	0.662	0.0390	0.0253
	1.00	0.0470	0.0151
	2.50	0.0995	0.0058
	6.13	0.1861	0.0035
Concrete	0.2	0.0023	0.0737
	0.662	0.0347	0.0197
	1.00	0.0503	0.0118
	2.50	0.0999	0.0051
	6.13	0.1717	0.0048
Iron	0.2	0.0272	-0.0100
	0.662	0.0430	0.0063
	1.00	0.0555	0.0045
	2.50	0.1009	0.0044
	6.13	0.1447	0.0077
Lead	0.2	0.0044	-0.0050
	0.662	0.0308	-0.0100
	1.00	0.0452	-0.0083
	2.50	0.0882	0.0001
	6.13	0.1126	0.0063

[a]From A. B. Chilton, C. M. Davisson, and L. A. Beach, *Transact. Am. Nucl. Soc.*, 8, 656 (1965).

neutrons, A_2 is on the order of 0.1, and the line-of-sight component may dominate. For thermal neutrons, A_2 is usually around 0.8, and the wall-scattered component often dominates for large Z/a.

Another way to use albedos is in a Monte Carlo calculation. Rather than tracking particles through many collisions in the shield, particles hitting the duct wall are reflected back into the duct at an energy and direction chosen according to the albedo. The albedo in effect expresses the results of measurements or previous calculations in which the tracking through the shield material was performed. It is assumed that the particles reemerge from the point at which they entered and that the albedo is insensitive to thickness or other details of the construction beyond the wall surface. In many problems the albedo Monte Carlo approach gives good results and saves an enormous amount of computer time.

Holes or ports are often closed with plugs or doors. Because of manufacturing tolerances, allowance for thermal expansion, or clearance for movement, there will be a gap between plug and hole. The plug and hole should have at least one step (see Figure 7.24) to block the line-of-sight component and reduce radiation streaming down the gap. Rotating shafts can also be provided with a step or collar. Experiments by Schamberger, Shore, et al. [37] on transmission of fast neutrons through offset air-filled ducts in water indicate that a single step is sufficient and should be located half-way through the shield, that the transmission is not very sensitive to the width of the slot or small thicknesses of material between the offset portions, and that little is gained by making the offset (step) more than twice the slot (gap) width. However, in practice it is advisable to keep the gap width less than 1 cm if feasible, and the step width might be some 5 cm, to compensate for construction irregularities. There is no line-of-sight component with a stepped duct. However, radiation can stream down the gap to the step, and some radiation can short-circuit the second half of the shield by streaming through the gap past the step. In addition, there will be a wall-scattered component, which can scatter around the step. It is difficult to calculate the radiation transmitted through the region of a stepped hole and plug. In general, a step will be sufficient for fast neutrons and primary γ rays

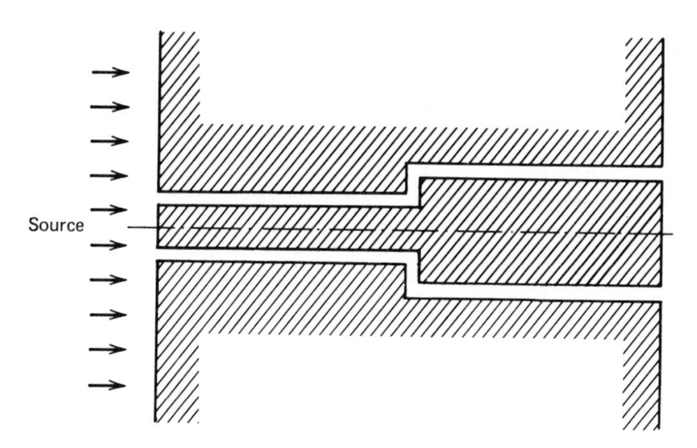

Fig. 7.24. Stepped cylindrical hole and plug with annular gap.

without requiring an increase in shield thickness, but thermal neutrons reflecting down the gap can produce secondary γ rays deep within the shield, which will require more shielding than if the penetration were absent.

When a hole cannot be plugged, such as in a gas-filled cooling pipe or a personnel access tunnel, the duct should be constructed with one or more bends, as shown in Figure 7.25. Then radiation has to be scattered at least once to escape, and if the albedo is small, the reduction in dose can be worthwhile. An estimate of the emergent dose can be made using the Simon-Clifford technique, in which the area irradiated at the end of the first leg is the source area for the second leg, and so on. Bent ducts and labyrinths or mazes can also be calculated by the albedo Monte Carlo method. Figure 7.26 shows the results of fast-neutron dose measurements and calculations performed with the AMC Monte Carlo code [38]. The source was a collimated beam from the Tower Shielding Reactor, incident at 45° on a small area at 4.6 m from the bend. Agreement between calculations and measurements is good. Greater reductions can be achieved with three-legged ducts (U- or Z-shaped).

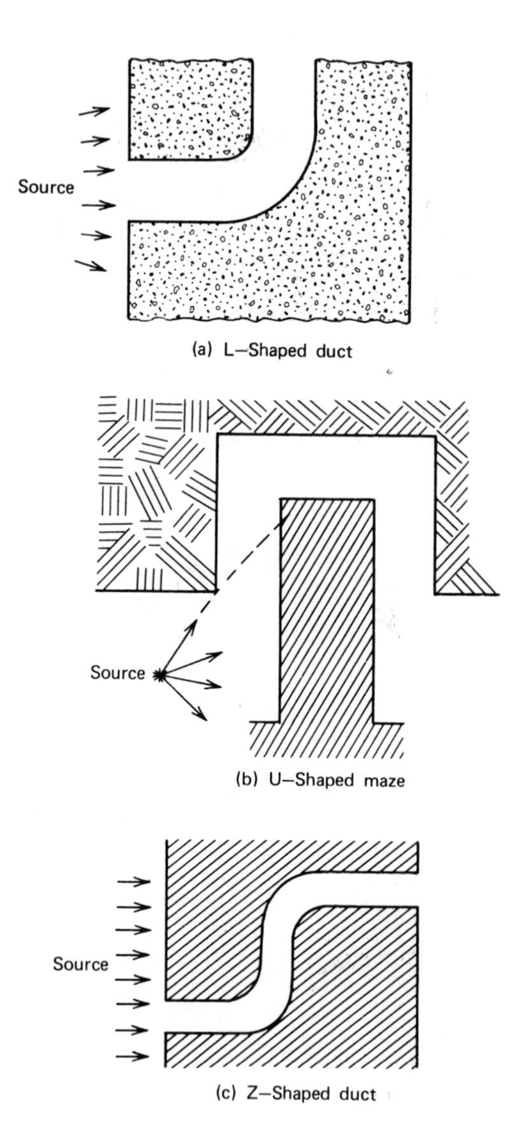

Fig. 7.25. Bent ducts and maze.

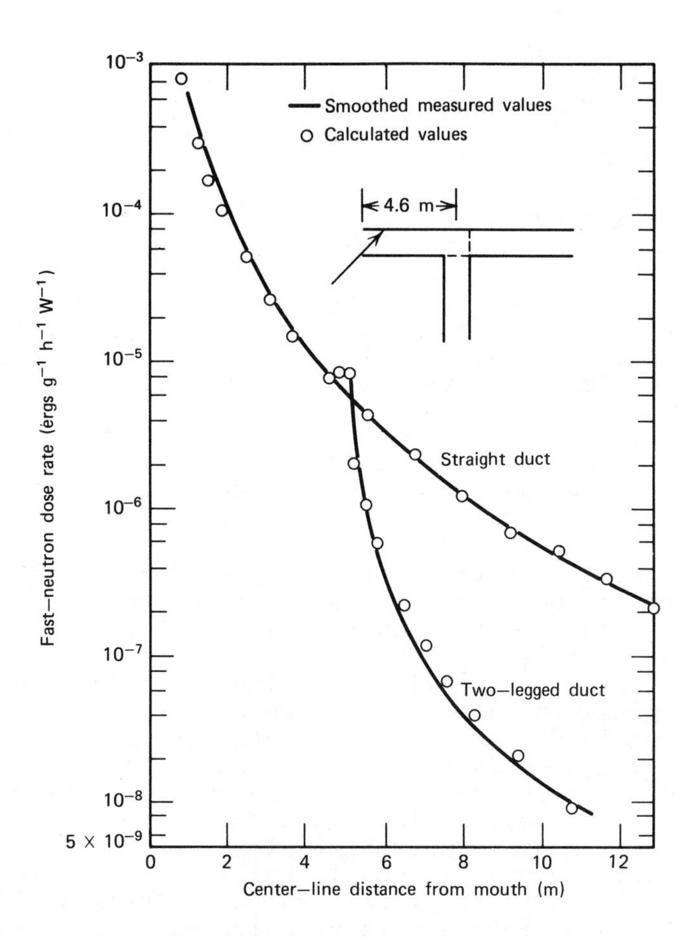

Fig. 7.26. Comparison of albedo Monte Carlo calculations and experiment for fast-neutron transmission through straight (one-legged) and L-shaped (two-legged) concrete duct, 0.91-m square (after Maerker and Muckenthaler [38]).

Electrical cables and small diameter cooling pipes are often twisted into a helix. The diameter of the helix should be a few times the diameter of the conduit or pipe. The pitch should be fairly large, to facilitate pulling the cable or to reduce pressure drop in the pipe, and to minimize the volume of the shield material displaced. Because the radiation has to make many reflections, the reduction in density through the volume of the penetration (helix) may be more significant than the opportunity to scatter down the pipe.

7.5 HEATING AND STRESS

The volumetric heat generation rate, $\dot{P}$ watts cm^{-3}, is obtained as a function of position in the shield by the methods discussed in Chapters 4 and 5. Usually one calculates the scalar flux density spectrum $\phi(\underline{r},E)$ and folds it with k(E), the flux density to kerma-rate factor,

$$\dot{P}(\underline{r}) = \int k(E)\ \phi(\underline{r},E)\ dE \tag{7.29}$$

The flux density is obtained from a transport calculation. It is not necessary to carry it out to deep penetrations, just in the inner region of the shield where the heating is greatest.

γ-Ray heating is usually most important for medium and high-Z elements, whereas neutron heating is most important for low-Z (small-A) elements, as indicated [39] in Table 7.12. Fast neutrons transfer energy mainly by elastic scattering, and the average energy imparted to the recoil nucleus goes as $A/(A+1)^2$. Inelastic scattering occurs mainly in heavy nuclei, where the recoil energy is small, but inelastic γ rays should be included in the γ-ray heating. Thermal neutrons generate capture γ rays, which contribute to the γ-ray flux density. Thermal neutrons also release appreciable energy in charged-particle-emitting reactions such as $^{10}B(n,\alpha)^7Li$ and $^6Li(n,\alpha)^3H$.

Heating (kerma) factors for γ rays are plotted in Figure 7.27, based on the energy-transfer coefficients in Hubbell [40] and Eq. (5.40). Kerma factors for neutrons [41] are plotted in Figures 7.28 and 7.29.

Given $\dot{P}(\underline{r})$, the problem is to calculate the temperature distribution in the shield and the total heat to be removed. This can be done by standard heat-transfer methods. For the simple case of an infinite slab of thickness Z, the heat conducted across 1 cm^2

TABLE 7.12. HEAT-GENERATION RATE FROM FLUX DENSITY OF 10^{13} cm^{-2} s^{-1} OF 1-MeV PHOTONS OR NEUTRONS

Material	Density ($g\ cm^{-3}$)	Photon ($W\ cm^{-3}$)	Neutron ($W\ cm^{-3}$)
Water	1.00	0.050	0.249
Graphite	1.67	0.073	0.045
Steel	7.85	0.324	0.006

of area in a plane at depth z is given by

$$\frac{dQ}{dz} = \dot{P}(z) \tag{7.30}$$

hence

$$Q(z) = \int_0^Z \dot{P}(z)\ dz \tag{7.31}$$

The heat flow rate and temperature gradient are related by

$$Q(z) = -K\left(\frac{dT}{dz}\right) \tag{7.32}$$

where K is the thermal conductivity. Then the temperature

$$T(z) = T_0 - \frac{1}{K}\int_0^Z Q(z)\ dz \tag{7.33}$$

and

$$\int_0^Z Q(z)\ dz = K(T_0 - T_Z)$$

The hottest plane occurs where $Q(z) = 0$.

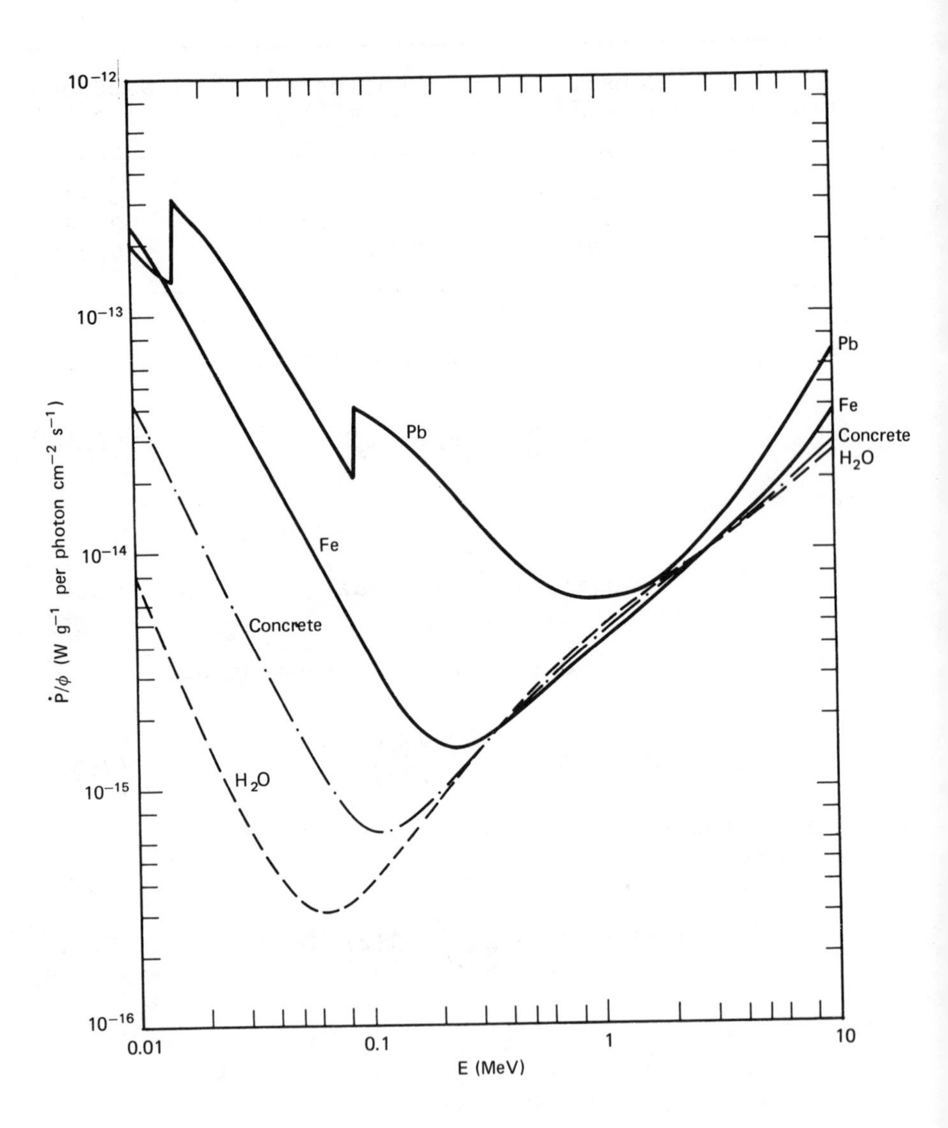

Fig. 7.27. Heating (kerma) factor for γ rays.

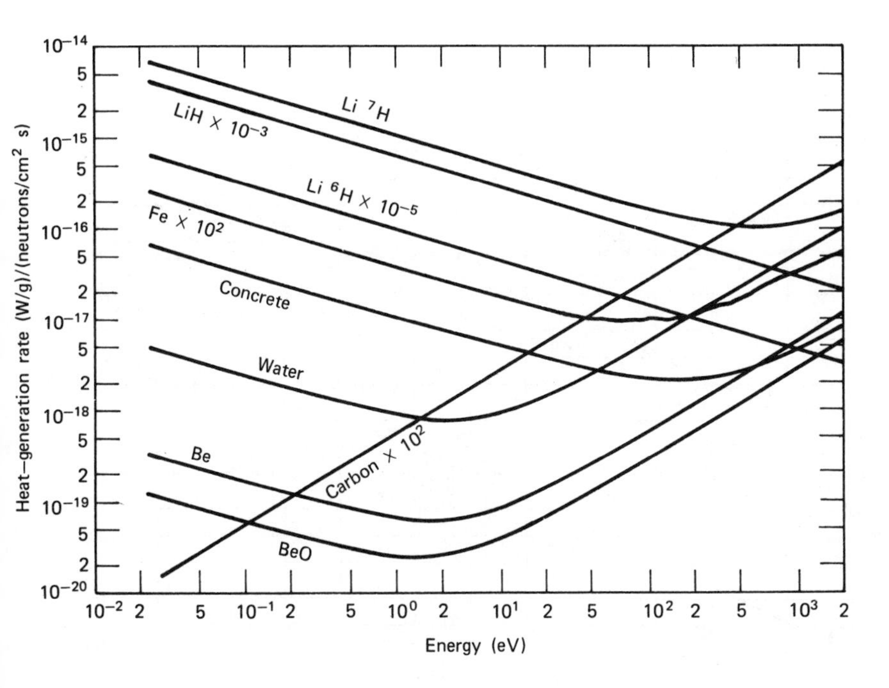

Fig. 7.28. Heating (kerma) factor for 0.02-eV to 1-keV neutrons (after Claiborne, Solomito, et al. [41]).

The maximum allowable temperature may be limited by loss of strength or melting (as in lead), boiling, or other change in properties (e.g., loss of water in concrete). Another consideration is thermal stress, caused by differential thermal expansion in a temperature gradient, if the material is constrained. Thermal stress is most serious when the thermal conductivity is low (resulting in a large temperature gradient) and strength is poor, as in concrete. Stress, especially tensile stress, can cause cracking, with loss of shielding effectiveness if radiation can stream through the cracks. The heat-generation and temperature limits for concrete given in Section 7.2 are sufficient to avoid excessive dehydration or thermal stress. Calculation of thermal stress in hollow cylinders is treated in the *Compendium* [42].

When predicted temperatures or gradients in the biological shield are too large, it is necessary to

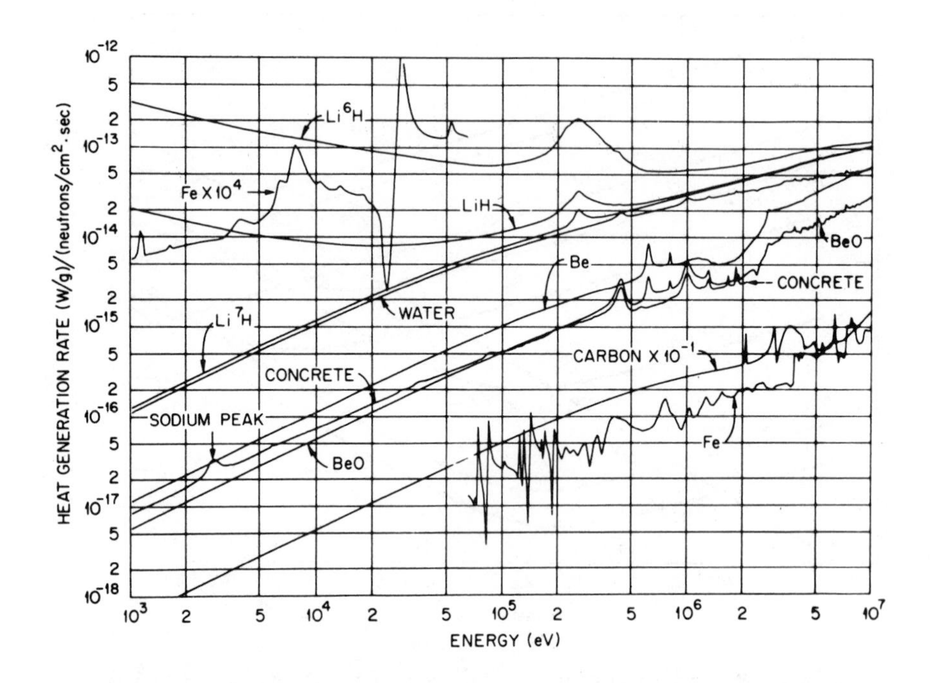

Fig. 7.29. Heating (kerma) factor for 1-keV to 10-MeV neutrons (after Claiborne, Solomito, et al. [41]).

insert a thermal shield. Reactor thermal shields are usually made of steel, divided into layers 2.5 to 5 cm thick, separated by passages for coolant. The steel may contain about 2% boron to absorb thermal neutrons and suppress activation and capture-γ-ray production in the steel. By absorbing most of the γ-radiation energy in the thermal shield, where the heat may be removed readily, cooling the biological shield is much facilitated.

7.6 TESTING PROCEDURES

Castings and other shield components that may contain voids should be inspected before installation, either by radiography or by scanning with a scintillation counter and source. Joints and bonds should be tested for radiation leakage as the shield is assembled because it is easier to make repairs then. Scanning requires a fairly small detector (e.g., 4 cm diameter), moved slowly across the joint while observing or recording the output of a count-rate meter. The scanner should be calibrated on a mockup containing slabs of the thickness being examined, with a known gap width. γ-Ray radiography for shield plates requires a strong, moderately high-energy source (e.g., 50 Ci of ^{60}Co). Proper radiographic inspection methods should be used, including calibration with a penetrameter (block containing holes of known diameter and depth).

Concrete shields for reactors or accelerators are usually constructed in either one layer or two (primary and secondary shields). It may be impractical to test such a shield for lack of a sufficiently strong source, until the reactor or accelerator is brought up to near full power or output. It is very important that the concrete be mixed and poured properly, to achieve design strength, density and hydrogen content, to prevent cracking, and to eliminate segregation and voids. The usual slump measurements and strength tests should be made. If anything, shield concrete should be somewhat mushy, so that it flows freely around obstacles and fills all the spaces (slump of 4 to 6 in.). The aggregate and trial batches should be tested for density. Water content is controlled by mixing proportions (including water mixed with coarse aggregate and sand) and damp curing. Water content may be measured by weight loss on heating (but not so hot that decomposition occurs), or by neutron moderation and scattering from test slabs [43]. A vibrator may be used in placing concrete, but care has to be taken to prevent segregation of the coarse aggregate, especially if it is dense. Methods of fabricating and testing concrete are discussed in other works [1,44].

The final step in shield engineering is to survey the neutron and γ-ray dose rates in accessible regions outside the shield. This is done as a radionuclide source is loaded into the shield and as a reactor is brought up to power [45] or an accelerator is raised to full beam current and energy. The area radiation monitors should be operating and measurements made while power is gradually increased, to detect any hot spots. A complete survey of the plant should be made at, say, 10% of full power, and again at full power.

A typical power plant is divided into radiation zones [46]. For example, zone 1 may be the unrestricted area for office workers and visitors, where the total dose rate cannot exceed 0.25 mrem/h. Zone 2 would be a restricted area that may be occupied by a radiation worker for 40 h per week; hence the maximum dose rate would be 2.5 mrem/h. Zone 3 might be a low-radiation zone where dose rates may be too high for 40 h/week occupancy, but less than 100 mrem/h. Zone 4 would be high-radiation areas (>100 mrem/h), where access is strictly controlled and radiation levels monitored. The radiation levels in zones 3 and 4 may be surveyed and working times planned accordingly. The purpose of the shield survey is to ensure that the allowable dose rates in each of the zones are not exceeded, and incidentally to tell the engineer if the shield was overdesigned, as a guide to future practice.

The instrumentation used for shield surveys is usually one or more γ-ray sensitive portable ion chambers (reading from at least 0.2 mrem/h) and a neutron-sensitive, γ-insensitive survey meter reading in millirem (usually a moderated BF_3 or ^3He proportional counter). Geiger-Müller counters may be used but should be checked against an ion chamber. Any streaming through cracks or other hot spots may be localized with a plastic or NaI scintillation counter. It is very important that the instruments be calibrated against a standard source, and be accurate to $\pm 15\%$.

The dose rate in certain areas may increase with operation because of accumulation of radioactivity. Obviously such areas must be surveyed periodically during operation until maximum levels are established.

There have been few publications of results of shield-radiation surveys. A few results, mostly for research reactors, are given in an RSIC report [47].

7.7 EXAMPLES

We now discuss six examples of shield engineering: (1) ^{60}Co γ-ray source shield, (2) ^{252}Cf source shield, (3) X-ray facility, (4) accelerator facility, (5) fusion reactor, and (6) fission reactor.

Cobalt-60 Source

Shipping and storage shields for ^{60}Co and other γ-ray sources are made of lead encased in steel, or steel alone. Minimum mass is achieved with a spherical shield, but construction is easier if the shield is cylindrical. Let us assume a point isotropic source of 100 mCi and neglect displacement by the source capsule. Cobalt-60 emits one 1.17-MeV and one 1.33-MeV γ ray per decay. These energies are close enough to use an average energy of 1.25 MeV. The strength of a 100-mCi ^{60}Co source is $S = 7.4 \times 10^9$ photons per second (1.25 MeV).

The unshielded flux density at 1 meter is $S/4\pi r^2 = 5.89 \times 10^9\ \gamma\ cm^{-2}\ s^{-1}$. Using the exposure factor from Figure 5.24, the exposure rate is 3.53×10^{-5} R s^{-1} or 127 mR per hour at 1 m. Suppose we want to reduce the exposure rate to 2 mR h^{-1} at 1 m. The transmission factor required is 2/127 = 0.016. Interpolating in Figures 7.4 and 7.5, we find that about 8 cm of lead or 13 cm of steel are required. The weight of a lead sphere of 8 cm radius is 24.3 kg. The weight of a steel sphere of 13 cm radius is 72.3 kg. The exposure rate at the surface of the steel sphere would be $2(100.13)^2 = 118$ mR h^{-1}.

An engineered shield for a radionuclide source includes a protective steel container, a lock to prevent accidental or unauthorized removal of the source, a handle or other means of manipulating the source without undue personnel exposure, and a warning label with the radiation symbol and measured dose rate, either at the surface or at 30 cm from the surface. Large sources ($\geq$ 1 Ci) are usually stored in a lead cask with a shutter and beam port for external irradiation in a collimated beam or a mechanism for inserting a specimen without exposing the experimenter, for internal irradiation. In applications such as γ radiography, a cable is used to extract and position the source remotely. Because mechanisms can fail, it is vital that the source-shield area be monitored with a radiation survey meter as it is approached, in the event the source has not retracted completely into the shield or the shutter has failed to close.

Californium-252 Source

Stoddard and Hootman [17] discuss a method of analyzing shield requirements for ^{252}Cf, and present the results of many discrete-ordinates calculations of the fast neutron, thermal neutron, primary γ ray, and capture-γ-ray dose rates in various materials, normalized to one neutron per second emitted by the source. Let us analyze a shield consisting of a 7.6-cm outer-diameter lead sphere (with a 1-cm-diameter cavity for the source) inside a boronated polyethylene sphere with an outer diameter of 100 cm. Thus the thickness of the polyethylene is 48.8 cm, and the thickness of the lead is 3.3 cm. From Figure 7.16 for unborated polyethylene, we obtain the dose-equivalent rates (multiplied by $4\pi r^2$) listed in Table 7.13. The lead will attenuate the ^{252}Cf γ rays by a factor of 0.1 but will have little effect on the other radiations. Adding, say, 10 mg cm^{-3} of natural boron to the polyethylene will reduce the capture-γ-ray dose by a factor of 0.3 (Figure 7.30) and the thermal neutron dose by a factor of 0.3 (polyethylene is similar to paraffin). Increasing the concentration of boron does not help much, and the capture-γ-ray dose predominates in this shield. The total $4\pi r^2 D$ is 6.1×10^{-4} per neutron s^{-1}, and at the surface of the shield ($r = 50$ cm), the dose-equivalent rate is 1.8×10^{-8} mrem h^{-1} per n s^{-1}. If the dose-equivalent rate is limited to 2.5 mrem h^{-1} at the surface, the maximum neutron emission rate is 1.3×10^8 n s^{-1}. Californium-252 emits 2.3×10^{12} n s^{-1} per gram (Table 2.3); hence the shield could accommodate up to 56 μg of ^{252}Cf.

X-Ray Facility

Design of shielded rooms for X- and γ-ray diagnostic radiography and therapeutic installations is treated [48] in NCRP Report No. 49, which should be consulted by anyone involved in such a design. Figure 7.31 shows an arrangement for an X-ray machine, with the useful beam directed toward a primary barrier (shield wall or floor). The secondary barrier is irradiated by leakage radiation from the source housing and by X-rays scattered by the patient and may be thinner than the primary barrier. Standard barrier materials are lead sheet or concrete.

The weekly exposure (R) in an occupied area at distance d_{pri} (meters) from the source is given by

$$X_u = \frac{\dot{X}_u}{(d_{pri})^2} AWUT \qquad (7.34)$$

where

TABLE 7.13. CALIFORNIUM-252 SHIELD

Radiation	Polyethylene $4\pi r^2 D$	Lead Factor	Boron Factor	Shield $4\pi r^2 D$
Fast neutron	1.0×10^{-4}	1.0	1.0	1.0×10^{-4}
Thermal neutron	1.0×10^{-5}	1.0	0.3	3.0×10^{-6}
Primary γ	2.2×10^{-3}	0.1	1.0	2.2×10^{-4}
Capture γ	1.3×10^{-3}	1.0	0.3	3.9×10^{-4}
			Total	6.1×10^{-4}[a]

[a] Units are $cm^2(mrem\ h^{-1})/(n\ s^{-1})$.

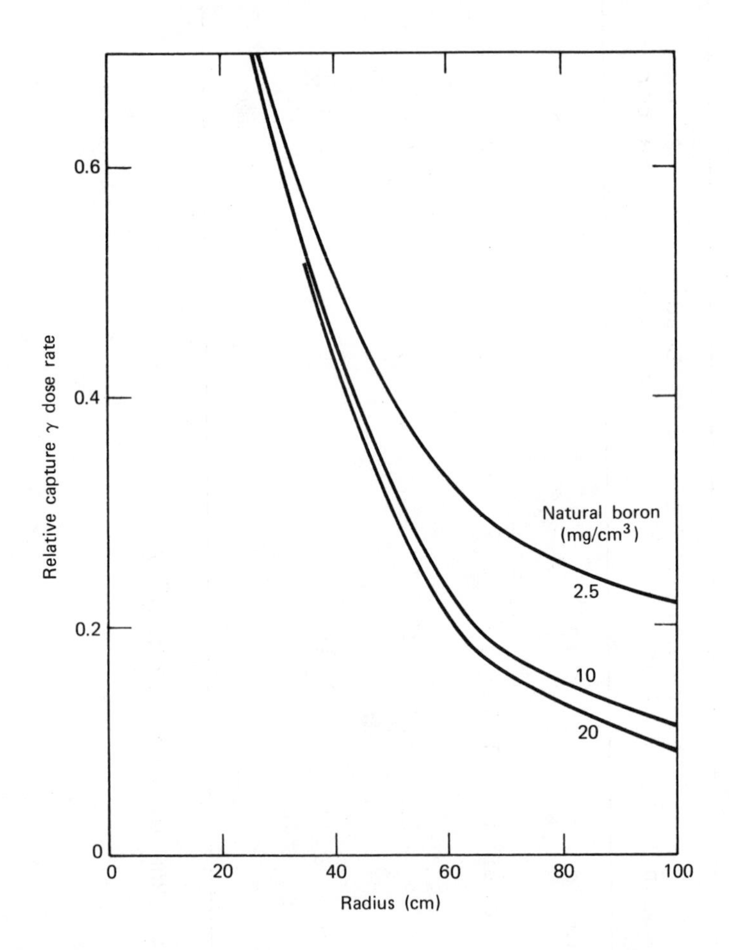

Fig. 7.30. Effect on the capture γ dose rate of adding boron to a spherical paraffin shield (after Stoddard and Hootman [17]).

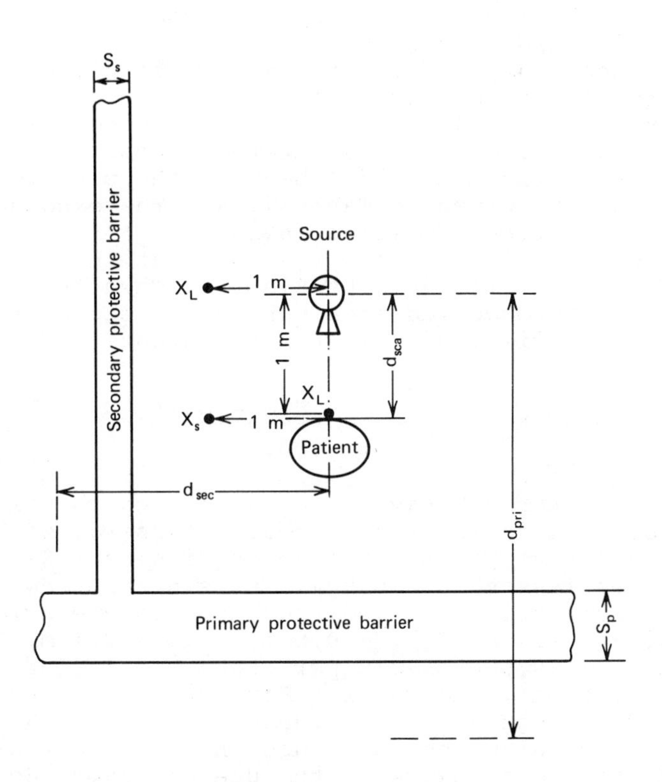

Fig. 7.31. Primary and secondary barriers in an X-ray facility (after Ref. 48).

$\dot{X}_u$ = exposure rate (R/min) at 1 m in unshielded useful beam, per mA of target current;
A = exposure attenuation (transmission) in the barrier;
W = workload, in mA-min per week;
U = use factor, fraction of week beam is directed toward barrier; and
T = occupancy factor for protected area.

For example, in a controlled area the maximum permissible weekly exposure is 0.1 R. Assume W = 20,000 mA-min, U = 0.25, T = 1, and d_{pri} = 2.1 m. Then K = $\dot{X}_u$A

$= 8.8 \times 10^{-5}$ R/mA-min. From Figure 7.8 this value of K requires $S_p = 0.45$ cm lead, for a 200-kVp X-ray machine. From Figure 7.32 we could use 37 cm of concrete instead of the lead.

The secondary barrier thickness depends on the intensity of the scattered plus leakage radiation. The scattered contribution can be evaluated using the experimental ratio, a, of scattered exposure: incident exposure (in the useful beam at the position of the patient) as given in Table 7.14. The exposure rate at the patient position is $\dot{X}_u/(d_{sca})^2$. The ratio, a, applies to a field area of 400 cm^2 at the patient. Let the actual field area be $F(cm^2)$. Then the weekly exposure in an occupied area behind the secondary protective barrier, d_{sec} (meters) distant, is given by

$$X_s = \frac{\dot{X}_u}{(d_{sca})^2} \frac{F\ a\ A\ WUT}{400\ (d_{sec})^2} \tag{7.35}$$

from the scattered X-rays. In evaluating the exposure transmission A it is assumed that the energy of the scattered X-rays is the same as the incident X-rays. At energies less than 500 keV, it is not worthwhile making a correction for the reduction in energy on scattering. Assume $d_{sca} = 0.5$ m, $d_{sec} = 2.1$ m, F = 400 cm^2, W = 20,000 mA-min per week, U = 1 (for scattered radiation), and T = 1. For 200 kVp X-rays scattered through 90°, a = 0.0019; hence $X = 34.5\ \dot{X}_u$ A. The housing should reduce the leakage radiation to 0.001 of the exposure in the useful beam. Hence for the same parameters

$$X_1 = \frac{(0.001)(20{,}000)}{(2.1)^2} \dot{X}_u\ A = 4.5\ \dot{X}_u\ A$$

Scattered plus leakage radiation requires K = (0.1)/(39) = 2.6×10^{-3}. Using Figures 7.8 and 7.32, we obtain a secondary barrier thickness $S_s = 0.28$ cm of lead or 20 cm of concrete for a 200-kVp machine.

Accelerator Facility

An existing facility for a Cockcroft-Walton accelerator (14-MeV neutron generator) is shown in Figure 7.33. A distance of 30 ft (915 cm) and a concrete wall 32 in. (81 cm) thick are used to reduce the dose rate at the

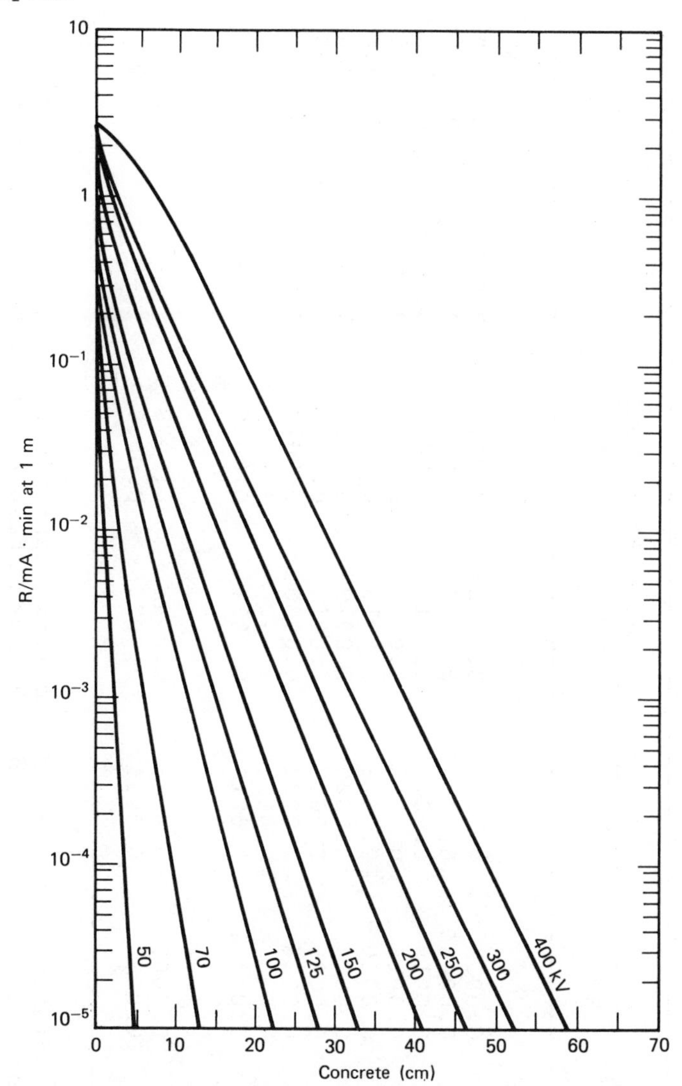

Fig. 7.32. Broad-beam transmission of X-rays through concrete, density 2.35 g/cm^3. At 50 to 300 kV: half-wave generator, tungsten reflection target; total beam filtration 1 mm aluminum at 50 kV, 1.5 at 70, 2 at 100, and 3 at 125 to 300. At 400 kV: constant potential generator, gold reflection target; 3 mm copper total beam filtration. Ordinate intercepts are 2.7 at 400 kV, 2.4 at 300 kV, 1.6 at 250 kV, 1.02 at 200 kV, 0.6 at 150 kV, 0.45 at 125 kV, 0.32 at 100 kV, 0.24 at 70 kV, and 0.19 at 50 kV (ICRP Publication 21).

TABLE 7.14. RATIO, a, OF SCATTERED:INCIDENT EXPOSURE[a]

Source	Scattering Angle (from Central Ray					
	30	45	60	90	120	135
X Rays						
50 kV	0.005	0.0002	0.00025	0.00035	0.0008	0.0010
70 kV	0.00065	0.00035	0.00035	0.005	0.0010	0.0013
100 kV	0.0015	0.0012	0.0012	0.0013	0.0020	0.0022
125 kV	0.0018	0.0015	0.0015	0.0015	0.0023	0.0025
150 kV	0.0020	0.0016	0.0016	0.0016	0.0024	0.0026
200 kV	0.0024	0.0020	0.0019	0.0019	0.0027	0.0028
250 kV	0.0025	0.0021	0.0019	0.0019	0.0027	0.0028
300 kV	0.0026	0.0022	0.0020	0.0019	0.0026	0.0028
4 MV	--	0.0027	--	--	--	--
6 MV	0.007	0.0018	0.0011	0.0006	--	0.0004
γ Rays						
^{137}Cs	0.0065	0.0050	0.0041	0.0028	--	0.0019
^{60}Co	0.0060	0.0036	0.0023	0.0009	--	0.0006

[a] Scattered radiation measured at 1 m from phantom when field area is 400 cm^2 at the phantom surface; incident exposure measured at center of field 1 m from the source but without phantom.

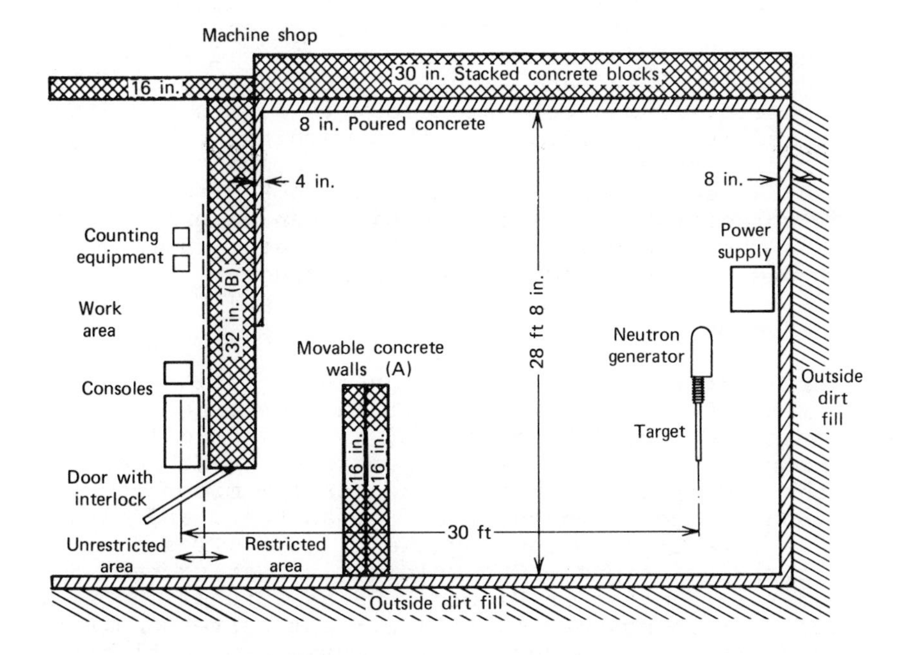

Fig. 7.33. Laboratory for a 14-MeV neutron generator emitting 4×10^{10} n s^{-1} (courtesy Texas Nuclear Corporation).

control console to an acceptable level. The accelerator generates 4×10^{10} n s^{-1} (14 MeV). Unshielded, the flux density at the console would be 3.8×10^3 n cm^{-2} s^{-1}, and the dose-equivalent rate would be 841 mrem h^{-1}, using the maximum dose-equivalent factor for 14-MeV neutrons from Table 5.14. The 81 cm of concrete should reduce the dose-equivalent rate to 4.5 mrem h^{-1}, based on the attenuation in Figure 7.12. This is acceptable considering the workload. The same accelerator would require over 120 cm of concrete shielding if the distance were reduced to 300 cm. Most of the dose equivalent comes from the fast neutrons. A thinner shield can be achieved if the first 50 cm or so immediately around the target is water, followed by concrete.

Figure 7.33 also shows a simple maze entrance. The door has little shielding ability, but is interlocked to shut off accelerator power when the door is opened. The accelerator cannot be started again until the door is closed and a reset button on the console is pressed. The accelerator room is a high radiation area and must be provided with a warning light, sign, and radiation monitor. As an added precaution, plugs can be wired in series in the interlock circuit. Each worker takes a plug with him when he enters the room. The accelerator cannot be operated until everyone has left and returned his plug.

Fusion Reactors

Figure 7.34 illustrates a design for a fusion reactor blanket [49], consisting of lithium contained in niobium and reflected by graphite, in cylindrical geometry. The plasma region generates 14-MeV neutrons in the $T(d,n)^4He$ reaction. The lithium moderates the neutrons, recovering their energy as heat, and also breeds more tritium fuel by the $^7Li(n,n'\alpha)T$ and $^6Li(n,\alpha)T$ reactions. The volumetric heating rate (W/cm^3) is plotted in Figure 7.35 as a function of distance, for an assumed neutron current corresponding to 10 MW/m^2 kinetic energy carried outward. Heating is produced by neutron scattering and by absorption of secondary γ rays. Discrete-ordinates calculations give 1.33 for tritium breeding (T atoms per fusion neutron), 0.90 from ^6Li and 0.43 from ^7Li. Additional

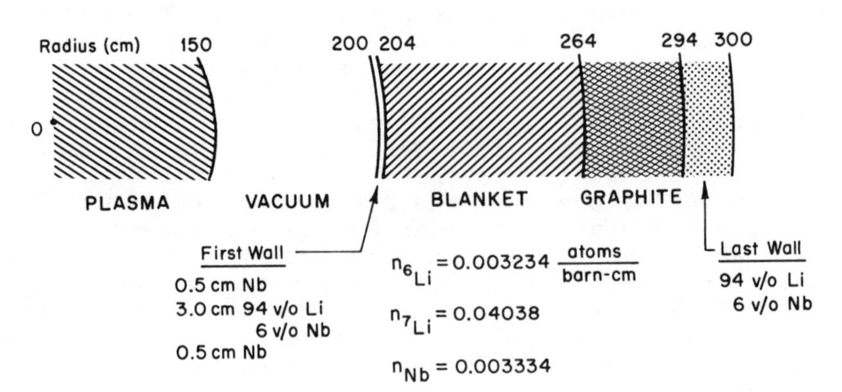

Fig. 7.34. Cylindrical fusion reactor blanket model (after Steiner and Blow [49]).

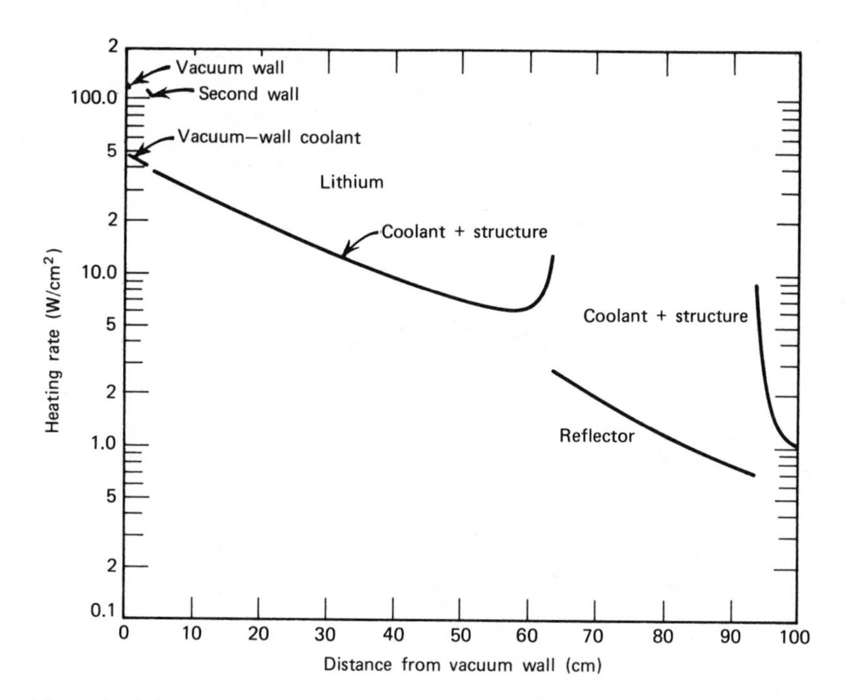

Fig. 7.35. Heating rate as a function of distance from vacuum wall in the cylindrical graphite-reflected Li-Nb blanket model, for 10-MW/m^2 neutron loading.

shielding is required to reduce nuclear heating if there is a cryogenic magnet.

The blanket and shield for an actual fusion reactor is likely to be more complex than in the example because of penetrations and noncylindrical geometry. A very complete study has been made [50] for a Tokomak device, considering radiation damage as well as heating and other parameters.

Fission Reactors

Shielding design for nuclear power plants requires computer calculations. Although some success has been achieved with approximate methods such as kernel integration with buildup (for γ rays) and removal diffusion (for neutrons), the trend is to apply two-dimensional discrete-ordinates calculations and three-dimensional Monte Carlo calculations, tested against experiment.

Figure 7.36 illustrates the general design of the concrete shielding for a pressurized water reactor (PWR) [51]. The reactor vessel is surrounded by about 6 ft (183 cm) of reinforced concrete, which also serves as a support. The dose rate in the reactor-vessel cavity, at the midplane of the core (position 1), is about 10^4 rad/h from γ rays plus 3×10^5 rem/h from

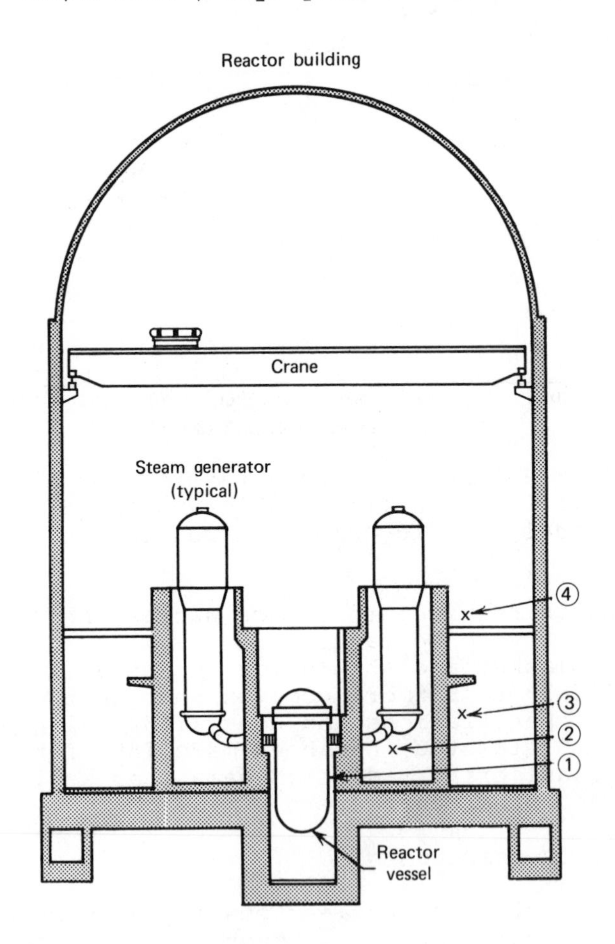

Fig. 7.36. Concrete shielding and primary loop components in a typical PWR power plant.

neutrons, during operation at a nominal 3000 MW (thermal). Figure 7.37 shows the calculated γ-ray and neutron flux densities as a function of distance through the primary shield. Outside the primary shield most of the dose during operation comes from the ^{16}N activity in the water or steam. The dose rate near the steam generator (position 2) is about 50 rad/h

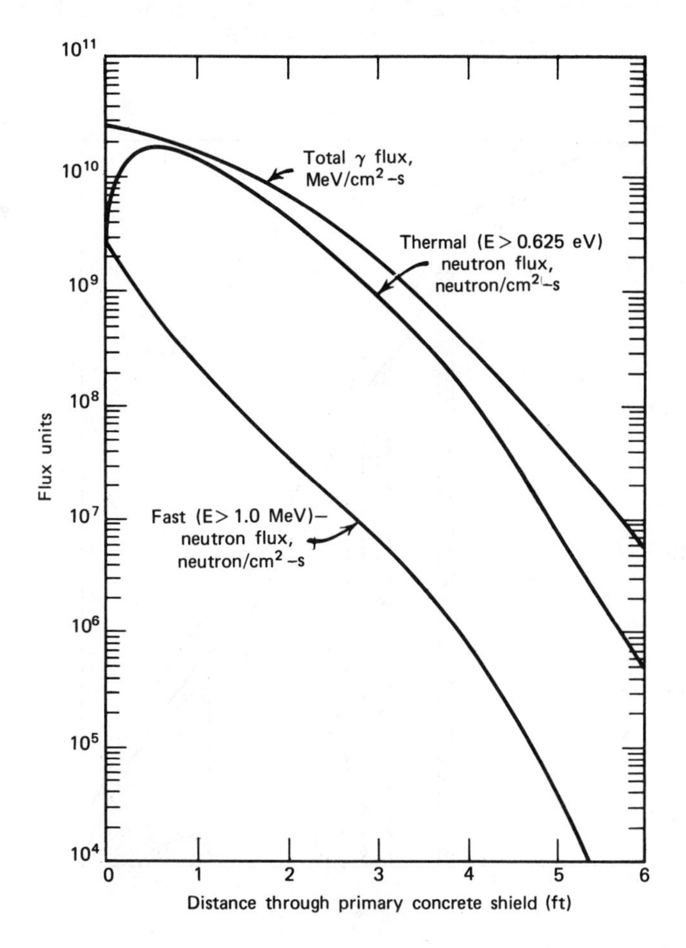

Fig. 7.37. Neutron and γ-ray flux densities in the primary concrete shield of a 3000-MWt PWR (after Locante, Lowder, et al. [46]).

from γ rays and 0.2 rem/h from neutrons. The steam generators, pumps, and piping are shielded by 2 to 3 ft of concrete, reducing dose rates outside (position 3) to about 0.2 rad/h (γ) and 0.2 rem/h (neutron). The dose rates on the operating deck (position 4) vary from 5 to 10 mrem/h.

After shutdown, and decay of the 7-s ^{16}N activity, the dominant source is ^{58}Co- and ^{60}Co-activated corrosion products, plated out in the primary loop. γ-Ray dose rates may be several rad/h in the annulus between the reactor vessel and primary shield. Another problem

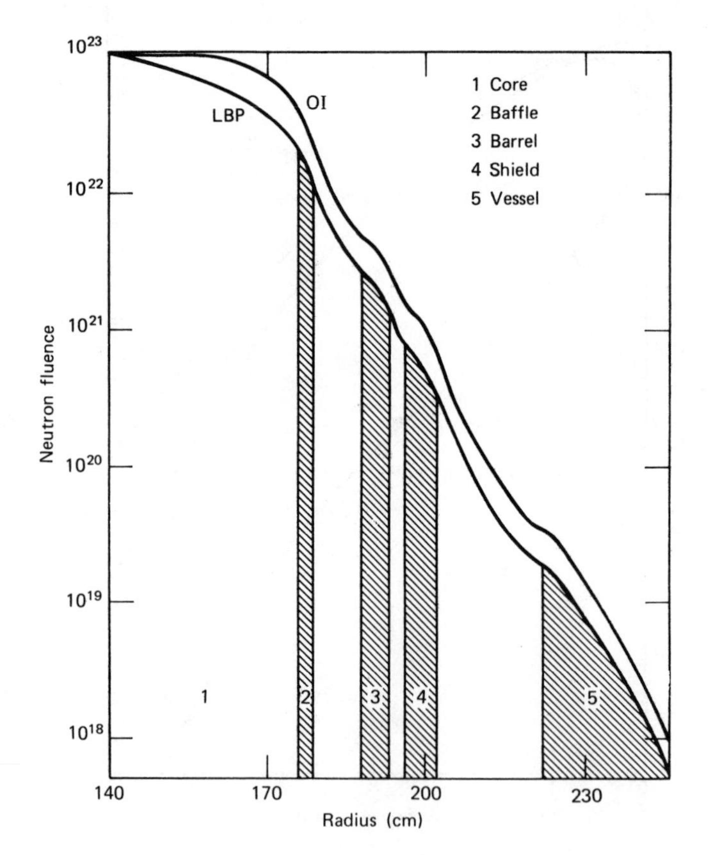

Fig. 7.38. Predicted fast neutron fluences over design life in a PWR with OI or LBP fuel-reloading schemes (after Sapyta and Simmons [53]).

is the streaming of neutrons in the annulus, partially thermalized in the concrete, leading to excessive activation of the reactor-vessel head and surrounding area, limiting access during refueling and increasing dose rates on the operating deck because of capture γ rays [52]. Additional shielding may be required above the annulus.

The shield designer may be called on to predict flux densities at the reactor control and safety ion chambers (in the annulus), the heating in the thermal shield, reactor vessel, and biological shield, and the fast neutron-fluence and spectrum for predicting radiation damage. Figure 7.38 plots on the fast-neutron fluence (>1 MeV) in the reactor vessel for the design 32 full-power-years service life of a 3000-MW PWR [53], as calculated by the ANISN discrete ordinates code. The abbreviation LBP refers to lumped burnable poison, and OI to out/in, fuel-reloading schemes and corresponding flux-density distributions as plotted in Figure 2.27. The maximum allowable fluence is 3.5 to 5.0×10^{13} n/cm^2, depending on the steel.

Analysis of reinforced ordinary concrete shielding for nuclear power plants is treated in an American National Standard [54]. Data are given for approximate design calculations using point-kernel, removal, and albedo-scatter methods, as well as guidelines on choice of computational procedures.

REFERENCES

1. R. G. Jaeger, Ed., *Engineering Compendium on Radiation Shielding*, Vol. II, *Shielding Materials*, Springer-Verlag, New York, 1975, Chapter 9.
2. R. C. Weast, Ed., *Handbook of Chemistry and Physics*, The Chemical Rubber Co., Cleveland, Ohio, periodically revised.
3. A. Hönig, "The Attenuation Factors for Broad Beam Gamma Radiation in Lead, Steel, Concrete, and Water," *Inzenyske Stavby*, 3, 117-130 (1968).
4. "Data for Protection Against Ionizing Radiation from External Sources: Supplement to ICRP Publication 15," ICRP Publication 21, Pergamon Press, Oxford, 1973.
5. H. L. Beck, "Gamma-Ray Spectrum from a Point ^{137}Cs Source in Infinite Water," in A. E. Profio, Ed., *Shielding Benchmark Problems*, Report ORNL-RSIC-25 (ANS-SD-9), 1968. See also M. Alberg, *Nucl. Sci. Eng.*, 30, 65-74 (1967).

6. J. D. Marshall and M. B. Wells, "A Comparison of Monte Carlo Calculated and Experimentally Determined Energy and Angle Spectra for ^{137}Cs," Radiation Research Associates Report RRA-M74 (1967).
7. Charles W. Garrett, "Gamma-Ray Dose Above a Plane Source of ^{60}Co on an Air/Ground Interface," ORNL-RSIC-25 (ANS-SD-9) (1968).
8. Herbert Goldstein and J. Ernest Wilkins, Jr., "Calculations for the Penetration of Gamma Rays," Report NYO-3075 (1954).
9. R. G. Jaeger, Ed., *Engineering Compendium on Radiation Shielding*, Vol. I, Springer-Verlag, New York, 1968, Chapter 4.
10. F. J. Allen and A. T. Futterer, "Neutron Transmission Data," *Nucleonics*, 21 (8), 120 (1963).
11. F. A. R. Schmidt, "The Attenuation Properties of Concrete for Shielding of Neutrons of Energy Less Than 15 MeV," Report ORNL-RSIC-26 (1970).
12. N. M. Schaeffer, Ed., "Reactor Shielding for Nuclear Engineers," TID-25951 (1973) pp. 458 and 460. The data were obtained from J. P. Nichols, USAEC Report ORNL-TM-1167 (1965) and F. B. F. Kam and F. H. S. Clark, *Nucl. Appl. Technol.*, 3, 433-435 (1967).
13. "Attenuation in Water of Radiation from the Bulk Shielding Reactor," ORNL-2518 (1958), quoted in ANL-5800 Reactor Physics Constants.
14. L. Harris, Jr., "Measurement of Fast Neutron Spectra in Water and Graphite," University of Michigan, Department of Nuclear Engineering Report 07786-1-T (1967).
15. A. E. Profio, R. J. Cerbone, and D. L. Huffman, "The Neutron Spectrum from a Fission Source in Water," *Nucl. Sci. Eng.*, 49, 232-236 (1972).
16. H. Ing and W. G. Cross, "Spectra and Dosimetry of Neutrons from Moderation of ^{235}U and ^{252}Cf Fission Sources in H_2O," *Health Physics*, 29, 839-851 (1975).
17. D. H. Stoddard and H. E. Hootman, "^{252}Cf Shielding Guide," AEC Savannah River Laboratory Report DP-1246 (1971).
18. A. E. Profio, R. J. Cerbone, and D. L. Huffman, "Fast Neutron Penetration in Paraffin and Lithium Hydride," *Nucl. Sci. Eng.*, 44, 376-387 (1971).
19. C. E. Burgart, "Neutron and Secondary Gamma Ray Fluence Transmitted Through a Slab of Borated

Polyethylene," Shielding Benchmark Problems, ORNL-RSIC-25 (ANS-SD-9), Suppl. 2 (1973).

20. G. P. Lahti, "Fission Neutron Attenuation in Lithium-6, Natural Lithium Hydride, and Tungsten," Report NASA TN D-4684 (1968).
21. A. E. Profio, R. J. Cerbone, and D. L. Huffman, "Experimental Verification of Neutron and Gamma-Ray Transport Calculations in Lithium Hydride and Tungsten," Air Force Report AFWL-TR-68-138 (1969).
22. A. E. Profio, H. M. Antuñez, and D. L. Huffman, "The Neutron Spectrum from a Fission Source in Graphite," *Nucl. Sci. Eng.*, 35, 91-103 (1969).
23. A. E. Profio, "Fast Neutron Spectrum from a Point Fission Source in Infinite Graphite," ORNL-RSIC-25 (ANS-SD-9) (1969).
24. E. A. Straker, "Neutron Spectrum from Point Fission and 14 MeV Sources in Infinite Air," ORNL-RSIC-25 (ANS-SD-9) (1969).
25. "A Review of Calculations of Radiation Transport in Air: Theory, Techniques, and Computer Codes," Report ORNL-RSIC-33 (CONF-711125) (1971).
26. W. E. Selph, "Albedos, Ducts, and Voids," in *Reactor Shielding for Nuclear Engineers*, N. M. Schaeffer, Ed., TID-25951 (1973), Chapter 7.
27. D. K. Trubey, "A Calculation of Radiation Penetration of Cylindrical Duct Walls," Report ORNL-CF-63-2-64 (1963).
28. R. E. Maerker and F. J. Muckenthaler, "Calculation and Measurement of the Fast-Neutron Differential Dose Albedo for Concrete," *Nucl. Sci. Eng.*, 22, 455-462 (1965).
29. F. J. Allen, A. Futterer, and W. Wright, "Dependence of Neutron Albedos on the Hydrogen Content of a Shield," Report BRL-1224, Ballistics Research Laboratories (1963).
30. R. L. French and M. B. Wells, "An Angular Dependent Albedo for Fast-Neutron Reflection Calculations," RRA-M31 (1963).
31. W. A. Coleman, R. E. Maerker, F. J. Muckenthaler, and P. N. Stevens, "Calculation of Doubly Differential Current Albedos for Epicadmium Neutrons Incident on Concrete and Comparison of the Reflected Subcadmium Component with Experiment," *Nucl. Sci. Eng.*, 27, 411 (1967).
32. Wade E. Selph, "Neutron and Gamma Ray Albedos," ORNL-RSIC-21 (DASA-1892-2) (1968).

33. A. Mockel, "Reflection and Transmission by a Strongly-Absorbing Slab," *Nucl. Sci. Eng.*, 22, 339 (1965).
34. M. B. Wells, "Differential Dose Albedos for Calculation of Gamma-Ray Reflection from Concrete," Radiation Research Associates Report RRA-T46 (1964).
35. A. B. Chilton and C. M. Huddleston, *Nucl. Sci. Eng.*, 17, 419 (1963); A. B. Chilton, C. M. Davisson, and L. A. Beach, *Trans. Am. Nucl. Soc.*, 8, 656 (1965); A. B. Chilton, "A Modified Formula for Differential Exposure Albedo for Gamma Rays Reflected from Concrete," *Nucl. Sci. Eng.*, 27, 481 (1967).
36. A. Simon and C. E. Clifford, "The Attenuation of Neutrons by Air Ducts in Shields," *Nucl. Sci. Eng.*, 1, 156-166 (1956).
37. R. D. Schamberger, F. J. Shore, and H. P. Sleeper, "BNL-2019 to BNL-2028," (1954). Quoted in B. T. Price, C. C. Horton, and K. T. Spinney, *Radiation Shielding*, Pergamon Press, Oxford, 1957, pp. 207-209.
38. R. E. Maerker and F. J. Muckenthaler, "Monte Carlo Calculations Using the Albedo Concept of Fast-Neutron Dose Rates Along the Center Lines of One- and Two-Legged Square Concrete Open Ducts, and Comparison with Experiment," ORNL-TM-1557 (1966); *Trans. Am. Nucl. Soc.*, 9, 147 (1966).
39. *Engineering Compendium on Radiation Shielding*, Vol. I, p. 440.
40. J. Hubbell, Report NSDRS-NBS 29 (1969).
41. H. C. Claiborne, M. Solomito and J. J. Ritts, "Heat Generation by Neutrons in Some Moderating and Shielding Materials," *Nucl. Eng. Design*, 15, 232-236 (1971).
42. *Engineering Compendium*, Vol. I, pp. 473-486.
43. K. Preiss and P. J. Grant, "The Optimization of a Neutron Scattering Water Content Gauge for Soils or Concrete," *J. Sci. Instrum.*, 41, 548 (1964).
44. American National Standard ANS-11.13/N101.6-1972, "Concrete Radiation Shields," American Nuclear Society (1972).
45. American National Standard ANS-18.9-1972, "Procedure for Testing Biological Shielding in Nuclear Power Plants," American Nuclear Society (1972).
46. J. Locante, W. M. Lowder, B. A. Engholm, and W. E. Kreger, "Panel Discussion: Engineering Problems

in Power Reactor Shielding," *Nucl. Technol.*, 26, 496-507 (1975).
47. "Compilation of Data on Experimental Shielding Facilities and Tests of Shields of Operating Reactors," Report ORNL-RSIC-24 (EACRP-U-37) (1968).
48. "Structural Shielding Design and Evaluation for Medical Use of X-Rays and Gamma Rays of Energies Up to 10 MeV," NCRP Report No. 49, National Council on Radiation Protection and Measurements, Washington, D. C. (1976).
49. D. Steiner and S. Blow, Culham Report CLM-P345, and D. Steiner, "The Nuclear Performance of Fusion Reactor Blankets," *Nucl. Appl. Technol.*, 9, 83-91 (1970).
50. M. A. Abdou, "Nuclear Design of the Blanket/ Shield System for a Tokomak Experimental Power Reactor," *Nucl. Technol.*, 29, 7 (1976).
51. J. Sejvar, "Normal Operating Radiation Levels in Pressurized Water Reactor Plants," *Nucl. Technol.*, 36, 48-55 (1977).
52. "Nuclear Reactor Shielding," in R. W. Roussin, L. S. Abbott, and D. E. Bartine, Eds., *Proc. 5th Internat. Conf. Reactor Shielding*, Knoxville, Tenn., April 18-23, 1977, Science Press, Princeton, N. J.
53. J. J. Sapyta and G. L. Simmons, "Shielding Design and Analysis Methods for Pressurized Water Reactors," *Nucl. Technol.*, 26, 508-515 (1975).
54. American National Standard ANSI/ANS-6.4-1977 (N403), "Guidelines on the Nuclear Analysis and Design of Concrete Radiation Shields for Nuclear Power Plants."

PROBLEMS

1. Compare water, polyethylene, and concrete for shielding a 10-mg source of ^{252}Cf, on a mass, thickness, and cost basis.

2. Derive expressions for the β-particle and γ-ray dose rates to the skin or whole body from submersion in a semiinfinite (for β) or infinite cloud of air containing radioactive gas at a concentration of χ(Ci m^{-3}) if the average β-particle energy per decay is $\bar{E}_\beta$ and the average γ-ray energy per decay is $\bar{E}_\gamma$ (MeV).

3. Design a compact concrete shield (not a room) for a 14-MeV neutron generator emitting 4×10^{10} n s^{-1}. Allow a 30-cm-diameter void around the target for sample irradiations. Consider how you would service the accelerator, insert and remove samples, and run cables to the accelerator from the console and power supply located outside the shield.

4. A straight cylindrical hole, 5 cm in diameter, runs perpendicularly through a concrete wall 50 cm thick. A plane, isotropic, 1.0-MeV γ-ray source emitting 10^7 photons/s per cm^2 into the forward hemisphere is located on one side. Calculate the dose rate on the hole axis on the opposite side. Compare with the dose rate without the hole.

5. Design a slab shield of ordinary concrete for a perpendicularly incident beam of fission neutrons at a flux density of 10^5 n/cm^2s, plus 1 MeV γ rays at 10^5 γ/cm^2s, if the dose-equivalent rate on the opposite face is not to exceed the occupational limit.

6. Calculate the maximum temperature in a 1-cm thick slab of lead irradiated by a perpendicularly-incident beam of 500 keV γ rays.

8

RADIOLOGY

8.1 DIAGNOSTIC RADIOLOGY

Roentgenography

The physics of image formation in diagnostic roentgenography is discussed in the literature [1-3]. Here we are concerned with contrast, transmission, and dose to the body as influenced by the attenuation and absorption of X-rays.

X-Rays are used to image bone and soft tissue, sometimes with the aid of a contrast medium such as iodine or barium compound. The geometry is shown schematically in Figure 8.1. The electron beam of the X-ray tube is focused to a small spot ($\sim$1 mm^2) on the anode, which is usually made of tungsten. The bremsstrahlung and fluorescence X-rays diverge from the focus, limited by a diaphragm and cone. The X-rays pass through a filter, which may be only the wall of the tube (intrinsic filtration), or include a filter added to absorb low-energy photons. These photons will contribute little, if anything, to the image because of absorption in the body but do contribute to the dose. The patient is located at a distance FSD (focus-to-skin distance) or SSD (source-to-skin distance), normally 50 to 100 cm. The imaging detector is located at distance FSD, usually immediately behind the patient. The detector may be silver halide emulsion (film) with or without fluorescent intensifying screens, a charged selenium plate (xeroradiography), or the fluorescent screen of an image-intensifier tube as used in fluoroscope. In any event, some minimum fluence or dose is required to produce a usable response in the detector.

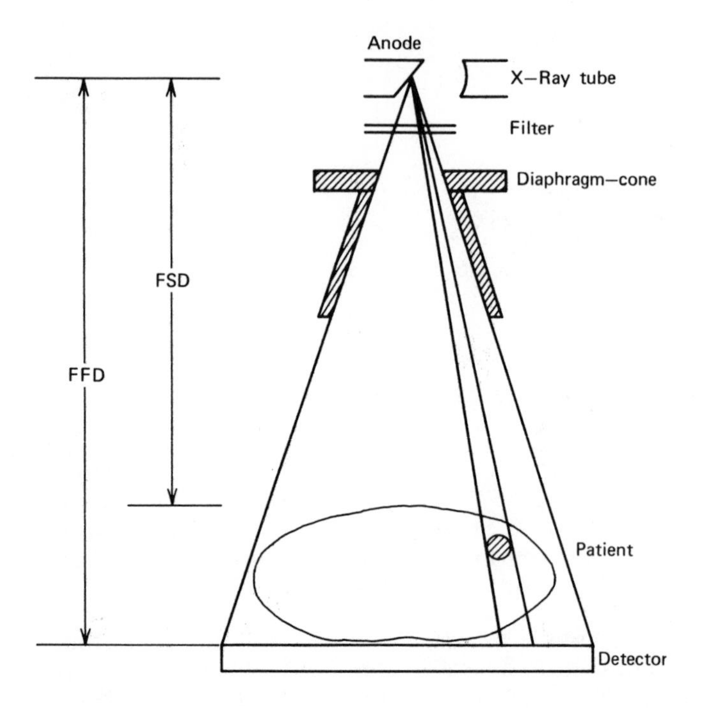

Fig. 8.1. Geometry for diagnostic radiology.

The desired image is produced by the differential attenuation of the uncollided photons, as indicated in Figure 8.2. The incident flux density ϕ_0 is the same for the first ray passing through a thickness T of tissue of linear attenuation coefficient μ_a and the second ray passing through tissue of thickness t and attenuation coefficient μ_b imbedded in the other tissue. The transmitted flux density (at a given energy) in the first ray is

$$\phi_1 = \phi_0 \exp -\mu_a T \quad (8.1)$$

and for the second ray

$$\phi_2 = \phi_0 \exp -\mu_a (T - t) \exp -\mu_b t \quad (8.2)$$

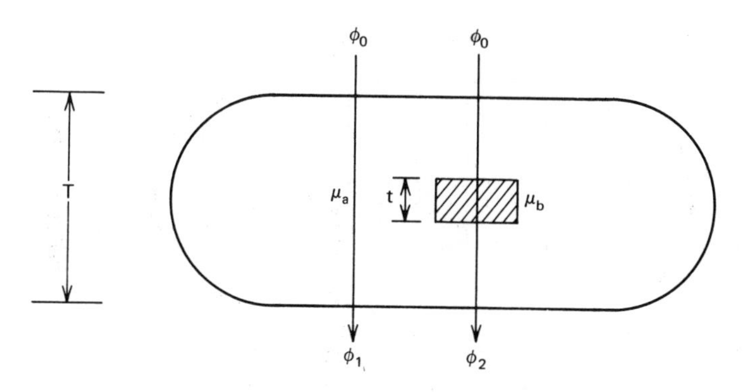

Fig. 8.2. Attenuation in object of attenuation coefficient μ_b and thickness t, imbedded in matrix of attenuation coefficient μ_a and thickness T.

The object contrast

$$C = \frac{\phi_2 - \phi_1}{\phi_1} = \exp -(\mu_a - \mu_b)t - 1 \qquad (8.3)$$

and if $(\mu_a - \mu_b)t << 1$,

$$C \simeq (\mu_b - \mu_a)t \qquad (8.4)$$

Object contrast depends on thickness and the difference in linear attenuation coefficients, which in turn depends on the elemental compositions and densities. In a polyenergetic beam the spectrum varies with depth because of preferential absorption of low-energy photons, but for most purposes the attenuation coefficients may be evaluated at the average energy.

Contrast is diminished by scattered photons, which tend to fill in the X-ray "shadow" of an attenuating object, as shown in Figure 8.3. The number of scattered photons detected can be reduced by introducing a collimator or grid (Bucky grid) of thin lead strips, spaced by strips of low-absorption material such as aluminum. The uncollided photons travel more or less parallel to the strips and some can still reach the film, but the obliquely scattered photons are absorbed. The grid casts a shadow, but the

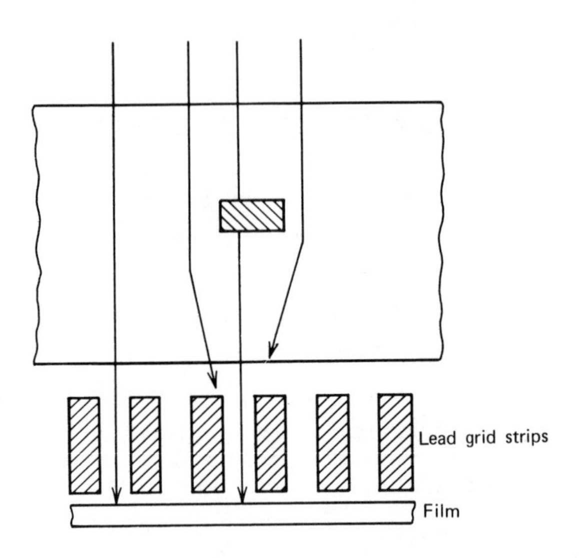

Fig. 8.3. Reduction of contrast by scattered photons and use of grid.

appearance is improved by making the strips very thin or by moving the grid perpendicular to the strips during the exposure. Since some uncollided photons are absorbed, the exposure and thus the dose to the patient must be increased.

Scattered photons tend to blur the image, reducing spatial resolution or sharpness as well as contrast. Blurring also occurs if the source, patient, or detector moves during the exposure, and involuntary patient or organ movement limits the duration of an exposure. Geometrical unsharpness or blurring arises from the finite size of the focal spot or source, resulting in a penumbra (region of partial shadow) instead of just the umbra (shadow) desired, as seen in Figure 8.4. The width of the penumbra at the detector is the geometric unsharpness

$$U = F \frac{\ell}{L} \tag{8.5}$$

where F is the diameter of the source, L is the distance from source to object point, and ℓ is the distance from object to the detector. For a point source

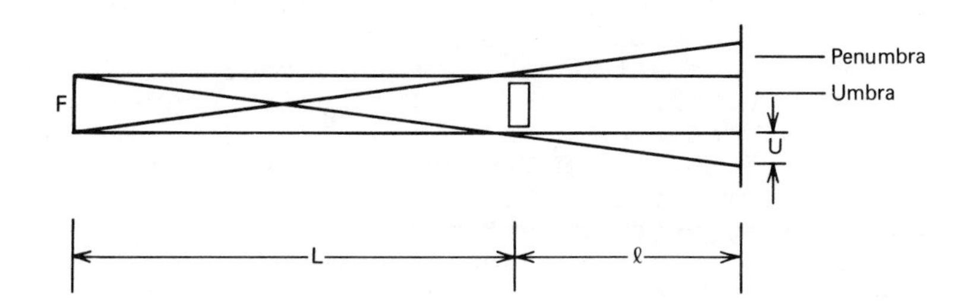

Fig. 8.4. Geometric unsharpness U for finite-source diameter F, source-to-object distance L, and object-to-detector distance ℓ.

U is zero. Geometric unsharpness is usually negligible in medical roentgenography because of the small focal spot and large FSD but can be significant in other applications.

X-Ray Dosimetry

The detector sensitivity determines the minimum fluence or dose required at the detector. For a medium speed film exposed between medium-speed fluorescent intensifying screens, an exposure of 1 mrad gives an optical density of 1 above base density and fog. X-Ray film exposed without intensifying screens may require 40 or 50 mrad.

The dose at the incident surface, or the maximum dose within the body, is larger because of the attenuation in the body. The dose transmission of the useful (uncollided) photons is

$$\frac{D}{D_0} = e^{-\mu T} \tag{8.6}$$

For example, if the effective energy of the X-ray beam is 60 keV and we approximate the body by water, $\mu = 0.2\ \text{cm}^{-1}$. Then if the thickness T = 20 cm and exit dose D = 1 mrad, the dose at the incident surface D_0 = 55 mrad.

Scattered photons contribute a large fraction of the absorbed dose in the body irradiated with a broad beam of X-rays or γ rays. Jones [4] measured the total (scattered and uncollided) dose with LiF TLDs as a function of position in an anthropomorphic phantom of tissue-equivalent material (including bones in head and trunk), irradiated with 11 monoenergetic γ-ray sources from 0.025 to 1.25 MeV. Results were reported later [5] in terms of the absorbed dose at the position of the testes, ovaries, gut mucosa, and bone marrow, normalized to one roentgen exposure at the FSD with no phantom in place (this is standard practice for electromagnetic radiation where the output of a particular machine is measured by the free-air exposure). The largest absorbed dose is to the testes because they are only lightly shielded by overlying tissue. The ratio of testes absorbed dose to free-air exposure (R) varies with energy. The maximum is 1.60 and occurs at 100 keV. Results are also given [5,6] normalized to the exposure measured by a film badge on the chest. This is of interest in radiation protection, but in medical practice the X-ray beam would be collimated and the testes and ovaries covered with a lead apron, unless a radiograph of the gonadal region were required.

Delafield [7] measured the absorbed dose in an elliptical water phantom approximating the trunk for low- and high-energy γ-ray sources incident in a broad beam parallel to the minor axis. Results are expressed as "percent depth dose" or percent of surface dose at depth, $100(D/D_0)$, where D is the absorbed dose at the surface of incidence, as plotted in Figure 8.5. If the absorbed dose (tissue) is measured in air (no phantom) at the FSD, the reading must be multiplied by the backscatter factor to correct for the increase in absorbed dose at the surface because of scattering in the phantom, or body. If the free-field exposure X_0 is measured instead of absorbed dose, an additional R-to-rad correction factor f_x is needed. Thus X_0 roentgens measured in air at the FSD with no phantom results in an absorbed dose at depth

$$D = X_0 f_x f_b p \tag{8.7}$$

where f_x = 0.87 rad per roentgen for muscle at low energies (Figure 5.26), f_b is the surface backscatter

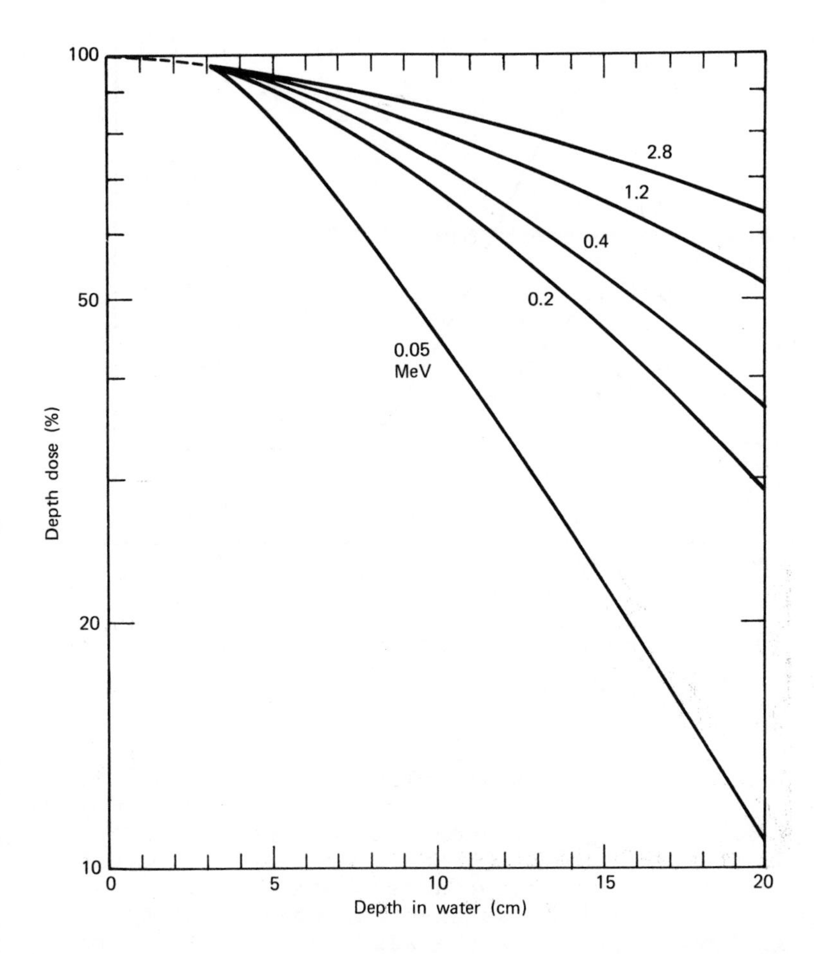

Fig. 8.5. Absorbed dose at depth in an elliptical water phantom as percentage of surface dose, for protons of different energies (after Delafield [7]).

factor plotted in Figure 8.6, and p is the fraction of surface dose at depth (percent depth dose from Figure 8.5, divided by 100).

Monte Carlo calculations of the absorbed dose in an elliptical water phantom have been published by Sidwell, Burlin, et al. [8]. The phantom approximates

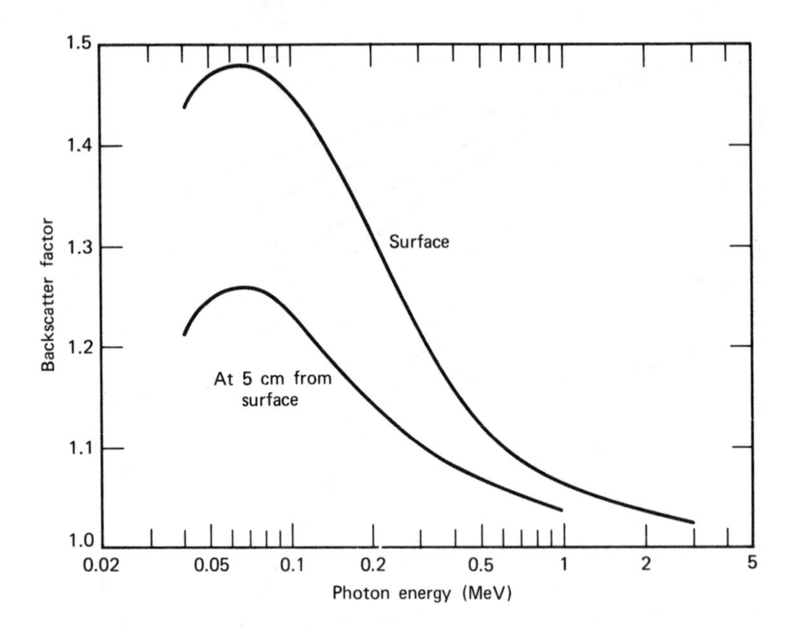

Fig. 8.6. Backscatter factor for correcting absorbed dose without phantom to absorbed dose at surface of a phantom (after Delafield [7]).

the trunk by an elliptical cylinder 60 cm high, with a major axis of 30 cm and a minor axis of 20 cm. Calculations were made for monoenergetic γ rays with energies from 0.02 MeV to 1.25 MeV, incident in a broad beam parallel to the minor axis. Some results are plotted in Figure 8.7, as absorbed dose (rad) per unit fluence (photons/cm^2) at the surface, as a function of depth along the minor axis. These curves may be used in lieu of Figures 8.5 and 8.6 and Eq. (8.7) because they are normalized to fluence instead of exposure.

Claiborne and Trubey [9] have used Monte Carlo and discrete-ordinates methods to calculate the absorbed-dose distributions for broad parallel beams of monoenergetic γ rays (0.01 to 16 MeV) incident on a slab of tissue of "standard man" composition. The results were used to obtain the maximum dose per unit incident fluence, plotted in Figure 5.33. Monte Carlo calculations in more realistic phantoms have been carried out by others [10,11], but the homogeneous

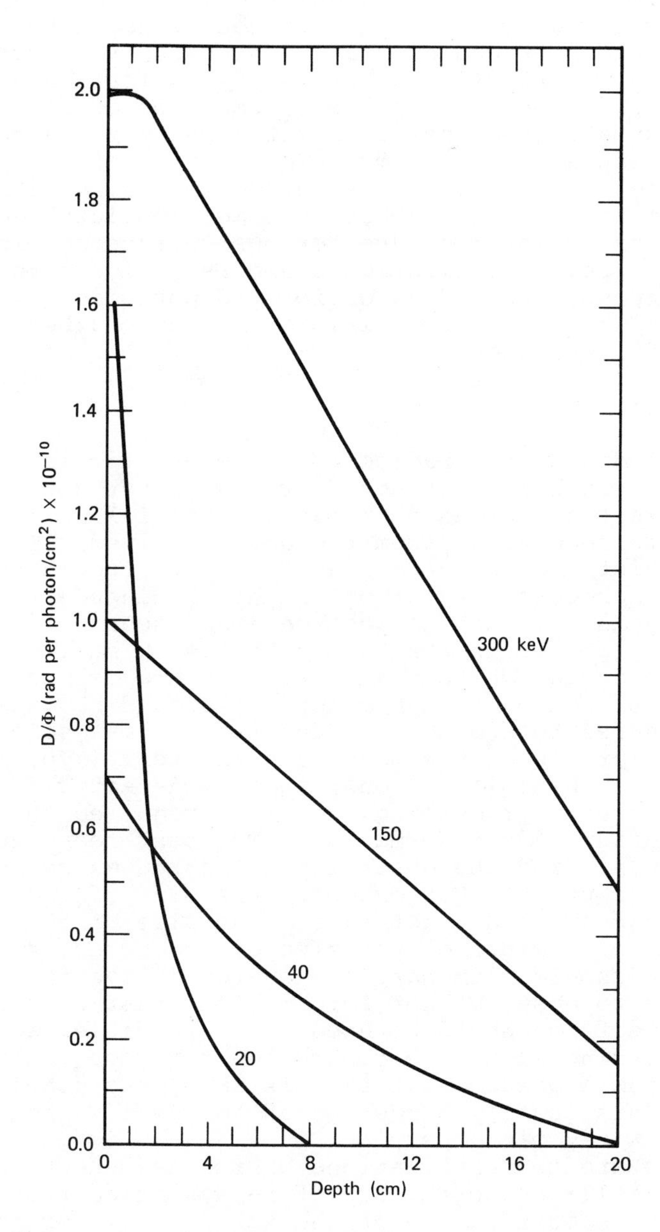

Fig. 8.7. Absorbed dose at depth as a function of energy for unit fluence at surface.

slab or elliptic cylinder calculations are adequate for estimating the dose in diagnostic roentgenography.

Measurement techniques are discussed in Chapter 5 and in ICRU reports [12]. The cavity ionization chamber is usually applied, for measurement of either exposure or absorbed dose in a water phantom may be measured with a small (5 mm diameter) ionization chamber, or the LiF thermoluminescent dosimeter, calibrated against an ionization chamber or some "absolute" device such as a calorimeter or the ferrous sulfate chemical dosimeter. Cavity ionization chambers designed to measure exposure should be calibrated periodically against an absolute, free-air, ionization chamber at a standards laboratory.

Benefit/Risk Ratio

In medical diagnostic radiography the benefit is the improvement in health expected from a better diagnosis. The risk is the increased probability of ill health (e.g., carcinogenesis) from the dose received. There are no strict limits on dose as there are from other man-made sources of radiation, although diagnostic roentgenography is the largest man-made source of radiation exposure in the United States. The physician is free to decide whether an X-ray picture is needed and this is good, although an effort should be made to keep exposures to a minimum. The radiologist selects the number or views, the machine parameters (kVp, mAs, filtration, SSD, collimation), and the detector (e.g., film and screen, processing, grid) as required to give adequate diagnostic information. The best technique should be used and the dose reduced as far as is reasonably achievable. For example, the size of the collimated beam should not exceed the size of the organs to be examined or the size of the detector, filtration should eliminate low-energy X-rays that contribute to dose but not information, fast-film-screen detectors should be used where feasible, and film processing should be controlled for good quality with minimum exposure. If this is done, the benefit: risk ratio is usually large enough to justify the taking of X-ray pictures.

There are two applications where the benefit:risk ratio is small; hence the use of X-rays is controversial. One is irradiation of the fetus, as in examination of the mother before delivery, because the fetus is much more radiosensitive than the adult.

Ultrasonography has largely displaced roentgenography in this application. The other problem area is mass screening of women for breast cancer by mammography. The incidence of breast cancer in women over 35 is about 3000 cases per year, per million women, detectable by mammography but not by palpation, and almost all curable when detected early. According to the BEIR report (Chapter 6), the risk of cancer induction is six cases per year, per million women exposed to one rad, and these cancers continue to be expressed for 20 years following a latent period of about 10 years. Thus the absolute risk is 120 cases, per million women exposed, per rad. The benefit ratio for one screening is (3000/120) = 25, if the average dose to the breasts is 1 rad. Older techniques require on the order of 3 rad per examination, but new screens and new processes such as electron mammography have reduced the average dose to the order of 0.3 rad per examination. Further reductions are limited by quantum statistics (mottle or noise).

Neutron Radiography

Thermal neutrons are used in industrial radiography, such as nondestructive inspection of explosives, plastics, elastomers, or water inside metallic components. Hydrogen attenuates thermal neutrons much more strongly than most metals do, which is the opposite of X-rays. Similar differences in attenuation coefficients occur in tissues because of differences in hydrogen concentration. However, thermal-neutron radiography has not been adopted in medicine because the large scattering and small transmission limit thickness to about 2 cm. Table 8.1 compares the attenuation coefficients for neutrons and 60-keV X-rays. Fast neutrons (2 MeV) have attenuation coefficients comparable to the X-rays, but contrast (difference in attenuation coefficients) is not as good. Note that bone is less attenuating than soft tissue for 2 MeV neutrons, whereas bone is more attenuating than soft tissues for 60-keV X-rays. The main disadvantage of fast-neutron radiography is the large dose equivalent delivered, unless the imaging detector is quite efficient. Because most neutron sources also emit γ rays the detector should be insensitive to γ rays. A detector that meets these criteria and also provides good spatial resolution has not yet been developed.

TABLE 8.1. ATTENUATION COEFFICIENTS FOR NEUTRONS AND X-RAYS

Radiation	μ (cm^{-1})		
	Muscle	Fat	Bone
60-keV X-rays	0.20	0.17	0.45
Thermal neutrons	3.43	3.46	2.33
0.2-MeV Neutrons	0.72	0.70	0.55
2-MeV Neutrons	0.29	0.26	0.25
14-MeV Neutrons	0.096	0.097	0.123

Neutron Dosimetry

Monte Carlo calculations of absorbed dose and dose equivalent have been made for broad parallel beams of monoenergetic neutrons (0.025 eV to 14 MeV) incident on a slab of tissue 30 cm thick, or on a cylinder of tissue 30 cm in diameter and 60 cm high [13-15]. The results in terms of maximum dose equivalent per unit fluence are listed in Table 5.13. The distribution of dose equivalent in the cylindrical phantom is plotted in Figure 8.8.

Measurements of absorbed dose and dose equivalent are discussed in Chapter 5 and in an ICRU report [16]. The preferred method is to use a tissue-equivalent ionization chamber to measure the absorbed dose, and the LET proportional counter to measure the absorbed dose as a function of linear energy transfer, from which the dose equivalent can be obtained. Tissue equivalence for neutron dosimetry implies the same composition or proportions of C, N, O, and H as in soft tissue, with the concentration of hydrogen as particularly important. The absorbed dose and LET should be measured in a tissue-equivalent phantom. It is desirable to measure the absorbed dose from

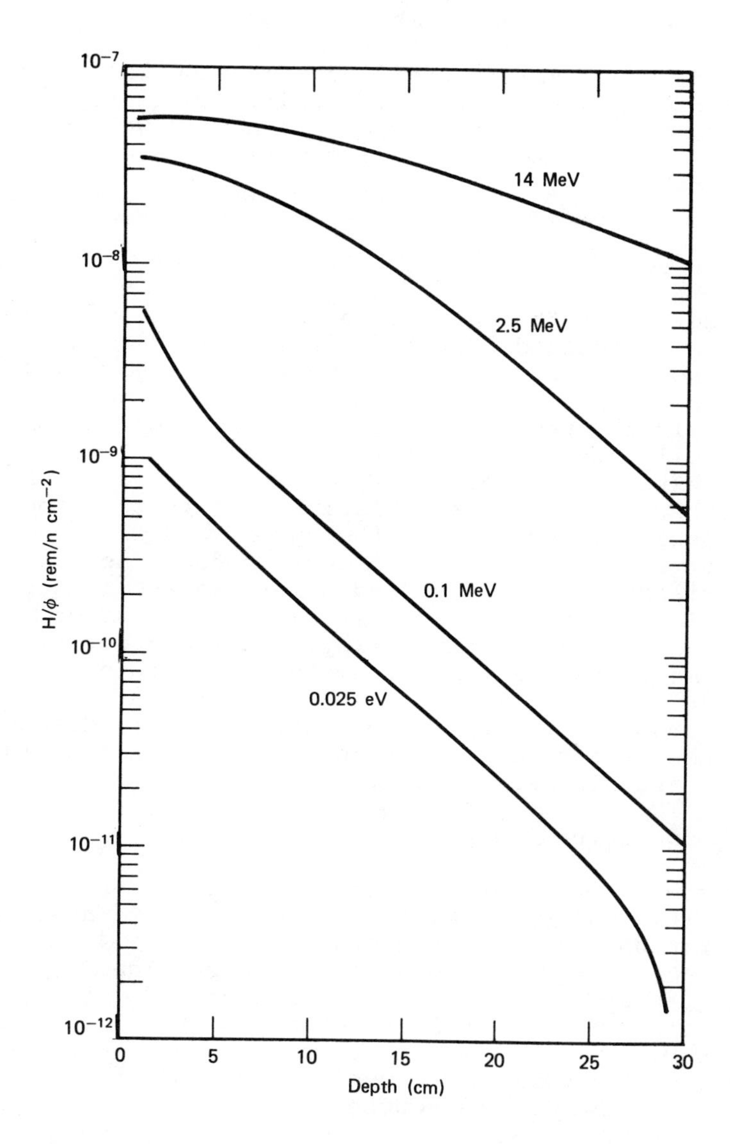

Fig. 8.8. Dose-equivalent vs depth in cylindrical phantom irradiated with monoenergetic neutrons (after Ref. 15).

γ rays alone, including the hydrogen capture-γ rays generated in the phantom. This requires a neutron-insensitive ionization chamber. The graphite chamber filled with CO_2 is satisfactory at neutron energies below 5 MeV, but an argon-filled magnesium-wall chamber is better at higher energies. Neutron dosimetry and dosimetry in mixed n - γ fields are not as well developed as X-ray dosimetry, and improvements are possible and desirable.

8.2 THERAPEUTIC RADIOLOGY

The objective of therapeutic radiology [1,17] is to deliver a prescribed dose to a well-defined volume (e.g., thousands of rads to a tumor) while giving as small a dose as feasible to the surrounding normal tissue. In teletherapy the body is irradiated with a narrow collimated beam of X-rays, γ rays, or neutrons from an external source. In brachytherapy (also called "plesiotherapy") the source is imbedded in the tumor or in a body cavity next to the tumor, and $1/r^2$ and tissue attenuation are relied on to reduce the dose to normal tissue. Skin, eye, and other organs may also be irradiated with a β-particle source in contact, which is a form of brachytherapy. In therapeutic nuclear medicine, radioactive material is injected and accumulates preferentially in certain organs. Photon and neutron teletherapy and brachytherapy are discussed here, and therapeutic and diagnostic nuclear medicine are discussed in Section 8.3.

Photon Teletherapy

The source in photon teletherapy may be an X-ray machine (typically 250 kVp), ^{60}Co or ^{137}Cs γ rays, or bremsstrahlung from an accelerator such as a 4- or 6-MV LINAC. The target or capsule is shielded by lead or tungsten as shown in Figure 8.9 except for the port to extract a collimated beam, or adjustable field size at the SSD. The collimator is often made of depleted uranium. The shutter is closed to shut off the beam, or in other designs the source is moved to a safe position in the shield.

The spatial distribution of the absorbed dose is usually specified by the dose along the central axis as a function of depth d and by an isodose map

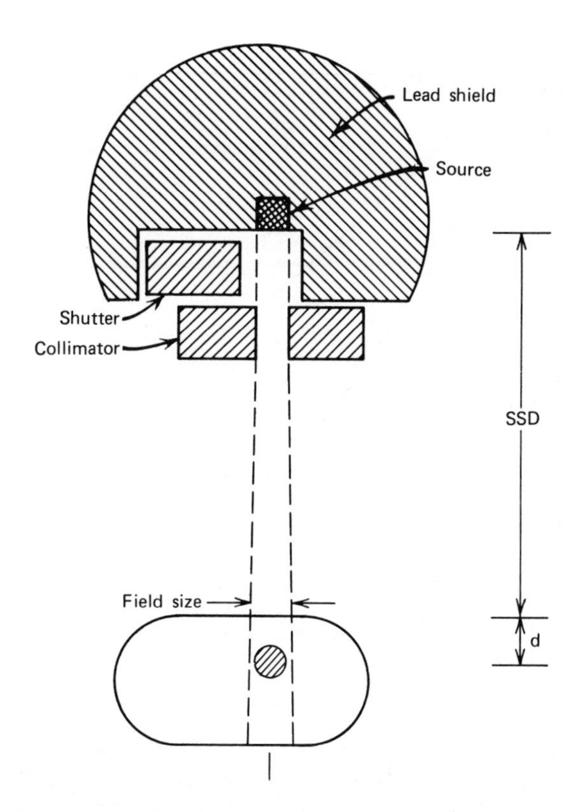

Fig. 8.9. Teletherapy source and irradiation geometry.

(contours of equal dose in a plane including the central axis, normalized to 100% at the point of maximum absorbed dose). Conversion between exposure (R) measured in free air and absorbed dose at the reference point is made using the R:rad and backscatter factors in Eq. (8.7) and correcting from the surface to the reference point.

The central-axis dose distribution depends on the spectrum of the incident photons, the SSD and source size (because of geometrical attenuation), and the field size (because of scattering). Figure 8.10 plots the percent depth dose for 250-kVp X-rays and 4-MV bremsstrahlung, for 50-cm SSD and a 10-cm by 10-cm square field. The dose from the 4-MV bremsstrahlung

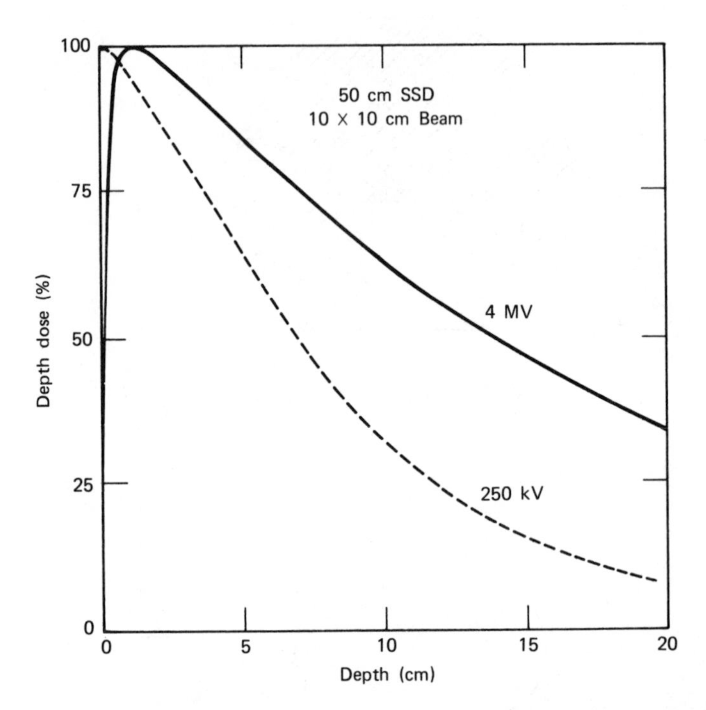

Fig. 8.10. Percent of maximum absorbed dose as a function of depth for 250-kVp and 4-MV X-rays, 50 cm SSD, 10 cm × 10 cm field, along central axis of tissue phantom.

drops less rapidly with depth than do the 250-kVp X-rays, which is advantageous in treating deep tumors and reducing the dose to tissue above the tumor. The 4-MV radiation also gives a desirable "skin sparing" effect because the absorbed dose reaches a maximum at about 1 cm depth instead of at the skin, as the secondary electrons come into equilibrium. A larger SSD reduces the variation with depth because of geometrical attenuation but also reduces the magnitude of the absorbed dose. The 250-kVp X-ray source generally uses an SSD of 50 cm, whereas ^{60}Co and accelerator sources often use an 80- to 100-cm SSD.

The influence of the field size is seen in Figure 8.11 for filtered 250-kVp X-rays. The percent depth dose at 10-cm depth is plotted as a function of

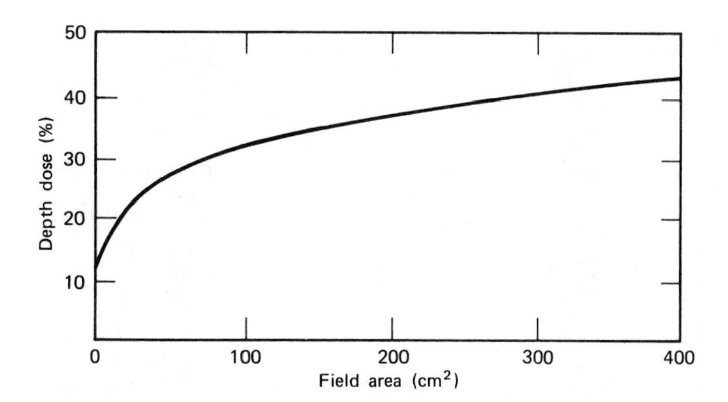

Fig. 8.11. Percent depth dose at 10 cm for 250-kVp X-rays as function of field area.

field area (for square or circular fields). The dose increases with increasing field area because a larger volume contributes scattered photons to the dose on axis. The backscatter at the surface also increases with field area. The increase from scattering is somewhat less at higher beam energies because there are fewer collisions and the particles are emitted predominantly in the forward direction.

Figure 8.12 is an example of an isodose chart. The dose decreases toward the edge of the beam because there is a smaller contribution from scattered photons and because of geometric and material attenuation in the diverging beam. There is then a rapid decrease at the edge of the umbra ($\pm$5cm) but not a step because of scattering, the penumbra from finite source size, and the penumbra from penetration of the walls of the collimator. Isodose charts are usually measured by traversing a small ion chamber in a water or tissue-equivalent liquid phantom. The contours may be drawn by connecting points of equal dose or may be traced automatically by means of servomechanism driving the ion chamber.

In practice, beams from more than one direction are used in teletherapy. Figure 8.13 is an isodose chart for the combined dose from four ^{60}Co beams

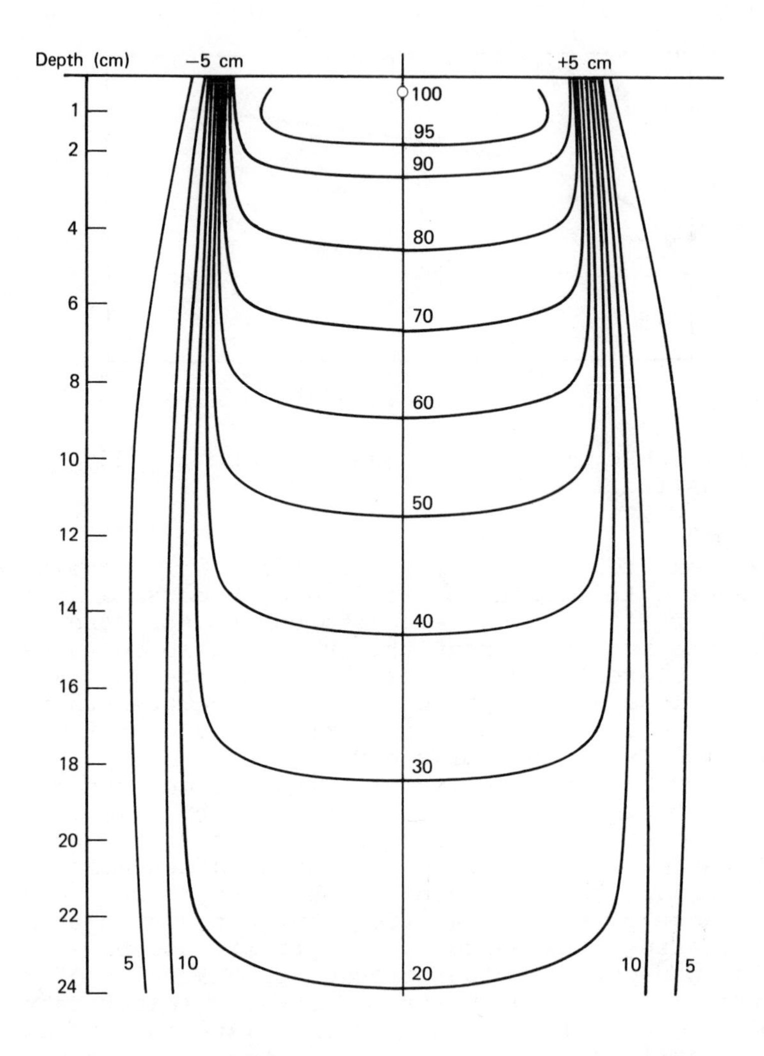

Fig. 8.12. Isodose map for ^{60}Co γ rays, 10 cm × 10 cm field at 80 cm SSD.

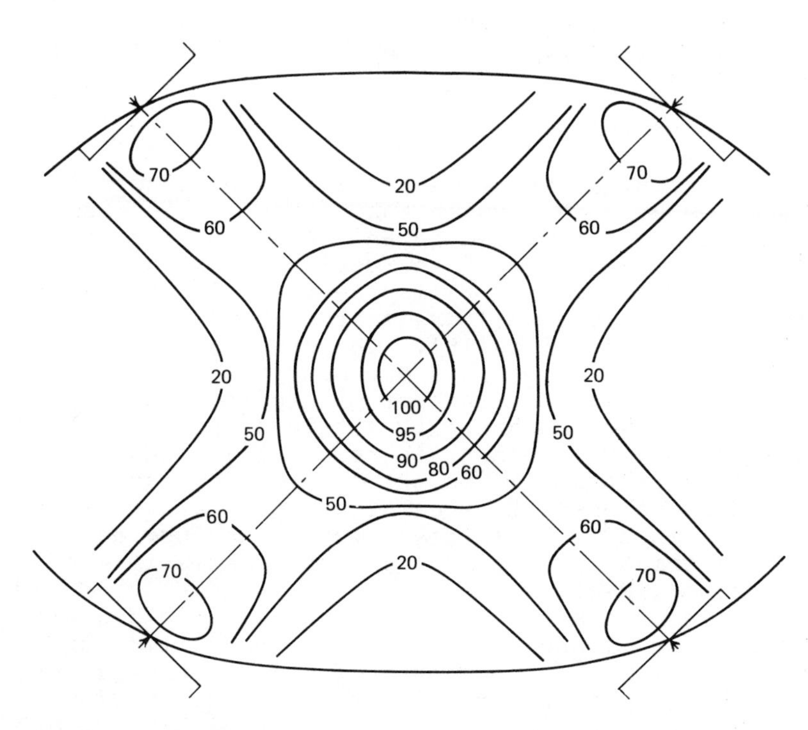

Fig. 8.13. Isodose map for four ^{60}Co γ-ray beams, 5 cm × 5 cm fields at 60 cm SSD (after M. Cohen and S. Martin, *Atlas of Radiation Dose Distributions*, IAEA, 1966).

intersecting at right angles. The percent dose is normalized to 100 at the point of intersection of the beam axes. Note that the dose is now maximum at depth and is smaller at the entry portals.

These results are for a homogeneous phantom. The dose at depth is decreased by attenuation in bone, increased because of smaller attenuation in low-density tissue such as lung, and modified by details of the body contours, hence varying thickness of tissue between the surface and the tumor. Additional attenuating material, called "bolus," may be added at the entry portal to equalize the distributions and match a precomputed or measured isodose chart. Bone is usually avoided by angling the beam, both to simplify the determination of dose and to minimize irradiation of radiosensitive tissues such as bone marrow and the spinal cord.

Neutron Teletherapy

Fast neutrons are being investigated for cancer teletherapy because the oxygen enhancement ratio (OER) is only about 1.6, compared to an OER of about 3 for X- and γ rays, as discussed in Section 6.7. Thus hypoxic or anoxic cells in tumors may be treated without excessive damage to the well-oxygenated normal tissues surrounding the tumor [18]. The greatest need is for intense sources of 7 to 15 MeV neutrons, capable of delivering at least 5 rad per minute at a SSD of 75 cm (i.e., source strength 10^{12} n s^{-1}). Cyclotrons operating on the $^{9}Be(d,n)^{10}B$ reaction and sealed-tube accelerators operating on the $^{3}H(d,n)^{4}He$ reaction with a mixed deuteron-triton beam (for long target life) are being developed with large source strengths, and clinical trials are in progress using existing cyclotrons [19-21].

The accelerator target is usually shielded with layers of steel and polyethylene. The collimator is usually made of steel because of its relatively large total cross section and inelastic scattering cross section for high-energy neutrons. Steel has the disadvantage of generating capture γ rays and becoming activated when irradiated with thermal neutrons. The beam from a cyclotron is usually fixed, but the shielding for sealed-tube generators may be sufficiently light to allow rotation, as in teletherapy with ^{60}Co sources.

The axial depth dose curve for 14-MeV neutrons in tissue is similar to that for ^{60}Co γ rays, as shown in Figure 8.14. An isodose map is given in Figure 8.15. Penumbra is larger, but beam definition is acceptable for neutrons. Neutrons from the $^{9}Be(d,n)$ reaction at 16-MeV deuteron energy have an average energy of 7 MeV. Penetration in tissue is intermediate between that for ^{60}Co and 250-kVp X-rays. Lower-energy neutrons are not suitable for teletherapy.

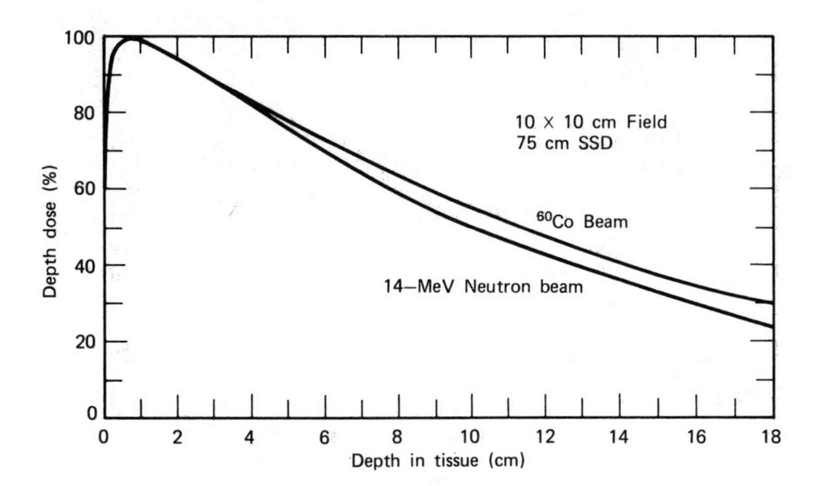

Fig. 8.14. Depth dose distribution of 14-MeV neutrons and ^{60}Co γ rays.

Brachytherapy

Radiation therapy may be carried out with implants of encapsulated radioisotopes in the form of a "needle" some 2 cm long and 3 mm in diameter [3,17]. The source may be inserted in a body cavity close to the tumor or surgically implanted within the tumor for a prescribed time. Radium in equilibrium with its daughters is often used for γ-ray brachytherapy. The needles contain 1 to 3 mg each of ^{226}Ra, and several may be arranged in a pattern to distribute the dose evenly. The exposure rate in air at 1 cm from a 1 mg source, filtered by 0.5 mm platinum (wall of the capsule), is 8.25 R/h. At 1 mm Pt filtration, the dose rate is 10% lower. The capsule absorbs the α and β particles and prevents contamination by the radium or radon daughter. Figure 8.16 is an isodose chart for a 1-mg radium needle (rotationally summetric about the axis). Reactor-produced radioisotopes, for example, ^{198}Au, are also used.

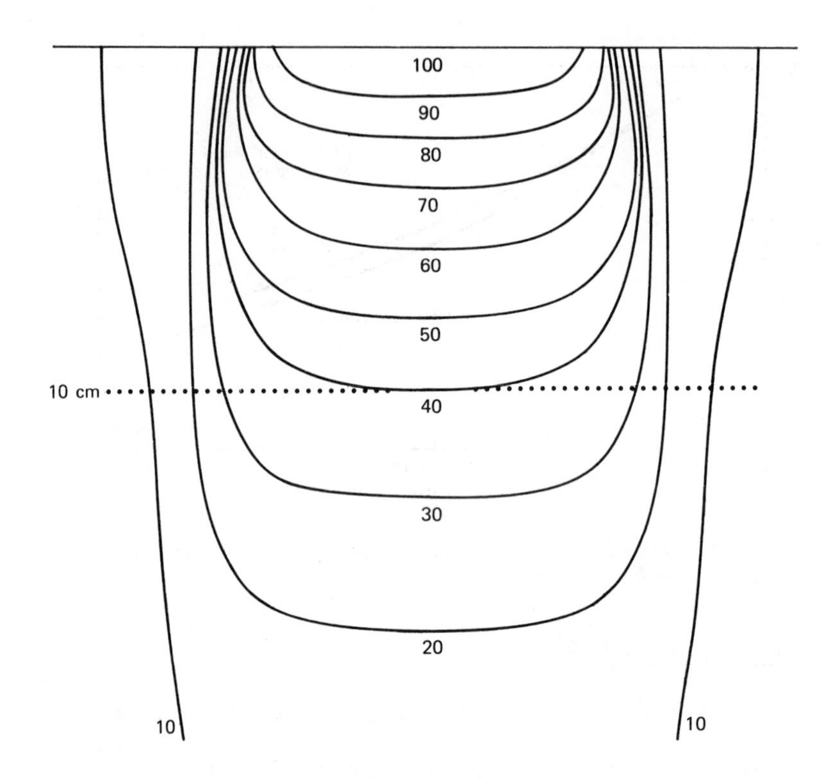

Fig. 8.15. Isodose map for 14-MeV neutrons, 10 cm × 10 cm field, and 55 cm SSD.

Californium-252 is being investigated for brachytherapy with fast neutrons (average energy 2 MeV) because their OER is smaller than for X-rays and there may be therapeutic advantage for anoxic tumors. Figure 8.17 displays isodose distributions for the neutrons and the γ rays emitted by ^{252}Cf, normalized to one microgram.

Figure 8.18 plots the percent depth dose from a β particle "applicator," using ^{90}Sr and its daughter ^{90}Y. The dose rate at the surface is typically 100 rad/min in treatment of eye tumors.

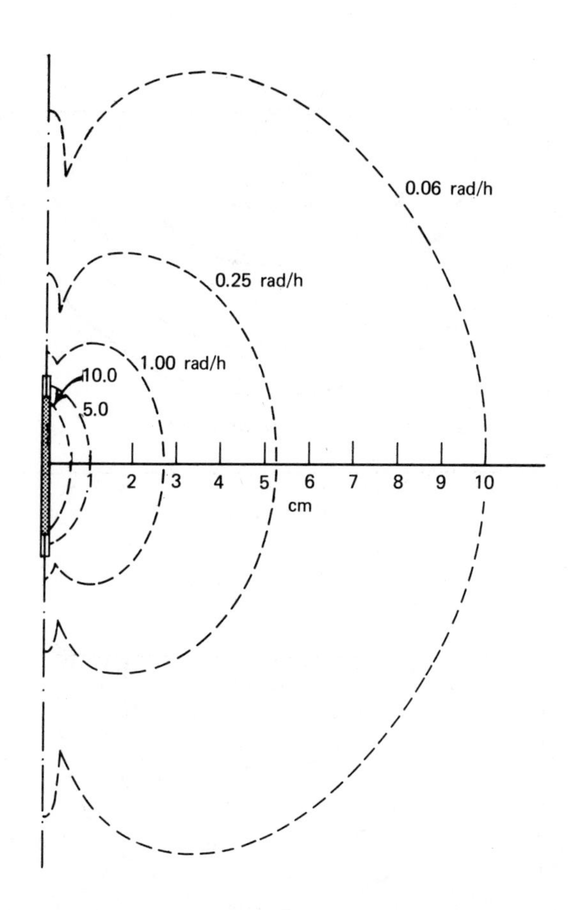

Fig. 8.16. Isodose curves for a 1-mg radium needle showing section of the cylindrically symmetric distribution [after J. Rose, *Am. J. Roentgenol.*, 97, 1032 (1966)].

8.3 NUCLEAR MEDICINE

Diagnostic Nuclear Medicine

The uptake of an element in an organ or the size and shape of an organ or lesion can be measured by counting or mapping the distribution of a radioisotope in

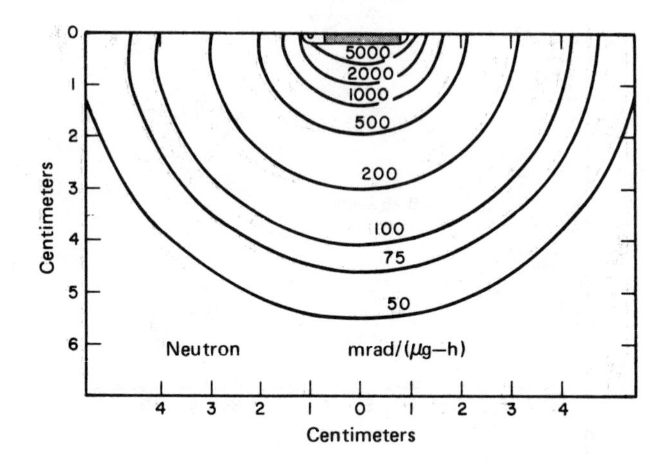

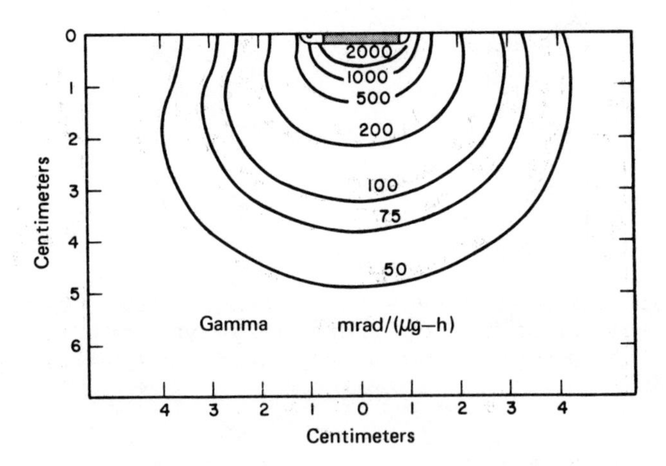

Fig. 8.17. Measured neutron and γ-ray dose rates (mrad/h μ^{-1}) in tissue-equivalent liquid around a ^{252}Cf source (after Californium-252 Progress, No. 9, Oct. 1971, U.S. Atomic Energy Commission).

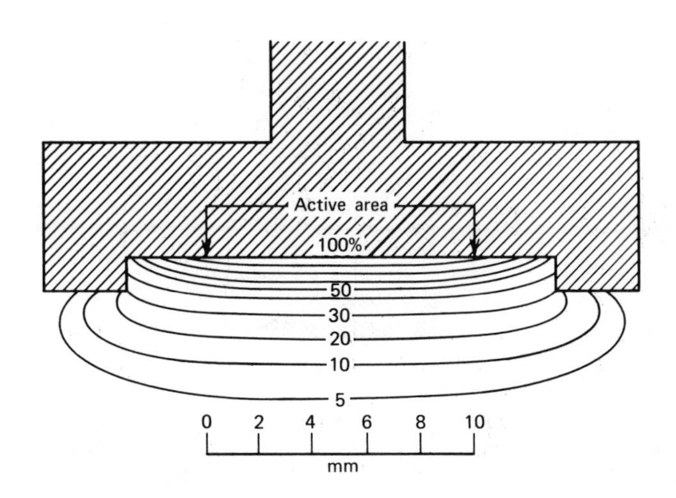

Fig. 8.18. Isodose distributions for ^{90}Sr + ^{90}Y applicator.

the organ [22]. The radioisotope, in suitable chemical form, is usually injected intraveneously but may be inhaled or ingested.

The uptake of iodine in the thyroid can be measured by ingesting ^{131}I as sodium iodide and counting the γ rays from the thyroid 24 h later. The count rate is compared with that from a solution, equal to the amount injected, and located in a plastic phantom to approximate the attenuation and scattering in the neck. The detector is a sodium iodide scintillation counter, shielded by about 5 cm of lead against background including radioiodine in other parts of the body, with a wide collimator to encompass the entire thyroid, as shown in Figure 8.19. About 30% of the ingested ^{131}I activity is present in the normal thyroid at 24 h after ingestion. Decreased or increased activity indicates possible hypothyroidism or hyperthyroidism. The activity decreases with time because of radioactive decay and elimination, with an effective half-life of 7.6 days.

The mapping or imaging of the spatial distribution of radioactivity in an organ is done either by scanning a well-collimated detector in a raster or X-Y pattern,

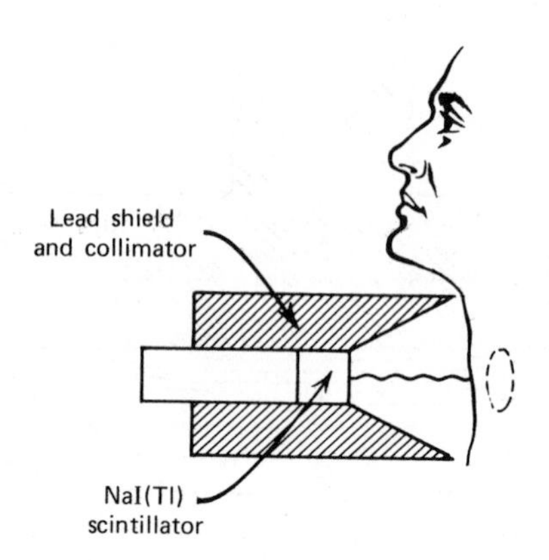

Fig. 8.19. Open-collimated scintillation detector for thyroid radioiodine uptake measurement.

or by using a "camera" consisting of a collimated, position-sensitive, γ-ray detector. A scanner may use a 7.6- to 10-cm-diameter NaI(Tl) scintillation counter, shielded by about 5 cm lead, with a multiple-hole collimator such as that illustrated in Figure 8.20. The "focusing" type of collimator contains seven to over 30 tapered and angled holes, such that γ rays emitted from a small volume at the "focus" are detected with good efficiency, whereas γ rays from other volumes are not. The number of holes depends on the scintillator size, diameter of the collimator holes, and thickness of the septum between the holes. The septum thickness has to be large enough to absorb nearly all the γ rays; the minimum thickness depends on the γ-ray energy but is typically 2 mm for 100 keV. A small hole diameter and a long collimator give better spatial resolution but lower geometric detection efficiency, and a compromise is sought. A typical lead collimator for 100 to 200 keV γ rays will have a hole diameter of 3 mm to 5 mm at the small end (entrance) and a collimator length of 76 mm. Resolution and collimator design are discussed by Gottschalk and Beck [23] and by Brownell [24]. The image is built up by scanning the detector across the field of interest and displaying the counts per unit area by the density of dots or in some other fashion.

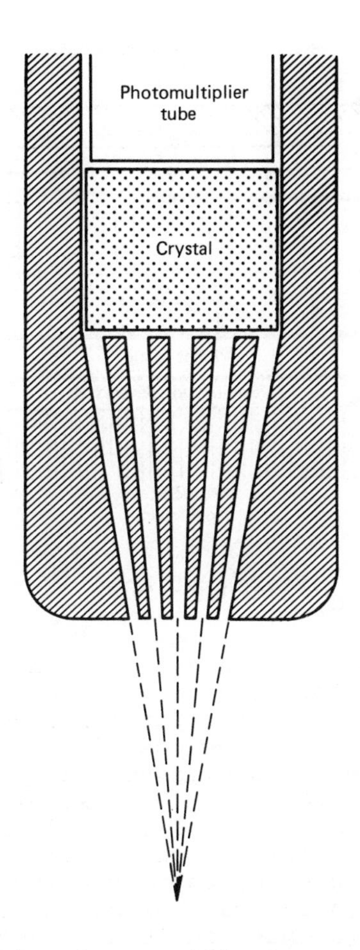

Fig. 8.20. Focusing-type collimator and shielded sodium iodide scintillation detector for scanning.

The most popular γ-ray imaging device is the Anger camera, shown schematically in Figure 8.21. The γ rays, collimated by thousands of small-diameter (∿3 mm) parallel holes in a 45-mm-long collimator, impinge in a NaI(Tl) crystal 29 cm in diameter and 1.3 cm thick. The scintillation light is detected by 19 photo-

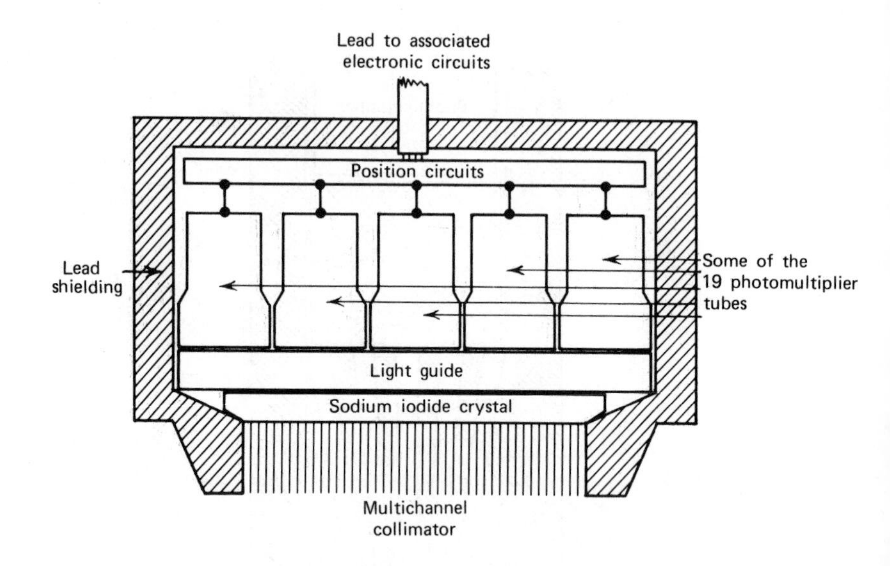

Fig. 8.21. Diagram of Anger γ-ray camera.

multiplier tubes. The intensity of light from a single pulse depends on the distance of the scintillation point from the photomultiplier tube, and this information is used to derive the position (X-Y coordinates) from the signal amplitude in several tubes. Spatial resolution is 0.5 to 1 cm. Each scintillation event is recorded as a momentary dot or flash of light at a corresponding X-Y position on an oscilloscope screen. The image is produced by photographing the dots on Polaroid film in a camera viewing the oscilloscope screen. The image thus consists of dots with a higher dot density in regions of higher activity.

Counting statistics limit the minimum activity per unit area that can be imaged properly. A typical scintiphoto will have 500,000 dots (counts) over the central 500-cm^2 area, or 1000 dots per cm^2. Counting time is limited to 10 to 30 min because of patient movement or throughput. The overall detection efficiency is quite small, on the order of 10^{-4} because of the small solid angle subtended by each collimator hole and the distance from organ to crystal even with the collimator in contact with the body. Thus the activity in an organ has to be fairly high, perhaps 10 μCi to 1 mCi. The organ and whole-body doses are then on the order of 0.1 to 1 rad. Dosimetry is discussed in Section 8.4.

Therapeutic Nuclear Medicine

Applications of internally deposited radioisotopes to therapy are very restricted because of the dose to other organs [1,22]. Iodine is concentrated in the thyroid gland, and ingestion of 3 to 6 mCi of ^{131}I can be used to treat hyperthyroidism by destroying part of the gland. Cancerous lesions of the thyroid cannot be treated because they take up less iodine than the normal tissue, but radioiodine has been considered for treatment of metastases.

Radioactive phosphorus, ^{32}P, is widely distributed in the body immediately after injection, but then concentrates in the bone, liver, and spleen. It is used in a dosage of about 4 mCi for treatment of polycythemia vera and leukemia, delivering about 60 rad to the bone marrow.

Colloidal gold (^{198}Au) is used for intracavity radiotherapy because it does not spread from the cavity.

8.4 INTERNAL DOSIMETRY

It is difficult to measure the dose delivered to an organ from an internally deposited radioisotope. Sometimes the activity can be derived from the count rate in an external γ-ray detector, or from the activity excreted in feces, urine, sweat, or exhaled breath. Usually dose is calculated from a known intake of activity, the transport and elimination of elements in the body, and the relationship between the activity deposited in an organ and the dose to that or surrounding organs. Intake may occur by inhalation, ingestion (as in food or drink), injection (as in nuclear medicine), or by absorption through the intact skin or a wound, as indicated in Figure 8.22.

The intake by inhalation is equal to the breathing rate times the activity concentration in air (μCi cm^{-3}). Table 8.2 lists breathing rates and other physiological intake and excretion data for the standard adult [25]. Part of the inhaled material may be deposited

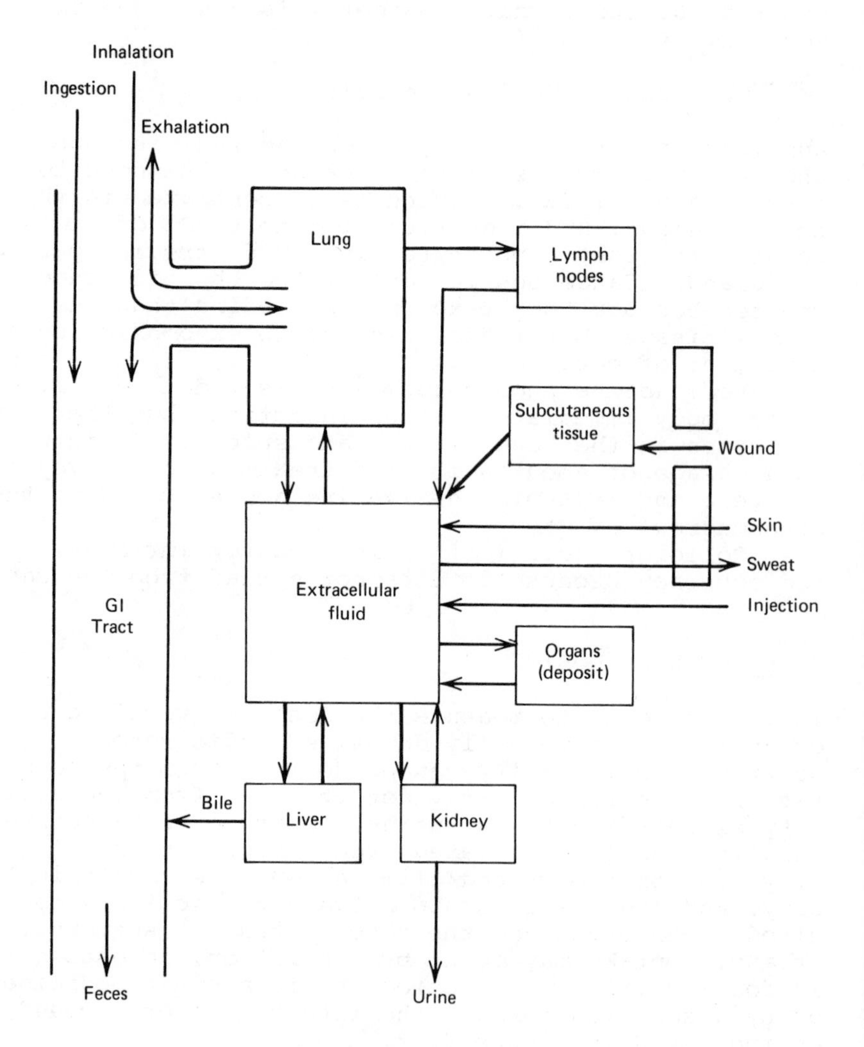

Fig. 8.22. Model of metabolic pathways (after ICRP Publication 10).

TABLE 8.2. INTAKE AND EXCRETION RATES FOR STANDARD MAN

Air inhaled in 8-h work period	1×10^7 cm^3	(1.25×10^6 cm^3 h^{-1})
Air inhaled in 18-h nonwork period	1×10^7 cm^3	(0.63×10^6 cm^3 h^{-1})
Average daily air intake	2×10^7 cm^3	(8.33×10^5 cm^3 h^{-1})
Water intake in 8-h work period	1100 cm^3	(137 cm^3 h^{-1})
Water intake in 16-h nonwork period	1100 cm^3	
Average daily water intake	2200 cm^3	(91.7 cm^3 h^{-1})

temporarily in the lung while the rest is exhaled. Material that is soluble in lung fluid is rapidly taken up by the blood. Particulates may be removed by ciliary action in the respiratory tract and then swallowed, thus entering the GI tract. The lung model originally defined by the ICRP is summarized in Table 8.3, giving the fractions going by the different pathways and the times for each, for soluble and relatively insoluble compounds. A more detailed model has been proposed [26] taking into account the probability that aerosol particles of different sizes will be lodged and then eliminated from various parts of the respiratory tract, and accounting for transport to the lumph nodes as well as in the GI tract. However, application of this model requires more knowledge of particle size distributions than is usually available. The fraction of inhaled activity that is deposited in a given organ is f_a. This will depend on the solubility of extracellular fluid (mostly blood), the organ in question, and the time scale. The ICRP publications and Table 8.4 give f_a for the "critical" organ, that is the organ having the greatest fraction of the maximum permissible dose (discussed in Section 6.6). The critical organ may be the whole body, lung, GI tract, kidney, or an organ that accumulates an element, such as strontium in the skeleton or iodine in the thyroid. International Commission on Radiological Protection Publication 2 considers continuous inhalation (40-h week or 168 h per week for nonoccupational exposure), and ICRP Publication 10 considers a single intake. For a single intake, f_a applies to the maximum in the organ deposition as a function of time after the intake. The dose from inert gases (except radon) is calculated for external irradiation from submersion in a cloud of contaminated air rather than the small dose to the lungs from inhaled material.

The ICRP model for solids in the GI tract is summarized in Table 8.5. The GI tract is divided into stomach, small intestine, upper large intestine, and lower large intestine. Mean residence times are given, along with average mass of contents and dimensions of the portions of the tract, for the purpose of calculating the dose to the walls of the tract. Soluble compounds are rapidly transferred to extracellular fluid (blood) and from blood to certain organs where the radioisotope is deposited. The fraction of ingested activity transferred to blood is listed as f_1 in the

TABLE 8.3. ICRP (1959) LUNG MODEL FOR AEROSOLS

Distribution	Readily Soluble Compounds (%)	Other Compounds (%)
Exhaled immediately	25	25
Upper-respiratory passages then swallowed	50	50[b]
Lung then rapidly to blood	25	
Lung then swallowed within 24 h		12.5[b]
Lung then to blood with 120-day half-life[a]		12.5

[a]Except retention half-life is 1 yr for Pu and 4 yr for Th.

[b]Total of 62.5% is swallowed.

ICRP reports and Table 8.4. The fraction transferred from blood to the critical organ is listed as f_2'. The fraction of ingested activity ingested in water and deposited in the organ is $f_W = f_1 f_2'$. Sometimes the transfer fractions are not known, but f_2, the fraction of activity in the total body that resides in the organ, is known, together with the total body activity.

Nearly 100% of tritiated water, HTO, is rapidly absorbed through the skin. Iodine vapor or solution and iodides are also absorbed through the skin, whereas most other materials are not. Radioactive material can enter through a wound, but as it is likely to be absorbed only slowly from the subcutaneous tissue, it is desirable to cleanse the wound as soon as possible and remove the contamination. Radioactive compounds injected intravenously enter the blood immediately and

TABLE 8.4. METABOLIC INFORMATION FROM ICRP PUBLICATIONS 2 AND 6

Radionuclide	Critical Organ	f_1	f_2
1 H-3 (oxide)	Body tissue	1.0	1.0
6 C-14 (CO_2)	Fat	1.0	0.1
11 Na-22	Total body	1.0	1.0
15 P-32	Bone	0.75	0.5
16 S-35	Testis	1.0	1.0
17 Cl-36	Total body	1.0	1.0
20 Ca-45	Bone	0.6	0.9
26 Fe-59	Spleen	0.1	0.02
27 Co-60	Total body	0.3	1.0
30 Zn-65	Total body	0.1	1.0
37 Rb-86	Total body	1.0	1.0
38 Sr-85	Total body	0.3	1.0
38 Sr-89	Bone	0.3	0.99
39 Sr-90	Bone	0.3	0.99
52 Te-132	Kidney	0.25	0.07
53 I-131	Thyroid	1.0	0.2
53 I-132	Thyroid	1.0	0.2
55 Cs-137	Total body	1.0	1.0
56 Ba-140	Bone	0.05	0.7
58 Ce-144	Bone	$<10^{-4}$	0.38
79 Au-198	Kidney	0.1	0.03
84 Po-210	Spleen	0.06	0.07
88 Ra-226	Bone	0.3	0.99
90 Th nat	Bone	$<10^{-4}$	0.9
92 U nat & U 238	Kidney	$<10^{-2}$	0.065
92 U-233	Bone	$<10^{-2}$	0.85
92 U-234	Bone	$<10^{-2}$	0.85
92 U-235	Kidney	$<10^{-2}$	0.065
93 Np-239	Bone	$<10^{-4}$	0.45
94 Pu-239	Bone	3×10^{-5}	0.9
94 Pu-241	Bone	3×10^{-5}	0.81

f_2'	f_w	f_a	ε	T_r (d)
1.0	1.0	1.0	0.010	4.5×10^3
0.025	0.025	0.02	0.27	2.0×10^6
1.0	1.0	0.75	1.6	950.0
0.5	0.375	0.32	3.5	14.3
1.0	1.0	0.75	0.056	87.1
1.0	1.0	0.75	0.26	1.2×10^8
0.9	0.54	0.5	0.43	164.0
0.02	2.0×10^{-3}	6×10^{-3}	0.34	45.1
1.0	0.3	0.4	1.5	1.9×10^3
1.0	0.1	0.3	0.32	245.0
1.0	1.0	0.75	0.70	18.6
1.0	0.3	0.4	0.33	65.0
0.7	0.21	0.28	2.8	50.5
0.3	0.09	0.12	5.5	1.0×10^4
0.07	0.02	0.03	0.96	3.2
0.3	0.3	0.23	0.23	8.0
0.3	0.3	0.23	0.65	0.097
1.0	1.0	0.75	0.59	1.1×10^4
0.7	0.035	0.19	4.2	12.8
0.3	3.0×10^{-5}	0.075	6.3	290.0
0.03	3.0×10^{-9}	9×10^{-3}	0.41	2.7
0.04	2.0×10^{-3}	0.01	55.0	138.4
0.1	0.03	0.04	110.0	5.9×10^5
0.7	7.0×10^{-5}	0.18	270.0	?
0.11	1.1×10^{-3}	0.028	43.0	?
0.11	1.1×10^{-3}	0.028	250.0	5.9×10^7
0.11	1.1×10^{-3}	0.028	240.0	9.1×10^7
0.11	1.1×10^{-3}	0.028	46.0	2.6×10^{11}
0.45	4.5×10^{-5}	0.11	0.98	2.33
0.8	2.4×10^{-5}	0.2	270.0	8.9×10^6
0.8	2.4×10^{-5}	0.2	14.0	4.8×10^3

TABLE 8.4. METABOLIC INFORMATION FROM ICRP PUBLICATIONS 2 AND 6 (Contd.)

Radionuclide	T_b[a] (d)		T_{eff} (d)
1 H-3 (oxide)	12.0		12.0
6 C-14 (CO_2)	40.0	(10)	40.0
11 Na-22	11.0		11.0
15 P-32	1155.0	(257)	14.1
16 S-35	90.0		44.3
17 Cl-36	29.0		29.0
20 Ca-45	1.8×10^4	(1.64×10^4)	162.0
26 Fe-59	600.0	(800)	41.9
27 Co-60	9.5		9.5
30 Zn-65	933.0		194.0
37 Rb-86	45.0		13.2
38 Sr-85	1.3×10^4		64.7
38 Sr-89	1.8×10^4	(1.3×10^4)	50.4
38 Sr-90	1.8×10^4	(1.3×10^4)	6.4×10^3
52 Te-132	30.0	(15)	2.9
53 I-131	138.0		7.6
53 I-132	138.0		0.097
55 Cs-137	70.0		70.0
56 Ba-140	65.0		10.7
58 Ce-144	1500.0	(563)	243.0
79 Au-198	280.0	(120)	2.7
84 Po-210	60.0	(30)	42.0
88 Ra-226	1.64×10^4	(900)	1.6×10^4
90 Th nat	7.3×10^4	(5.7×10^4)	7.3×10^4
92 U nat & U 238	15.0	(100)	15.0
92 U-233	300.0	(100)	300.0
92 U-234	300.0	(100)	300.0
92 U-235	15.0	(100)	15.0
93 Np-239	7.3×10^4	(3.9×10^4)	2.33
94 Pu-239	7.3×10^4	(6.5×10^4)	7.2×10^4
94 Pu-241	7.3×10^4	(6.5×10^4)	4.5×10^3

[a] T_b in parenthesis is for total body.

TABLE 8.5. ICRP (1959) GI-TRACT MODEL

Portion	Mass of Contents (g)	Time Food Remains (h)	Effective Radius (cm)
Stomach	250	1	10
Small intestine	400	4	10
Upper large intestine	220	13	5
Lower large intestine	135	24	5

are taken up by other organs in a matter of minutes to hours. Thus only f_2' is needed for injected materials.

The radioactive material initially deposited in an organ is depleted by radioactive decay and biological elimination. The material is transferred to extracellular fluid (blood or lymph) and thence to an organ of excretion. Many substances are excreted in the urine, and the kidney may be subjected to an appreciable dose. Other compounds are broken down or trapped in the liver and excreted in feces through the bile. The concentration in sweat is approximately equal to the concentration in blood, but excretion is variable. Tritiated water vapor may be exhaled. The organ content or burden (in μCi) may be obtained from

$$\frac{d\,q(t)}{dt} = -\lambda_r t - E(t) \tag{8.8}$$

where $\lambda_r = (0.693/T_r)$ is the radioactive decay constant and $E(t)$ is activity eliminated per unit time, at time t. If we normalize to q_0 (μCi) in blood at $t = 0$, for a single intake we can write

$$q(t) = q_0 R(t) \tag{8.9}$$

where the retention function

$$R(t) = \exp(-\lambda_r t) \left[1 - \int_0^t \exp(\lambda_r t)\ Y(\tau)\ d\tau \right] \quad (8.10)$$

and $Y(\tau) = [E(\tau)/q_0]$ is the normalized elimination function. Very often the elimination can be represented by an exponential function with a biological half-life of T_b. Then

$$R(t) = \exp \frac{-0.693}{T_{eff}} \quad (8.11)$$

where

$$T_{eff} = \left(T_r^{-1} + T_b^{-1}\right)^{-1}, \qquad \lambda_{eff} = \lambda_r + \lambda_b \quad (8.12)$$

Values of T_r, T_b, and T_{eff} are given in Table 8.4. For some substances the biological elimination is represented better by a power law, $E(t) = ct^{-d}$, where c and d are constants. Expressions for R(t) and $Y(\tau)$ are given in the references [25].

If radioactive material is inhaled, ingested, or otherwise taken into the body at a continuous rate P (μCi per second), the body content is obtained from

$$\frac{dq_b(t)}{dt} = P - \lambda_{eff}\ q_b(t) \quad (8.13)$$

and the organ content

$$q(t) = f_2 q_b(t) = f_2 P \frac{1 - \exp(-\lambda_{eff}\ t)}{\lambda_{eff}} \quad (8.14)$$

The dose-equivalent rate to an organ, is

$$\dot{H} = \frac{kq\varepsilon}{m} \quad (8.15)$$

where

$k = (3.7 \times 10^4 \text{ dis s}^{-1} \ \mu\text{Ci}^{-1})(1.60 \times 10^{-8} \text{ g-rad MeV}^{-1})$,

m = mass of organ (g); and

$\varepsilon = \Sigma\, a_i E_i F_i (\text{RBE})_i n_i$, where E_i is the energy of the i-th particle or γ ray emitted, a_i is the fraction of decays emitting radiation of this energy and type, F_i is the fraction of the energy absorbed in the organ, $(\text{RBE})_i$ is the relative biological effectiveness for converting from rad to rem, and n_i is a modifying factor correcting for nonuniform activity, or nonequilibrium with daughter activities, relative effectiveness compared to ^{226}Ra + daughters when comparing permissible bone content to 0.1 μCi ^{226}Ra, and so on.

The effective dose equivalent ε is listed for some radioisotopes and organs in Table 8.4. For other situations one can obtain the energies and decay fractions from a compilation such as the *Table of Isotopes* [27] or the Medical Internal Radiation Dose (MIRD) pamphlets [28]. The MIRD pamphlets give values corresponding to F, as well as the absorbed dose in various organs from activity deposited in a given organ, S. These factors are calculated by the Monte Carlo method in the mathematical model or phantom shown in Figure 1.2. The organs are represented by somewhat simplified geometrical shapes and sizes corresponding to an adult man (or woman for dose to the ovaries), with proper density. Previously the fraction of γ-ray energy absorbed in the organ was estimated from the effective radius and unit density. It is usually a good assumption that all of the α- and β-particle energy is absorbed in the organ containing the activity (F = 1). For small or thin organs F is less than unity and may be calculated by the methods discussed in the MIRD pamphlets.

The cumulative dose equivalent received over a time span t_1 to t_2 is

$$H = \int_{t_2}^{t_1} \dot{H}\, dt \tag{8.16}$$

which can also be expressed as

$$H = \frac{kQ\varepsilon}{m} \tag{8.17}$$

where Q is the time integral of the organ content. If Q is expressed in (μCi-days), ε in MeV, and m in grams, then k = 51.2 and H is in rem. Two limiting situations are of interest. For a single intake, the "dose commitment" is obtained if $t_1 = 0$ and $t_2 = \infty$. Maximum permissible concentrations in air and water are based on continuous intake, such that $\dot{H}$ at the end of 50 years ($t_1 = 0$, $t_2 = 50$ years) should not exceed the maximum permissible dose rate (rem per year) from Section 6.6. For radioisotopes with effective half-lives T_{eff} small compared to 50 years, the organ content will be in equilibrium (annual uptake equal to annual elimination including decay). For isotopes with long effective half-lives the maximum permissible dose will be reached at the end of 50 years. Some MPC (maximum permissible concentration) values are listed in Table 6.6.

8.5 EXTERNAL DOSIMETRY

Most of the discussion has centered on the dose from parallel collimated beams incident on the body. This is conservative for most radiation protection purposes, because neutrons or photons incident other than perpendicularly do not penetrate as far into the body. The American National Standard ANSI/ANS-6.1.1-1977 (N666), available from the American Nuclear Society, gives flux:dose factors based on the maximum dose equivalent from broad parallel-beam irradiation [9,15]. For accurate results, the standard recommends using the fitted expression

$$\ln DF(E) = A + B \ln E + C(\ln E)^2 + D(\ln E)^3 \tag{8.18}$$

where DF is the dose factor, rem/h per particle/cm^2s. The coefficients A, B, C, and D are listed for γ rays in Table 8.6 and for neutrons in Table 8.7 (energy is in MeV).

There is another situation where a more or less isotropic incident distribution is of interest, namely, when the body is immersed in a large cloud of radioactive material in air. This problem is discussed by Slade [29]. Assume equilibrium between energy emitted and energy absorbed in a given volume in infinite air.

TABLE 8.6. COEFFICIENTS FOR γ-RAY FLUX:DOSE FACTOR

Photon Energy (MeV)	A	B	C	D
0.01 to 0.03	-2.0477 +01	-1.7454		
0.03 to 0.5	-1.3626 +01	-5.7117 -01	-1.0954	-2.4897 -01
0.5 to 5.0	-1.3133 +01	7.2008 -01	-3.3603 -02	
5.0 to 15.0	-1.2791 +01	2.8309 -01	1.0873 -01	

TABLE 8.7. COEFFICIENTS FOR NEUTRON FLUX:DOSE FACTOR

Neutron Energy (MeV)	A	B	C	D
2.5 -08 to 1.0 -07	-1.2514 +01			
1.0 -07 to 1.0 -02	-1.2210 +01	1.7165 -01	2.6034 -02	1.0273 -03
0.01 to 0.1	-8.9302	7.8440 -01		
0.1 to 0.5	-8.6632	9.0037 -01		
0.5 to 1.0	-8.9359	5.0696 -01		
1.0 to 2.5	-8.9359	-5.5979 -02		
2.5 to 5.0	-9.2822	3.2193 -01		
5.0 to 7.0	-8.4741	-1.8018 -01		
7.0 to 10.0	-8.8247			
10.0 to 14.0	-1.1208 +01	1.0352		
14.0 to 20.0	-9.1202	2.4395 -01		

Then the external dose from β particles may be estimated from

$$D_\beta = 0.457\ \overline{E}_\beta\ \chi \tag{8.19}$$

where

D_β = β-particle dose rate (rad/s);
$\overline{E}_\beta$ = average energy of β particles (MeV/dis); and
χ = concentration of isotope in cloud (Ci/m^3).

and 0.457 is a units conversion factor.

The external dose rate from immersion in an infinite cloud of γ-ray-emitting radioisotope may be estimated from

$$D_\gamma = 0.507\ \overline{E}_\gamma\ \chi$$

where

D_γ = dose rate from γ rays (rad/s);
$\overline{E}_\gamma$ = average energy of γ rays (MeV/dis); and
χ = concentration of isotope (Ci/m^3).

and 0.507 is a units conversion factor times a correction from air to tissue absorption. For a person on the ground, hence in a semiinfinite cloud, the dose rate is halved. The concentrations may be calculated by the method of Section 4.6 if emission rates are known.

REFERENCES

1. W. J. Meredith and J. B. Massey, *Fundamental Physics of Radiology*, 2nd ed., Williams and Wilkins, Baltimore, 1972.
2. Michel M. Ter-Pogossian, *The Physical Aspects of Diagnostic Radiology*, Harper and Row, New York, 1967.
3. William R. Hendee, *Medical Radiation Physics*, Year Book Medical Publishers, Chicago, 1970.

4. A. R. Jones, "Measurement of the Dose Absorbed in Various Organs as a Function of the External Gamma-Ray Exposure," Report AECL-2240, Atomic Energy of Canada, Ltd. (1964).
5. A. R. Jones, "Proposed Calibration Factors for Various Dosimeters at Different Energies," *Health Physics*, 12, 663-671 (1966).
6. "Data for Protection Against Ionizing Radiation from External Sources: Supplement to ICRP Publication 15," ICRP Publication 21, Pergamon Press, Oxford (1971).
7. H. J. Delafield, "Gamma Ray Exposure Measurements in a Man Phantom Related to Personnel Dosimetry," AERE-R4430, Atomic Energy Research Establishment, Harwell (1963). Quoted in ICRP Publication 21.
8. J. M. Sidwell, T. E. Burlin, and B. M. Wheatley, "Calculations of the Absorbed Dose in a Phantom from Photon Fluence and Some Applications to Radiological Protection," *Br. J. Radiol.*, 42, 522-529 (1969).
9. H. C. Claiborne and D. K. Trubey, "Dose Rate in a Slab Phantom from Monoenergetic Gamma Rays," *Nucl. Appl. Technol.*, 8, 450-455 (1970).
10. N. A. Frigerio and R. F. Coley, "Depth Dose Determinations. III. Standard Man Phantom and Various Gamma Sources," *Phys. Med. Biol.*, 18, 187-194 (1973).
11. L. Koblinger and P. Zarand, "Monte Carlo Calculations on Chest X-Ray Examinations for the Determination of the Absorbed Dose and Image Quality," *Phys. Med. Biol.*, 18, 518-531 (1973).
12. "Radiation Dosimetry: X-Rays Generated at Potentials of 5 to 150 kV," ICRU Report 17 (1970); "Radiation Dosimetry: X-Rays and Gamma Rays with Maximum Photon Energies Between 0.6 and 50 MeV," ICRU Report 14 (1969); "Measurement of Absorbed Dose in a Phantom Irradiated by a Single Beam of X or Gamma Rays," ICRU Report 23 (1973); International Commission on Radiation Units and Measurements, Washington, D. C.
13. W. S. Snyder and C. Neufeld, "On the Passage of Heavy Particles Through Tissue," *Radiat. Res.*, 6, 67 (1967); reprinted in NBS Handbook 63.
14. G. A. Auxier, W. S. Snyder, and T. D. Jones, "Neutron Interactions and Penetration in Tissue," in *Radiation Dosimetry*, 2nd ed., Vol. I, F. H. Attix and W. C. Roesch, Eds., Academic Press, New York, 1968, Chapter 6.

15. "Protection Against Neutron Radiation," NCRP Report No. 38, National Council on Radiation Protection and Measurements, Washington, D. C., 1971.
16. "Neutron Dosimetry for Biology and Medicine," ICRU Report 26 (1977); International Commission on Radiation Units and Measurements, Washington, D. C.
17. H. E. Johns and J. R. Cunningham, *The Physics of Radiology*, 3rd ed., Charle C. Thomas Publishers, Springfield, Ill., 1969.
18. Eric J. Hall, *Radiobiology for the Radiologist*, Harper and Row, Hagerstown, Md., 1973.
19. William Duncan, "Fast Neutrons in Radiotherapy," in K. E. Halnan, Ed., *Recent Advances in Cancer and Radiotherapeutics: Clinical Oncology*, Churchill Livingstone, Edinburgh, 1972.
20. Fundamental and Practical Aspects of the Application of Fast Neutrons in Clinical Radiotherapy, Proc. Meet. Radiobiol. Inst. TNO, Rijswijk, Netherlands, June 22-24, 1970; published in *Eur. J. Cancer*, 7, (2/3) (May 1971).
21. Particle Radiation Therapy, Proc. Internat. Workshop, October 1-3, 1975, Key Biscayne, Fla., American College of Radiology, Philadelphia, (1976).
22. W. M. Blahd, *Nuclear Medicine*, 2nd ed., McGraw-Hill, New York, 1971.
23. Alexander Gottschalk and Robert N. Beck, *Fundamental Problems in Scanning*, Charles C. Thomas Publishers, Springfield, Ill., 1968.
24. Gordon L. Brownell, "Theory of Radioisotope Scanning," in *Medical Radioisotope Scanning*, Vol. I, International Atomic Energy Agency, Vienna, 1964.
25. "Report of Committee II on Permissible Dose for Internal Radiation (1959)," ICRP Publication 2, Pergamon Press, Oxford. See also revisions in ICRP Publications 6, 10, and 10a.
26. W. S. Snyder, "Internal Exposure," in K. Z. Morgan and J. E. Turner, Eds., *Principles in Radiation Protection*, Wiley, New York, 1967.
27. C. M. Lederer, J. M. Hollander, and I. Perlman, *Table of Isotopes*, 6th ed., Wiley, New York, 1967.
28. Medical Internal Radiation Dose (MIRD) Committee, Society of Nuclear Medicine, New York: Pamphlet 1 discusses the model and method, Pamphlets 2 and 3

discuss energy deposition from photons, Pamphlet 5 gives absorbed fractions for photons, Pamphlets 7 and 8 discuss absorbed dose and absorbed fractions for electrons, Pamphlet 10 gives the decay schemes and energies for selected radionuclides, and Pamphlet 11 gives the effective dose equivalents.

29. D. H. Slade, "Meterology and Atomic Energy 1968," available as report TID-24190 from the National Technical Information Service, Springfield, Va., 22151.

PROBLEMS

1. Construct the shape of the X-ray spectrum from a 250-kV constant potential X-ray tube with tungsten anode and total filtration equivalent to 2 mm of aluminum.

2. Calculate the transmission, contrast, and maximum dose equivalent for 120-kVp X-rays, a 1-cm-diameter tumor similar in attenuation to water, in an abdomen 20 cm thick. Do you think this tumor can be imaged successfully? Assume an output of 25 R/min at 1 m.

3. Calculate the central-axis adsorbed-dose rate in tissue at 10 cm depth for a ^{60}Co teletherapy source at 80 cm SSD and 100 cm^2 field area if the exposure rate in free air at the SSD is 20 R/min.

4. Calculate the central-axis absorbed-dose rate at 10 cm depth in tissue from a 14-MeV neutron generator emitting 10^{12} n s^{-1} at a SSD of 75 cm.

5. What is the dose to the adult thyroid (m = 20 g) from ingestion of 50 μCi of ^{131}I? What is the chance of inducing cancer from this dose?

6. What is the dose to the adult thyroid from inhalation of air containing 10^{-7} μCi/ml of soluble ^{131}I compound for 8 h occupational exposure?

INDEX